Fundamentals and Applications of Gas Phase Ion Chemistry

NATO ASI Series

Advanced Science Institute Series

A Series presenting the results of activities sponsored by the NATO Science Committee, which aims at the dissemination of advanced scientific and technological knowledge, with a view to strengthening links between scientific communities.

The Series is published by an international board of publishers in conjunction with the NATO Scientific Affairs Division

A Life Sciences	Plenum Publishing Corporation
B Physics	London and New York
C Mathematical and Physical Sciences	Kluwer Academic Publishers
D Behavioural and Social Sciences	Dordrecht, Boston and London
E Applied Sciences	
F Computer and Systems Sciences	Springer-Verlag
G Ecological Sciences	Berlin, Heidelberg, New York, London,
H Cell Biology	Paris and Tokyo
I Global Environment Change	

PARTNERSHIP SUB-SERIES

1. Disarmament Technologies	Kluwer Academic Publishers
2. Environment	Springer-Verlag / Kluwer Academic Publishers
3. High Technology	Kluwer Academic Publishers
4. Science and Technology Policy	Kluwer Academic Publishers
5. Computer Networking	Kluwer Academic Publishers

The Partnership Sub-Series incorporates activities undertaken in collaboration with NATO's Cooperation Partners, the countries of the CIS and Central and Eastern Europe, in Priority Areas of concern to those countries.

NATO-PCO-DATA BASE

The electronic index to the NATO ASI Series provides full bibliographical references (with keywords and/or abstracts) to about 50,000 contributions from international scientists published in all sections of the NATO ASI Series. Access to the NATO-PCO-DATA BASE is possible via a CD-ROM "NATO Science and Technology Disk" with user-friendly retrieval software in English, French, and German (©WTV GmbH and DATAWARE Technologies, Inc. 1989). The CD-ROM contains the AGARD Aerospace Database.

The CD-ROM can be ordered through any member of the Board of Publishers or through NATO-PCO, Overijse, Belgium.

Series C: Mathematical and Physical Sciences – Vol. 521

Fundamentals and Applications of Gas Phase Ion Chemistry

edited by

Keith R. Jennings
Department of Biological Sciences,
University of Warwick,
Coventry, U.K.

Kluwer Academic Publishers

Dordrecht / Boston / London

Published in cooperation with NATO Scientific Affairs Division

Proceedings of the NATO Advanced Study Institute on
Fundamentals and Applications of Gas Phase Ion Chemistry Chemistry
Grainau, Germany
7 - 18 August 1995

A C.I.P. Catalogue record for this book is available from the Library of Congress.

ISBN 0-7923-5463-X

Published by Kluwer Academic Publishers,
P.O. Box 17, 3300 AA Dordrecht, The Netherlands.

Sold and distributed in North, Central and South America
by Kluwer Academic Publishers,
101 Philip Drive, Norwell, MA 02061, U.S.A.

In all other countries, sold and distributed
by Kluwer Academic Publishers,
P.O. Box 322, 3300 AH Dordrecht, The Netherlands.

Printed on acid-free paper

Printed in the Netherlands

CONTENTS

PREFACE

This Volume presents the proceedings of the 1995 Advanced Study Institute entitled "Fundamentals and Applications of Gas Phase Ion Chemistry" held at Grainau, Germany during the period 7th-18th August, 1995. The Institute brought together over 80 scientists from many countries and, for the first time in this series, we were delighted to be able to welcome representatives from a number of countries in Eastern Europe.

This was the sixth in a series of NATO-sponsored Institutes devoted to various aspects of the physics and chemistry of gas phase ions. Although there continued to be a strong emphasis on various fundamental aspects of the subject, there was a broadening of the topics covered compared with previous Institutes with several sessions being devoted to discussing various types of application. This reflects the great expansion of the use of mass spectrometry in the investigation of biochemical and biological problems and in its use in different branches of the chemical industry such as in the investigation of polymers and other materials.

As usual, the Institute was built around the many invited lectures but, in addition, a poster session allowed many participants to display their own work and a number of more specialised topics formed the basis of lively workshops organised by the participants during the Institute. The Volume therefore contains not only chapters based on most of the lectures but summaries of the more important points brought out in the workshops are also included. The diversity of subject matter covered is a clear indication of how our increased understanding of gas phase ion chemistry is now being applied in many areas of pure and applied science, far beyond what could be envisaged at the first meeting held in Biarritz in 1974.

I should like to express my sincere thanks to Hans-Friedrich Grützmacher for his invaluable help with much of the local planning at Grainau, to Terry McMahon for dealing with all enquiries and applications in North America and to other members of the Organising Committee for their assistance in planning and organising the Institute. It also a pleasure to record our thanks to the staff of the Hotel am Badersee, Grainau, for their efficient and courteous assistance throughout the meeting and to Dr. Luis V. da Cunha, Director of the ASI Programme of the Scientific Affairs Division of NATO, for his continuing support of this series of Institutes. Finally I should to thank Dr. Elizabeth Kelly for her invaluable help in editing this Volume.

Keith R. Jennings,
University of Warwick,
Coventry, CV4 7AL,
United Kingdom.
March, 1998.

INTERMEDIATES IN IONIC GAS PHASE ORGANIC REACTIONS: I. WEAKLY BONDED SPECIES

H.E. AUDIER [(a)], P. MOURGUES [(a)], J. TORTAJADA [(b)] and D. BERTHOMIEU [(a)]

(a) Laboratoire des Mécanismes réactionnels, CNRS, Ecole Polytechnique, 91128 Palaiseau, France.
(b) Lab. de Chimie Structurale Organique, Université Paris VI, 4 Place Jussieu, 75252 Paris Cedex 05, France.

When the mechanism of an unimolecular or bimolecular ionic reaction is studied in the gas phase, a great number of methods exists in order to determine the structure of the reaction products [1]. In contrast, determination of the intermediates is often difficult. These intermediates can be ionized molecules or classical carbocations which are easy to generate and to study. However, generally, these intermediates are original species such as ionized carbenes, distonic ions (following article) or "complexes". The knowledge of the behaviour of these species is necessary to understand the reaction mechanisms. In the present article, complexes as reaction intermediates will be presented.
Through the years 1976 to 1982, several authors (De Puy, Bowen, Williams, Morton, Longevialle, etc.) have proposed that "complexes" can be key-intermediates in the gas-phase reactions of organic ions [2]. Depending on the putative structure of these complexes but also on the authors, different concepts were used in the literature : ion-neutral complexes, electrostatic complexes, H-bonded species, proton bound dimers, etc... For this reason, we will give, in introduction, a very general definition of these "complexes" which will be discussed further: they are stable or transient weakly bonded species (WBS) in which an ion and one or several neutrals remain in interaction but are not bonded by a classical covalent bond.

K. R. Jennings (ed.), Fundamentals and Applications of Gas Phase Ion Chemistry, 1–25.

1. Reactions within "complexes".

During the last twenty years, the gas phase chemists have used the concept of "complex" to explain experimental results. "Complexes" were considered as flexible structures allowing a great number of reactions, such as transfers of H^+, H^-, $H^\bullet$, CH_3^+, H_2O, formation of new bonds, isomerization of the cation, etc... The double well model of Brauman [3] is the base of the interpretations proposed (Eq.1 to 3). Since they maintain in interaction an ion and a neutral [4], "complexes" play the same role in the gas-phase than solvent cage effect in solution, as it has been suggested by McAdoo and Morton [4f].

$$ABC^+ \Leftrightarrow [A^+, BC] \Leftrightarrow BCA^{+\cdot} \qquad (1)$$

$$ABH^+ \Leftrightarrow [A, BH^+] \Leftrightarrow [AH^+, B] \qquad (2)$$

$$AB^+ \Leftrightarrow [A^+, B] \Leftrightarrow [A^+, B'] \Leftrightarrow AB'^+ \qquad (3)$$

In this section we will give several "historical examples" and therefore first arguments suggesting the intermediacy of "complexes" in these kinds of reactions. In the further sections we will give evidences, for the same systems, of the intermediacy of these complexes.

1.1. REARRANGEMENTS

1.1.1. *The Bowen-Williams rearrangement*

The reactions of the metastable $CH_3CH_2OCH_2^+$ cation and protonated propionaldehyde yield the same products. Bowen and Williams [5] have proposed an isomerization mechanism involving protonated formaldehyde weakly bonded to ethylene as key-intermediate (Scheme 1). Labelling is in agreement with this mechanism.

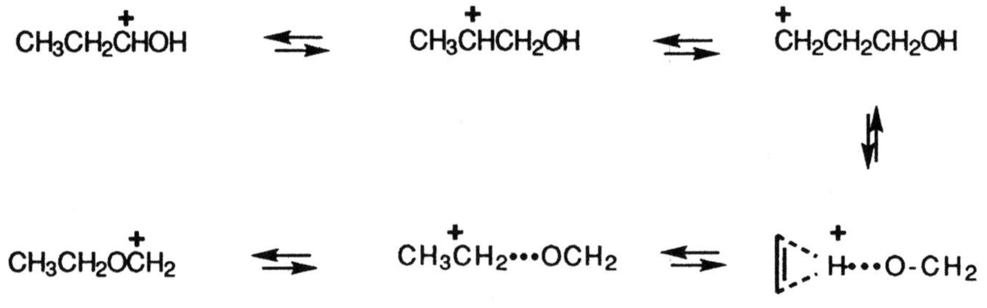

Scheme 1

1.1.2. *The McLafferty rearrangement*

It is well known that McLafferty rearrangement begins with a γ-H transfer to an ionized carbonyl group, leading to an intermediate distonic ion which yields by simple cleavage an enol$^{+\bullet}$ with alkene loss [6a]. At low energy, these moieties can remain weakly bonded and return, through formation of a new bond, to the reactants (Scheme 2). This reversible internal return explains the carbon atoms permutation observed prior to dissociation [6b-d].

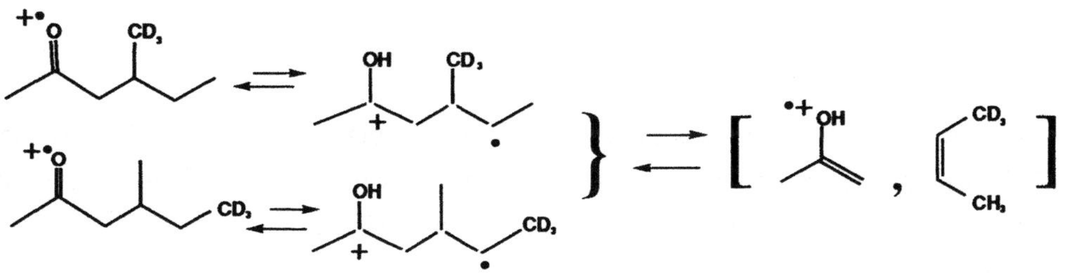

Scheme 2

1.1.3. *Apparent 1,3 H-transfer*

The reaction of the metastable $CH_3COCH_2CO^+$ cation only yields the CH_3CO^+ fragment-ion ($T_{0.5}$ = 1 meV)[7]. Labelling indicates that the dissociation is preceded by a permutation of the two CH_2-C=O moieties (Scheme 3).

$CH_3CO^*CH_2CO^+$	→	CH_3CO^+ (50 %)	and	$^*CH_3CO^+$ (50 %)
$CH_3COCD_2CO^+$	→	CH_3CO^+ (45 %)	and	CHD_2CO^+ (55 %)
$CD_3COCH_2CO^+$	→	CD_3CO^+ (95 %)	and	CH_2DCO^+ (5 %)

Scheme 3

From the data it has been concluded [7] that there is an intermediate or transient symmetric structure which may be either a [ketene ... H^+ ... ketene] complex, or alternatively two interconverting complexes (Eq. 4) leading to an apparent 1,3-H transfer. This latter proposal is conforted by the strong isotope effect observed for H-transfer when $CD_3COCH_2CO^+$ reacts (Scheme 3), which shows that H-transfer is rate determining.

$$CH_3CO^*CH_2CO^+ \rightarrow [CH_3CO^+, {}^*CH_2CO] \Leftrightarrow [CH_2CO, {}^*CH_3CO^+] \qquad \underline{4}$$

1.2. ELIMINATIONS AND H-EXCHANGE

The 1,2 elimination observed in protonated alcohols, amines or alkyl ethers is often preceded by a specific but extensive H-exchange consistent with the intermediacy of two interconverting complexes α and β (Scheme 4) [8].

$$\begin{array}{llcl} \text{iso-}C_3H_7OH^+R \rightarrow & [\text{iso-}C_3H_7^+, ROH] & \Leftrightarrow & [C_3H_6, ROH_2^+] \\ & \alpha \downarrow & & \beta \downarrow \\ & \text{iso-}C_3H_7^+, + HOR & & ROH_2^+ + C_3H_6 \end{array}$$

Scheme 4 (R= H, CH_3, C_2H_5)

A similar phenomenon is observed in the unimolecular dissociation of protonated alkylbenzenes [9], as for instance isopropylbenzene (Scheme 5) [9c]. However, in this case the H-exchange is incomplete for reasons that will be discussed further.

$$\begin{array}{llcl} C_6H_6\text{-iso-}C_3H_7^+ \rightarrow & [C_6H_6, \text{iso-}C_3H_7^+] & \Leftrightarrow & [C_6H_7^+, C_3H_6] \\ & \alpha \downarrow & & \beta \downarrow \\ & \text{iso-}C_3H_7^+ + C_6H_6 & & C_6H_7^+ + C_3H_6 \end{array}$$

Scheme 5

1.3. CATALYZED ISOMERIZATION OF THE ION

Metastable protonated isopropylbenzene gives the $C_6H_7^+$ and iso-$C_3H_7^+$ fragments in a ratio 9/1 (Scheme 5). In contrast protonated sec-butylbenzene only yields a $C_4H_9^+$ fragment. If a sec-$C_4H_9^+$ fragment is formed, the data can not be understood since the thermochemistry of the reactions for isopropyl and sec-butylbenzene are almost the same in such an hypothesis. That is why Harrisson [9a] proposed that, within a complex, the sec-butyl cation isomerizes into the much more stable tert-butyl structure (Scheme 6).

$$\begin{array}{lcl} C_6H_6\text{-sec-}C_4H_9^+ \rightarrow [C_6H_6, \text{sec-}C_4H_9^+] & \rightarrow & [C_6H_6, \text{tert-}C_4H_9^+] \\ & & \downarrow \\ & & \text{tert-}C_4H_9^+ + C_6H_6 \end{array}$$

Scheme 6

In several small systems (Eq. 5-6) [10a-b], it has been shown that the neutral can catalyze the isomerization of the ion within a complex, as it has been reviewed by Bohme [10c].

$COH^+ + H_2 \rightarrow [COH^+ ... H_2] \rightarrow [CHO^+ ... H_2] \rightarrow CHO^+ + H_2$ 5

$HCN^{+\bullet} + CO \rightarrow [HCN^{+\bullet}, CO] \rightarrow [CNH^{+\bullet} + CO] \rightarrow CNH^{+\bullet} + CO$ 6

As it will be developed in the following article, the CID spectra demonstrate that a water molecule catalyzes the isomerization of ionized methanol into its α-distonic ion counterpart [11] through a catalyzed 1,2-H shift (Eq. 7). Similarly, methanol catalyzes an 1,3-H transfer, isomerizing ionized acetaldehyde into its more stable enolic structure [12].

$CH_3OH^{+\bullet} + H_2O \rightarrow [CH_3OH^{+\bullet}, H_2O] \rightarrow [^\bullet CH_2OH_2^+, H_2O] \rightarrow {}^\bullet CH_2OH_2^+ + H_2O$ 7

$[CH_3CHO^{+\bullet}, CH_3OH] \rightarrow [CH_2CHOH^{+\bullet}, CH_3OH] \rightarrow CH_2CHOH^{+\bullet} + CH_3OH$ 8

2. Experimental evidences for the intermediacy of "complexes".

Proving the intermediacy of "complexes" requires several complementary experimental methods and, when possible, experiment has to be confronted with calculations.

2.1. LONG DISTANCE INTERACTIONS

The first (and early) way is to choose a model in which two functional groups are far away enough to consider that their direct interaction is impossible. For instance De Puy has studied a conjugated and linear anion (Eq. 9) which eliminates water. Since a direct H-transfer to the oxygen is impossible for steric reasons, De Puy has supposed the intermediacy of a [neutral, OH^-] complex in which the OH^- anion can move and abstract a proton from the neutral [13].

$RCH = CHCH\text{-}N{=}N\text{-}OH^- \rightarrow [RCH{=}CHCH{=}N^+{=}N^-, OH^-]$ 9

At the same time Botter and Longevialle studied ionized 3,20-diamino pregnane. At high energy this ion yields a $CH_3CHNH_2^+$ (m/z 44) product formed by simple cleavage. At low energy it eliminates successively CH_3CHNH and NH_3 while the di-ND_2 ion loses specifically CH_3CHND and then ND_3. This result has been explained by the existence of the intermediate complex shown in scheme 7 [2d] [14].

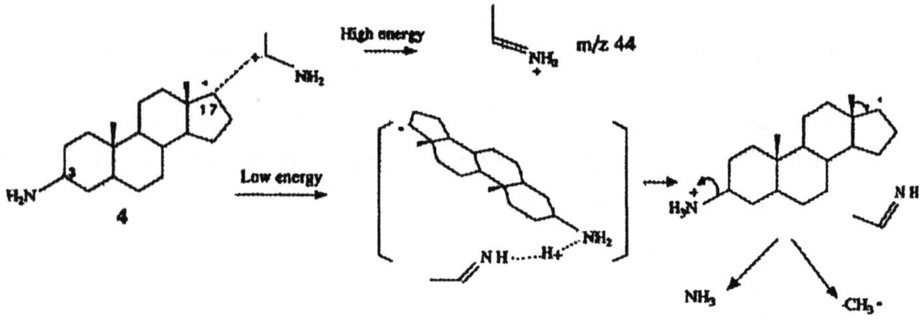

Scheme 7

2.2. STEREOCHEMISTRY

The study of the stereospecificity of a reaction is a very good way to prove the intermediacy of complexes. For instance, sec-butylphenylether$^{\bullet+}$ leads to the formation of a phenol$^{\bullet+}$ product-ion with butene elimination. From the study of the structure of the neutrals (produced as a mixture), Morton [4a-b] concluded that the fragmentation begins with a simple cleavage, and gives a [$C_6H_5O^{\bullet}$, sec-$C_4H_9^+$] "complex" in which a non selective H^+ abstraction by the radical leads to the observed products (Scheme 8).

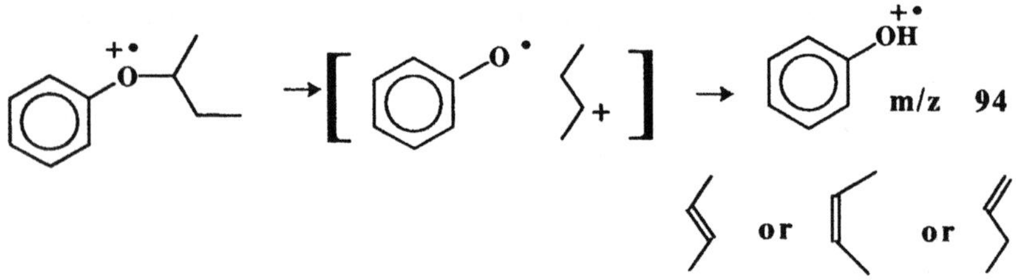

Scheme 8

In contrast other authors [4c] showed that, in the unimolecular reaction of metastable ionized sec-butylphenylether, the transfer of one secondary H of the chain was highly favored. It was concluded that, at least partially, the reaction pathway was concerted. This was later established by Morton from the study of stereolabelled ions [4d]. The fragmentation of metastable ions shows that the tertiary hydrogen is poorly transfered, while there is a significant difference between the threo and erythro stereoisomers labelled with one deuterium in position 3 (Figure 1).

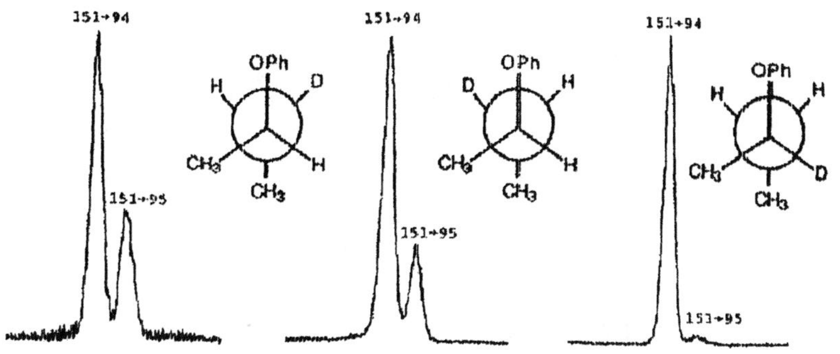

Figure 1: MIKE spectra of isomeric sec-butylphenylethers

The detailed results show the occurence of two mechanisms: (i) the first one is a concerted process, and is therefore concerned by the stereochemistry of the secondary H, while (ii) the second one gives rise to the formation of intermediate complexes in which the initial stereochemistry is lost.

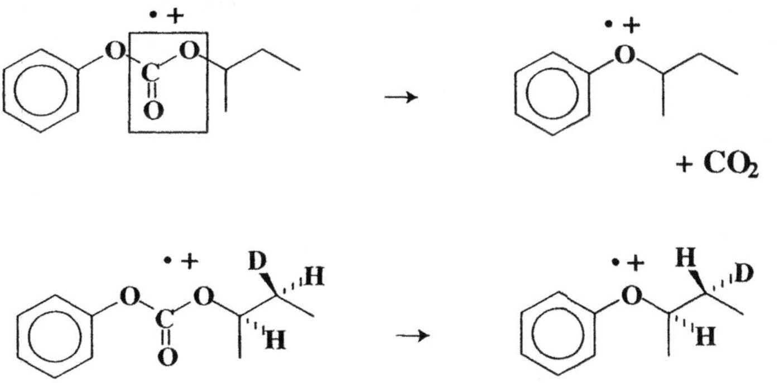

Scheme 9

The possibility to distinguish between the threo et erythro 3-d, s-butyl phenyl ether has been used to clarify the mechanism of CO_2 loss from ionized phenyl s-butyl carbonates, which leads to the formation of phenyl s-butyl ether radical cation. Labelling by O^{18} shows that the oxygen bound to the ring remains in the product-ion while the erythro 3-d carbonate$^{\bullet+}$ yields the threo s-butyl phenyl ether, and vice-versa (Scheme 9) [4e].The configuration inversion during the reaction is explained by the initial formation of a complex (Scheme 10) followed by a back side attack. Therefore, there is rotation of a moiety with respect to the other within the complex.

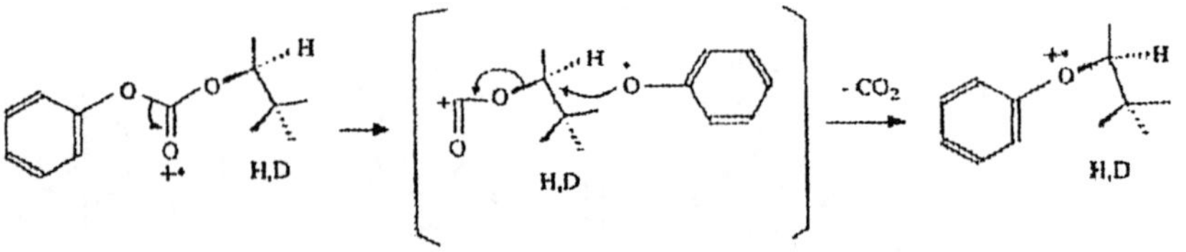

Scheme 10

2.3. ION MOLECULE REACTIONS

Comparison between the unimolecular reaction of an ion and that of the isomeric adduct formed by ion-molecule reaction can lead to interesting conclusions.

2.3.1. *Evidence for the stability of "π-complexes" [arene, alkyl cation$^+$].*

The adduct [benzene, iso-$C_3H_7^+$], formed in the ion source by reaction of iso-$C_3H_7^+$ with benzene, leads to the same MIKE and CID spectra than protonated isopropylbenzene. Even if a π-complex is involved in the fragmentation pathway of protonated isopropylbenzene, it cannot be evidenced since it interconverts rapidly with the covalent structure [9e].

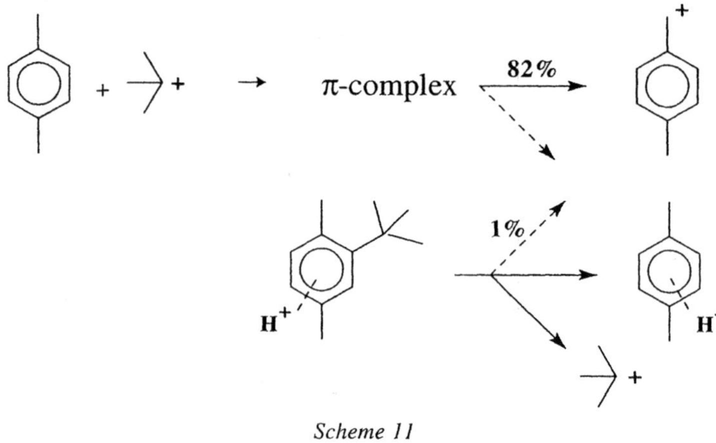

Scheme 11

Therefore it was necessary to find a model in which addition of the cation to the ring is hindered. For instance Cacace [9f] has proved by the radiolytic technique that the rate constant of the reaction of the tert-$C_4H_9^+$ cation with paraxylene is zero. For this reason, the behaviour of the metastable [tert-$C_4H_9^+$, paraxylene] adduct and that of its corresponding covalent structure are completely different (Scheme 11)[9e].

This shows that the π-complex formed in the ion-source by ion-molecule reaction has a life-time sufficient to survive up to the 2nd FFR.

2.3.2. *Stability of the [CH_3^+, CH_3OH] complex*

In the source of a double focusing mass spectrometer the reaction of CH_3^+ with CH_3OH yields $^{13}CH_3OHCH_3^+$ which loses CH_4 and $^{13}CH_4$ in a 1/1 ratio (Scheme 12).

$$^{13}CH_3^+ + CH_3OH \rightarrow {}^{13}CH_3OHCH_3^+ \rightarrow \begin{cases} {}^+CH_2OH + {}^{13}CH_4 & 50\ \% \\ {}^{+13}CH_2OH + CH_4 & 50\ \% \end{cases}$$

Scheme 12

In contrast, the reaction of $^{13}CH_3^+$ with CH_3OH, performed in the cell of a FT-ICR, leads to a higher ratio of $^{13}CH_4$ loss (Scheme 13, [15a]).

$$^{13}CH_3^+ + CH_3OH \rightarrow \begin{cases} {}^+CH_2OH + {}^{13}CH_4 & 57\ \% \\ {}^{+13}CH_2OH + CH_4 & 43\ \% \end{cases}$$

Scheme 13

From these results it has been deduced that the reaction leads for about 85 % to the symmetric $^{13}CH_3OHCH_3^+$ structure and for 15 % to the [$^{13}CH_3^+$, CH_3OH] complex which specifically eliminates $^{13}CH_4$ (Eq. 10).

$$^{13}CH_3^+ + CH_3OH \rightarrow [{}^{13}CH_3^+, CH_3OH] \rightarrow CH_2OH^+ + {}^{13}CH_4 \qquad \underline{10}$$

2.4. LIGAND EXCHANGE

In a FT-ICR cell, the observation of a ligand exchange between an ion and a neutral molecule is always a noteworthy result, but the interpretation is not straightforward. This reaction can occur as well for long bond ions (Eq. 11) as for complexes (Eq. 12).

$$A^+\cdots B + C \rightarrow A^+\cdots C + B \qquad \underline{11}$$

$$[A^+, B] + C \rightarrow [A^+, C] + B \qquad \underline{12}$$

Both kinds of reactions can be illustrated by the ligand exchange occuring in the ions involved in the Bowen-Williams and the McLafferty rearrangements (Scheme 14) [16] as well as for the substitution of the β-distonic ions [11] [16].

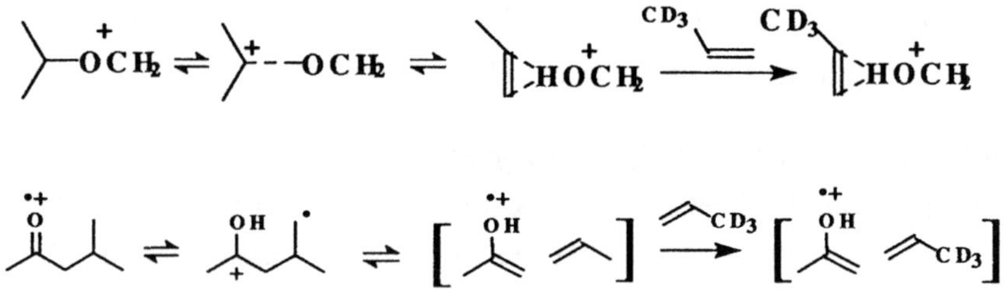

Scheme 14

2.5. STRUCTURAL STUDIES BY HPMS

The study at various temperatures by pulsed high pressure mass spectrometry (PHPMS) [17a-b] of the equilibrium constant for the association of tert-butyl cation with small neutrals [17c] shows unambiguously that, in many cases, the complex exists as two isomeric forms, their relative proportions depending strongly on temperature. The fitting of the van't Hoff plots (presenting in the methanol case a sharp break, figure 2) according to the coexistence of two structures leads to the conclusion that a covalently bound complex prevails at low temperatures, whereas at higher ones the complex is an electrostatically bound adduct, presenting lower -ΔH° and-ΔS° values of enthalpy and entropy for association.

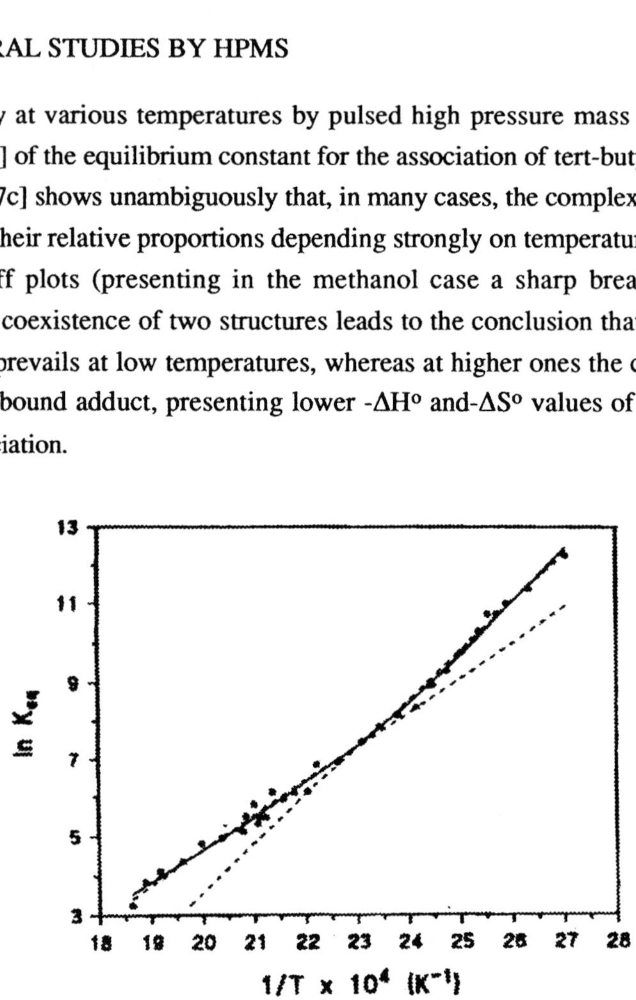

Figure 2: van't Hoff plot for the tert-$C_4H_9^+$ / CH_3OH system

2.6. OTHER METHODS

The intermediacy of WBS. can also be evidenced by other experimental methods [18], such as neutralization-reionization (NRMS) [18a-e], based on the fact that, in contrast with molecular ions, neutral species corresponding to complexes are rarely long-living, and therefore dissociate before reionization. Other methods such as "angle resolved distribution" of "osculating complexes", are based on asymmetric angular distribution of fragments when the life-time of the complex is of same order that its reorientation time. Crossed beam experiments [18f] or rf-octopoles ions guides designs[18g] are generally used for such studies.

3. Complexes as stable species

3.1. ENERGY MINIMA

When a bond of an organic ion is stretched, the energy increases, the system becomes flexible and can be stabilized by formation of a stable complex. Conversely, when an ion reacts with a neutral, an intermediate complex is often formed prior to the creation of a covalent bond (Figure 3) [2d].

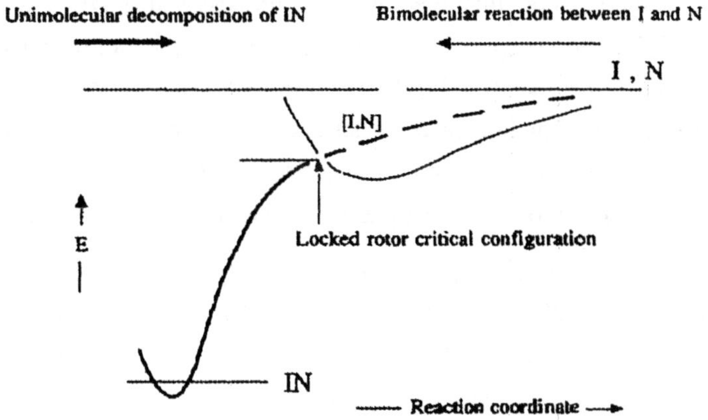

Figure 3: Mode of formation of complexes

Between the ion and a neutral, different kinds of interactions exist which can be repulsive or attractive (interactions charge/dipole, charge/induced dipole, induced dipole/induced dipole, charge exchange, etc.). Complexes can correspond to deep minima on the

Potential Energy Surface (PES). In such cases, the interaction energy is less than in a classical covalent bond but can overpass 50 kcal mole^{-1} in H-bonded species .
However complexes most often correspond to shallow minima on the PES and to rather flexible structures. In such cases, a theoretical approach based on calculations is necessary in order to verify that the complex lies in an energy minimum, to know its geometry or its interaction energy, and to evaluate the different components of this interaction energy.

3.2. COMPONENTS OF THE INTERACTION ENERGY

The partitioning scheme of Morokuma (1976) for the analysis of the components of the interaction energy is based on the Hartree-Fock approximation [19]. No correlation is included, so dispersion interactions are not taken into account, leading only to semi-quantitative results. The stabilization energy is defined as the sum of five terms (Eq. 13):

$$E_{STAB} = E_{ES} + E_{P} + E_{EX\text{-}REP} + E_{CT} + E_{MIX} \qquad \underline{13}$$

E_{ES} = classical electrostatic interaction between occupied MO's without any mixing of MO's; E_P = polarization coming from the mixing of occupied MO and vacant MO of each molecule; $E_{EX\text{-}REP}$ = exchange-repulsion arising from the mixing of electrons and delocalization between both entities A and B; E_{CT} = charge-transfer coming from the mixing of electrons and delocalization between the occupied MO's of A and the vacant MO's of B and vice versa; E_{MIX} = difference between total ΔE and the sum of the four terms, taking in account higher order interactions.

The interest of the method can be illustrated by the examples of the [H_3O^+, isobutene] **a** and the [H_2O, tert-$C_4H_9^+$] **b** complexes (Figure 4). The first one can be considered as the interaction of one hydrogen atom of H_3O^+ with the π system of neutral isobutene: the charge transfer term is important and, since the structure is compact, the repulsion forces are strong. In contrast, in the second system the electrostatic term is highly dominant and E_{CT} negligible (Table 1): **b** can be considered as a pure electrostatic complex [8f, 20].

	E_{STAB}	E_{ES}	E_P	$E_{EX\text{-}R}$	E_{CT}	E_{MIX}
[H_3O^+, C_4H_8]	-28	-19.2	-16.7	+20.7	-14.9	+2.1
[t-$C_4H_9^+$, H_2O]	-12.8	-15	-1.5	+4.7	-1.6	+0.7

Table 1 (kcal mol^{-1})

This energy decomposition can also be done by semi-empirical methods such as the method of Claverie which takes into account dipersion interactions. The interaction energy E_{STAB} is calculated as a sum of five terms : electrostatic E_{EL}, polarization E_{POL}, dispersion

E_{DISP}, exchange repulsion E_{REP} and charge exchange E_{CT}. The interaction energy is

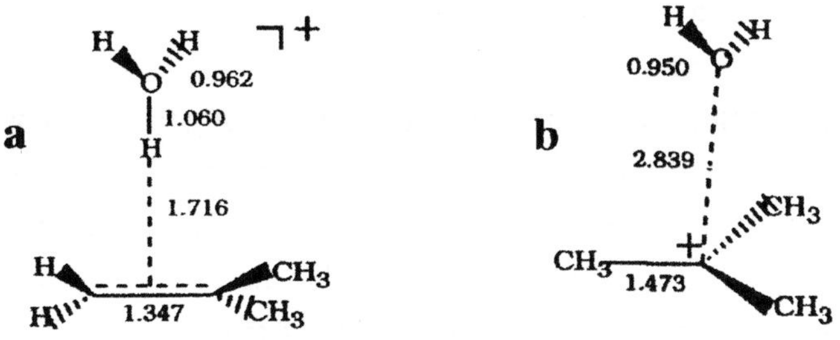

Figure 4: Geometries of the [H_3O^+, isobutene] and [H_2O, tert-$C_4H_9^+$] complexes

calculated from frozen fragments and it is assumed that in van der Waals complexes the geometries of the two subunits are not modified. A simulated annealing calculation allows exploring the PES. The minima found are then optimized.
For instance, experiment shows that [arene, alkyl cation] π-complexes [9e] correspond to stable species. This was confirmed by calculation which indicates that E_{STAB} lies between 11.7 and 15.5 kcal/mole depending on the arene and the cation. Table 2 shows that: (i) only E_{POL} increases from benzene to toluene, which is in agreement with the respective polarizability of benzene and toluene; (ii) E_{DISP} is rather important and represents from 26 % to 35 % of the attractive energy; (iii) charge transfer is negligible: these complexes are classical electrostatic complexes.

	E_{STAB}	E_{EL}	E_{POL}	E_{REP}	E_{DISP}	E_{CT}
[Benzene, iso-$C_3H_7^+$]	-13.9	-8.4	-5.7	+5.4	-5	-1
[Benzene, tert-$C_4H_9^+$]	-11.7	-6.5	-4.3	+4.1	-5	-1
[p-Toluene, iso-$C_3H_7^+$]	-15.5	-8.6	-6.9	+6.0	-6	-1
[p-Toluene, tert-$C_4H_9^+$]	-12.9	-6.5	-5.0	+4.6	-6	-1

Table 2 (in kcal mole^{-1})

Bader analysis. The nature of the bonding interactions in the different complexes can be investigated by a topological analysis of the electron charge density distribution ρ and its Laplacien $\nabla^2\rho$. As has been shown by Bader and co-workers [21], $\nabla^2\rho$ identifies regions of the space where the electronic charge is locally depleted (in this case $\nabla^2\rho$ is positive), or reinforced, and in this case $\nabla^2\rho$ is negative.

The case $\nabla^2\rho > 0$ is typically associated with interactions between closed-shell systems (ionic bonds, hydrogen bonds, van der Waals associations...), while $\nabla^2\rho < 0$ characterizes covalent bonds.

In figure 5 are represented the Laplacian of the charge density for the complexes [H_3O^+, isobutene] **a** and [H_2O, tert-$C_4H_9^+$] **b**, and the contour maps of $\nabla^2\rho$. Full lines indicate decreases, and dotted lines increases in electronic density. The charge transfer (H bond) appears clearly for **a** (Fig. 5a) while for **b** (Fig. 5b), the two moieties are completely separated [20], in contrast with the covalent structure of protonated tert-butanol.

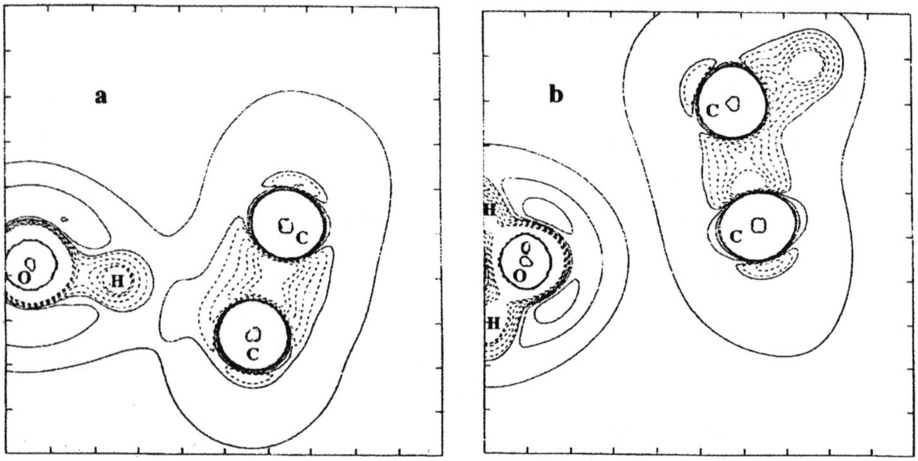

Figure 5: Laplacian of the charge density for [H_3O^+, isobutene] and [H_2O, tert-$C_4H_9^+$]

Different kinds of "complexes"? From the geometry of the complex and the distance between the ion and the neutral moiety, the interaction energy, the components of this energy and the Bader analysis, it is often possible to propose dominant characteristics for a complex. It is proposed to distinguish between two limit cases (Table 1, Fig. 4, 5, 6):

- Electrostatic complexes (ions **b** and **g**) in which the charge transfer is negligible. The distance between both entities is important (3Å or more) and the Bader analysis clearly shows the existence of two separated entities. These complexes are generally flexible allowing a facile reorientation [2e]. The interaction energy lies in most cases between 5 and 15 kcal mol^{-1} and can be roughly evaluated when the dipole moment and the induced dipole moment of the neutral molecule are known [2a].
- Charge transfer complexes in which a significant part of the interaction energy comes from a charge transfer, which often creates a rigidity in the system. These complexes are

compact and their Bader diagrams resemble those of covalent species. Since the distance between the entities is rather short, the attractive and repulsive electrostatic forces are strong. Therefore the stability of these species depends on the relative orientation of the weak bond coming from charge transfer and of the dipole moment of the neutral. In favorable cases, the interaction energy can reach 40 kcal mol^{-1} or more. Ion **d** (Fig. 6) illustrates the case of a strong π-interaction. In contrast, it has been shown above that the so-called "π-complexes" [arene, cation] do not correspond to a π-interaction [9e]. Solvated radical cations **f**, **h** and **i** examplify the so called "H-bonded species" particularly studied by Terlouw and by Schwarz [22]. Finally, "proton bound dimers" [$M...H^+...M'$] is a third classic case where the interaction energy can be easily evaluated [23].
However, it should be noted that there are not only two kinds of "complexes", but a continuum between these limit cases, which raises a question of vocabulary.

3.3. CONSEQUENCES OF THE PLURALITY OF STRUCTURES

3.3.1. *Plurality of structures*

When an ion interacts with a neutral, a plurality of stable structures can correspond to the plurality of stabilization factors of a complex. To a same pair correspond (Figure 6): three stable structures **c**, **d** and **e** for the CH_2OH^+/ethylen system, intermediates in the Bowen-Williams rearrangement [5c]; one H-bonded complex **f** and one electrostatic complex **g** for the $^{\bullet}CH_2CH_2OH^+$ / H_2O system [11]; two H-bonded species **h** and **i** for the $CH_3CH_2OH^{\bullet+}$ / H_2O system, (and similarly **h'** and **i'** for $CH_3OH^{\bullet+}$ / H_2O, not shown) [11]; three stable structures **j**, **k** and **l** for the CH_3CO^+ / CH_2CO system [7].

3.3.2. *Isomerization between different structures.*

Isomerization between different structures is often necessary to explain rearrangements. For instance, in the Bowen-Williams rearrangement, the interconversion between **c** and **d** happens through rotation of the ion with respect to the neutral (Fig. 7) [5c]. Similarly the PES of the $CH_3COCH_2CO^+$ cation (Scheme 3, Figure 8) [7], illustrates the key role of the isomerization between the complexes **j**, **k** and **l**.

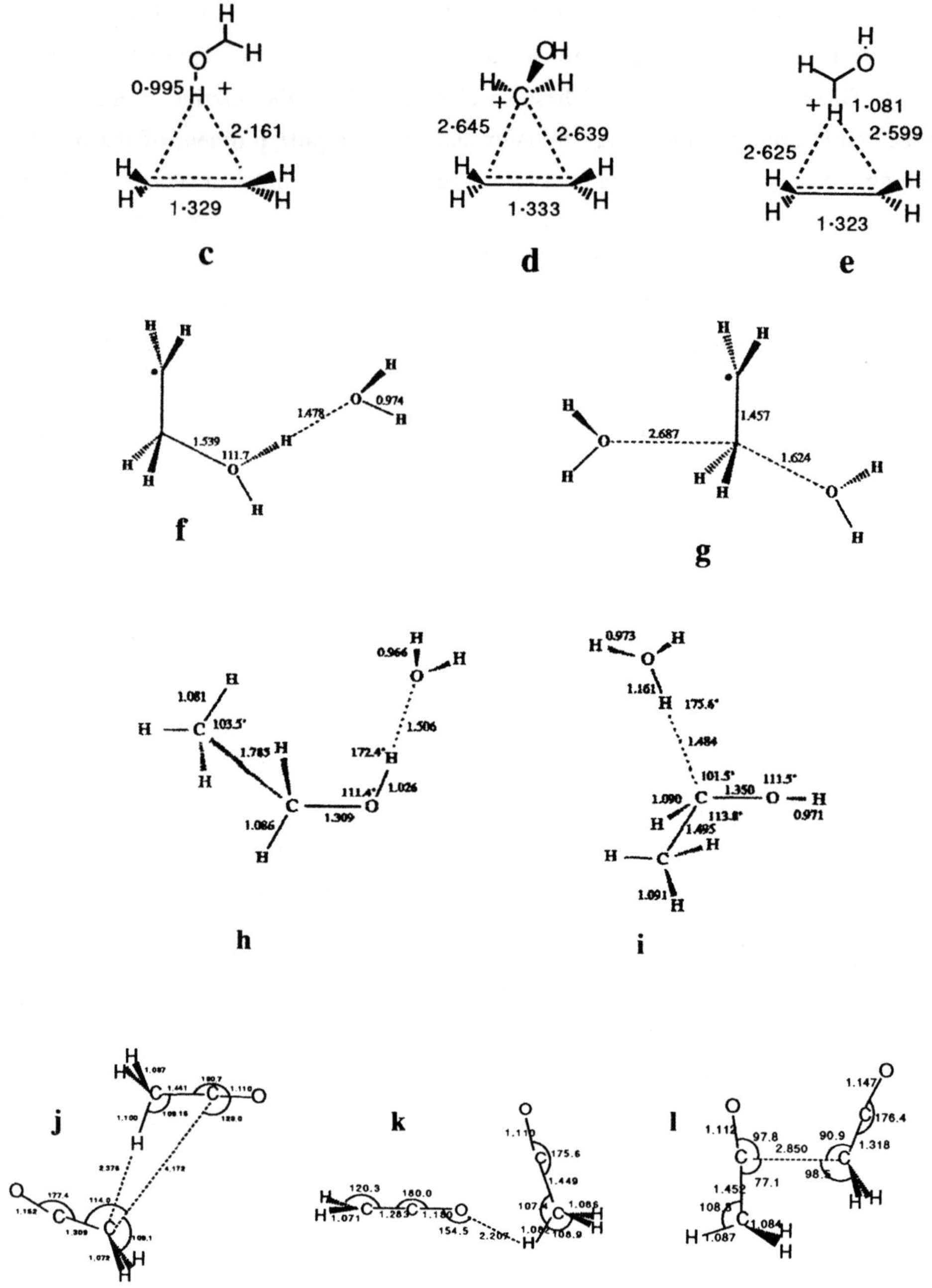
0·995
2·161
1·329
c
2·645
2·639
1·333
d
1·081
2·625
2·599
1·323
e
f
g
h
i
j
k
l

Figure 6

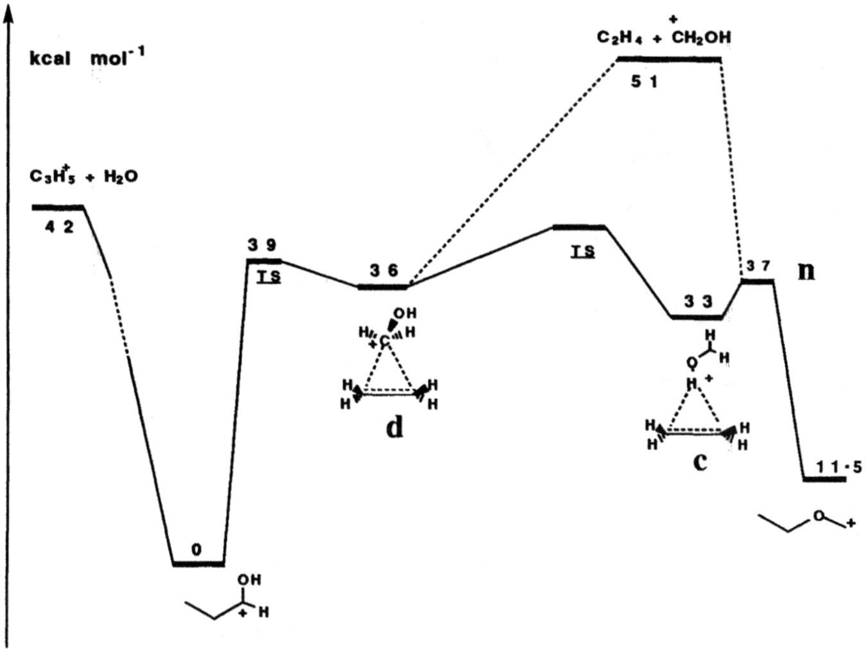

Figure 7: Energy profile of the Bowen-Williams rearrangement

3.3.3. *Structures of dissociation precursors.*

When several structures correspond to a same I^+ / N system, each structure can be the direct precursor in a dissociation or in a specific reaction. This is examplified by the fragmentation of protonated dialkyl ethers: the [iso-$C_3H_7^+$, ROH] complex yields iso-$C_3H_7^+$ while [C_3H_6, ROH_2^+] yields ROH_2^+(Scheme 4).

Similarly when the distonic ion $^{\bullet}CH_2CH_2OH_2^+$ reacts with a molecule M, the H-bonded complex **f** is the direct precursor in the protonation of the molecule, while the electrostatic complex **g** is the precursor of a nucleophilic substitution [11].

The consequences of this phenomenon are particularly interesting when, in contrast with the preceding examples, the isomeric structures do not interconvert (or interconvert very slowly). It is the case of the reaction of $CH_3CH_2OH^{+\bullet}$ with H_2O which gives two non-interconverting complexes **h** and **i** (Fig. 6), each of them having they own forthcoming [11]: **h** loses a methyl radical (Eq.14) while **i** leads to the α-distonic ion (Eq.15).

$$[CH_3CH_2OH^{\bullet+}\cdots OH_2] \rightarrow {}^{\bullet}CH_3 + [CH_2OH^{+}\cdots OH_2] \qquad \underline{14}$$

$$[CH_3{}^{\bullet}CHOH\,, H_3O^+] \rightarrow [CH_3{}^{\bullet}CHOH_2^{+}\cdots OH_2] \rightarrow CH_3{}^{\bullet}CHOH_2^+ + H_2O \qquad \underline{15}$$

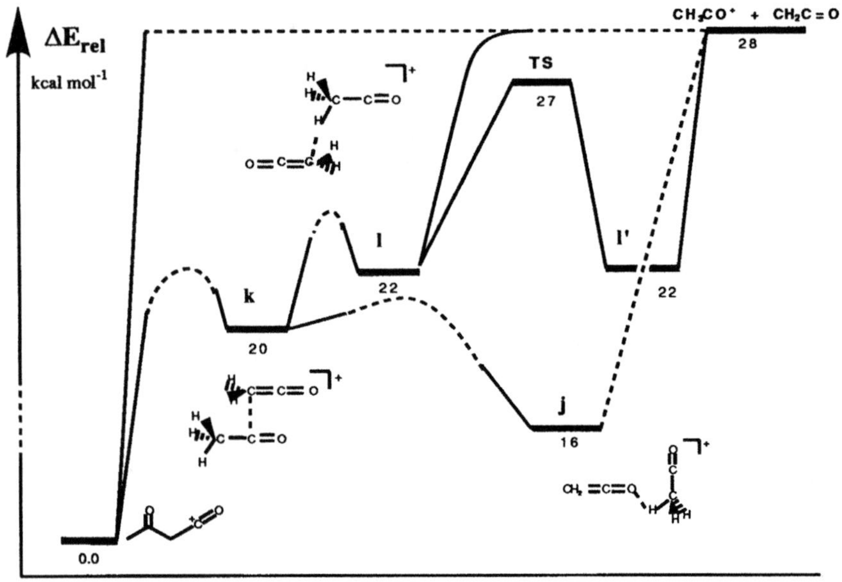

Figure 8: Energy profile of the $CH_3COCH_2CO^+$ cation

As it will be detailed in the following paper, the reaction between $CH_3OH^{+\bullet}$ and H_2O leads to the same kind of complexes **i'** and **h'**: **i'** leads to the α-distonic ion but, in this case, **h'** is unreactive, being a cul-de-sac in the reaction pathway.

3.3.4. *Cul-de-sac and reactive structures.*

Very stable structures are often cul-de-sacs in the rearrangement pathways. In the systems precedently studied, ions **i'** and **j** (Figure 8), which correspond to very stable structures, do not intervene in the rearrangement mechanisms.

Conversely, the complexes which operate in rearrangement processes requiring a great change of the geometry of the system, they generally lie rather close to the transition state energy, as illustrated by the energy profiles shown in the figures 7 and 8.

A qualitative explanation of this observation can be that, for an intermediate complex, there is generally a potential competition between its dissociation with a very favorable entropic factor, and one or several rearrangements corresponding to unfavorable entropic factors. Therefore, to be observed, a rearrangement involving a significant change in the geometry of the system must have a rather small activation energy compared to that of its dissociation, which means that the reactive complex must lie in energy only some kcal mol^{-1} under the transition state for rearrangement. It is the case in figures 7 and 8.

4. Complexes as transient species.

4.1. A STEPWISE OR A CONCERTED PROCESS ?

As a consequence of this last observation, it is sometimes difficult to distinguish between a stepwise and a concerted process. For instance, the H-exchange preceding the 1,2 elimination in protonated alcohols, ethers, amines or arenes [8-9] can be explained by the intermediacy of two complexes α and β (see first section, Schemes 4 and 5). However, depending on the structure of the reactant ion, the α and β complexes i/ correspond to minima on the PES or not, ii/ interconvert or not.

The unimolecular reaction of *protonated isopropyl benzene* (Fig. 9a) corresponds to a typical double well process. The [C_6H_6, iso-$C_3H_7^+$] α and [$C_6H_7^+$, C_3H_6] β intermediate complexes have been shown, by experiment and calculation, to be energy minima [9d]. The existence of a strong primary isotope effect [9c] shows that the H-transfer requires a significant energy barrier, in agreement with the calculated significant change of geometry when α → β. For this reason, H exchange is not complete.

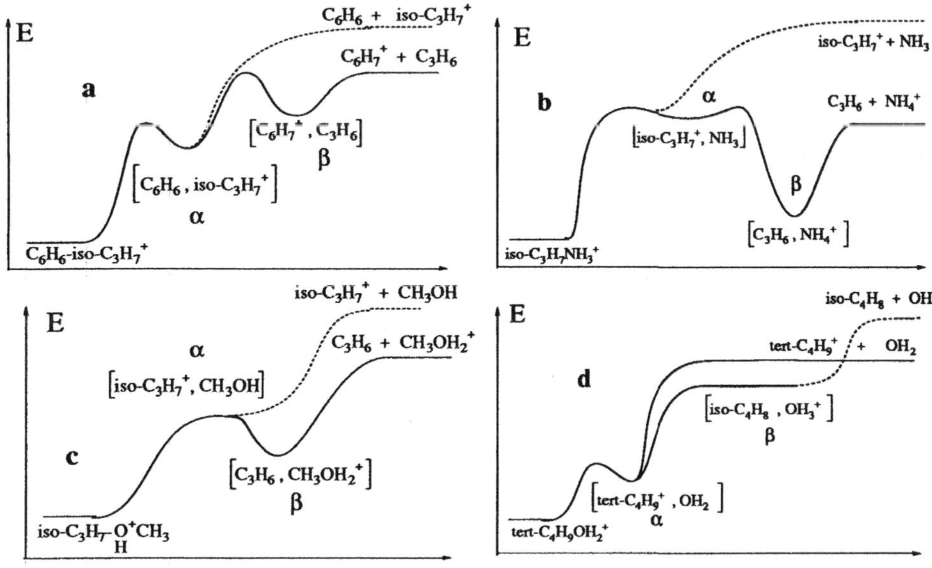

Figure 9

Fragmentation of metastable *protonated isopropylamine* (Fig. 9b)[8e] and *isopropyl methyl ether* (Fig. 9c)[8c] corresponds to degenerated forms of the double well model. In the first case, the [C_3H_6, NH_4^+] β-complex is clearly a deep energy minimum while the [iso-$C_3H_7^+$, NH_3] α-complex is a shallow minimum and there is not a great change in the geometry (and therefore no significant barrier) when α → β. This minimum lies rather high in energy and above the final state ($NH_4^+ + C_3H_6$). Therefore the reaction iso-$C_3H_7NH_3^+ \rightarrow [C_3H_6, NH_4^+]$ is irreversible and its transition state is very close to the α-complex .

In contrast, for protonated *isopropyl methyl ether*, it has not been possible to localize an energy minimum for the [iso-$C_3H_7^+$, CH_3OH] α-complex. The iso-$C_3H_7OH(CH_3)^+$ ion isomerizes to its β-complex [C_3H_6, $CH_3OH_2^+$] by a concerted process through a broad transition state possessing the structure of a weakly bonded species between an isopropyl cation and a methanol molecule [8c].

The unimolecular reaction of *protonated tert-butanol* (Fig. 9d) corresponds to a second degenerated form of the double well model. The [tert-$C_4H_9^+$, H_2O] α-complex is now an energy minimum while, in contrast, the [isobutene, H_3O^+] β-structure corresponds to a plateau which lies in energy below the most stable final state. The slow H-exchange observed when tert-$C_4H_9^+$ reacts with D_2O shows that both α and β structures operate in the H-exchange reaction, what means that β has a significant life-time [8f].

4.2. TRANSITION STATES

The ambiguous situation described for protonated isopropylamine and isopropyl methyl ether is rather current. For instance, protonated dimethyl ether loses methane (Eq. 10) via a stepwise process involving the intermediacy of a complex **m**, which has been shown to

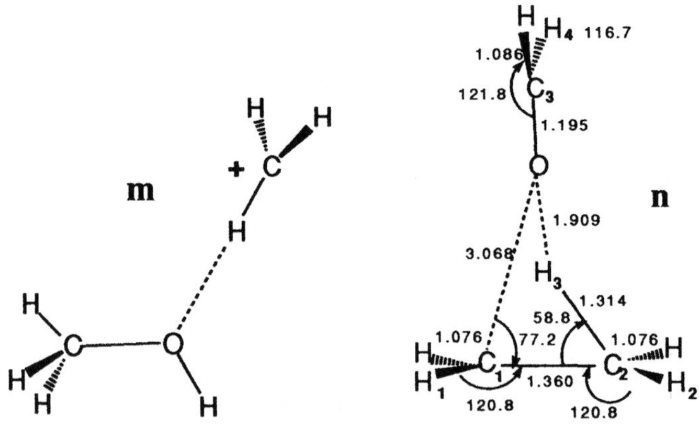

Figure 10

be stable by calculation (Figure 10) [15a]. Other authors have calculated the transition state leading to methane loss via a concerted mechanism: they found that its geometry strongly resembles that of **m** [15b].

Similarly, in the Bowen-Williams rearrangement, the concerted isomerization of $CH_3CH_2OCH_2^+$ into $[C_2H_4 \cdots HOCH_2^+]$ **c** (Fig. 7) occurs through the transition state **n** whose geometry is a $CH_3CH_2^+$ cation weakly bonded to OCH_2 (Fig. 10) [5c]. Even in this extreme example, for which it is clear that **n** is an energy maximum, is it so sure that its life-time is not significant ?

4.3. WHICH CRITERIUM: ENERGY MINIMUM OR LIFETIME ?

Is it necessary for a complex to be an energy minimum to operate in a rearrangement ? Probably not. From this point of view, in the case of protonated isopropyl methyl ether, the stretching of a C-O bond up to 2.7 Å leads to a critical configuration allowing one partner to flip over relative to the other within the [iso-$C_3H_7^+$, CH_3OH] α-complex (Fig. 11). This critical configuration corresponds to an entropic bottleneck and, energetically, to the onset of a plateau. When the C-O bond is further stretched, the energy remains almost constant up to 4 Å and the ion can turn freely within the complex with respect to the neutral. At any point of this plateau, a H-transfer yields, almost without energy barrier, but with an entropic bottleneck, the more stable [C_3H_6, $CH_3OH_2^+$] β-complex (Fig. 9c). Therefore, the α-complex lies between two entropic bottlenecks and may have a significant life-time, even if it doesn't lie at an energy minimum [8c]. These aspects have been developed by Morton for different kinds of complexes [2e].

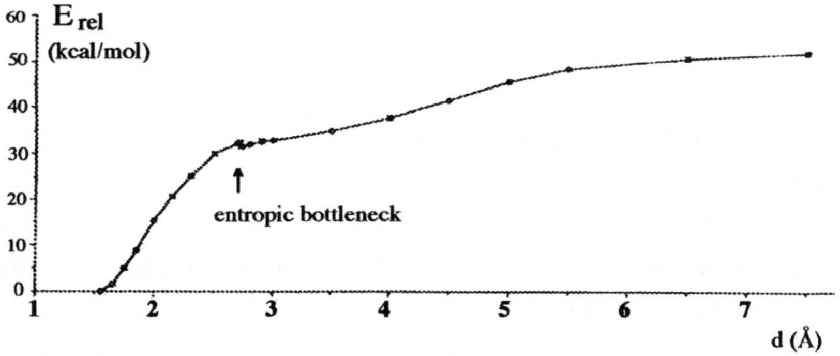

Figure 11 Relative energy of protonated methyl isopropyl ether according to the C_3H_7—O elongation.

5. Conclusion

- Complexes are key intermediates in rearrangement processes and in a great number of reactions (H-exchange, group transfer, reorganization of the initial ion). The data can be generally rationalized on the basis of the double well model.
- A great number of complementary experimental methods can be used to demonstrate the intermediacy of complexes. Experimental results must be confronted to calculation to verify if the complex corresponds to an energy minimum or not, to know its geometry, its energy and the components of the interaction energy.
- From these calculations, it is possible to distinguish between two limit kinds of complexes: (i) *electrostatic complexes* in which the charge transfer is negligible and the distance between both entities is large; these complexes are generally flexible, allowing a facile reorientation; (ii) *charge transfer complexes* in which a significant part of the interaction energy comes from a charge transfer which often creates a rigidity of the system; these complexes are compact and the attractive and repulsive electrostatic forces are strong; (iii) between these limit cases, there is a *continuum of situations*.
- The most stable structures do not necessarily correspond to reactive structures. These often lie rather close to the transition state. For this reason it is sometimes difficult to distinguish between a stepwise and a concerted mechanism.
- It is not necessary for a complex to be an energy minimum to operate in a rearrangement. When a complex lies between two entropic bottlenecks, its life-time can be significant, even if it does not correspond to an energy minimum.
-Finally, the question of vocabulary must be raised: have we to speak about "complexes", of "weakly bonded species" or of different kinds of weakly bonded species ?

6. References

1. McLafferty, F.W. and Turecek, F. (1993) *Interpretation of Mass Spectra*, University Science Book, Mill Valley, California.
2. a) McAdoo, D.J. (1988) Ion-Neutral Complexes in Unimolecular Decompositions, *Mass Spectrom. Rev.* **7,** 363-393; b) Hammerum, S. (1988) Formation and Stabilization of Intermediate Ion-Neutral Complexes in Radical Cation Dissociation Reactions, *J. Chem. Soc. Chem. Comun.*, 858-859; c) Bowen, R. D. (1991) Ion-Neutral Complexes *Acc. Chem. Res.* **24**, 364-371; d) Longevialle, P. (1992) Ion-Neutral Complexes in the unimolecular reactivity of organic cations in the gas phase,

Mass Spectrom. Rev. **11**, 157-192; e) Morton, T.H. (1992) The Reorientation Criterion and Positive Ion-Neutral Complexes, *Org. Mass Spectrom.* **27**, 353-368.

3. a) Farneth, W.E., Brauman, J.I. (1976) Dynamics of Proton Transfer Involving Delocalized Negative Ions in the Gas Phase, *J. Am. Chem. Soc.* **98**, 7891-7898; b) Brauman, J.I. (1979) in Ausloos, P. (ed.), *Kinetics of Ion-Molecule Reactions*, Plenum Press, New York, p. 153.

4. a) Morton, T.H. (1980) Ion-Molecule Complexes in Unimolecular Fragmentations of Gaseous Cations, *J. Am. Chem. Soc.* **102**, 1596-1602; b) Morton, T.H. (1982) Gas Phase Analogues of Solvolysis Reactions, *Tetrahedron* **38**, 3195-3243; c) Sozzi, G., Audier, H.E., Mourgues, P., Milliet, A. (1987) Alkyl phenyl ether radical cation in the gas phase, *Org. Mass Spectrom.* **22**, 746-747; d) Kondrat, R.W., Morton, T.H. (1988) Stereochemical differentiation in the decomposition of *sec*-butyl phenyl ether molecular ions, *Org. Mass Spectrom.* **23**, 555-557; e) Audier, H. E., Morton, T. H. (1991) Backside displacement in the unimolecular gas phase decarboxylation of Alkyl Phenyl Carbonate Radical Cations, *J. Am. Chem. Soc.* **113**, 9001-9003; f) McAdoo, D.J. and Morton, T.H. (1993) Gas-Phase Analogues of Cage Effects, *Acc. Chem. Res.* **26**, 295-302.

5. a) Bowen, R.D., Stapleton, B.J. and Williams, D.H. (1978) Non-concerted Unimolecular Reactions of Ions in the Gas Phase: Isomerisation of weakly Coordinated Carbenium Ions, *J. Chem. Soc. Chem. Commun.*, 24-26; b) Bowen, R.D. and Williams, D.H. (1980) Unimolecular Reactions of Isolated Organic Ions, *J. Am. Chem. Soc.* **102**, 2752-2756; c) Bouchoux, G., Penaud-Berruyer, F., Tortajada, J., Audier, H. E. and Mourgues, P. (1997) Reactivity of $C_3H_7O^+$ cations : a combined MS-MO approach, *J. Mass Spectrom.* **32**, 188.

6. a) Budzikiewicz, H., Djerassi C. and Williams D.H. (1993), *Mass Spectrometry of Organic Compounds*, Holden-Day, San Francisco; b) Bouchoux, Tortajada, J., G., Dagaut, J. and Fillaud, J. (1987) Unimolecular dissociations of the 2-hexanone$^{\bullet+}$ metastable ion, *Org. Mass Spectrom* **22**, 451-457; c) Bouchoux, G. (1988) Keto-enol tautomers and distonic ions: the chemistry of $C_nH_{2n}O^{+\cdot}$ radical-cations, Part I : *Mass Spectrom. Rev.* **7**, 1-39, Part II : *Mass Spectrom. Rev.* **7**, 203-255 ; d) Audier, H.E., Denhez, J.P. and Mourgues, P. (1989) Competitive isomerization processes of 5-methyl 2-hexanone radical cations, *Rapid Commun. Mass Spectrom.* **3**, 273-275.

7. Tortajada, J., Berthomieu, D., Morizur, J.P. and Audier, H. E. (1992) [Acylium, Ketene] ion-neutral complexes in the gas phase, *J. Am. Chem. Soc.* **114**, 10874-10880.

8. a) Harrison, A.G. (1987) Hydrogen Interchange Prior to the fragmentation of Protonated Molecules, *Org. Mass Spectrom.* **22**, 637-641; b)Terlouw, T.K., Weiske, T., Schwarz, H. and Holmes, J.L. (1986) Study of Ion structures Produced by the reaction of 2-Propyl Cations with Water and Methanol; Covalently Bound v. Hydrogen-bridged Adducts, *Org. Mass Spectrom.* **21**, 665-671; c) Audier, H.E., Monteiro, C., Berthomieu, D. and Tortajada, J. (1991) Complexes ion-molecule

intermédiaires dans la fragmentation des éthers protonés, *Int. J. Mass Spectrom. Ion Proc.* **104**, 145-161; d) Bouchoux, G., Hoppilliard, Y. (1990) Role of ion neutral complexes during acid-catalyzed dehydration of ethanol in the gas phase, *J. Am. Chem. Soc.* **112**, 9110-9115; e) Audier, H.E. and Morton, T.H. (1993) Rearrangements in Metastable Ion Decompositions of Protonated Methylamines *Org. Mass Spectrom.* **28**, 1218-1224; f) Berthomieu, D. and Audier, H. E. (1997) Intermediate Complexes in the Gas Phase Reactions of tert-butyl Cation with Water. *Eur. J. Mass Spectrom.* **3**, 19.

9. a) Hermann, J.A., Harrison, A.G. (1981) Effect of Protonation Exothermicity on the Chemical Ionization Mass Spectra of Some Alkylbenzenes, *Org. Mass Spectrom.* **16**, 423-427; b) Holman, R.W., Gross, M.L. (1989) Interconversion of Gas-Phase Arenium Ions and π-Complexes, *J. Am. Chem. Soc.* **111**, 3560-3565; c) Denhez, J.P., Monteiro, C., Mourgues, P. and Audier, H.E. (1991) Propylation en phase gazeuse: étude des ions dipropylbenzène protonés *New. J. Chem.* **15**, 273-278; d) Berthomieu, D., Audier, H.E., Denhez, J.P., Monteiro, C. , Mourgues, P. (1991) [C_6H_6, $C_3H_7^+$] and [$C_6H_7^+$, C_3H_6] ion neutral complexes intermediates in the reactions of protonated propylbenzenes. *Org. Mass Spectrom.* **26**, 271-275; e) Berthomieu, D., Brenner, V., Ohanessian, G., Denhez, J.P., Millié, P. and Audier, H.E. (1995) Ion-Neutral Complexes of Protonated Alkylbenzenes, *J. Phys. Chem.* **99**, 712-720; f) Giacomello, P. and Cacace, F. (1976) Gas-Phase Alkylation of Xylenes by t-$C_4H_9^+$ Ions, *J. Am. Chem. Soc.* **98**, 1823-1828; g) Attina, M., Cacace, F. (1992) Concerning sterical hindrance to deprotonation of gaseous polyalkyl benzenium ions, *Int. J. Mass Spectrom. Ion Processes* **120**, R1-R4; Kuck, D. (1990) Gas phase ion chemistry of protonated alkylbenzenes, *Mass Spectrom. Rev.* **9**, 583-630.
10. a) Wagner-Redecker, W., Kemper, P.R., Jarrold, M.F. and Bowers, M.T. (1985), *J. Phys. Chem.* **89**, 1121; b) Petrie, S., Freeman, C.G., Maut-Ner, M., McEwan, M.J. and Ferguson, E.E. (1990) Experimental Study of HCN^+ and HNC^+ Ion Chemistry, *J. Am. Chem. Soc.* **112**, 7121-7126; c) Bohme, D.K. (1992) Proton transport in the catalyzed gas-phase isomerization of protonated molecules, *Int. J. Mass Spectrom. Ion Processes* **115**, 95-110.
11. This book, following article.
12. Audier, H.E., Mourgues, P. and Van der Rest, G. to be published.
13. Bierbaum, V.M., De Puy, C.H. and Shapiro, R.H. (1977) Gas Phase Reactions of Anions with Nitrous Oxide and Carbon Dioxide, *J. Am. Chem. Soc.* **99**, 5800-5802.
14. Longevialle, P. and Botter, R. (1980) Evidence for Intramolecular Interaction between Ionic and Neutral Fragments in the Mass Spectrometer, *J. Chem. Soc. Chem. Commun.*, 823-825.
15. a)Audier, H.E. , Koyanagi, G.K., McMahon, T.B. and Thölmann, D. (1996) Evidence for non-covalent gas phase ion-neutral complexes of the form [CH_3^+, CH_3OH] and [CH_3^+, CH_3OCH_3], *J.*

Phys. Chem. **100**, 8220; b) Nguyen, M.T., Vanquikenborne, L.C. and Bouchoux, G. (1993), On the energy barrier for 1,2-elimination of methane from the dimethyloxonium cation, *Int. J. Mass Spectrom. on Processes* **124**, R11-R14.

16. Audier, H.E., Mourgues, P., Leblanc, D. and Hammerum, S. (1993) Bimolecular alkene displacement reactions in the gas phase. *C. R. Acad. Sci.* **317**, serie II, 27.
17. a) Kebarle, P. (1988) in J.M. Farrar and W.H. Saunders Jr. (eds.), *Techniques of Chemistry*, Vol. 20, Wiley, New York; b) McMahon, T.B. (1996) Experimental Approaches to the Unimolecular Dissociation of Gaseous Cluster Ions in N. Adams and L.M. Babcock (eds.), *Advances in Gas Phase Ion Chemistry*, Vol. 2, pp. 41-85; c) Norman, K. and McMahon, T.B. (1996) Covalent and Electrostatic Complexes in Association Reactions of *tert*-Butyl cation with Small Organics, *J. Am. Chem. Soc.* **118**, 2449-2457.
18. a) Polce, M.J. and Wesdemiotis, C. (1996) Stabilities and isomerization proclivities studied by dissociation and neutralization-reionization, *J. Am. Soc. Mass Spectrom.* **7**, 573-589; b) Holmes, J.L., Lossing, F.P., Terlouw, J.K. and Burgers, P.C. (1982) $[CH_2OH_2]^{•+}$ and Related Stable Gas-Phase Ion-Dipole Complexes, *J. Am. Chem. Soc.* **104**, 2931-2932; c) Wedesmiotis, C. and McLafferty, F.W. (1987) *Chem. Rev.* **87**, 485; d) Terlouw, J.K. and Schwarz, H. (1987) *Angew. Chem. Int. Ed. Engl.* **26**, 805; e) Holmes, J.L. (1989) The neutralization of organic cations, *Mass Spectrom. Rev.* **8**, 513-539; f) Levine, R.D. and Bernstein, R.B. (1987), *Molecular Reaction Dynamics and Chemical Reactivity*, Oxford University Press, Oxford; g) Gerlich, D. (1992) in C.Y. Ng and T. Baer (eds.), *Adv. Chem. Phys.* **82**, Wiley & Sons, New York, p. 1.
19. Kitaura, K. and Morokuma, K. (1976) A New Energy Decomposition Scheme for Molecular Interactions within the Hartree-Fock approximation, *Int. J. of Quantum Chemistry* **325**, 340.
20. Tortajada, J., conference at the Workshop on Weakly Bonded Species, Palaiseau (1996).
21. Bader, R.F.W. (1990) *Atoms in Molecules. A quantum theory*, Oxford University Press, Oxford.
22. a) Burgers, P.C., van den Berg, K.J., Visser, H. and Terlouw, J.K. (1990) Captodative substituted $C_4H_7O_2^+$ carbenium ions and their relationship with the proton-bound complex $CH_3CHO...H^+...CH_2CH_2O$, *Int. J. Mass Spectrom. Ion Phys.* **101**, 83-100; b) Heinrich, N., Drewello, T., Burgers, P.C., Morrow, J.C., Schmidt, J., Kulik, W., Terlouw, J.K. and Schwarz, H. (1992) Unimolecular Chemistry of methyl formate cation radical. A combined Theoretical and Experimental Study, *J. Am. Chem. Soc.* **114**, 3776-3783.
23. Larson, J.W. and McMahon, T.B. (1982), Formation, thermochemistry, and Relative Stabilities of Proton-Bound Dimers of Oxygen n-Donors Bases from ion Cyclotron Resonance Solvent-Exchange Equilibria Measurements, *J. Am. Chem. Soc.* **104**, 6255-6261.

INTERMEDIATES IN IONIC GAS PHASE ORGANIC REACTIONS: II. DISTONIC IONS

H.E. AUDIER, J. FOSSEY, D. LEBLANC,
P. MOURGUES and V. TROUDE
Laboratoire des Mécanismes réactionnels, CNRS,
Ecole Polytechnique, 91128, Palaiseau, France.

Distonic ions are stable open shell ionic species which intervene in a number of chemical processes in the gas phase. They are often well defined intermediates in the unimolecular decomposition of radical cations or in their bimolecular reactions.

1. Distonic Ions as Stable Species

1.1. DEFINITION

A distonic ion is a radical cation which would arise, formally, from ionization of a zwitterion or of a diradical [1-2]. Distonic ions are mostly isomers of molecular ions in which the charge and the radical are not borne by the same atoms. For instance, to ionized methanol $CH_3OH^{+\bullet}$ corresponds ${}^\bullet CH_2OH_2^+$, to ionized oxirane corresponds ${}^\bullet CH_2OCH_2^+$. Depending on the relative position of the charge and of the radical, they are refered as α, β, γ or δ-distonic ions:

$CH_3CH_2OH^{\bullet +}$	$CH_3{}^\bullet CHOH_2^+$	${}^\bullet CH_2CH_2OH_2^+$
Molecular ion	α-distonic ion	β-distonic ion

Mass spectrometrists have always postulated many such ions in order to rationalize their mass spectra [3]. However, for many years the question whether these distonic ions were

K. R. Jennings (ed.), Fundamentals and Applications of Gas Phase Ion Chemistry, 27–53.

stable or transient species was not addressed. It is only in the 70'ies that distonic ions were generated and studied for themselves [4-5]. In 1982-84 Radom showed by *ab initio* calculations that α and β-distonic ions generally correspond to minima on the potential energy surface [2] [6-7] and therefore are stable species. He proposed the definition given at the beginning of this paragraph [1-2]. Since 1982, the unimolecular and bimolecular reactions of distonic ions [1] [8] have been studied by several laboratories. Distonic ions can also be formed in condensed phase and characterized by ESR [1]. They are also intermediates in various reactions in solution (the Hoffman-Löffler-Freytag reaction [9]) or in biological processes.

1.2. FORMATION

Since distonic ions are stable species, it is often possible to specifically generate and characterize these ions. They are generally prepared by fragmentation of a radical cation, either by a ring opening or an irreversible H-transfer, followed by a simple cleavage (Scheme 1). Ion-molecule reactions can also be used to synthetize other distonic ions [8b].

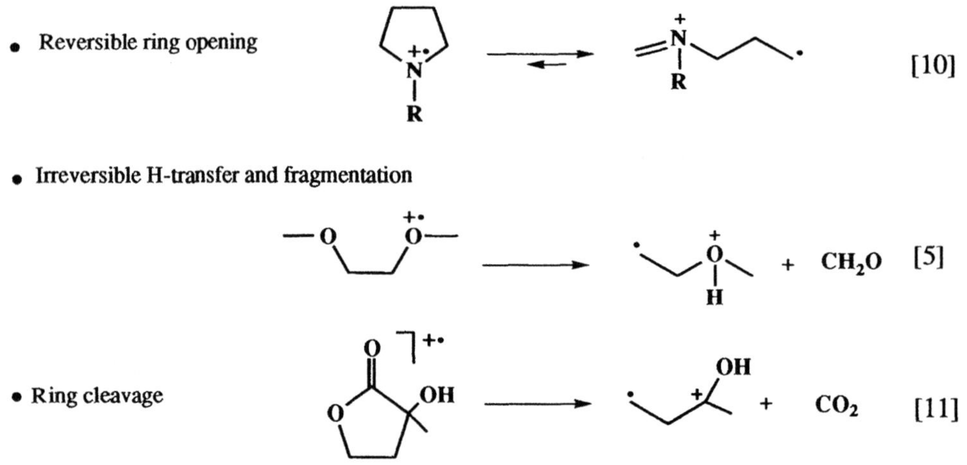

Scheme 1 : Different routes to distonic ions.

1.3. CHARACTERIZATION

i/ The most convenient method to characterize a stable distonic ion is to perform its CID spectrum which often allows to distinguish a molecular ion and its distonic counterparts (e.g. $CH_3OH^{\bullet+}$ (a) *versus* $^{\bullet}CH_2OH_2^+$ (b); figure 1 [12]). The formation of multiply charged ions give rise to the narrow peak m/z 16 (32^{++}).

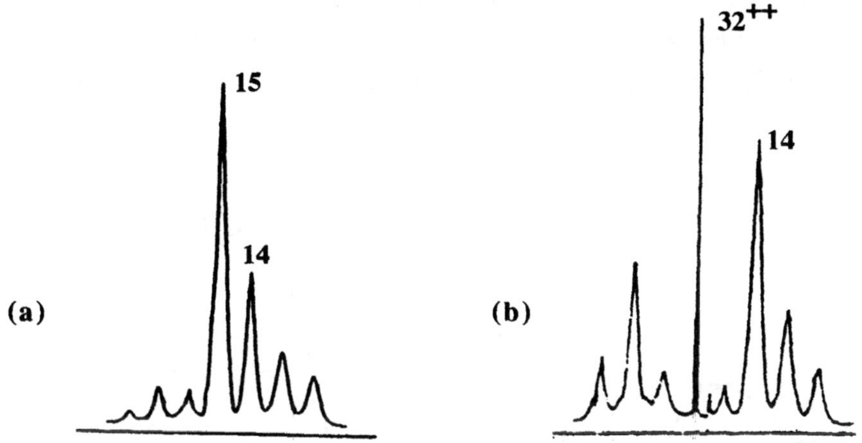

Figure 1: CID Spectra of (a) $CH_3OH^{+\bullet}$ and (b)$^{\bullet}CH_2OH_2^+$ [12].

ii/ Neutralization-reionization [13] can also be used. In contrast with molecular ions (Eq. 1), no long-lived neutral species correspond to distonic ions. Within the microsecond time window of the NRMS experiments diradical dissociates (Eq. 2). When a small fraction of the diradical escapes from decomposition, the "recovery peak" or "survivor ion" is significantly smaller for the distonic ion than for his tautomers after a neutralization-reionization process. Such property probably results from the less favorable Franck-Condon factor of the reionization of a diradical vis-à-vis the reionization of a closed-shell molecular ion [14].

$$CH_3CH_2OH^{+\bullet} \xrightarrow{\text{Neutralization (Xe)}} CH_3CH_2OH \xrightarrow{\text{Reionization } (O_2)} CH_3CH_2OH^{+\bullet} \qquad (1)$$

$$^{\bullet}CH_2CH_2OH_2^+ \xrightarrow{\text{Neutralization}} CH_2CH_2 + H_2O \xrightarrow{\text{Reionization}} CH_2CH_2^{+\bullet} + H_2O^{+\bullet} \qquad (2)$$

iii/ The use of bimolecular reactions can be a very sensitive method: for each distonic ion it is generally possible to find a reaction characteristic of its structure (see further).

1.4. ENERGETICS

The ΔH_f of a distonic ion [15] can be determined by experiment (e.g. Appearance Energy measurements) or calculated. Radom has shown [7] that distonic ions can be more stable than the molecular ion counterparts (Table 1) and that a significant barrier separates both species (Figure 2). Therefore, unimolecular isomerization is only rarely observed.

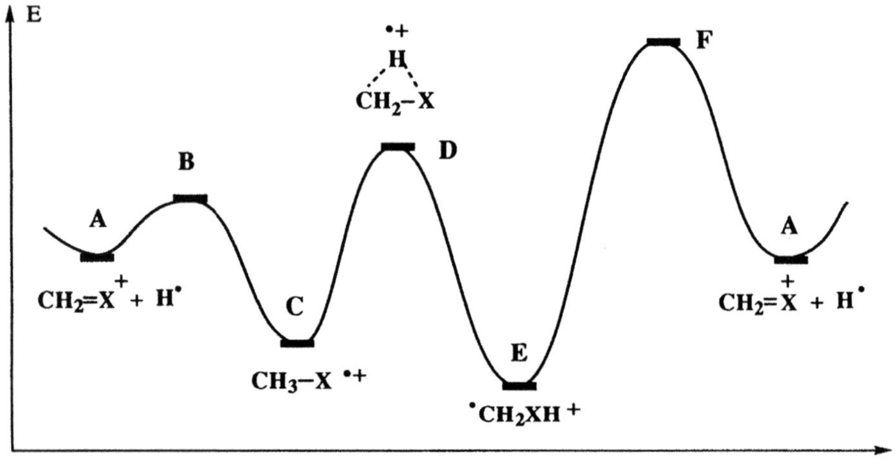

Figure 2: Calculated energy profile [7]

X	A	B	C	D	E	F
NH_2	29	35	2	42	0	48
OH	26	28	11	38	0	48
F	23	23	11	31	0	46

Table 1: Calculated energy values (kcal mol^{-1}) [7]

2. Early works on bimolecular reactions of distonic ions

The introduction of ICR techniques [16] to study ion-molecule reactions has open an extensively used field allowing the study of bimolecular reactions of distonic ions, and therefore the differentiation of ionic isomeric structures.

2.1. PROTON TRANSFER

The first historical example which can be related to bimolecular reactivity of a distonic species is the proton transfer to more basic neutrals from dialkyl-N-nitrosamines radical cations, which do have a distonic structure [17], although the exact structure was not ascertained. Some years later, Morton and Beauchamp [5] show that proton transfer could characterize the $^{\bullet}CH_2CH_2OHCH_3^+$ ion. It will be shown in the following part that proton transfer to the neutral is a characteristic of different kinds of distonic ions.

2.2.REACTIONS OF CYCLOPROPANE $^{\bullet+}$

Another early work in that field was the attempt to characterize the open distonic form of cyclopropane radical cation [18] by the study of its gas-phase reaction with NH_3: transfer of $CH_2^{\bullet+}$ occurs on ammonia, what could be indicative of the existence of $^{\bullet}CH_2CH_2CH_2^+$. However it was later shown that the cyclopropane radical cation indeed exists with a cyclic structure [19], and that in this case addition of NH_3 most probably occurs before ring opening and elimination of neutral ethylene (Eq. 3).

$$\Delta^{\bullet+} + NH_3 \rightarrow {}^{\bullet}CH_2CH_2CH_2NH_3^+ \rightarrow {}^{\bullet}CH_2NH_3^+ + CH_2CH_2 \qquad \underline{3}$$

2.3. $CH_2^{\bullet+}$ TRANSFER

Conversely, the open structure of oxirane $^{\bullet+}$ is a well defined, stable ion $^{\bullet}CH_2OCH_2^+$, which presents a very characteristic reactivity, transferring $CH_2^{\bullet+}$ to a great number of neutral molecules [8] [20]. By far the most studied through the 70'ies , this process was later used to prepare well-defined α-distonic ions such as $C_6H_5N^+\text{-}CH_2^{\bullet}$ (Eq. 4).

$$^{\bullet}CH_2OCH_2^{+} + C_6H_5N \rightarrow C_6H_5N^{+}\text{-}CH_2^{\bullet} + CH_2O \qquad \underline{4}$$

2.4. $CH_2CH_2^{\bullet +}$ TRANSFER

Another early study was devoted to the β-distonic ion $^{\bullet}CH_2CH_2OHCH_3^{+}$, which was shown to transfer ionized ethylene to a number of nitriles. The failure of such a transfer with other neutrals studied at this time by Busch et al. [21] prompted the authors in designing that reaction as specific to nitriles. But it was later shown [22] that such $CH_2CH_2^{\bullet +}$ transfer occurs onto a wide range of neutral compounds (dimethylether, ethanol, methanol, propene, ...), provided that their proton affinity is not too high. Nevertheless, the reactivity observed towards nitriles could well correspond to a different mechanism for the transfer of $CH_2CH_2^{\bullet +}$. It cannot occur by a radical mechanism because the nitrogen would become hypervalent [23].

2.5. THE AMBIDENT STRUCTURE OF THE $^{\bullet}CH_2CH_2OCH_2^{+}$ β-DISTONIC ION

Transfer of ionized ethylene onto neutrals is also a typical reaction of β-distonic ions, such as $^{\bullet}CH_2CH_2OCH_2^{+}$. The reactivity of this latter ion (formed by loss of CH_2O from 1,4-dioxane radical cation) reflects well the ambident distonic structure [24]. Protonation of the neutral, as well as formylation of pyridine at the nitrogen atom, were rationalized as reactions occurring at the ionic site while transfer of ionized ethylene on acetonitrile was considered to involve the CH_2CH_2 part of the distonic ion (Scheme 2).

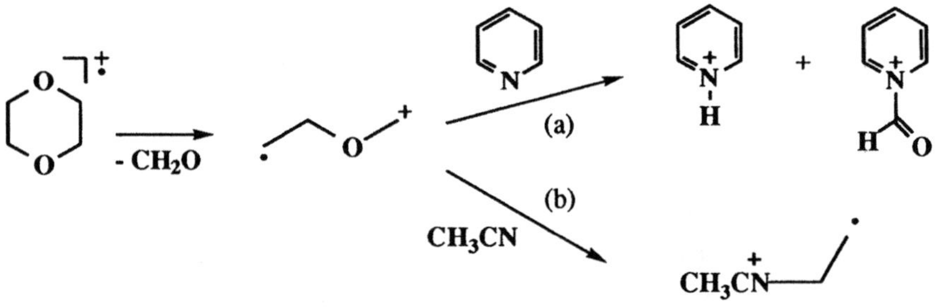

Scheme 2: $^{\bullet}CH_2CH_2OCH_2^{+}$ as an electrophilic reactant (route a) or a $CH_2CH_2^{+\bullet}$ transfer agent (route b). [24].

2.6. THE RADICAL SITE

New reactions have recently been introduced, such a transfer of $CH_3S^{\bullet}$ from dimethyldisulfide, which can be thought as occurring at the radical site of the ion [27]. For instance, such an abstraction takes place with the γ-distonic open structure of cyclobutanone, $^{\bullet}CH_2CH_2CH_2CO^+$, whereas this reaction is not observed with other isomers. In a similar way, $CH_3S^{\bullet}$ is abstracted by the already discussed distonic ion $^{\bullet}CH_2CH_2OCH_2^+$, and not by cyclic alkylether radical cations. Unfortunately this reaction appears not to be very general, charge exchange or protonation being often the only reactions observed with many even so well-defined distonic structures. This problem was tentatively overcome by the same group using dimethyl diselenide [28].

In the following part, this talk will be focussed on the bimolecular reactions of α and β-distonic ions isomers of alcohols$^{+\bullet}$, ethers$^{+\bullet}$ and amines$^{+\bullet}$. It will be shown that it is often possible to distinguish the reactions at the charge site and at the radical site of a distonic ion. However, the problem can be more complicated for α-distonic ions since appropriate neutrals can catalyse the reversible isomerization into their isomer molecular ions.

3. Reactions of α-distonic ions of alcohols, ethers or amines.

3.1. ISOMERIZATION BETWEEN MOLECULAR IONS AND α-DISTONIC IONS

It has been precedently shown that the unimolecular interconversion between molecular ions and α-distonic ions is spontaneously impossible. However, in the presence of basic neutral molecules, the reactions of the molecular ion and its α-distonic isomer often yield the same products, in the same ratio and at the same rate. This strongly suggests that, within a complex, the neutral base N catalyzes [29-30] the following isomerization (Eq. 5) which will now be discussed.

$$[^{\bullet}CH_2XH^+, N] \rightleftharpoons [CH_3X^{+\bullet}, N] \qquad \underline{5}$$

- *The reaction only yields the most stable isomer: the methanol case.*

Unimolecular reactions of ionized methanol were studied on a double focusing mass spectrometer (B/E geometry) with a chemical ionization ion-source. When the ion-source pressure is rather low (10^{-6} τorr), $CH_3OH^{+\bullet}$ gives the characteristic CID spectrum (Figure 1a). Injection of water into the ion source leads to a change in this CID spectrum. The CID spectrum becomes (b) which is characteristic of the $^{\bullet}CH_2OH_2^+$ structure (Figure 1b). This is not a collision induced process since there is no change in the CID spectrum of $CH_3OH^{+\bullet}$ when there is a pressure of 10^{-5} τorr of N_2 in the ion-source.

What happens is that H_2O catalyzes the isomerization of $CH_3OH^{+\bullet}$ into $^{\bullet}CH_2OH_2^+$. Correspondingly, since the α-distonic ion is more stable than the molecular ion, the introduction of water in the ion-source does not change the CID spectrum of $^{\bullet}CH_2OH_2^+$ (Eq. 6).

$$CH_3OH^{+\bullet} + H_2O \rightarrow [CH_3OH^{+\bullet}, H_2O] \downarrow [^{\bullet}CH_2OH_2^+, H_2O] \rightarrow {}^{\bullet}CH_2OH_2^+ + H_2O \qquad \underline{6}$$

- *Role of the base*

This proton transport (Eq. 5 and 6) is only operative for a subset of neutral bases determined largely by the proton affinity (PA) of the base involved. When PA[N] is too high and above PA[$^{\bullet}CH_2OH$] at the oxygen site (e.g. methanol), the proton transfer to the neutral corresponds to the most stable final state. This process is exothermic and proceeds at, or close to, the collision rate. Thus, in the case of methanol with no labelling involved, it is not immediately apparent whether or not proton transfer is accompanied by isomerization of the radical cation to the distonic ion. Conversely, when PA[N] is too low (e.g.: N_2O) and under PA[$^{\bullet}CH_2OH$] at the carbon site, the base is not strong enough to abstract and transport the proton within the [$CH_3OH^{+\bullet}$, N] complex. In such cases, the isomerization is no longer observed.

- *other systems*

The isomerization of $CH_3NH_2^{+\bullet}$ (spectrum c) into the more stable ion $^{\bullet}CH_2NH_3^+$ (spectrum e; figure 3) can similarly be catalyzed by acetone. This neutral is efficient since PA [acetone] lies only some kcal mol^{-1} under that of the $^{\bullet}CH_2NH_2$ radical [29b].

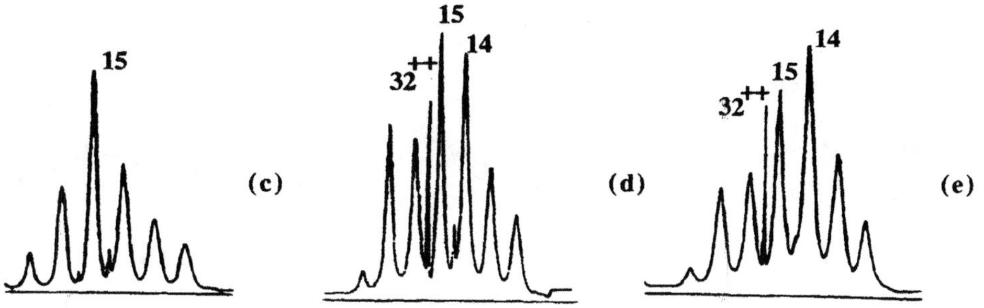

Figure 3 : CA spectra of $CH_3NH_2^{+\bullet}$ (spectrum c), $^{\bullet}CH_2NH_3^{+}$ (spectrum e) and of $CH_3NH_2^{+\bullet}$ after isomerization catalyzed by acetone(d) .

In contrast, the α-distonic ion $^{\bullet}CH_2BrH^{+}$ is less stable than $CH_3Br^{+\bullet}$. Therefore H_2O isomerizes $^{\bullet}CH_2BrH^{+}$ into $CH_3Br^{+\bullet}$ while $CH_3Br^{+\bullet}$ is unreactive towards H_2O [29b].

-The interconversion between solvated ions leads to H-exchange.
In the cell of an ICR, both $CH_3OH^{+\bullet}$ and $^{\bullet}CH_2OH_2^{+}$ rapidly exchange two H with D_2O and slowly the two others [29]. The rapid exchange confirms the isomerization of the [$CH_3OH^{+\bullet}$, H_2O] adduct into [$^{\bullet}CH_2OH_2^{+}$, H_2O] which undergoes subsequent H/D exchange with D_2O. More generally, by similar processes, it is possible to exchange all the α-hydrogens of a molecular ion by reaction with an appropriate labelled neutral.

$CH_3OCH_3^{+\bullet}$	exchanges	6H	with	CCl_3CH_2OD
$CH_3NHCH_3^{+\bullet}$	"	7H	"	ND_3
$CH_3CH_2NH_2^{+\bullet}$	"	4H	"	ND_3
$CH_3Br^{+\bullet}$	"	3H	"	D_2O
$CH_3CH_2Br^{+\bullet}$	"	2H	"	D_2O

3.2. A MECHANISM BY PROTON TRANSPORT

High-level *ab initio* calculations (UMP2/6-31G**) support the formation, in the first step of the reaction between methanol$^{+\bullet}$ and water, of two stabilized H-bonded adducts [31].

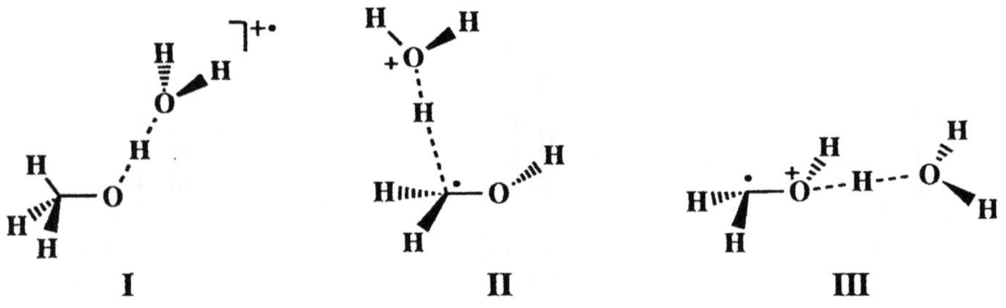

Figure 4

A molecule of water can approach ionized methanol either by the oxygen side or by the carbon side. In the first case, the stable H-bonded adduct **I** (Figure 4) is formed, with a stabilization energy of 29.7 kcal mol^{-1} compared to the reactants. In ion **I** the hydroxylic hydrogen of methanol is bonded to the oxygen of water. In the second case, the creation of a H-bond between the C(α)-bonded hydrogen of the ion and the oxygen of water leads to the transient adduct **II** [H_3O^+, $^{\bullet}CH_2OH$] which isomerizes into the stable H-bonded species **III** [$^{\bullet}CH_2OH_2^+$... OH_2] in which the α-distonic ion $^{\bullet}CH_2OH_2^+$ is H-bonded to a water molecule. Adduct **III** has the greatest interaction energy : 38.3 kcal mol^{-1}.

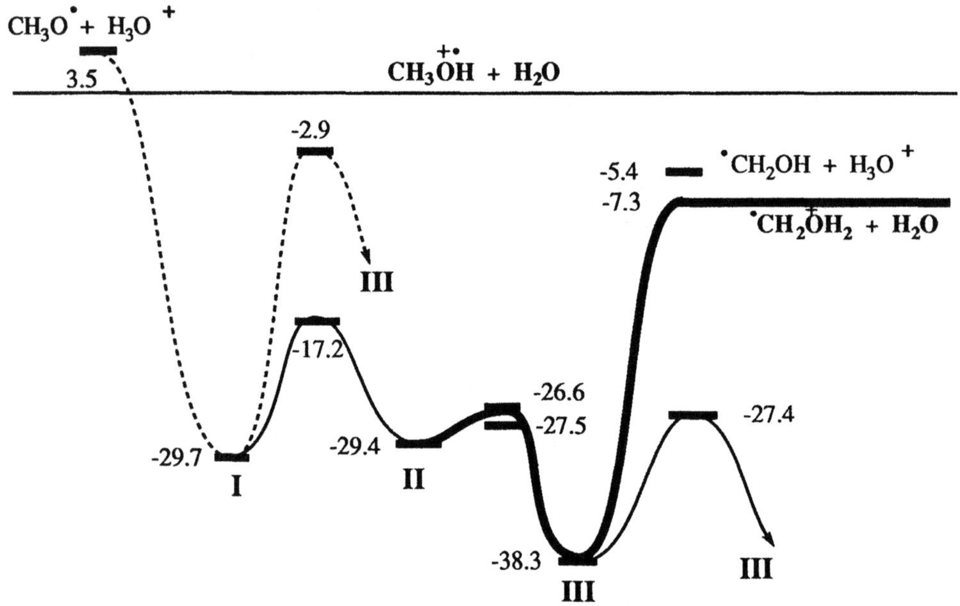

Figure 5 : Energy profile (kcal mol^{-1}) for the water-catalyzed interconversion of $CH_3OH^{+\bullet}$ to $^{\bullet}CH_2OH_2^+$ involving O-H hydrated (I and III) and C-H hydrated (II) species [31].

The reversible isomerization (**II**⇔**III**) occurs *via* two different transition states (three or five-membered ring) with in both cases a very low barrier. **III** can also easily isomerize by migration of the water molecule between the two O-bonded hydrogens. Finally **III** gives $^{\bullet}CH_2OH_2^+$ by simple cleavage and is the reactive structure for H/D exchange reactions (bold curve in Figure 5).

The rapid exchange of two hydrogens when ionized methanol reacts with D_2O is explained by (i) the interconversion **II**⇔**III**, followed by successive reactions of the such formed $^{\bullet}CH_2OH_2^+$ ion.

A significant energy barrier separates **I** and **II**. Adduct **I** is almost unreactive and appears to be a cul-de-sac in the reaction pathway. Since the isomerization (**III** ⇔ **I**) requires a significant energy barrier, the exchange of the two others α-H is slow (Figure 5).

3.3. REACTIONS OF α-DISTONIC IONS AND OF MOLECULAR IONS

Molecular ions $M^{+\bullet}$ and their α-distonic counterparts often react similarly with a neutral N since [$CH_3X^{+\bullet}$, N] $\rightleftharpoons$ [$^{\bullet}CH_2XH^+$, N]. For both ions, three kinds of reactions are observed : i/ Protonation of the neutral giving NH^+, ii/ $H^{\bullet}$ abstraction from the neutral giving MH^+, iii/ Loss of radical from the [$M^{+\bullet}$, N] encounter complex. Therefore, the problem is to determine the reactive structure (molecular or distonic ion ?) for each kind of reaction.

-Protonation of the neutral

Experiments show that the α-distonic ion is the reactive structure. When $CD_3OH^{+\bullet}$ reacts with CD_3OH in an ICR cell, protonation of the neutral by H^+ or D^+ is observed in a 1/1 ratio at the beginning of the reaction. Since the protonation is exothermic, the $^{\bullet}CD_2OHD^+$ distonic ion, formed by isomerization of $CD_3OH^{+\bullet}$, protonates rapidly the neutral either by H^+ or by D^+. The back reaction giving $CD_3OH^{+\bullet}$ or $CD_2HOD^{+\bullet}$ is very slow (Scheme 3) [29].

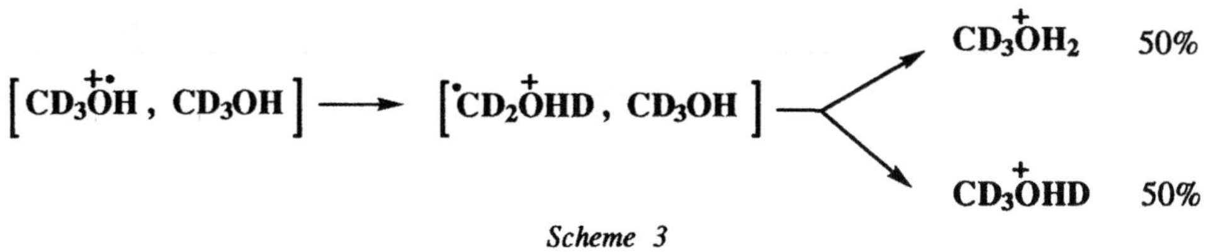

Scheme 3

The reaction of $CD_3NH_2^{+\bullet}$ with CD_3NH_2 leads similarly to a protonation by H^+ or D^+ in a 2/1 ratio since in the intermediate distonic ion $^{\bullet}CD_2NH_2D^+$ there are two hydrogens and one deuterium bonded to nitrogen. Finally, $CD_3NHCD_3^{+\bullet}$ protonates CD_3NHCD_3 by H^+ or D^+ in a 1/1 ratio (Scheme 4). Therefore, in the studied cases, the reactive form for protonation is the α-distonic ion.

$$\left[CD_3\overset{+\bullet}{N}H_2 \, , \, CD_3NH_2 \right] \longrightarrow \left[\overset{\bullet}{C}D_2\overset{+}{N}H_2D \, , \, CD_3NH_2 \right]$$

$$\left[CD_3\overset{+\bullet}{N}HCD_3 \, , \, CD_3NHCD_3 \right] \longrightarrow \left[\overset{\bullet}{C}D_2\overset{+}{N}HDCD_3 \, , \, CD_3NHCD_3 \right]$$

Scheme 4

The deprotonation method leads to measure the PA at the heteroatom of the α-radical $^{\bullet}CH_2X$. When

$^{\bullet}CH_2XH^+ + N \rightarrow {}^{\bullet}CH_2X + NH^+$ then PA [N] > PA [$^{\bullet}CH_2X$]

$^{\bullet}CH_2XH^+ + N \rightarrow$ nothing then PA [N] < PA [$^{\bullet}CH_2X$]

The results [32] are summarized in table 2, which shows that PA at the heteroarom of the $^{\bullet}CH_2X$ α-radical are between 12 and 16 kcal mol^{-1} less than that of corresponding CH_3X saturated species.

Radical	PA [$^{\bullet}CH_2X$] kcal mol^{-1}	Molecule	PA [CH_3X] kcal mol^{-1}
$^{\bullet}CH_2OH$	166	CH_3OH	182
$CH_3CHOH^{\bullet}$	174	CH_3CH_2OH	188
$^{\bullet}CH_2OCH_3$	180	CH_3OCH_3	192
$^{\bullet}CH_2NH_2$	202	CH_3NH_2	217
$CH_3CHNH_2^{\bullet}$	207	$CH_3CH_2NH_2$	219

Table 2

- H• abstraction from the neutral

Both $CH_3NH_2^{+\bullet}$ and $^\bullet CH_2NH_3^+$ ions react with diethylether or with acetone either by protonation of the neutral or by H• abstraction (Scheme 5). In order to know the reactive structure for this last reaction, two methods are possible:

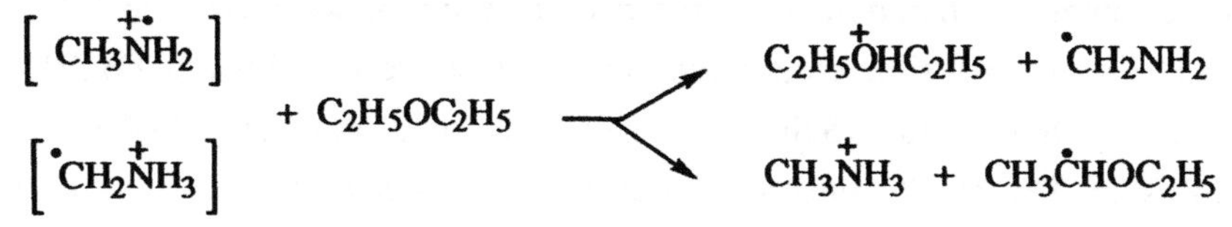

Scheme 5

The first one is to study reactions with labelled neutrals. By reaction with $D_5C_2OC_2D_5$, $CH_3NH_2^{+\bullet}$ would give the $CH_3NH_2D^+$ product-ion by D• abstraction while by the same reaction, $^\bullet CH_2NH_3^+$ would give $CH_2DNH_3^+$. Since the CID spectrum of the labelled protonated methylamine yields a pure CH_3^+ fragment, it can be concluded that the molecular ion is the reactive structure in the H• (or D•) abstraction reaction.

The second method is to use neutrals with low PA. It has been precedently shown that when PA [N] is significantly lower than PA [$^\bullet CH_2X$], isomerization between $CH_3X^{+\bullet}$ and $^\bullet CH_2XH^+$ does not take place. In such cases we observed that molecular ions react by H• abstraction but not the α-distonic ions (Scheme 6). Therefore the radical cation is the reactive form for H• abstraction.

$$CH_3NH_2^{+\bullet} + CH_3I \longrightarrow CH_3NH_3^+ + {}^\bullet CH_2I$$
$$^\bullet CH_2NH_3^+ + CH_3I \not\longrightarrow CH_3NH_3^+ + {}^\bullet CH_2I$$

$$CH_3OH^{+\bullet} + CH_4 \longrightarrow CH_3OH_2^+ + {}^\bullet CH_3$$
$$^\bullet CH_2OH_2^+ + CH_4 \not\longrightarrow CH_3OH_2^+ + {}^\bullet CH_3$$

Scheme 6

- Loss of a radical from the encounter complex

The reaction of $CH_3CD_2OH^{+\bullet}$ with water highlights the presence of two distinct reaction channels [33]. On the one hand a rapid exchange of one α-D takes place. This exchange involves the encounter complex **II'** (scheme 7) which isomerizes into the solvated α-distonic ion [$CH_3{}^\bullet CDOHD^+...H_2O$] **III'** . This intermediate dissociates to give the product-ions $CH_3{}^{\bullet}CDOHD^+$ and, after exchange with H_2O, $CH_3{}^{\bullet}CDOH_2{}^+$, but cannot eliminate a CH_3 radical (Scheme 7).

On the other hand, the encounter complex eliminates the terminal methyl group. The reaction is not preceded by H/D exchange. This indicated the existence of a distinct reaction channel which may involve hydrogen bonding between water and the original ethanol OH group. By simple cleavage this intermediate **I'** yields a $[CD_2OH^+ ... OH_2]$ product-ion with $^{\bullet}CH_3$ loss (Scheme 7) which only leads to a pure CD_2OH^+ fragment upon collision. This result, confirmed by calculations, indicates that a significant energy barrier separates the adducts **I'** and **II'**.

The behaviours of the [methanol$^{\bullet +}$ / H_2O] and [ethanol$^{\bullet +}$ / H_2O] systems are very similar except in one part : the energetic and entropic factors favour the $^{\bullet}CH_3$ loss in the ethanol case. For this reason, **I'** is no more a cul-de-sac in the reaction pathway and the molecular ion is the reactive structure in the $^{\bullet}CH_3$ loss reaction.

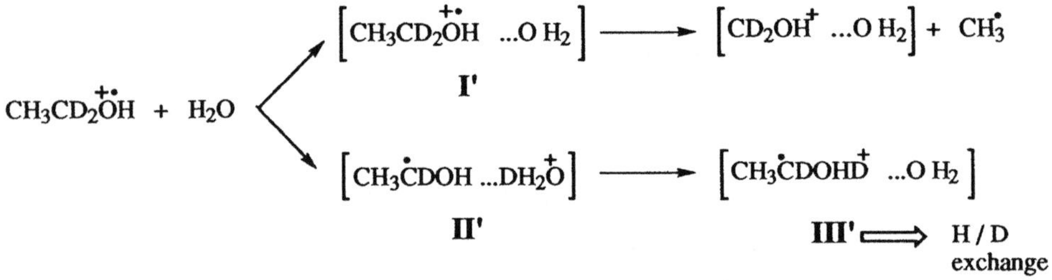

Scheme 7

When the reactant ion does not possess acidic hydrogens, the isomerization of the molecular ion into the α-distonic ion is generally the only open channel, as illustrated by the diethyl ether$^{+\bullet}$ + water reaction. Interaction between the H_2O molecule and the α–hydrogen atom results in isomerization to the α-distonic radical cation; this conversion is strongly exothermic, owing to the hydrogen bonding in the product adduct. By C-O bond cleavage, this intermediate yields a H-bonded product-ion wich $C_2H_5{}^{\bullet}$ loss [33] (Scheme 8).

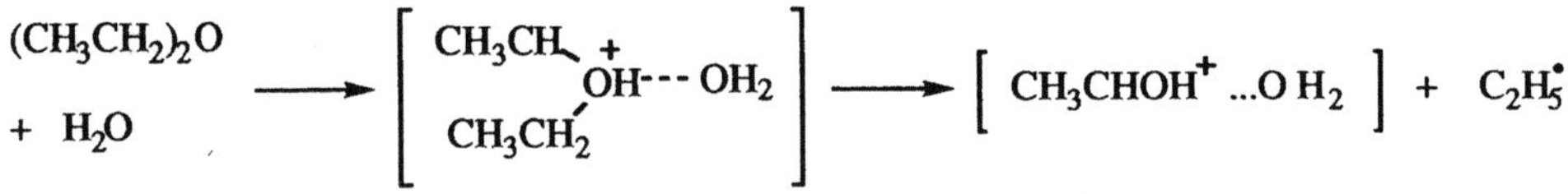

Scheme 8

In conclusion, H-bonding plays a determining role in these reactions, since it stabilizes the initial adduct and the final products. Depending on the structure of the reactants, the reactive structure in the radical loss reaction can be either the molecular ion or the α-distonic ion.

4. Reactions of β-distonic ions of alcohols, ethers and amines.

Small β-distonic ions isomerize neither into α-distonic ions nor into molecular ions. It is therefore possible to distinguish : i/ the reactions at the positive site, ii/ the reactions at the radical site, iii/ the substitution or addition reactions involving both sites. These reactions will be studied and compared for three β-distonic ions $^{\bullet}CH_2CH_2OH_2^+$ (**1**), $^{\bullet}CH_2CH_2OHCH_3^+$ (**2**) and $^{\bullet}CH_2CH_2NH_3^+$ (**3**).

4.1. STRUCTURES OF NAKED AND SOLVATED β-DISTONIC IONS.

-Naked ions

The structures of ions **1** and **3**, obtained by *ab initio* calculations, are represented figure 6 [22b] [34]. Compared to protonated ethanol (d = 1.547 Å), the C-O bond in ion **1** is longer (d = 1.604 Å). For ion **3** the length of the C-N bond is almost identical of that in protonated ethylamine (1.538 *versus* 1.519 Å). More important is the positive charge repartition. In ion **1**, 56 % of the charge is borne by the CH_2CH_2 moiety while in ion **3** only 36 % of the charge lies on this moiety. The α-carbon bears an important part of the positive charge in ion **1** (37 %) while this part is significantly less for ion **3** (22 %). Finally, the calculated spin density indicates that the β-carbon presents a pure radical character in all cases.

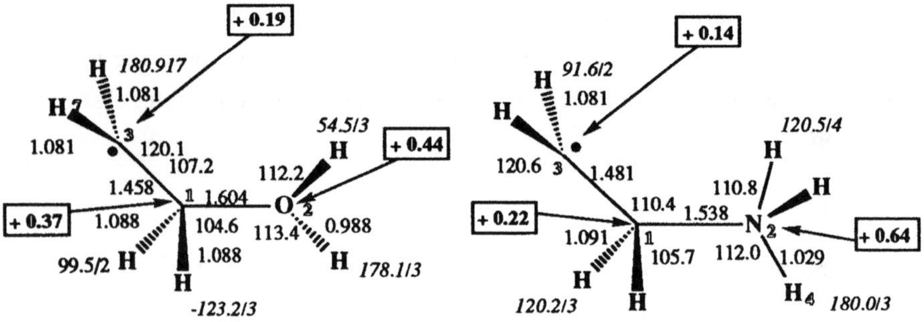

Figure 6 : EUMP2/6-31G optimized geometries for ions 1 and 3. Bond lengths in Å, angles and dihedral (italics) in degrees, net charges in framed numbers.*

-Solvated ions

When a molecule of H_2O approaches the oxygen side of ion **1**, calculations show that the H-bonded complex C_1 is formed. C_1, characterized by the existence of a H-bond between the hydroxylic hydrogen of the ion and the oxygen of the water molecule (Figure 7), is highly stabilized (~ 29 kcal mol^{-1}).

The attack of water on the radical carbon in the β-position gives a repulsive three electrons bond, and no critical point (minimum or transition structure) is found in this region. When the molecule of water approaches the α-carbon, the energy of the system decreases up to the ion-neutral complex C_2. In this complex electrostatic interactions exist between the charge borne by the α-carbon and the oxygen of water (Figure 7). The interaction energy is about 9 kcal mol^{-1}.

More generally, two similar complexes are found when ions **1**, **2** or **3** interact with a variety of neutrals [22b] [34]. The barrier for interconversion between both structures is low [35] but it is interesting to determine the role of each of them in the reactivity of the distonic ion.

Complex C_1

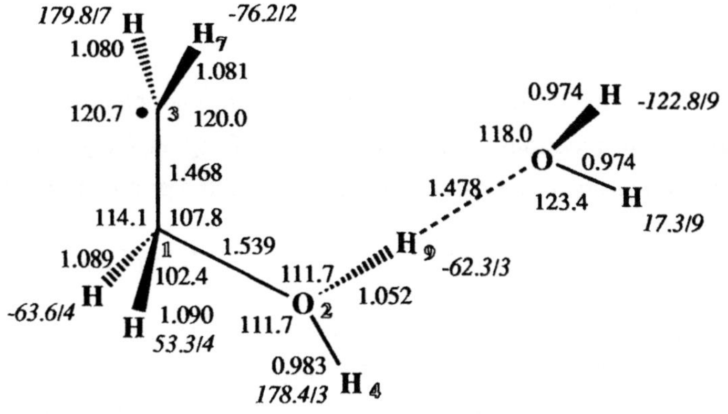

Complex C_2

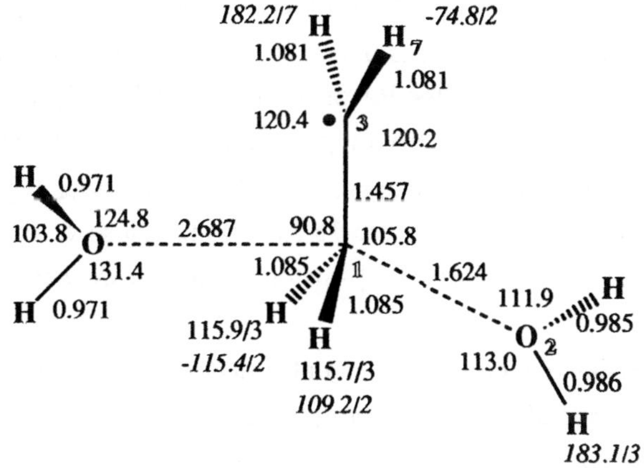

Figure 7 : EUMP2/6-31G optimized geometries for complexes C_1 and C_2.*

4.2. REACTIONS AT THE POSITIVE SITE

Protonation of the neutral is the most characteristic reaction at the positive site (Eq. 7). The reactions of labelled ions (i.e. $^{\bullet}CH_2CH_2OD_2^+$) indicates that the heteroatom bonded hydrogens are specifically involved in the proton transfer.

$$^{\bullet}CH_2CH_2XH^+ \ + \ N \quad \rightarrow \quad ^{\bullet}CH_2CH_2X \ + \ NH^+ \qquad \underline{7}$$

The H-bonded complexes whose structures are similar to C_1 (Figure 7) are the direct precursors for this reaction. For instance, in the corresponding complex of the ($^{\bullet}CH_2CH_2OH_2^+$/ CH_3OCH_3) system, the O-bonded hydrogen of the ion is already near the oxygen of CH_3OCH_3 (1.53 Å). Therefore, the simple cleavage of the complex may directly yield, with a very small energy barrier (about 1 kcal mol^{-1}), protonated dimethyl ether [22b].

The question arises to know if the β-distonic ions $^{\bullet}CH_2CH_2XH^+$ are more or less acidic than the corresponding saturated species $CH_3CH_2XH^+$. In other words, what is the influence of the radical on the PA of β–radicals $^{\bullet}CH_2CH_2X$? Experimental results and calculations show that the PA of β–radicals are almost the same than those of the corresponding saturated compounds [22b].This behaviour highly contrasts with that of α-radicals $^{\bullet}CH_2X$ whose PA are 15 kcal under those of the corresponding saturated compounds [32]. Therefore, the β-distonic ions are weaker acids than their α-distonic isomers.

The H-bonded complexes can also undergo C-heteroatom bond cleavage. A first example corresponds to the reactions, recently studied, of ion **2** with tert-butanol [36]. The such formed H-bonded complexes eliminate isobutene to yield solvated distonic ions (Eq. 8).

$$^{\bullet}CH_2CH_2O(H)CH_3^+ + \text{tert-}C_4H_9OH \rightarrow [^{\bullet}CH_2CH_2O(H)CH_3^+, H_2O\,] + \text{iso-}C_4H_8 \qquad \underline{8}$$

4.3. NUCLEOPHILIC SUBSTITUTION

The $C_2H_4^{+\bullet}$ transfer to the neutral yields in many cases another distonic ion (Eq. 9) [22] [24] [38]. The reaction can also be considered as a substitution of the XH group of the ion $^{\bullet}CH_2CH_2XH^+$ by the neutral .

$$^{\bullet}CH_2CH_2OH_2^+ + CH_2O \rightarrow {}^{\bullet}CH_2CH_2OCH_2^+ + H_2O \qquad \underline{9}$$

Extent of the reaction

This process is often considered as a characteristic reaction of β-distonic ions $^{\bullet}CH_2CH_2XH^+$. In fact this reaction is not general. If $^{\bullet}CH_2CH_2OH_2^+$ reacts by this way with a great number of neutrals [38], the transfer of $C_2H_4^{+\bullet}$ from $^{\bullet}CH_2CH_2NH_3^+$ is exceptional.

Thermochemistry can, in part, explain these differences. For instance, the reaction of $^{\bullet}CH_2CH_2OH_2^+$ with CH_2O leads to an exothermic substitution and is observed. By contrast, the substitution reaction is endothermic with $^{\bullet}CH_2CH_2NH_3^+$ and not observed.

Thermoneutral substitutions

However thermochemistry is not the only factor, as shown by the ICR study of thermoneutral substitutions. The rate of the thermoneutral reaction (substitution of XH by labelled *XH) is very different for the various ions [34]. Substitution of H_2O by D_2O is rapid for the $^{\bullet}CH_2CH_2OH_2^+$ ion ($k_{exp}/k_{coll} = 0.3$). Substitution of CH_3OH by CD_3OD in $^{\bullet}CH_2CH_2OHCH_3^+$ is slow ($k_{exp}/k_{coll} = 0.05$) while that of C_2H_5OH by C_2D_5OD in $^{\bullet}CH_2CH_2OH^+C_2H_5$ is very slow. Substitution of NH_3 by $^{15}NH_3$ is not observed in $^{\bullet}CH_2CH_2NH_3^+$ (Eq. 10 to 12).

$$^{\bullet}CH_2CH_2OH_2^+ + H_2{}^{18}O \rightarrow {}^{\bullet}CH_2CH_2{}^{18}OH_2^+ + H_2O \quad \text{rapid} \qquad \underline{10}$$

$$^{\bullet}CH_2CH_2OHCH_3^+ + CD_3OD \rightarrow {}^{\bullet}CH_2CH_2ODCD_3^+ + CH_3OH \quad \text{slow} \qquad \underline{11}$$

$$^{\bullet}CH_2CH_2NH_3^+ + {}^{15}NH_3 \rightarrow \text{no reaction} \qquad \underline{12}$$

It can be observed that the reaction rate correlates with the density of positive charge on the α-carbon, suggesting that the reaction mechanism corresponds to a nucleophilic substitution. The calculated energy profile confirms this statement.

Energy profile

When a water molecule interacts with the α-carbon of $^{\bullet}CH_2CH_2OH_2^+$, an energy minimum is calculated for the electrostatic complex C_2 (Figure 7). A nucleophilic substitution can take place within this C_2 complex via a symmetric transition state

(Figure 8) which lies in energy 5.7 kcal mol^{-1} under that of the reactants. Therefore the athermic nucleophilic substitution of ion **1** is easy [34].

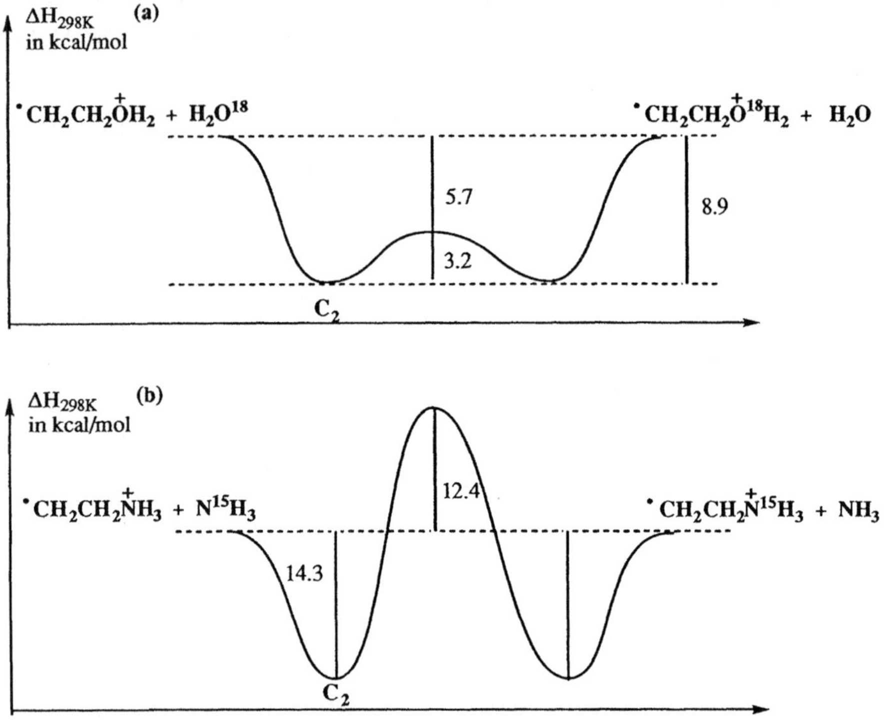

Figure 8 : Energy diagram for reactions between ion $^{\bullet}CH_2CH_2OHCH_3^+$ *and* H_2O *(a) and* $^{\bullet}CH_2CH_2NH_3^+$ *and* NH_3 *(b).*

In contrast, in the $^{\bullet}CH_2CH_2NH_3^+$ / NH_3 system (Figure 8), the transition state for substitution is now 12.4 kcal mol^{-1} above the reactant energy. Therefore experiments and calculations show that substitution does not occur in this case.

In conclusion, the electrostatic complex C_2 is the reactive structure in the substitution reaction while C_1 operates in the proton transfer. It will now be shown that both structures are involved for the other reactions.

4.4. REACTIONS AT THE RADICAL SITE

Although proton or ionized ethylene transfer to neutral species are the main reactions of the β–distonic ions **1** to **3** with most of neutral reactants, reaction at their radical site could be a characteristic reaction of such ions, the spatial separation of the two potentially reactive sites opening the possibility for an ambident behavior in the gas phase. And effectively, reactions such as hydrogen atom abstraction, iodine abstraction or addition to a carbonyl compound have been observed, but have been found to be dependent both on the distonic ion and on the neutral target. However, none of the ions **1** to **3** exhibit conventional radical reactivity with CH_3SSCH_3 [27], O_2 or N_2O. With dimethyl disulfide the four ions lead mostly to a charge exchange reaction, a slow $CH_3S^•$ transfer being observed only for ion **3**.

H• abstraction

H• abstraction from the neutral is the most important reaction at the radical site (Eq. 13).

$$^•CH_2CH_2XH^+ \; + \; RH \; \rightarrow \; CH_3CH_2XH^+ \; + \; R^• \qquad \underline{13}$$

When observed, this reaction always occurs at the radical site of the β-distonic ions, as shown by CID studies of the protonated species obtained. For instance, in the reaction depicted in Eq. 14, the $CH_2DCH_2NH_3^+$ product-ion loses NH_3 under collision giving the product ion $C_2H_4D^+$ (m/z 30), while no $C_2H_5^+$ ion (m/z 29) is formed. This contrasts strongly with the reaction of the ethylamine molecular ion; H• abstraction by this species is shown by CID to occur exclusively at the nitrogen atom. Similar results are obtained with other β–distonic ions. This shows that there is no appreciable interconversion of the β–distonic ion and the corresponding molecular ions or vice versa.

$$^•CH_2CH_2NH_3^+ \; + \; D_5C_2OC_2D_5 \; \rightarrow \; CH_2DCH_2NH_3^+ \; + \; CD_3{}^•CDOC_2D_5 \qquad \underline{14}$$

I• abstraction

Iodine compounds are suitable reactants to trap radicals . Because of the low value of its C-I Bond Dissociation Energy (BDE= 48 kcal mol^{-1}), allyl iodide was chosen to characterize the radical site of the β-distonic ions: I• abstraction was observed with ions **2** and **3** (Eq. 15). Similarly, Br• abstraction occurs when ion **2** reacts with $CBrCl_3$. However, in spite of the fact that these reactions are highly exothermic, these halogen abstractions occur at a low efficiency, suggesting a significant energy barrier on the reaction coordinate. For example, the bromine abstraction reaction of ion **2** with $CBrCl_3$ (BDE= 54 kcal mol^{-1}) is -46 kcal mol^{-1} exothermic with a total reaction efficiency of 0.2%.

$$^{\bullet}CH_2CH_2XH^+ + CH_2CHCH_2I \rightarrow ICH_2CH_2XH^+ + CH_2CHCH_2^{\bullet} \qquad \underline{15}$$

Existence of an energy barrier

Nevertheless, atom or $H^{\bullet}$ abstraction is far from being a general reaction for the ions **1** to **3** . For instance, although the thermochemistry of $H^{\bullet}$ abstraction by $^{\bullet}CH_2CH_2OH_2^+$ **1** is particularly favorable (HA= 106 kcal mol^{-1}), this ion does not abstract $H^{\bullet}$ from any of the neutral molecules so far examined. In this case, the competing reactions (protonation, ionized ethylene transfer and charge exchange) have been shown to be very easy , often occuring at the collision rate, which suggests that $H^{\bullet}$ abstraction, in contrast, requires overcoming an energy barrier. This is further supported by the existence of strong primary isotope effects: when ion **2** reacts with CH_3OCH_3, the products resulting of $C_2H_4^{\bullet+}$ transfer and $H^{\bullet}$ abstraction are in a 1.4 ratio; the same ratio is about 5 with CD_3OCD_3. A similar observation can be done for the reactions with CH_3COCH_3 (ratio=2) and CD_3COCD_3 (ratio=7).

Reactive structure

The structure and the isomerization of the encounter complex may play a determining role, since ions **2** and **3** generally react by $H^{\bullet}$ abstraction with alcohols and ethers but never with acetonitrile. For instance, whereas the thermochemistry of the $H^{\bullet}$ abstraction reaction is almost the same when the $^{\bullet}CH_2CH_2OHCH_3^+$ ion **2** reacts with dimethyl ether or acetonitrile, the reaction is observed with the first reactant only. This can be understood if the structures of the reacting complexes are considered. For the ($^{\bullet}CH_2CH_2OHCH_3^+/CH_3OCH_3$) system, the distances between one hydrogen of the neutral and the radical site of the ion is rather small as well for the H-bonded complex (3.949 Å) as for the electrostatic complex (3.323 Å). In contrast, for the ($^{\bullet}CH_2CH_2OHCH_3^+/ CH_3CN$) system, these distances are larger as well for the H-bonded complex (5.949 Å) as for the electrostatic complex (6.025 Å). Therefore, the energy barrier for $H^{\bullet}$ abstraction in the latter case is probably higher than in the first one. The role of the reacting complexes in this process also explains the absence of $H^{\bullet}$ abstraction from propane, never observed in spite of its exothermicity. The complexes [β–distonic ion, propane] are poorly stabilized and the energy barrier for H• abstraction may prevent the reaction from occuring.

4.5. REGIOSPECIFIC ADDITIONS

The $^{\bullet}CH_2CH_2OHCH_3^+$ ion **2** reacts with CH_2O by a specific attack of the carbon atom of formaldehyde at the radical carbon of the distonic ion [37]. The such formed intermediate ion eliminates water after a 1,5 H transfer (Scheme 9). Ion **2** also reacts with acetaldehyde by a similar addition process followed by H_2O and $CH_3^{\bullet}$ losses.

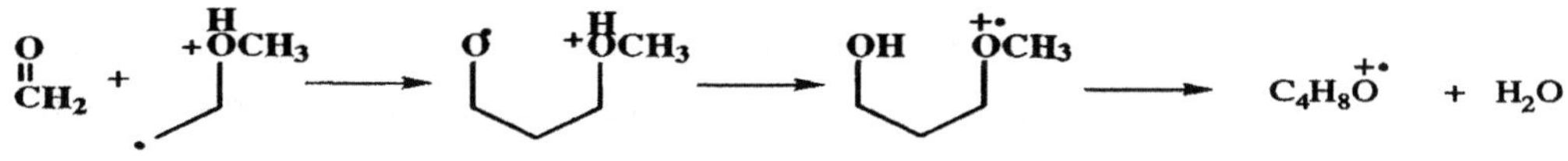

Scheme 9

In fact, the reaction involves both sites of the distonic ion. The regiospecificity is the consequence of the formation, in the first step of the reaction, of a H-bonded complex involving an acidic hydrogen of the ion followed by a C-C bond formation.
A similar reaction has been described for the $^{\bullet}CH_2CH_2OCH_2^+$ distonic ion [38]. This ion reacts with CH_2O (and CD_2O) by exchange of the OCH_2 group and by addition. The product ion resulting of this addition (followed by a rapid H• loss) is shown by CID to be the $[M - H]^+$ ion of 1,3-dioxane$^{+\bullet}$ and not of 1,4-dioxane$^{+\bullet}$. This indicates that the reaction is regiospecific and involves the two sites of the distonic ion. It begins with an attack at the positive site of the cation by the oxygen of CH_2O followed by ring-closure (Scheme 10).

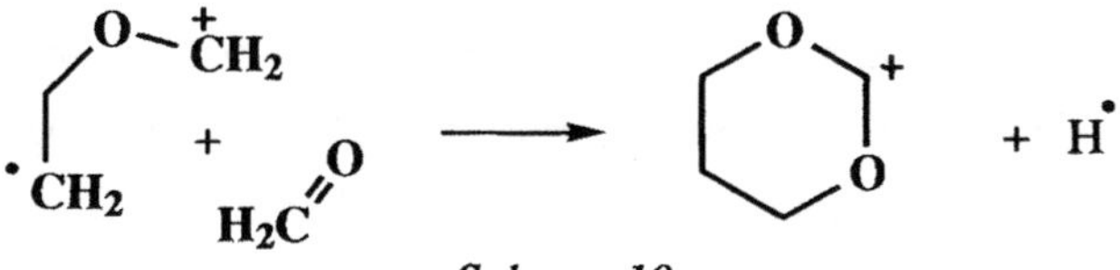

Scheme 10

5. Conclusions

Distonic ions form a very interesting group of chemical reactants in the gas phase. They are often easy to prepare and their chemistry is very rich, and they are useful to generate other ions. Moreover, a good knowledge of their behaviour is necessary to understand the mechanisms of unimolecular reactions of radical cations.

If the distonic ions of alcohols, ethers or amines are now considered, several characteristic can be selected.

(i) The behaviour of β-distonic ions derived from alcohols, ethers or amines is completely different from that of the corresponding α-distonic isomers:

- whereas appropriate neutral molecules can catalyse the isomerization of α-distonic ions into their molecular ion counterparts, such a process is not observed for β-distonic ions, unless an indirect pathway is open.
- α-distonic ions are much more acidic than the corresponding saturated species, in contrast with β-distonic ions,
- atom abstraction by the radical site is exceptional with α-distonic ions, but frequent for β-distonic ions.
- α-distonic ions $^{\bullet}CH_2XH^+$ rarely transfer $CH_2^{\bullet+}$ to the neutral molecule while $CH_2CH_2^{\bullet+}$ transfer is the most interesting reaction of the β-distonic ions $^{\bullet}CH_2CH_2XH^+$.

(ii) Therefore, in contrast with α-distonic ions, β-distonic ions react as ambident species exhibiting an expected reactivity as well at the positive site (protonation of the neutral) as at the radical site (atom abstraction). The third kind of reaction, $C_2H_4^{\bullet+}$ transfer to the neutral molecule, is in fact a nucleophilic substitution depending on the density of the positive charge borne by the α-carbon of the ion, and of the nucleophilic character of the neutral reactant.

(iii) Calculations show that the interaction between a β-distonic ion and a molecule yields two stable complexes: a H-bonded complex highly stabilized (up to 35 kcal/ mol) and an electrostatic complex less stabilized (between 8 and 15 kcal/mol). Each of them can either isomerize into the other or lead to specific reactions, mainly protonation of the neutral for the H-bonded complex and ionized ethylene transfer for the electrostatic complex.

6. References

1. Hammerum, S. (1988) Distonic Radical Cations in gaseous and condensed phase, *Mass Spectrom. Reviews* **7**, 123-202.
2. Radom, L., Bouma, W.J., Nobes, R.H. and Yates,B.F. (1984) *Pure Appl. Chem.* **56**, 1831.
3. Budzikiewicz, H., Djerassi, C. and Williams, D.H. (1967) *Mass Spectrometry of Organic Compounds*, Holden-Day, San Francisco.
4 Billets, S., Jaffé, H.H. and Kaplan, F. (1970) Rearrangements of Molecular Ions of Dialkyl-N-nitrosamines, *J. Am. Chem. Soc.* **92**, 6964-6965.
5. Morton, T.H. and Beauchamp, J.L. (1975) The Low Energy EI Mass Spectra of α,ω-Dialkoxyalkanes Examined by ICR Spectroscopy, *J. Am. Chem. Soc.* **97**, 2355-2361.
6. a) Bouma, W.J., Nobes, R.H. and Radom, L. (1982) The methylenoxonium radical cation $[CH_2OH_2]^{\bullet+}$: A surprisingly stable isomer of the methanol radical cation, *J. Am. Chem. Soc.* **104**, 2929-2930; b) Bouma, W.J., Nobes, R.H. and Radom, L. (1983) Unusual Low-Energy Isomers of the Ethanol and Dimethyl Ether Radical Cations, *J. Am. Chem. Soc.* **105**, 1743-1746.
7. Yates, B.F., Bouma, W.J. and Radom, L. (1987) Ylides and Ylidions: a Comparative Study of Unusual Gas-Phase Structures, *J. Am. Chem. Soc.* **109**, 2250-2263.
8. a) Stirk, K.M., Kiminkinen, L.K.M. and Kenttämaa, H.I. (1992) Ion-Molecule Reactions of Distonic Radical Cations, *Chem. Rev.*, **92**, 1649; b) Smith, R.L., Chou, P.K. and Kenttämaa, H.I. (1996) Structure and Reactivity of Selected Distonic Radical Cations, in T. Baer, C-Y. Ng and I. Powis (eds) *The Structure, Energetics and Dynamics of Organic Ions*, John Wiley & Sons, Chichester, pp. 197-261.
9. Wolff, M.E. (1963) Cyclisation of N-halogenated amines (the Hoffman-Löffler reaction), *Chem. Rev.* **63**, 55-64.
10. Zahorszky, U.I. (1983) Evidence for extensive recombination of the ring-opened to the original molecular ions of 2-substituted piperidines and pyrrolidines, *J. Chem. Soc. Perkin 2* 1655-1660.
11. Polce, M.J. and Wesdemiotis, C. (1996) Stabilities and isomerization proclivities studied by dissociation and neutralization-reionization, *J. Am. Soc. Mass Spectrom.* **7**, 573-589.
12. Holmes, J.L., Lossing, F.P., Terlouw, J.K. and Burgers, P.C. (1982) $[CH_2OH_2]^{\bullet+}$ and Related Stable Gas-Phase Ion-Dipole Complexes, *J. Am. Chem. Soc.* **104**, 2931-2932.

13. a) Wedesmiotis, C. and McLafferty, F.W. (1987) *Chem. Rev.* **87**, 485-500; b) Terlouw, J.K. and Schwarz, H. (1987) *Angew. Chem. Int. Ed. Engl.* **26**, 805-815; c) Holmes, J.L. (1989) The neutralization of organic cations, *Mass Spectrom. Rev.* **8**, 513-539.

14. Polce, J.M. and Wedesmiostis, C. (1993) *J. Am. Chem. Soc.* **115**, 10849-10856.

15. Splitter, J.S. (1994), Mass-Spectral Intermediate-Ion Structures, in J.S. Splitter and F. Turecek (eds.), *Applications of Mass Spectrometry to organic stereochemistry*, V.C.H. Publishers, New York, pp. 39-81.

16. a) Beauchamp, J.L., Anders, L.R. and Baldeschwieler, J.D. (1967) The Study of Ion-Molecule Reactions in Chloroethylene by Ion Cyclotron Resonance Spectroscopy, *J. Am. Chem. Soc.* **89**, 4569-4577; b) Baldeschwieler, J.D. (1968) *Science* **159**, 263.

17. Billets, S., Jaffé, H.H. and Kaplan, F. (1970) Rearrangements of Molecular Ions of Dialkyl-N-nitrosamines, *J. Am. Chem. Soc.* **92**, 6964-6965.

18. Gross, M.L. and McLafferty, F.W. (1971) Identification of $C_3H_6^{+\bullet}$ Structural Isomers by Ion Cyclotron Resonance Spectroscopy, *J. Am. Chem. Soc.* **93**, 1267-1268.

19. Sack, T.M., Miller, D.L. and Gross, M.L. (1985) The Ring Opening of Gas-Phase Cyclopropane Radical Cations, *J. Am. Chem. Soc.* **107**, 6795-6800.

20. Bouma, W.J., MacLeod J.K. and Radom, L. (1978) Experimental Proof of the Existence of a Fourth Stable Gas Phase $C_2H_4O^{+\bullet}$ Isomer: the Open Ethylene Oxide Ion, *J. Chem. Soc. Chem. Commun.*, 724-725.

21. Busch, K.L., Nixon, W.B., Clinton, B. and Bursey, M.M. (1978) Design of an Ion-Molecule Reaction Specific for Neutrals Nitriles, *J. Am. Chem. Soc.* **100**, 1621-1622.

22. a) Audier, H.E., Mourgues, P., Leblanc, D. and Hammerum, S. (1993) Bimolecular ionic alkene displacement reactions in the gas phase, *C. R. Acad. Sci. Paris,* **317**, Série II, 27-31; b) Audier, H.E., Mourgues, P., Leblanc, D., Troude, V. and Hammerum, S. (1996) Bimolecular reactions of β-distonic ions $^{\bullet}CH_2CH_2XH^+$ (X= OH, OCH_3, OC_2H_5, NH_2), to be published.

23. Kiminkinen, L.K.M., Stirk, K.G. and Kenttämaa, H.I. (1992) The distonic isomer of the unstable radical cation of dimethyl propyl phosphate, *J. Am. Chem. Soc.* **114**, 2027-2031.

24. Wittneben, D. and Grützmacher, H.F. (1990) Unimolecular and bimolecular reactions of the β –distonic ion $^{\bullet}CH_2CH_2OCH_2^+$ and its deuterated derivatives, *Int. J. Mass Spectrom. Ion Processes* **100**, 545-563.

25. Baumann, B.C., MacLeod, J.K. and Radom, L. (1980) Experimental proof for the gas-phase existence of the C-C ring opened trimethylene oxide cation, *J. Am. Chem. Soc.* **102**, 7927-7928.

26. Bouma, J.W., MacLeod, J..K. and Radom, L. (1980) Structure and stabilities of $C_3H_6O^{\bullet+}$ isomers. An ab initio M.O. study, *J. Am. Chem. Soc.* **102**, 2246-2252.

27. Stirk, K.M., Orlowski, J.C., Leeck, D.T. and Kenttämaa, H.I. (1992) The Identification of Distonic Radical Cations on the Basis of a Reaction with Dimethyl Disulfide. *J. Am. Chem. Soc.* **114**, 8604-8606.

28. Beasley, B.J., Smith, R.L. and Kenttämaa, H.I. (1995) A new reagent for structure-specific ion-molecule reactions. Dimethyl Diselenide, J. Mass Spectrom. **30**, 384-385.

29. a) Mourgues, P., Audier, H.E., Leblanc, D. and Hammerum, S. (1993) Bimolecular reactions of distonic ions: proton transfer and hydrogen atom abstraction with $^{\bullet}CH_2OH_2^+$, *Org. Mass Spectrom.* **28**, 1098-1100; b) Audier, H.E., Leblanc, D., Mourgues, P., McMahon, T.B. and Hammerum, S. (1994) Catalysed isomerisation of simple radical cations in the gas phase, *J. Chem. Soc. Chem. Commun.*, 2329-2330.

30. de Visser, S.P., de Koning, L.J. and Nibbering, N.M.M. (1997) Chemical and thermodynamic properties of methyl chloride dimers radical cations in the gas phase, submitted for publication.

31. Gauld, J.W., Audier, H., Fossey, J. and Radom L. (1996) Water catalyzed interconversion of conventional and distonic radical cations: methanol and methyleneoxonium radical cations, *J. Am. Chem. Soc.* **118**, 6299-6300.

32. a) Szulejko, J.E. and MacMahon, T.B. (1993) Progress towards an absolute gas-phase proton affinity scale, *J. Am. Chem. Soc.* **115**, 7839-7848; b) Audier, H.E., Fossey, J., Mourgues, P., Leblanc, D. and Hammerum, S. (1996) Deprotonation of α-distonic ions. Proton affinities of the α-radicals, *Int. J. Mass Spectrom. Ion Processes*, **157/158**, 275-282.

33. Audier, H.E., Fossey, J., Mourgues, P., McMahon, T.B. and Hammerum, S. (1996) Formation of proton-bound dimers as the driving force for alkyl radical loss in the gas phase reactions of radical cations, *J. Phys. Chem.* **100**, 18380-18386.

34. Audier, H.E., Fossey, J., Leblanc, D. and Mourgues, P. (1996) Thermoneutral nucleophilic substitution of β-distonic ions $^{\bullet}CH_2CH_2XH^+$ in the gas phase (X= OH, OCH_3, OC_2H_5, NH_2), *Bull. Soc. Chim. Fr.* **133**, 59-64.

35. Zipse, H. (1995) The Addition of Water to Ethylene and Butene Radical Cation. Models for the Reaction of Alkene Radical Cations with Nucleophiles, *J. Am. Chem. Soc.* **117**, 11798-11806.

36. Troude, V., Van der Rest, G., Mourgues, P. and Audier, H.E. (1997) Preparation and Reactivity of Solvated Distonic Ions and Ionized Enols in the Gas Phase, *J. Am. Chem. Soc.* **119**, 9287-9288.

37. Troude, V., Leblanc, D., Mourgues, P. and Audier, H.E. (1995) Regiospecific addition of CH_2O at the radical site of the $^{\bullet}CH_2CH_2OHCH_3^+$ distonic ion, *J. Mass Spectrom.* **30**, 1747-1751.

38 Mourgues, P., Audier, H.E. and Hammerum, S. (1994) A regiospecific cycloaddition: successive reactions of $^{\bullet}CH_2CH_2OH_2^+$ with CH_2O, *Rapid Commun. Mass Spectrom.*, **8**, 53-55.

Collisions of Ions with Surfaces

R.G. Cooks and S.A. Miller
Aston Laboratory for Mass Spectrometry, Department of Chemistry, Purdue University, West Lafayette, IN, 47907-1393

Abstract

Collisions of ions with surfaces may be elastic, inelastic or reactive, the latter category including charge changing and atom and group transfer processes. Inelastic collisions, which lead to the dissociation of the projectile, are used to characterize both biological ions and smaller organic ions and hence the neutral compounds for which they are surrogates. This process, termed surface-induced dissociation, is compared to the analgous gas-phase collision-activated dissociation experiment. Partitioning of the projectile translational energy into projectile vibrational energy, projectile electronic excitation, surface internal energy, and scattered ion kinetic energy in ion/surface collisions is illustrated for several projectiles and surfaces. Surfaces chosen to minimize projectile neutralization, which is competitive with the other collision processes, are discussed as is the effect of surface nature on energy partitioning and ion/surface reactivity. Reactive collisions, which include simple chemical sputtering of ions from the surface as well as the transfer of an atom or groups of atoms to the projectile ion, are reviewed. Chemical sputtering and diagnostic ion/surface reactions are shown to represent a means of surface analysis and reactive ion/surface collisions are shown to cause chemical modification of self-assembled monolayer surfaces. Mechanisms for both reactive and inelastic collisions are considered. Instrumentation to study the fundamental aspects of these processes in the energy range of a few tens of eV, is described.

K. R. Jennings (ed.), Fundamentals and Applications of Gas Phase Ion Chemistry, 55–114.

Introduction

Low energy polyatomic ions can scatter from surfaces in elastic, inelastic, and reactive collision processes.[1] Elastic collisions are not of great chemical interest but inelastic collisions, which lead to projectile excitation and hence to surface-induced dissociation (SID), have received attention due to the useful ion structural information obtained from the product ion spectra.[2] Surface-induced dissociation, analogous to the gas-phase process known either as collisionally activated dissociation (CAD) or collision-induced dissociation (CID), leads to a large internal energy deposition with a narrow distribution in the projectile ion and produces high quality mass spectra for both small and large organic ions.[3,4] Accompanying SID are ion/surface reactions which can be used as a means of surface analysis via diagnostic reactions between the projectile and the outermost atoms of a surface.[5,6,7] Ion/surface reactions have opened a field with interesting mechanisms[8,9,10,11,12,13,14] and these remarkable reactions display both parallels to and differences from gas-phase ion/molecule reactions.[15] Ion/surface reactions offers a new means of chemically-selective surface modification.[16]

The characteristic feature of the ionic collision processes discussed here is that the component of the kinetic energy in the direction of the surface normal is on the order of 10 - 100 eV. Thus, ions whose mass is 100 Daltons travel the distances of interest, ca. 1 nm, in 10^{-13} s. The low energies restrict penetration to the upper region (for example, a few carbon atoms of a self-assembled monolayer, SAM, surface) of a surface and the short times restrict the types of possible reactions.

1. Surface-Induced Dissociation

An experiment of Heiland illustrates the processes of elastic and inelastic scattering.[17] A beam of D_3^+ of 5 keV is scattered specularly from a gold surface using an angle of incidence (from the normal) of 85° and the kinetic energy of the scattered ions is measured. Three principal peaks are observed at approximately 5, 3.3, and 1.6 keV, corresponding to D_3^+, D_2^+, and D^+, respectively. This experiment, therefore, demonstrates the occurrence of elastic scattering (D_3^+) as well as inelastic scattering (D_2^+ and D^+). The latter ions are products of surface-induced dissociation. Note that there is additional information in the velocity distributions of the scattered ions which is not of immediate interest here.

A more typical SID experiment employs two mass analyzers, one to mass-select the reactant ion and the second to mass-analyze the scattered ionic products. Figure 1 displays SID spectra of the polyatomic molecular ion of p-methyl phenetole at two energies. It illustrates the control which can be exercised over the degree of fragmentation by adjusting the collision energy. The fragments observed are the same as occur in the electron impact (EI) mass spectrum and can be accounted for by simple unimolecular fragmentations of the excited projectile ion. This and similar data suggest a model for polyatomic ion SID which is analogous to that which occurs in CAD, i.e. delayed fragmentation separated in time and space from the collision event. As we shall see below, this model is not followed in all cases. One can also see from Figure 1 that, even though many of the SID processes are identical to those observed in CAD, significant amounts of internal energy can be lodged in the projectile by SID. This makes accessible high energy product ions such as $C_2H_5^+$ and/or CHO^+ at 29 Thomson (Th).[18] Note too, the complete disappearance of the projectile ion in the 30 eV collision spectrum. It is not possible for some ions to "miss" their target as it is in CAD where large impact parameter events corresponding to minimal excitation are much more common than are small impact parameter, high energy deposition collisions.[19]

SID product ions are often structurally diagnostic. Among many examples of this effect, the polynuclear aromatic hydrocarbons provide an excellent demonstration of the greater number and abundance of structurally diagnostic fragment ions in SID, compared to CAD spectra for a given collision energy.[14,20] The pyrene molecular ion fragments to give abundant ions with two to sixteen carbon atoms upon collision at a hydrocarbon-covered surface. The fragment ions are more than an order of magnitude more abundant than the residual parent ion in the case of 100 eV collisions (45° angle of incidence, 90° scattering angle). By contrast, single-collision conditions with argon as the target at the same collision energy do not result in any skeletal fragmentation; rather CAD yields the parent ion as the base peak with only the fragment ions due to loss of $H^{\cdot}$ and H_2 having significant abundance. These and some other characteristics of SID (some of which will be discussed further below) are summarized in Table 1.

2. Ion/Surface Reactions

The SID spectrum of the benzene molecular ion, 78 Th, with a deuterated-SAM (D-SAM) surface shows three ions of low abundance in addition to the expected fragments (Figure 2). These unexpected ions occur at 80, 92, and 93 Th and are due to addition of $D^{\cdot}$ and to $CD_3^{\cdot}$ addition followed by the loss of HD and D_2. Clearly, ion/surface reactions, processes which are formally analogous to ion/molecule reactions, also occur in the energy range which yields SID products. The projectile serves as one reagent, a characteristic which distinguishes ion/surface reactions from the interfacial reactions which accompany desorption ionization. These latter

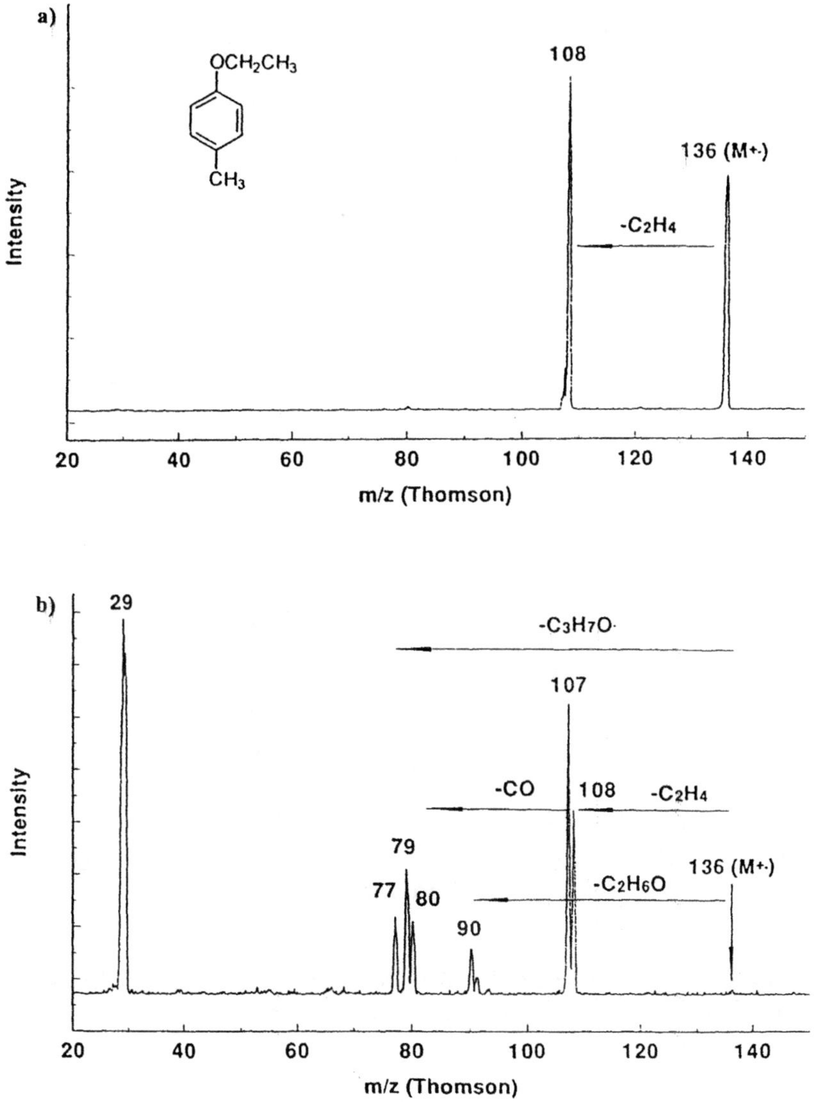

Figure 1: SID product ion spectra of the p-methyl phenetole molecular ion (m/z 136) upon collision with a F-SAM surface at (a) 15 eV and (b) 30 eV. Some of the dissociation channels are shown explicitly in (b).

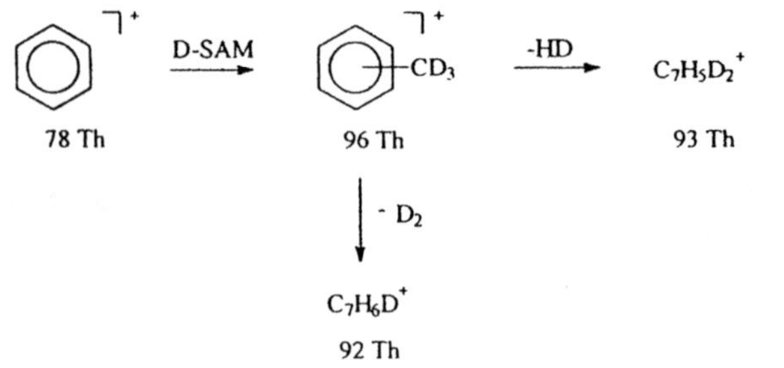

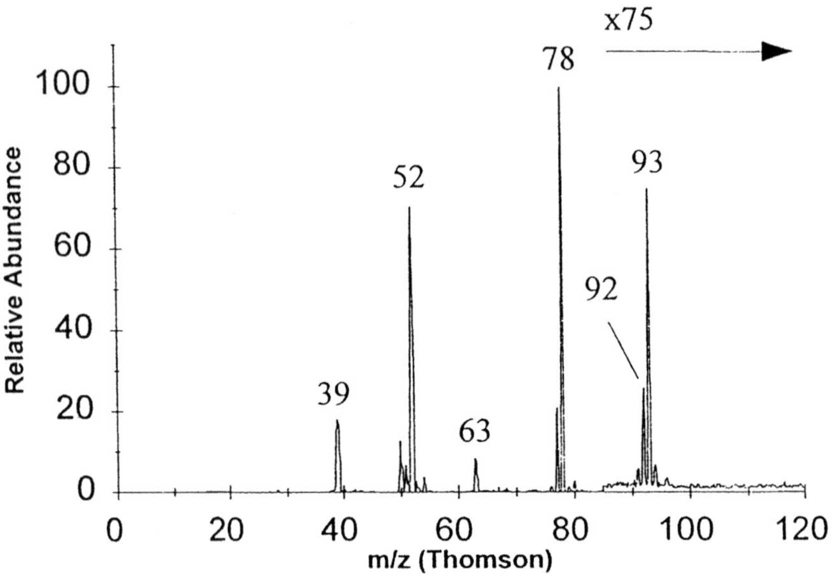

Figure 2: Product ion spectrum generated by 30 eV collisions of the benzene molecular ion (m/z 78) upon a D-SAM surface. Note the reaction products at m/z 80, 92, and 93.

Table 1:
Characteristics of Surface-Induced Dissociation (SID)

(1) Internal energy deposition (V) is variable depending upon the collision energy (T).
(2) Internal energy deposition can be made large, more than 10 eV is readily accessible.
(3) The distribution of internal energies is relatively narrow (~ 4 eV FWHM).
(4) The efficiency of energy transfer (T→V) is about 13% for hydrocarbon surfaces and 19% for fluorinated surfaces.
(5) The SID efficiency (total scattered ions vs. incident ions) is not well known, but is more than 10% in many systems.
(6) SID is applicable to large biomolecules.

Table 2:
Ion/Surface Reaction Types

Type 1: Reactions in which new bonds involve both projectile ion and target surface.[69]

Example: hydrogen atom abstraction by benzene radical cation.

Type 2: Reactions in which new bonds involve only the target surface. Well known from molecular SIMS, e.g. reduction, transalkylation, etc.[21]

Example: Fullerene cage expansion upon collision with 1.7 keV $Cr(CO)_6^{+\cdot}$ projectile ions.[70] It is interesting to note that C_2 addition is a process also seen when C_{60} is the projectile, viz. by a Type 1 ion/surface reaction.[50]

Type 3: Reactions in which new bonds involve only the projectile ion.[23]

Example: $NiH_2^{+\cdot}$ formation from nickelocenium cation upon collision with an F-SAM surface.[23]

processes include a wide variety of chemical reactions where the projectile merely serves to deliver energy to the system[21] Formally it is possible to organize ion/surface reactions into three types, depending on the location of the new bonds which are formed. This classification, with an example of each type, is shown in Table 2.

The fact that ion/surface reactions are worth careful attention is demonstrated dramatically by the spectra recorded when W^+ is scattered from a fluorinated self-assembled monolayer (F-SAM) or the surface of a liquid polymer, perfluoropolyether surface (Figure 3).[22] Ion/surface reaction products dominate these spectra and, given the low projectile kinetic energy, the observation of WF_n^+ ions demonstrates the presence of fluorine within the first few angstroms of the surface (the projectile penetration depth). Note that even the minor features of this spectrum are readily explained in terms of reactive collisions (for example, the ions at 234 Th and 253 Th are assigned as WCF_2^+ and WCF_3^+). Table 3 records some of the features of ion/surface reactions and includes some characteristics which will be discussed below.

The surface-induced dissociative rearrangement is an example of a recently recognized class of ion/surface reactions (Type 3, Table 2), e.g..

$$C_5H_5NiC_5H_5^{+\cdot} \xrightarrow{\text{F-SAM}} NiH_2^{+\cdot} + C_{10}H_8.$$

This particular reaction occurs over a relatively narrow range of collision energies (and hence internal energies, as befits a rearrangement). Variation in the intensity of $NiH_2^{+\cdot}$ with collision and electron impact energies suggests that the mechanism involves the excitation of a low-energy vibrational mode of the molecular ion, resulting in cyclopentadienyl ring fusion to form $C_{10}H_8$ with the simultaneous transfer of two hydrogen atoms to $Ni^{+\cdot}$.[23]

Table 3:
Characteristics of Ion/Surface Reactions

(1) Odd electron projectile ions (open shell) are more likely to react.
(2) Multiple atom transfers are common for metal ions and fluorocarbon surfaces.
(3) Polyatomic group transfers are common for polyaromatic hydrocarbons and hydrocarbon surfaces.
(4) Ion/surface reactions often occur at the threshold for chemical sputtering.
(5) Both exothermic and endothermic reactions occur.
(6) Analogous gas-phase ion/molecule reactions often occur.
(7) Ion/surface reaction products usually have low relative abundances < 10%.

Table 4:
Ion/Surface Collision Processes

Elastic Collisions	$AB^+ + SX \rightarrow AB^+ + SX$
Inelastic Collisions	$AB^+ + SX \rightarrow AB^{+\bullet} + SX$; $AB^{+\bullet} \rightarrow A^+ + B$
Reactive Collisions	$AB^+ + SX \rightarrow ABX^+ + S$
Surface Modification	$AB^+ + SX \rightarrow AX^+ + SB$
Charge Exchange	$AB^+ + SX \rightarrow AB + SX^+$
Chemical Sputtering	$AB^+ + SX \rightarrow AB + S + X^+$
Other Charge Changing Collisions	$AB^+ + SX \rightarrow AB^{+2} + SX + e^-$ $AB^- + SX \rightarrow AB^+ + SX + 2e^-$

Where AB^+ is the projectile and S is a surface covered by adsorbate X.

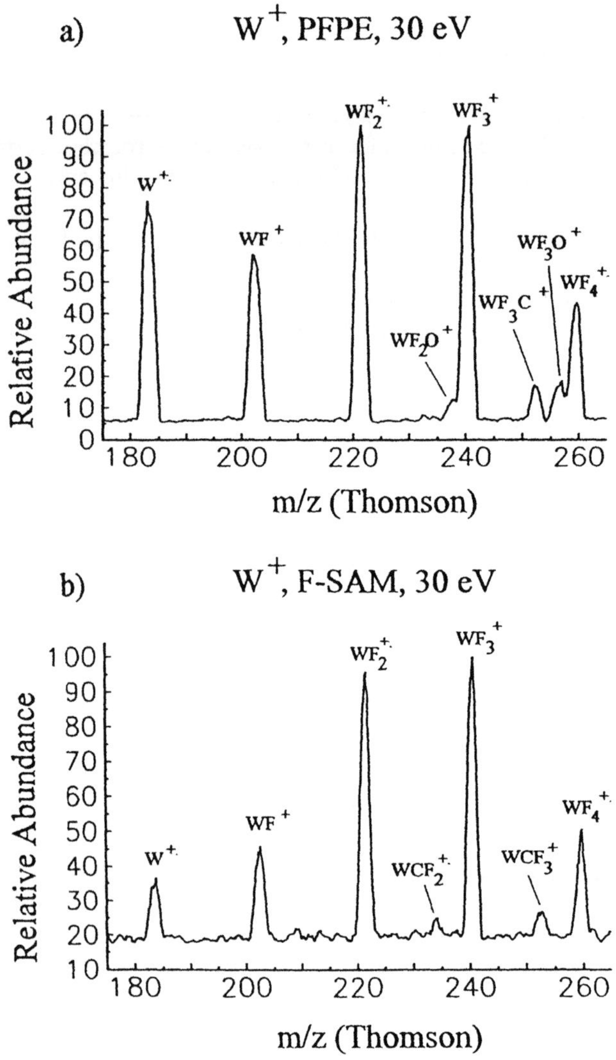

Figure 3: Product ion spectrum produced from 30 eV collisions of $^{186}W^{+\cdot}$ upon (a) liquid perfluoropolyether surface and (b) upon a F-SAM surface.

3. Surface Effects

The nature of the surface plays a crucial role in both SID (where it affects scattered ion yields and internal energy deposition) and in ion/surface reactions (where it is a reagent and so affects the observed reactions as well as the scattered ion yields). A key to success in the types of experiments described here is to minimize the extent of neutralization at the surface. This is illustrated for a metal, Re, and a metal oxide, ReO_2 which has a higher work function and causes correspondingly less neutralization (Figure 4).[24] Self-assembled alkanethiol monolayers have rapidly become the surfaces of choice for SID because they minimize neutralization[25] and also provide a well-defined chemical surface (the organic chemists' single crystal) for ion/surface reactions.[3] Note that they can be prepared with various terminal functional groups, a feature of particular interest in the study of ion/surface reactions.[3] Further evidence of the importance of neutralization in ion/surface collision phenomena is provided by considering collisions at a fluorocarbon surface by the rare gas ions $Ne^{+\cdot}$, $Ar^{+\cdot}$, $Kr^{+\cdot}$, and $Xe^{+\cdot}$, the recombination energies (RE) of which are respectively 21.56 eV, 15.76 eV, 14.00 eV, and 12.13 eV.[26] Abundant scattered projectiles are observed in all cases except for that of $Kr^{+\cdot}$, no doubt due to near-resonant neutralization of $Kr^{+\cdot}$ by the F-SAM (ionization energy (IE) estimated to be ca. 13.4 eV, e.g. C_3F_8 = 13.38 eV[26]). More recently, the same property of a low degree of neutralization (hence high ion yields) has been found using some liquid polymer surfaces. For example, $W(CO)_6^{+\cdot}$ fragments upon 30 eV collisions at an eight micron thick, liquid perfluoropolyether surface to yield $W(CO)_3^{+\cdot}$ as the most abundant product and $W(CO)_2^{+\cdot}$, $W(CO)_4^{+\cdot}$, and $W(CO)_5^{+\cdot}$ as the only other product ions.[6] There is virtually no difference between this perfluoropolyether and a F-SAM surface in collisions with many different projectile ions, either in SID or ion/surface

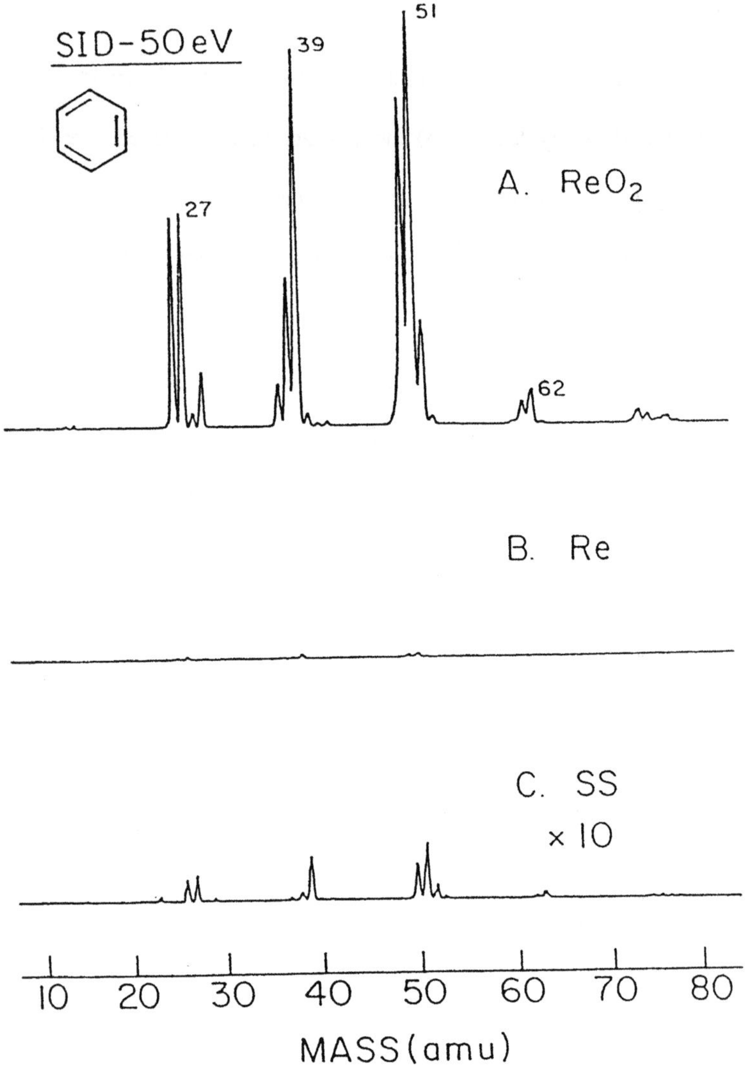

Figure 4: Product ion spectra produced from 50 eV collisions of the benzene molecular ion upon three different surfaces, ReO_2, Re, and stainless steel (SS). (Reproduced with permission from reference 24.)

reactions (see Figure 3).[6]

The chemical nature of the surface has a significant effect on energy partitioning during the ion/surface collision.[3,27] An F-SAM is more effective at converting translational to internal energy than is the corresponding hydrocarbon surface (Figure 5).[3] Figure 5 reveals that pyrazine molecular ion fragments by the sequential loss of HCN molecules; the loss of one molecule of HCN produces the base peak in the F-SAM spectrum. The hydrocarbon SAM surface is as reactive towards this projectile, showing additions of $D^{\cdot}$ and $CD_3^{\cdot}$ with a corresponding fragmentation due to HCN loss from $(M+D)^+$. The F-SAM surface is not reactive toward the pyrazine projectile, i.e. there is no fluorocarbon pickup. The F-SAM also produces higher secondary ion yields than does the hydrocarbon surface. For the benzene molecular ion, the F-SAM shows five times higher secondary ion yields than does the hydrocarbon-SAM (H-SAM)[25] presumably due to resonant electron transfer in the latter case (IE of the H-SAM is estimated to be ca. 9.6 eV, close to the value of benzene, 9.25 eV[26]).

4. Charge Exchange and Other Processes

The discussion so far has been restricted to SID and ion/surface reactions, although the occurrence of electron transfer has been noted above in connection with ion neutralization. A fuller listing of ion/surface collision processes is given in Table 4 and we now proceed to discuss charge transfer and the related process of chemical sputtering. Interest in charge exchange is rapidly growing, given the increasing use of multiply-charged ions especially in connection with electrospray ionization. The C_{60}^{+2}/F-SAM spectrum (Figure 6) illustrates the substantial contribution to the scattered ion signal which may be made by dissociative charge exchange.[10]

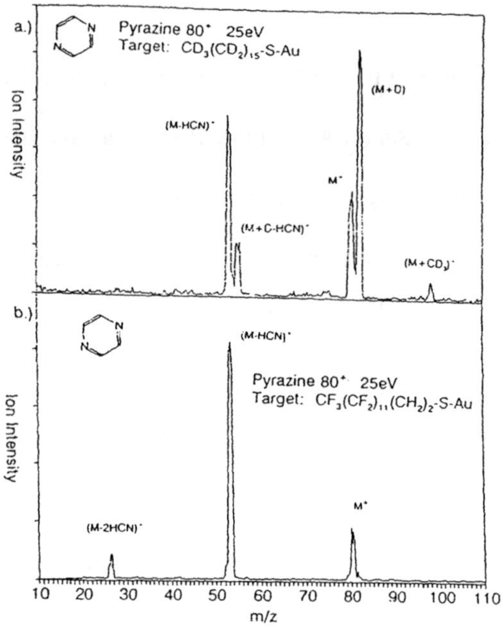

Figure 5: Product ion spectrum resulting from 25 eV collisions of pyrazine (m/z 80) upon (a) D-SAM surface and (b) upon a F-SAM surface. (Reproduced with permission from reference 80.)

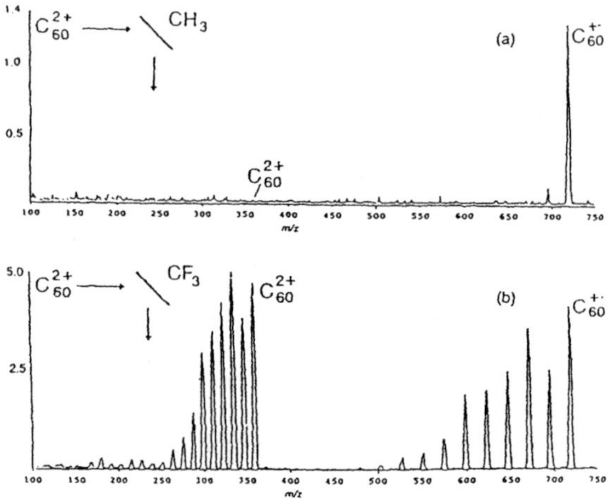

Figure 6: Ion/surface collision spectra obtained by colliding C_{60}^{2+} at 250 eV with (a) H-SAM surface and (b) F-SAM surface. (Reproduced with permission from reference 10.)

Charge exchange can cause ionization of the surface and can result in dissociation of surface bonds and the ejection of ions. This process, known as chemical sputtering,[28] occurs at low collision energies for suitable projectile ions, such as $Xe^{+\cdot}$. For example, 60 eV collisions of $Xe^{+\cdot}$ upon a F-SAM surface results in the production of the following surface-derived ions with abundances of at least 10%: CF^+ (70%), $CF_2^{+\cdot}$ (50%), CF_3^+ (100%), $C_2F_4^{+\cdot}$ (10%), $C_2F_5^+$ (15%), and $C_3F_5^+$ (10%). Note that the information provided is similar to that yielded by conventional momentum transfer sputtering (the usual static SIMS experiment[29]) except that the experiment appears to be more gentle in that it only yields information on the groups present at the outermost monolayers of a surface. Some characteristics of chemical sputtering are summarized in Table 5.

5. Instrumental Requirements and Their Realization

A fundamental requirement for low energy ion/surface collision experiments is that the primary and secondary ions are mass analyzed, that is one performs a MS/MS experiment. Time-of-flight instruments are a special case in which one time measurement can be used to select a projectile ion and measure the product ions in a single mass analyzer. Ultra high vacuum conditions ($< 10^{-9}$ Torr) are required for clean single crystal surfaces, however, SAM surfaces allow this requirement to be relaxed and more modest vacuum conditions ($< 10^{-6}$ Torr) suffice. Low ion doses, preferably static conditions (a beam size of 4 mm^2 with an ion beam current of less than 1 nA is typical in these experiments), result in longer lifetimes (days) of the SAM surfaces. A variable collision energy with a normal component of kinetic energy corresponding to more than 10 eV (e.g. a 15 eV collision at 45° angle of incidence) is also useful because it allows one to change the internal energy deposited (compare Figure 2). Other useful capabilities include kinetic energy analysis of the scattered product ions and the ability to vary the incident angle independently of the scattering angle.

Detection of surface modifications produced by low energy ion beam/surface collisions requires independent methods which can characterize the surface before and after ion/surface collisions. Many techniques are available to characterize the composition and structure of clean and adsorbate-covered surfaces. Electron spectroscopy for chemical analysis (ESCA) and surface infrared spectroscopy have been used to characterize SAMs.[30,31] Other techniques, such as high resolution electron energy loss spectroscopy (HREELS), Auger electron spectroscopy (AES), atomic force microscopy (AFM), and scanning tunneling microscopy (STM) are also producing interesting data.[32] Less traditional but effective techniques for examining organic surfaces, namely time-of-flight secondary ion mass spectrometry (TOF-SIMS)[33] and chemical sputtering,[28] can be used to recognize surface modifications. TOF-SIMS and chemical sputtering have the inherent disadvantage of being destructive to the surface but they do allow determination of experimental conditions for efficient surface modification.

All or some of the above capabilites have been realized in instruments of a wide-variety of types. The instrument designs can be divided into on-axis and off-axis geometries. On-axis geometries have the exit slit or aperture of one mass analyzer in-line with the entrance slit or aperture of the second mass analyzer. The on-axis group can be further divided into large deflection angle/low collision energy and glancing deflection angle/high collision energy types. The tandem quadrupole with the analyzers arranged at 90° is the classic case of the off-axis geometry.[34] An example of an off-axis, large deflection angle/low collision energy arrangement in which the MS/MS requirement is fulfilled with a Wien filter/quadrupole combination is shown in

Table 5:
Characteristics of Chemical Sputtering

(1) Chemical sputtering results from the ejection of groups from the surface by charge exchange.
(2) Chemical sputtering can be exothermic or endothermic.
(3) Typically onset collision energy for chemical sputtering is 20 eV for near specular collision geometry.
(4) Xenon cation is a good chemical sputtering agent.
(5) Chemical sputtering has a high cross section.
(6) Chemical sputtering is highly surface sensitive.

Table 6
Some SID and Ion/Surface Reaction Instrument Types

On-Axis Geometry

<u>Large deflection angle/Low collision energy</u>

1. Multi-sector instruments[71]

<u>Glancing deflection angle/High collision energy</u>

1. Hybrid sector/TOF[72]
2. Wien/Wien filter combination[73]
3. Hybrid sector/quadrupole[74,75]
4. Multi-sector instruments[76]

Off-Axis Geometry

<u>Large deflection angle/Low collision energy</u>

1. Tandem time-of-flight (TOF)[77,78,79]
2. Hybrid sector/quadrupole[2,80]
3. Wien filter/quadrupole instruments[81,82]

Trapping Instruments

1. Ion traps[83]
2. Fourier transform ion cyclotron resonance mass spectrometers (FTICRMS).[84,85,86]

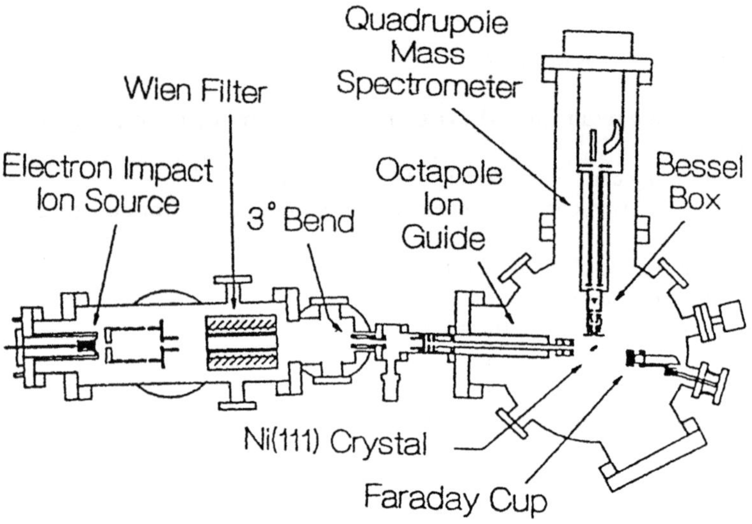

Figure 7: Tandem mass spectrometer utilizing a Wien filter/quadrupole combination for the study of ion/surface collision phenomena. (Reproduced with permission from reference 81.)

Figure 7.[82] Some instruments are summarized in Table 6.

Characterization of ion/surface reaction products (an MS/MS/MS experiment) would certainly be an advantage in future instruments. Such an instrument would require extremely high sensitivity due to the low product ion abundances involved. An ion trap mass spectrometer as the post-surface analyzer would provide both the multiple stages of MS and the sensitivity.

6. Applications to Ion Structural Characterization

The information from an SID experiment can be used for ion characterization, particularly if the measurement is made as a function of collision energy. The results can then be represented in the form of an energy resolved mass spectrum (ERMS) or breakdown curve of which an example is shown in Figure 8.[27] Note the ease with which these data allow one to recognize that the F-SAM surface is harder, *viz.* imparts more internal energy, than the D-SAM surface.

Numerous examples of isomeric ion distinction have been described using ion/surface collisions.[35,36,37,38] In some cases only SID processes are observed but the occurrence of ion/surface reactions is a valuable ancillary tool for distinguishing between isomeric ions. Consider $C_5H_6^{+\cdot}$, 66 Th, generated from various precursors. Two SID product ion spectra are shown in Figure 9 which represents the extremes of fragmentation for two of the isomers studied. It is suggested that a mixture of two structures are involved, a cyclopentadiene structure, which resists fragmentation, and an acyclic isomer. This interpretation is strongly supported by the plot of the degree of fragmentation which correlates linearly with the degree of charge stripping, a process characteristic only of the linear form.[36]

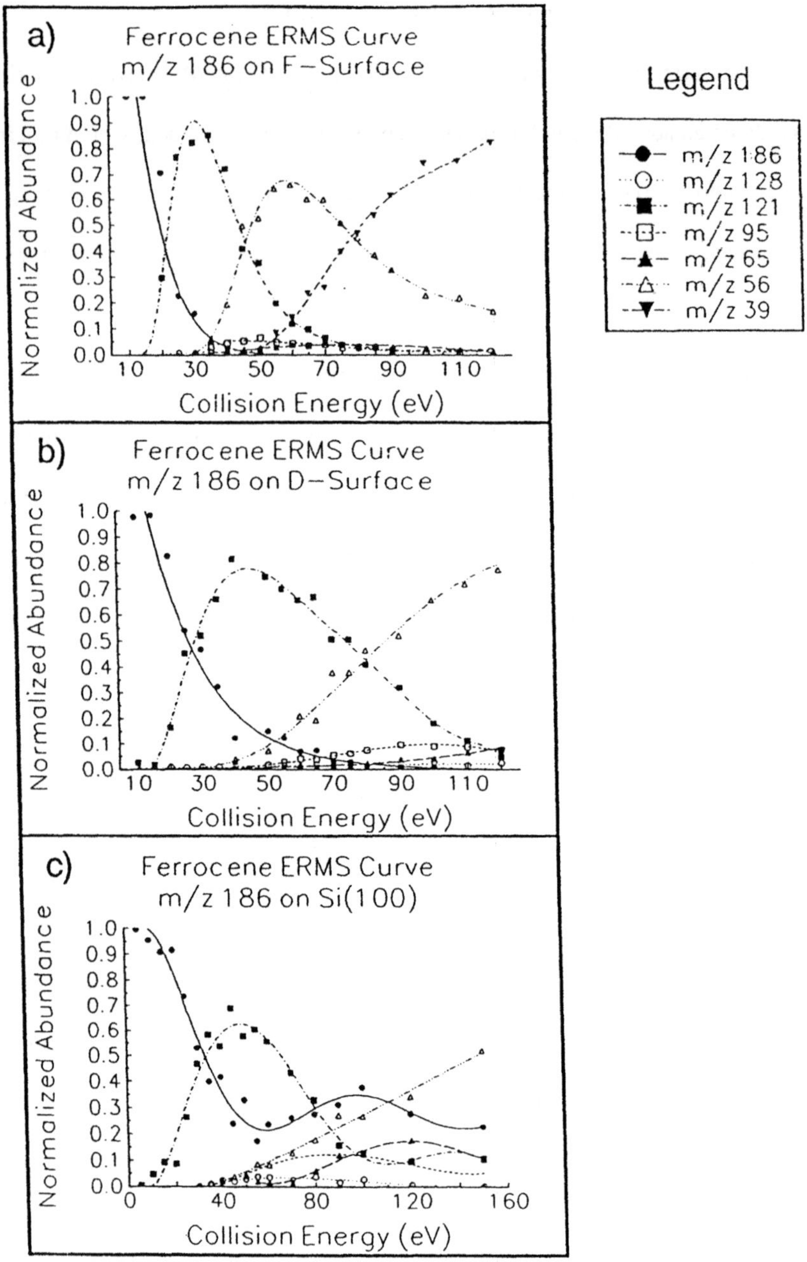

Figure 8: Energy resolved mass spectra plots of the ferrocene molecular ion (m/z 186) for collisions upon (a) a F-SAM surface, (b) a D-SAM surface, and (c) a Si(100) surface. (Reproduced with permission from reference 27.)

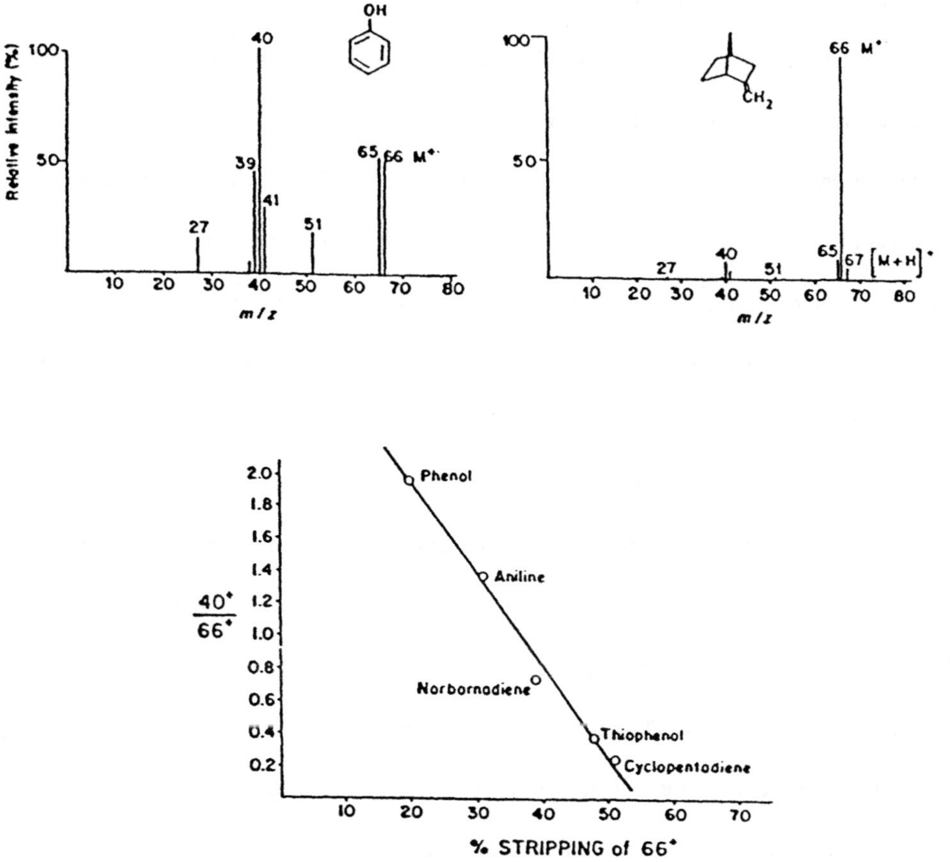

Figure 9: SID spectra for two $C_5H_6^{+\cdot}$ (m/z 66) isomers produced from two different precursors upon collision with a stainless steel surface at a 25 eV collision energy. A plot of m/z 40 relative to m/z 66 as a function of percent charge stripping is also shown for several precursors. (Adapted with permission from reference 36.)

7. Applications to Biomolecule Structure Elucidation

Surface-induced dissociation can be used analogously to gas-phase CAD to elucidate molecular structures.[2] Here, we briefly consider its use for peptide sequencing.[39,40,41] Note that the performance achieved in SID depends, just as it does in CAD, on the quality of the mass analyzers used (hence comparisons need to be made with this factor in mind) as well as on the particular choice of surface. The latter choice, as already shown, influences both the ion yield and the internal energy deposition. Fluorocarbon surfaces are preferred over hydrocarbon surfaces on both counts.

The overall quality of SID spectra on fluorocarbon surfaces can be compared with CAD spectra by examining the melittin data shown in Figure 10.[40,42] Note that the resolution in this SID experiment is lower than in CAD but in other respects, the data are comparable. In general, spectra taken using a fluorocarbon surface, show significant improvement in SID performance over hydrocarbon surfaces. The fluorinated surface gives higher ion yields and hence improved mass resolution and, because of its greater internal energy deposition, it also produces the characteristic d- and w- ions in amounts comparable to those observed in high energy CAD spectra.[40] SID efficiencies for a pentapeptide and a octapeptide were measured to be 13 % and 20%, respectively while 7 keV CAD efficiencies average at just 1%.[41] These results suggest that SID is a viable activation method for studying biological compounds, especially peptides.

8. Internal Energy Deposition in Ion/Surface Collisions

The importance of the distribution of internal energies, P(ϵ), acquired by a population of ions is that it controls the subsequent (unimolecular) behavior of the population and hence the

a)

b)

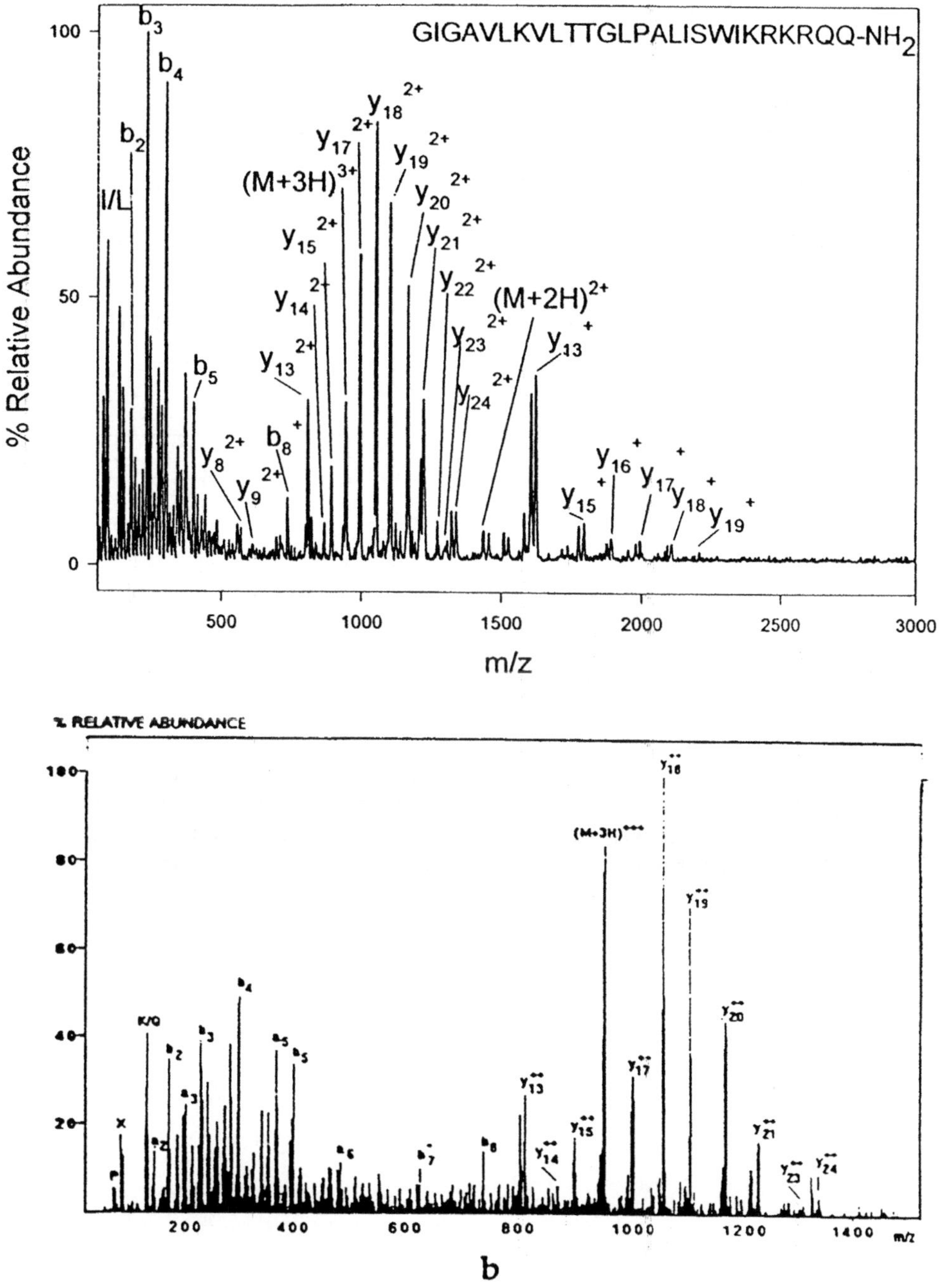

Figure 10: (a) SID spectrum obtained for the $[M+3H]^{3+}$ ion (m/z 949.8) of melittin at a collision energy of 135 eV on a fluorinated SAM surface and (b) CAD spectrum of $[M+3H]^{3+}$ ion (m/z 949.8) of melittin recorded at a collision energy of 90 eV. (Adapted with permission from reference 42 and Prof. V. Wysocki.)

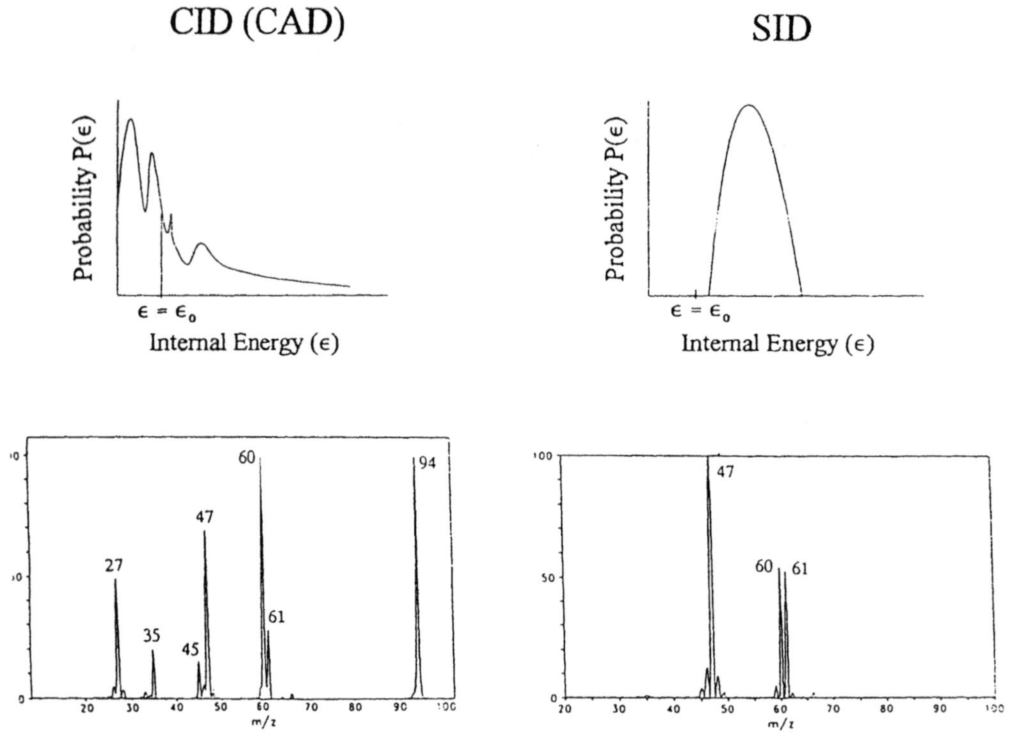

igure 11: Comparison of two hypothetical P(ε) curves for CAD and SID with their orresponding 25 eV collision energy spectra of the molecular ion, 1,2-ethanedithiol (m/z 94). he CID spectrum was taken using Kr as the collision gas under multiple collision conditions and ie SID spectrum was taken using a D-SAM surface. (Adapted with permission from reference 3.)

fragmentation pattern. The energy distributions associated with collisionally-activated dissociation and surface-induced dissociation are characteristically different, viz. SID distributions are narrower.[1,43] One can infer this by making a comparison of the 25 eV SID and CAD (Kr, multiple collisions) spectra of 1,2-ethanediol (Figure 11). In the CAD spectrum, the presence of all the ions which occur in the SID spectrum suggests that the internal energies of the parent ions in the latter experiment are also represented in the former. However, the additional presence of undissociated parent ions in the CAD spectrum demands that the internal energy distribution of this population include some ions of lower internal energy as well. Thus, the SID internal energy distribution must be narrower than the CAD distribution.

More accurate estimates of the distribution of internal energy are possible by examining the dissociation of characteristic "thermometer" ions which dissociate by a well-defined set of prompt sequential reactions of known energy requirement.[44] If these reactions are simple bond cleavages to which the "sudden death" approximation is applicable (assuming that all ions with the necessary energy for fragmentation, will fragment), then the kinetic behavior represented by the abundances of ions in the mass spectrum can be used to measure the distribution of internal energies, $P(\epsilon)$. The argument used is derived from the fact that a mass spectrum can be calculated from a knowledge of the energetics of fragmentation and the distribution of internal energies. In other words, from a knowledge of the unimolecular rate constants involved in the set of competing and consecutive reactions occurring from an energized parent ion, one can calculate a breakdown curve, the normalized plot of relative ion abundances expressed explicitly as a function of internal energy.[45] Convoluting this data with the distribution of internal energies, $P(\epsilon)$, gives the mass spectrum. Arguing in reverse, it is possible to deduce $P(\epsilon)$ by deconvolution of the mass spectrum if one knows the breakdown curve. This approach becomes extremely simple if the parent ion is especially chosen to have a linear fragmentation series in which the thresholds for production of particular fragment ions represent points of complete conversion of the precursor ion into the fragment ion. For example, the fragmentation pattern and thermochemical data for $W(CO)_6^{+\cdot}$ are included in Figure 12. The threshold values will include the kinetic shift and hence are best represented by experimental values derived from a measured appearance energy. The concept is outlined in Figure 12. The $P(\epsilon)$ curve is produced by plotting the abundance of each fragment ion corrected for the 'bin' width of appearance energies at the average internal energy for its formation. The average internal energy of the excited projectile ion is then determined by the weighted average of the $P(\epsilon)$ curve.

This method of determining $P(\epsilon)$ can be illustrated in the case of $W(CO)_6^{+\cdot}$ by two experiments, charge exchange with $Ar^{+\cdot}$ (Figure 13) and SID. The gas-phase charge exchange of $Ar^{+\cdot}$ with $W(CO)_6$ to yield $W(CO)_6^{+\cdot}$ provides a test of the method. Under single-collision conditions this process should give a single-valued internal energy of $W(CO)_6^{+\cdot}$, the expected value being given by $RE(Ar^{+\cdot}) - IE(W(CO)_6) = 15.7$ eV - 8.5 eV, i.e. approximately 7.2 eV (note the IE of $W(CO)_6$ is an average value).[46] The extent to which $W(CO)_6^{+\cdot}$ fragments can be used to measure its internal energy distribution and the result, 7.1 eV (Figure 13), is in agreement with the prediction.

Application of this method to metal carbonyl ions activated by surface collisions allows calculation of the average internal energy deposition and hence the percent translational to vibrational energy (% T -> V) conversion.[44] This energy partitioning percentage is constant over a wide range of collision energies for many metal carbonyls as it is also for the $SiEt_4^{+\cdot}$

ION	m/z	IE/AE (eV)	E_o (eV)
$W(CO)_6^{+\cdot}$	352	8.5	0.0
$W(CO)_5^{+\cdot}$	324	9.7	1.2
$W(CO)_4^{+\cdot}$	296	11.9	3.4
$W(CO)_3^{+\cdot}$	268	13.7	5.2
$W(CO)_2^{+\cdot}$	240	16.0	7.5
$WCO^{+\cdot}$	212	18.6	10.1
$W^{+\cdot}$	184	21.5	13.0

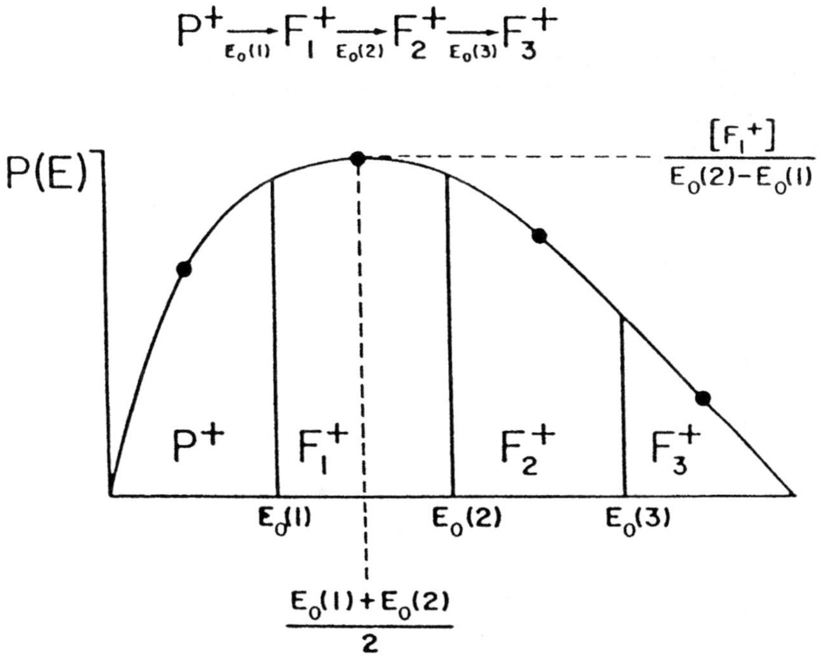

Figure 12: Ionization energies (IE), appearance energies (AE), and activation energies (E_o) of the molecular ion and fragments of $W(CO)_6^{+\cdot}$ and the construction of a P(ϵ) curve from the experimental ion abundances (●). (Adapted with permission from Wysocki, et. al *Int. J. Mass Spectrom. Ion Processes* **1987**, *75*, 181-208.)

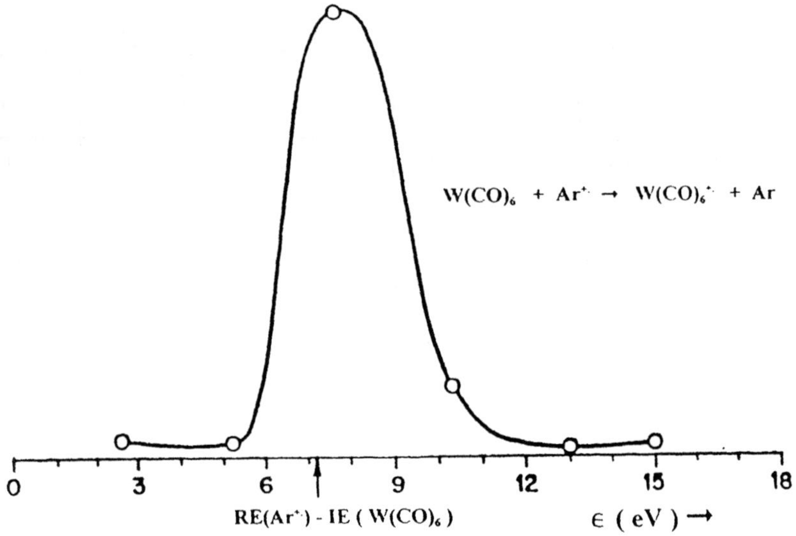

Figure 13: P(ε) curve developed from charge exchange data for $Ar^{+\cdot}$ and $W(CO)_6$ with an average internal energy deposition near the value of 7.2 eV calculated from {RE($Ar^{+\cdot}$) - IE ($W(CO)_6$)}. (Adapted with permission from Wysocki, et. al *Int. J. Mass Spectrom. Ion Processes* **1987**, *75*, 181-208.)

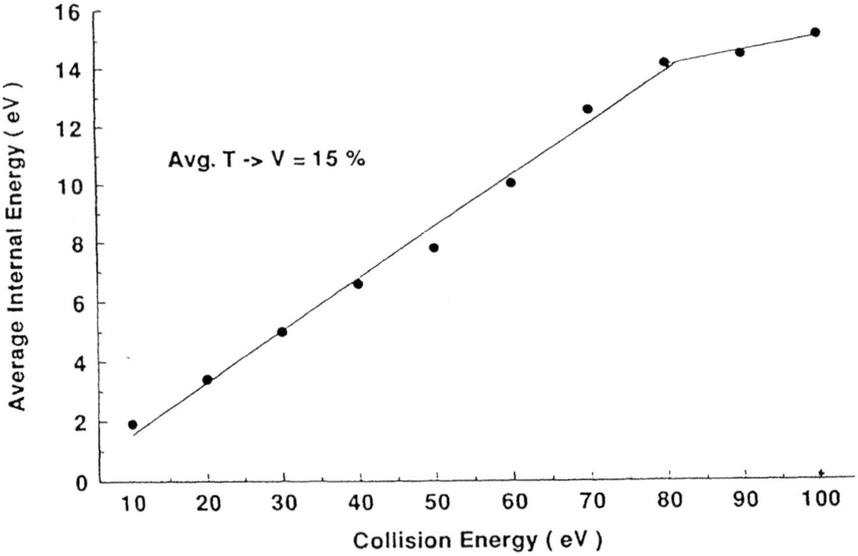

Figure 14: A plot of the average internal energy deposited into the molecular ion, tetraethylsilane, after colliding with a stainless steel surface as a function of collision energy. The translational to vibrational energy conversion is given by the slope of the line which corresponds to 15% conversion. (Adapted with permission from reference 77.)

Table 7:
Estimated Translational to Internal Energy Conversion Efficiency for Surfaces using Various Thermometer Molecules and Collision Energies

Ion	Target	%T->V Energy Conversion	
$Cr(CO)_6$	F-SAM (25 eV)	19%	
$W(CO)_6$	F-SAM (30 eV)	18%	F Targets
$W(CO)_6$	F-SAM (50 eV)	20%	
$W(CO)_6$	PFPE (50 eV)	18%	
ferrocene	F-SAM (25-70 eV)	24% (20%)	
$Cr(CO)_6$	D-SAM (25 eV)	12%	
ferrocene	D-SAM (25-70 eV)	15% (13%)	
$Cr(CO)_6$	Fc (25 eV)	11%	Hydrocarbon-like
n-butyl-benzene	Fc (25 eV)	12%	Targets
$Cr(CO)_6$	SS (25 eV)	11%	
$W(CO)_6$	SS (25-100 eV)	13%	
$Fe(CO)_5$	SS (25-60 eV)	13%	
Et_4Si	SS (10-100 eV)	12%	

Where PFPE is perfluoropolyether, Fc is a ferrocene terminated self-assembled monolayer, and SS is stainless steal with background hydrocarbon adsorbates.

thermometer ion (Figure 14). This behavior is suggestive of an impulsive excitation process with fixed effective target mass. Note the plateau in Figure 14 as the internal energy reaches about 14 eV. This plateau may be attributed to the inability of the projectile ion to adsorb more internal energy or it may be the result of secondary processes, such as chemical sputtering or ion/surface reactions. Wysocki and co-workers have used benzene as a thermometer molecule and shown that it adsorbs larger quantities of internal energy resulting in a larger linear range of T->V conversion.[47]

The T -> V conversion efficiency depends largely on the nature of the target .[37] Table 7 shows a collection of data for which it is evident that fluorocarbon surfaces convert some 18-20% of the laboratory translational energy of the ion into internal energy while a number close to 13% is appropriate for hydrocarbon surfaces. It should be emphasized that these data are estimates that refer to a particular set of scattering conditions employing near specular scattering, and where necessary, they have been corrected for the initial internal energy of the projectile.

The internal energy required to dissociate biomolecules depends on their structure. In particular, peptides which contain strong basic groups such as arginine are more difficult to dissociate than those which do not contain such basic groups. Wysocki et. al have established a linear relationship between the gas phase basicity of the most basic amino acids in a pentapeptide and the collision energy required for its fragmentation.[48] In spite of this valuable information, very little is known of the actual magnitudes of the internal energies which are deposited, by SID or any other means, into biological ions.

The conversion of translational into internal energy in fullerene collisions is of special interest since a simple mechanism has been proposed, *viz.* resilience of the ball leading to a maximum uptake of 14 eV of internal energy.[49] In this case, the % T -> V conversion is a constant until the internal energy increases to the maximum value. Results for $C_{60}^{+\cdot}$ are available for different surfaces, experimental conditions, and investigators, and they all show efficient conversion in the low tens of percent of translational to internal energy transfer.[50,51,52]

A description of the partitioning of the translational energy of the projectile ion upon collision with a surface requires that the kinetic as well as the internal energy of the recoiling ion be described. The kinetic energy measurement of the scattered ions has not been made for many systems but the available results make it clear that only a small fraction of the total energy is disposed of in this way. For example, C_{60}^{+} ions scattered from various surfaces have been examined and the scattered intact C_{60}^{+} ions show the velocity distributions reproduced in Figure 15.[53] Note that increasing target mass is associated with lower scattered ion velocities and that just 2 - 10 % of the precollision energy is partitioned into scattered ion translational energies in these cases.

For polyatomic ions other than $C_{60}^{+\cdot}$ much less data is available as this is a difficult measurement at the low collision energies which are of particular interest. However, data from Hanley and co-workers[54] as well as our own unpublished results (e.g. $C_6H_6^{+\cdot}$) on fluorocarbon surfaces, show that in at least some systems, these energies vary with the fragment ion examined and that they correspond to a few eV of energy. Clearly, the bulk of the energy supplied in inelastic collisions must be lodged in the surface itself since only 10 - 25 % goes into internal excitation of the projectile and less than this into the scattered ion translational energy.

An important conclusion emerges from the magnitude of the post-collision velocities: if fragmentation does not occur directly upon impact (and there is certainly evidence in some small

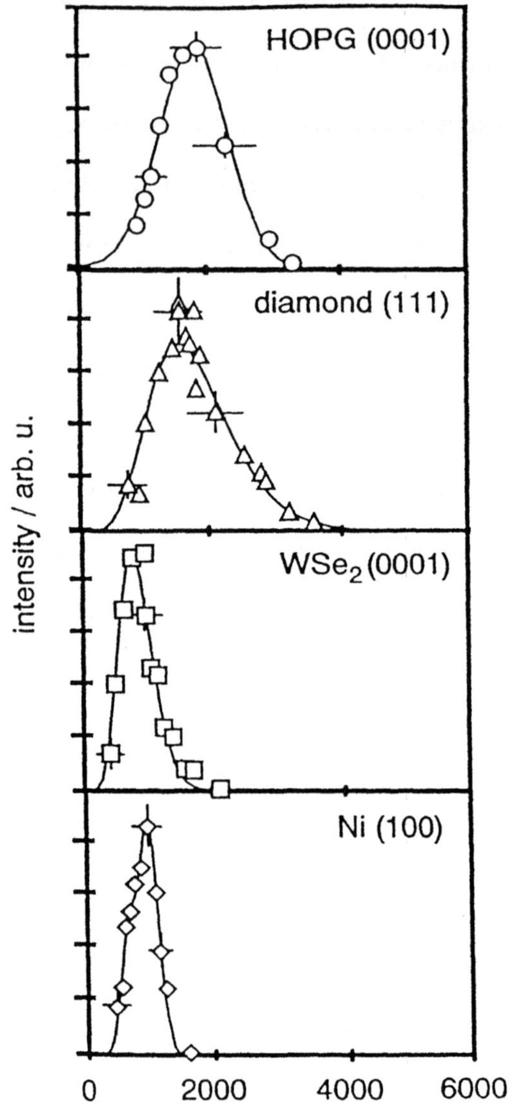

Figure 15: Velocity distributions of $C_{60}^{+\cdot}$ ions scattered from four different surfaces with an impact energy of 270 eV for the HOPG, diamond, and tungsten diselenide surfaces and 150 eV for the nickel surface. (Reproduced with permission from reference 53.)

systems for such direct dissociation at the interface) then it will occur unimolecularly and often only after the ion has traveled a significant distance from the surface. Consider the case of an ion of mass 200 Dalton and collision energy of 100 eV. The average internal energy deposited into the scattered projectile from a "hard" surface such as the F-SAM will be about 20 % (i.e., 20 eV), and the scattered translational energy will be perhaps 3 eV (3%), the normal component of which will correspond to ca. 2 eV, for a normal component velocity of 1×10^5 cm/s. The maximum rate constant for a statistical unimolecular fragmentation is on the order of 3×10^{11} s^{-1} (after which non-ergodic processes occur in conflict with the evidence from the SID spectra of many organic ions that fragmentation is due to conventional statistical fragmentation). Hence, the maximum value for the average ion lifetime is the reciprocal of the rate constant (i.e., 3 ps) and by this time the ion is 3 nm from the surface. It is therefore expected that activation at the surface will be followed by delayed gas phase dissociation for the many SID processes in which slower dissociation of larger ions occurs.

Angular effects are also known for ion/surface reactive scattering at SAM surfaces as illustrated in Figure 16.[22] Here the incoming angle is fixed while the total scattering angle is varied. As the data show, more glancing collisions are less effective in setting up the complex collision sequences required to abstract multiple F atoms by a single W^+ projectile.

An alternative method of altering the angle of interaction of the ion and a SAM surface has been reported by Wysocki et al. who examined the effect of alkyl chain length on the occurrence of hydrogen atom abstraction by $C_6H_6^{+\cdot}$.[55] They observed an alternating pattern of reactivity explained by the odd-C number chains having hydrogens directed away from the surface (Figure 17).

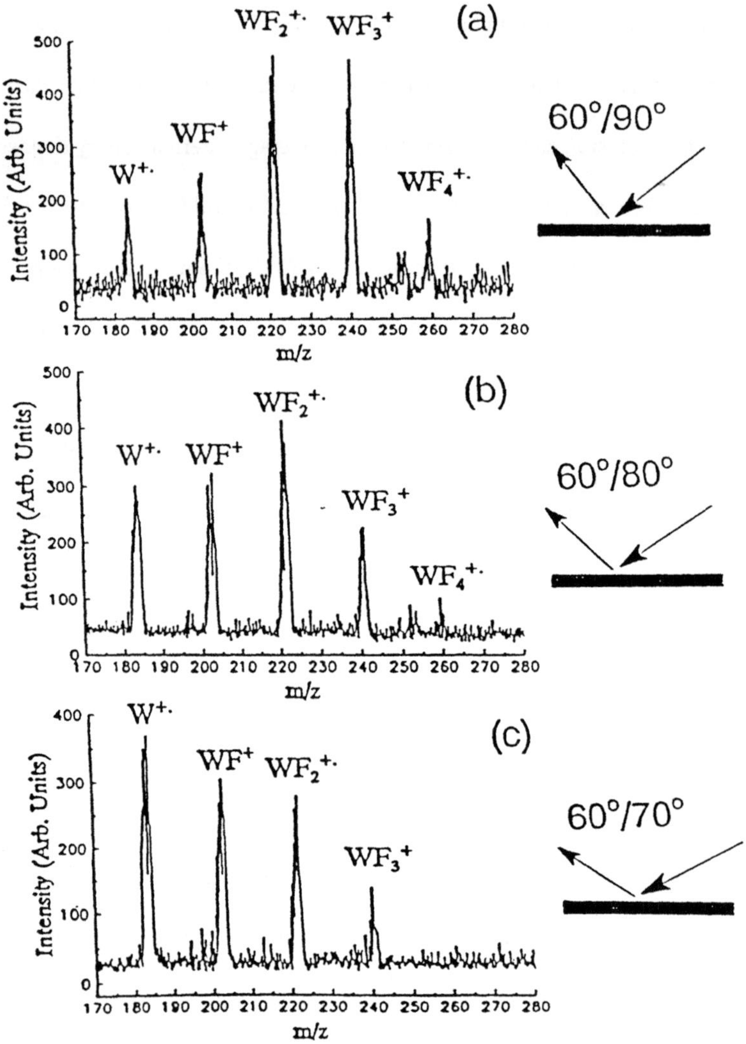

Figure 16: Ion/surface reaction spectra resulting from collisions of $^{186}W^{+\cdot}$ upon a F-SAM surface with a 30 eV collision energy at various incident and scattering angles. (Reproduced with permission from reference 22.)

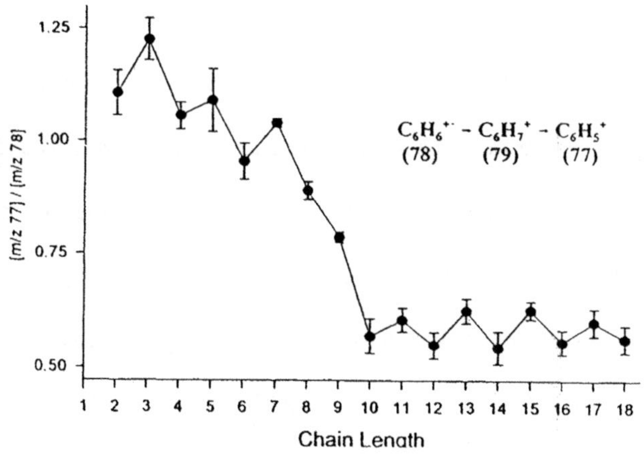

Figure 17: A plot of the relative reactivity of benzene ($[C_6H_6 + H - H_2]^+ / [C_6H_6]^{+\cdot}$) upon 30 eV collisions with H-SAM surfaces of varying chain length. (Reproduced with permission from Wysocki et. al, reference 55.)

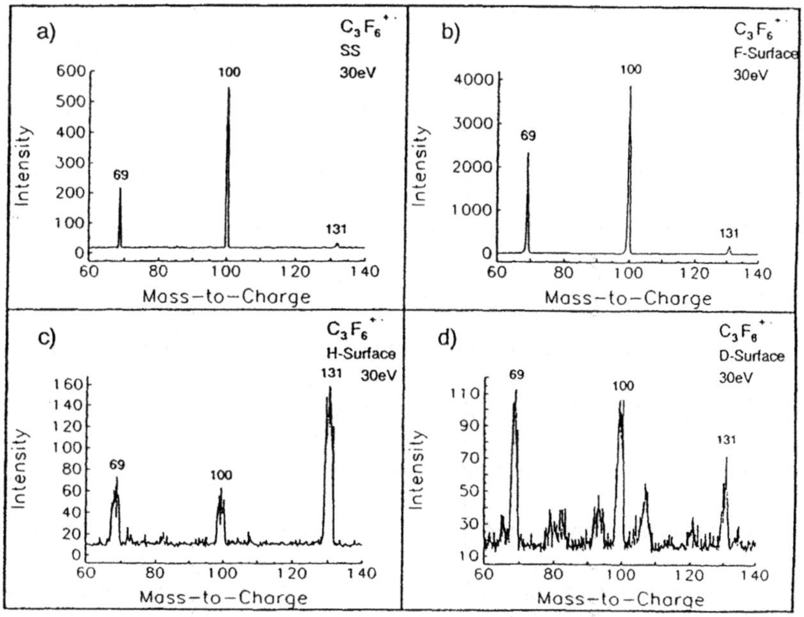

Figure 18: SID spectra resulting from 30 eV collisions of the molecular ion, $C_3F_6^{+\cdot}$, at four different surfaces: stainless steel, F-SAM, H-SAM, and a D-SAM surface. (Reproduced with permission from reference 58.)

9. Excited State Effects

The state of an ion (electronic and vibrational) will affect its behavior on collision with a surface, including the extent to which it is activated and its propensity to undergo charge transfer and chemical reactions. Although very few experiments of this type have been reported, interesting effects have been observed in both small molecular ions, e.g. NO^+,[56] and in large biomolecules.[40] Jacobs and coworkers have shown that simply increasing the number of vibrational quanta in NO^+ from $v = 2$ to $v = 6$ doubles the yield of O^- scattered from collisions of NO^+ on a GaAs single crystal surface by making accessible regions of the potential energy hypersurface which favor dissociative charge transfer.[57]

Similarly, whether a given protonated peptide is generated by electrospray ionization (ESI) or by liquid secondary ion mass spectrometry (LSIMS) has been shown to greatly affect its degree of fragmentation in SID by Wysocki and coworkers.[40] This is because of the very large difference in internal energies deposited upon formation of protonated ions by these two techniques. Even the apparently innocuous change of a SAM surface from the H-SAM to its analog D-SAM can have dramatic effects on the SID spectra of certain ions, as shown for $C_3F_6^{+\cdot}$ in Figure 18.[58] In this case, comparisons with other surfaces and examination of the effects of collision energy demonstrate that $C_3F_6^{+\cdot}$ can be excited into an electronically excited state from which it dissociates without converting to the ground state by $F^\cdot$ loss to give $C_3F_5^+$, 131 Th.[58] This process is measurably favored by the light $-CH_3$ terminus of the H-SAM, even when compared to the $-CD_3$ terminal group of a D-SAM surface. At increased collision energy, the electronic excitation process yields to direct vibrational excitation, the dominant product of which is $C_2F_4^{+\cdot}$, 100 Th.

One must note that this is an exceptional case, chosen for study because it was known from CAD work that access to a long-lived excited state is possible in $C_3F_6^{+\cdot}$.[58] At present, SID mechanisms are best described as being due to impulsive excitation, as reported in the case of $C_2H_3S^+$,[37] and described semi-quantitatively by Hanley et al.[59] In addition, while there is evidence in some systems, especially ions with a small number of atoms (and possibly in the case of C_{60}), for dissociation at the interface, the body of evidence favors delayed dissociation as already described.

10. Ion/Surface Reactions: A Survey

The rich variety of ion/surface reactions already known is summarized by the representative examples collected in Table 8. Note the pick-up of atoms (F, H, C), both singly and multiply, with or without loss of groups from the projectile. A more complex ion/surface reaction is demonstrated by the abstraction of multiple hydrogen atoms from a hydrocarbon surface by the highly unsaturated ion $C_2HN^{+\cdot}$. Unique product ions are associated with the characteristic transfer of a methyl group and associated H_2 elimination. Figure 19 shows fluorine abstraction by $Xe^{+\cdot}$, a case in which the resulting fluoride is relatively weakly bound. The XeF^+ bond energy of 2 eV places this value as an upper bound on T -> V conversion and it is noteworthy that this product is only observable over a relatively narrow range of collision energies. These collision energies are sufficiently high to transfer the 3.2 eV necessary to drive the reaction but sufficiently low to provide less than 2 eV internal energy to the product (note that less than 10% of the lab collision energy is lodged in the translational energy of the scatted ion, and 3.2 eV is required to drive the reaction). The bulk of the energy brought into the system by

Table 8:
Examples of Ion/Surface Reactions

Projectile	Surface	Products	Reference(s)
Mo^+	F-SAM	MoF_n^+ (n = 1 - 4)	22
$W(CO)_6^{+\cdot}$	Alkane	$WCOH^+$, $W(CO)_2H^+$	71
C^+	F-SAM	CF^+	64
$SiCl_4^{+\cdot}$	F-SAM	$SiCl_2F^+$	16,64
$C_6H_6^{+\cdot}$	Alkane	$[C_6H_6+H-H_2]^+$,etc.	25
$PAH^{+\cdot}$	Alkane	$PAH(C_2H_3)^+$, etc.	14
$HCCN^{+\cdot}$	H-SAM	H_nCCNH^+ (n = 1 - 3)	69,62
$Xe^{+\cdot}$	F-SAM	XeF^+	60
$C_{60}^{+\cdot}$	C_{60}	$C_{62}^{+\cdot}$, etc.	87
$Si_{12}^{+\cdot}$	Graphite	$Si_9C^{+\cdot}$	88
$Ni(C_5H_5)_2^{+\cdot}$	F-SAM	$NiH_2^{+\cdot}$	23
$OCNCO^+$	F-SAM	FCO^+, $F_2CN^{+\cdot}$	Unpublished results
$OCNCO^+$	D-SAM	DCO^+	Unpublished results
$OCNCS^+$	F-SAM	$NCSF^+$, FCS^+, FCO^+	This work

Table 9:
Methods Used to Study Ion/Surface Reaction Mechanisms

(1) Threshold measurements
(2) Thermochemical considerations
(3) Gas phase analogs (ion/molecule reactions)
(4) Correlation with chemical sputtering
(5) Isotopically labeled surfaces
(6) Subsequent surface analysis
(7) Scattered ion kinetic energy
(8) Scattered ion angular distribution

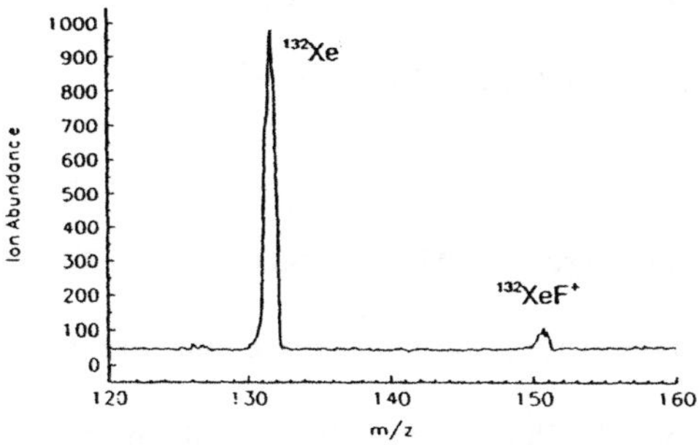

$$Xe^{+\cdot} + C_3F_8 \rightarrow XeF^+ + C_3F_7^{\cdot} \quad \Delta H_{rxn} = 3.2\ eV$$

XeF^+ Bond Energy = 2.0 eV

Collision Energy	40 eV
Heat of Reaction	3.2 eV (8%)
Internal Energy of Scattered Product	< 2 eV (5%)
Kinetic Energy of Scattered Product	< 4 eV (10%)
Internal Energy Distributed in FSAM	34 eV (77%)

Figure 19: Ion/surface reaction spectrum of 40 eV collisions of $^{132}Xe^{+\cdot}$ with a F-SAM surface including the thermochemistry related to the reaction and the estimated kinetic energy partitioning. (Reproduced with permission form reference 60.)

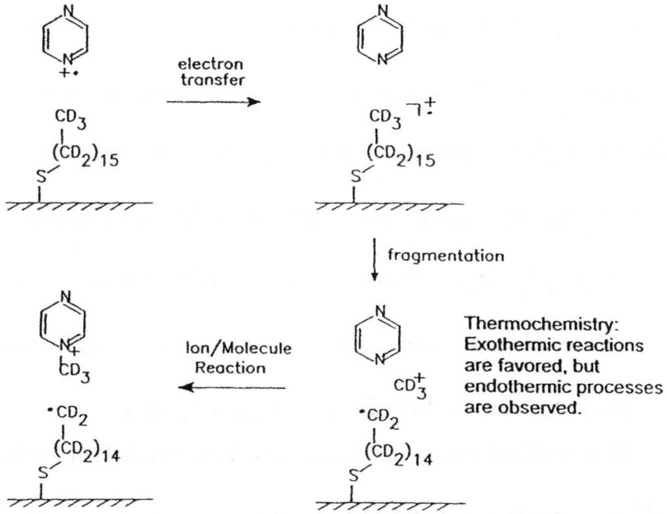

Figure 20: Electron transfer mechanism suggested for some ion/surface reactions illustrated in the case of ionized pyrazine and a deuterated surface.

the collision process is clearly deposited into the target.[60] This is appears to be the case in most surface-induced dissociations and ion/surface reactions.

11. Studies of Ion/Surface Reaction Mechanisms

These considerations naturally bring us to the topic of the mechanisms by which ion/surface reactions proceed. The tools used to elucidate mechanisms are still being developed but some are summarized in Table 9.

Examination of a large number of systems which undergo H and alkyl radical abstraction by the projectile ion shows that they appear to do so by such an electron transfer mechanism. The evidence is summarized in Table 10 and the mechanism is illustrated in Figure 20.[60] The requirement for product formation is not simply a high proton (or alkyl cation) affinity of the projectile. That this is a necessary but not sufficient condition is established by considering the case of ferrocene which has a high proton affinity but whose low ionization energy precludes electron transfer and which therefore does not react.[27]

Evidence for the charge exchange mechanism comes from examination of analogo ion/molecule reactions. This had been commented on earlier[1] and has been the subject of re work by Bernasek. In experiments on the ion/surface reactions of benzene with a hydrocarb surface, this group[61] suggests that the formation of the ion at 91 Th is the result of neutralis of benzene via an electron transfer with the surface, followed by a reaction with a chemical sputtered product, $C_3H_5^+$ and finally by C_2H_2 and H_2 loss.

This mechanism was tested using gas phase ion/molecule reactions in a triple quadrupole mass spectrometer in which benzene is the neutral collision gas and various

Table 10:
Evidence for Electron Transfer Mechanism

The following general trends, which have been observed for the collision of hydrocarbon-like ions with hydrocarbon surfaces, are consistent with a mechanism involving electron transfer:

IE(proj) << IE(surf)	SID No reaction Minimal sputtering
IE(proj) ≈ IE(surf)	SID Reactions Sputtering
IE(proj) >> IE(surf)	Some SID Some reaction Sputtering

In addition:

* For a given projectile, the collision energy threshold for ion/surface reaction often matches that for chemical sputtering.
* Groups abstracted from the surface during ion/surface reactions are similar to those produced by chemical sputtering.

hydrocarbon ions are allowed to react with benzene.

Multiple fluorine abstractions, like many ion/surface reactions, are strongly influenced by the associated thermochemistry. Most of these reactions are endothermic but there is nevertheless strong thermochemical control. In other words, the abundances of the metal fluoride ions and the degree of fluorination observed, correlate with the reaction thermochemistry. Some thermochemical data are collected in Table 11, all for reactions which are observed when metal ions collide with F-SAM surfaces. The thermochemical data display, separately, the cases in which two fluorines are abstracted (i) from adjacent alkyl chains with concomitant C-C single bond formation, (ii) a single chain with concomitant C=C double bond formation, or (iii) without bond formation at the surface. While experimental evidence is not available to distinguish these cases, H_2 abstraction from H-SAM surfaces by C_2HN^+ has been examined using an intimate mixture of deuterated and non-deuterated thioalkanes.[62] The results establish that the hydrogens are derived from the same chain in this case, presumably concomitantly with the thermochemically advantageous C=C bond formation.

Formation of multiple fluorine bonds occurs but at higher collision energies when polyatomic ions containing the metal of interest are used as the projectile, e.g. $W(CO)_3^+$.[22] This finding is part of the evidence that fragmentation precedes or accompanies bond formation in these cases, most likely in the course of a single collision sequence.

The above description of the ion/surface reaction mechanism is one in which a projectile ion excites and has several atom-atom encounters within a small local volume of a surface and then rapidly departs with a low but non-thermal kinetic energy. It is supported by measurements of the kinetic energies of the product ions[63] and leads to estimated interaction times on the order of a few ps.[54] It is further supported by the data which show that the abstracted groups are derived from a single chain in the case of hydrocarbon surfaces (see above). The single-chain origin is thermodynamically favored and indicates that the lifetimes of the interaction-complex is sufficient to allow this choice.

Much of the evidence for ion/surface reaction mechanisms is indirect, and is typified by the experiments in which analogies are sought with ion/molecule reactions. Many such cases have been examined and clear correspondence in the observed phenomena is often found. For example, Figure 21 shows multiple F atom abstractions by W^+ in single-collisions with a gaseous perfluoroalkane, C_6F_{14}. The mass-selection of a collision-complex surrogate, $W(CO)_6C_6F_{13}^+$, and the demonstration that this ion yields WF_n^+ fragments upon CAD, shows that the necessary bond-making reactions can occur simply by energizing a unit consisting of the projectile of interest, $W(CO)_6^{+\cdot}$, and a fluorocarbon unit.[15]

Single fluorine atom abstraction by PCl^+ occurs in both surface and gas-phase collisions as shown in Figure 22, although the process is characteristically much more favorable in the case of surface collisions. Significantly, Figure 22 shows that target modification occurs in the gas-phase (formation of $C_6F_{12}Cl^+$) in analogy to the surface modifications discussed in the next section.[64]

To conclude this section on ion/surface reaction mechanisms, one notes that the two principal mechanisms are (i) charge transfer exemplified by H^- and alkyl radical abstraction by radical cations and (ii) oxidative addition of fluorine atoms to metal cations.

12. Surface Modification by Low Energy Ion Beams

Low energy ion/surface collisions provide the ability to effect chemical modification of a

Table 11:
Thermochemical Values (kcal/mol) for Some Ion/Surface Reactions

$M^+ + F\text{-}SAM \rightarrow MF_n^+ + F\text{-}SAM$ (with the following new bonds)

With single bond formation (adjacent alkyl chains with concomitant C-C bond formation)

WF^+	11	MoF^+	36	CrF^+	18
WF_2^+	-48	MoF_2^+	-5	CrF_2^+	38
WF_3^+	-37	MoF_3^+	37	CrF_3^+	-1
WF_4^+	-99				

With multiple bond formation (C=C bond formation on single chain)

WF^+	11	MoF^+	36	CrF^+	18
WF_2^+	-34	MoF_2^+	9	CrF_2^+	52
WF_3^+	-23	MoF_3^+	41	CrF_3^+	13
WF_4^+	-71				

Sequential (no bond formation at the surface)

WF^+	11	MoF^+	36	CrF^+	18
WF_2^+	19	MoF_2^+	62	CrF_2^+	108
WF_3^+	26	MoF_3^+	90	CrF_3^+	159
WF_4^+					

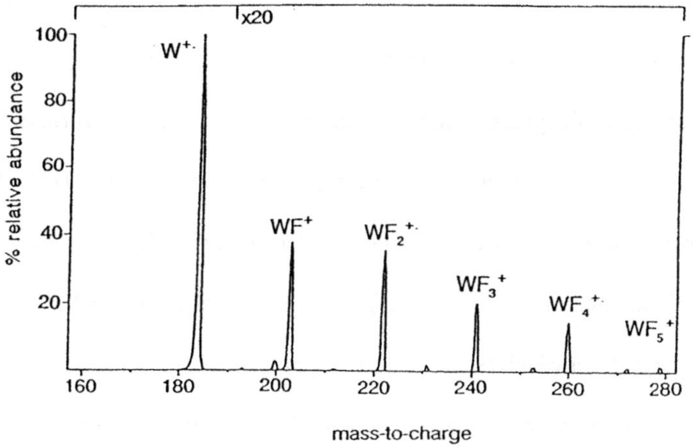

Figure 21: Ion/molecule reaction spectrum generated by 10 eV collisions of $^{184}W^{+\cdot}$ with 0.4 mTorr pressure of C_6F_{14}. (Reproduced with permission from reference 22.)

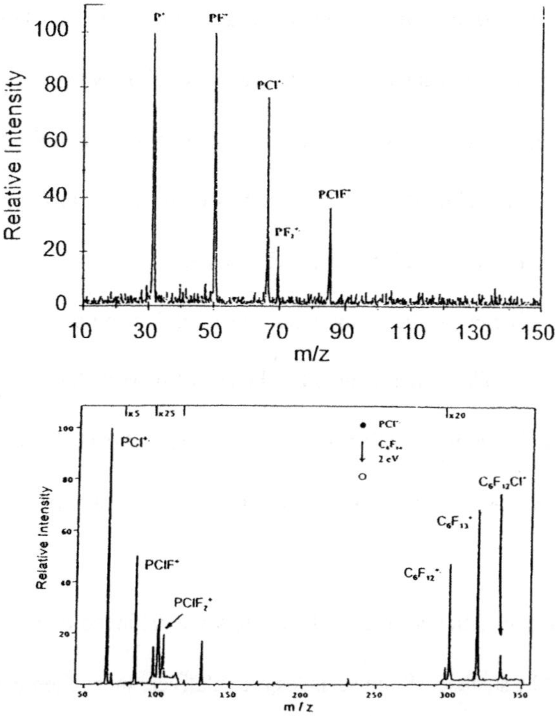

Figure 22: Comparison of the ion/surface and ion/molecule reaction product spectra of $^{31}P^{35}Cl^+$ (m/z 64) upon 40 eV collisions with a F-SAM surface and 2 eV collisions with C_6F_{14} under multiple collision conditions, respectively. (Reproduced with permission from reference 64.)

surface. Modification of surfaces may be achieved, *inter alia*, by solution treatment or by vapor deposition, plasma etching,[65] molecular beam scattering,[66] and ion beam implantation.[67] These methods offer a variety of degrees of control over the type and extent of surface alteration. Surfaces have also been modified using intermediate energy ion beams (500 eV - 1keV),[68] however, at these energies, chemical selectivity and specificity is difficult to attain.

Chemical modification using low energy ion beams is accomplished by mass selection of an appropriate reagent ion which reacts by donating and/or accepting an atom or group of atoms to/from the uppermost layers of the surface.[16] It has been suggested earlier in this paper that, when an ion reacts with a SAM surface, a free radical site may be left on the surface and be available for further modification. Three chemical pathways are available for this radical: i) it may react with a neutral species near or at the interface (which may be a product of SID or an adsorbed molecule); ii) it may initiate a reaction with neighbors in the SAM surface causing polymerization and radical propagation; iii) it may eliminate an atom or group of atoms from the surface causing unsaturation within the SAM surface.

A general scheme for the study of surface modification by low energy reactive ions is shown in Scheme 1. This scheme includes the use of *in-situ* chemical sputtering in a BEEQ mass spectrometer[80] as a means of prior analysis of the surface followed by a period of ion/surface reaction and final characterization by *in-situ* chemical sputtering and/or external analysis by TOF-SIMS.

The surface modifications described below are achieved by reacting a variety of projectile ions with an F-SAM surface. Such surfaces are relatively inert chemically but, as has been demonstrated, they react with many projectiles including bare transition metals, halogen

Scheme 1:
Surface Modification Methodology

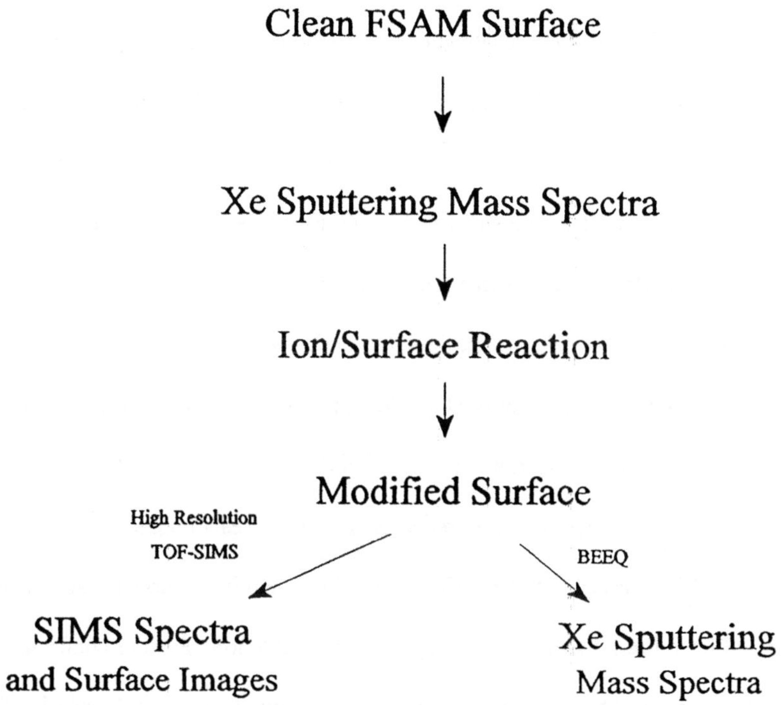

containing metals, halomethanes, $OCNCO^+$, and $OCNCS^+$. A number of halogenated projectiles which react with F-SAMs are summarized in Table 12. Note especially that these halogenated ions yield product ions which are indicative of an exchange process involving the ion and the surface, *viz.* transhalogenation.

Table 12 shows that $SiCl_4^{+\cdot}$ reacts with the F-SAM by dissociative halogen abstraction but not by simple halogen abstraction (i.e., no peak is observed corresponding to one-to-one transfer of Cl-for-F; $SiCl_4^{+\cdot}$ must dissociate in order to produce the observed ion/surface products). Nonetheless, Cl is lost and F is picked up by the projectile. In order to determine whether the lost Cl is incorporated into the surface (thus modifying it), chemical sputtering and high resolution TOF-SIMS experiments were performed. A 60 eV $^{132}Xe^{+\cdot}$ chemcial sputtering spectrum of a clean F-SAM surface produces major chemical sputtering (>20% relative intensity) peaks at 31, 50, and 69 Th. Other peaks are attributed to background hydrocarbons adsorbed to the surface and various minor fragments of the F-SAM. The F-SAM was then reacted with $Si^{35}Cl_2{}^{37}Cl_2^{+\cdot}$ at 60 eV for 2 hours giving scattered product ions listed in Table 12. A subsequent 60 eV $Xe^{+\cdot}$ chemical sputtering spectrum acquired after the 2 hour reaction period revealed two new peaks at 85 Th and 87 Th assigned to $CF_2{}^{35}Cl^+$ and $CF_2{}^{37}Cl^+$ and in a nearly 1:1 abundance ratio as was expected from the 1:1 isotopic labeling of the projectile ion. Further confirmation that the F-SAM was modified by a Cl-for-F exchange was obtained by external analysis of the surface with high resolution TOF-SIMS using a 15keV beam of Ga^+. This experiment also resulted in the expected 1:1 ratio of $CF_2{}^{35}Cl^+$ and $CF_2{}^{37}Cl^+$.

The ions $OCNCO^+$ and $OCNCS^+$ demonstrate additional chemical phenomena associated with low energy ion collisions with a F-SAM surface. The collision of $OCNCS^+$ with a F-SAM at

Table 12:
Ions Which Undergo Halogen Transfer Reactions with FSAM Surfaces

Projectile	Transhalogenation Products	Dissociative Transhalogenation Products
CCl^+	CF^+ (100)*,#	n/a†
CCl_2^+	$CClF^+$ (3)	CF^+ (40)#
CCl_3^+	CCl_2F^+ (5)	$CClF^+$ (2)
CBr^+	CF^+ (40)#	n/a
$CHBr^+$	CHF^+ (25)	CF^+ (60)#
CHI^+	CHF^+ (35)	CF^+ (55)#
CH_2Br^+	CH_2F^+ (50)	CF^+ (30)#
CH_2I^+	CH_2F^+ (70)	CF^+ (25)#
CH_3Br^+	n/a	CH_2F^+
CH_3I^+	n/a	CH_2F^+
CH_3Cl^+	n/a	CH_2F^+
$AlCl^+$	AlF^+ (25)	n/a
$AlCl_2^+$	$AlClF^+$ (70), AlF_2^+ (35)	AlF^+ (55)
$AlCl_3^+$	$AlCl_2F^+$ (5)	$AlClF^+$ (70), AlF_2^+ (15), AlF^+ (25)
$SiCl^+$	SiF^+ (100)	n/a
$SiCl_2^+$	$SiClF^+$ (15)	SiF^+ (100)
$SiCl_3^+$	$SiCl_2F^+$ (95)	$SiClF^+$ (15), SiF^+ (55)
$SiCl_4^+$	n/a	$SiCl_2F^+$ (55), $SiClF^+$ (10), SiF^+ (15)
$GeCl^+$	GeF^+ (100)	n/a
$GeCl_2^+$	n/a	GeF^+ (55)
$GeCl_3^+$	$GeCl_2F^+$ (15)	$GeClF^+$ (5), GeF^+ (30)
$GeCl_4^+$	n/a	$GeCl_2F^+$ (5), GeF^+ (10)

Projectile	Transhalogenation Products	Dissociative Transhalogenation Products
$SnCl^+$	SnF^+ (100)	n/a
$SnCl_2^+$	n/a	SnF^+ (20)
$SnCl_3^+$	$SnCl_2F^+$ (5)	$SnClF^+$ (5), SnF^+ (40)
$SnCl_4^+$	n/a	$SnCl_2F^+$ (2), $SnClF^+$ (2), SnF^+ (20)
$PbCl^+$	PbF^+ (20)	n/a
PCl^+	PF^+ (100)	n/a
PCl_2^+	$PClF^+$ (100), PF_2^+ (85)	$PClCF_2^+$ (8), $PClCF^+$ (5), $PClF_2^+$ (3), PF^+ (80)
PCl_3^+	PCl_2F^+ (5), $PClF_2^+$ (2)	$PCl_2F_2^+$ (1), $PClF^+$ (90), PF_2^+ (40), PF^+ (35)
$SbCl^+$	SbF^+ (100)	n/a
$SbCl_2^+$	$SbClF^+$ (95), SbF_2^+ (45)	SbF^+ (70)
$SbCl_3^+$	n/a	$SbClF^+$ (15), SbF_2^+ (10), SbF^+ (20)
$BiCl^+$	BiF^+ (80)	n/a
$BiCl_2^+$	$BiClF^+$ (35), BiF_2^+ (5)	BiF^+ (45)
$BiCl_3^+$	n/a	$BiClF^+$ (30), BiF^+ (25)
$SeCl^+$	SeF^+ (80)	$SeCF_2^+$ (20), $SeCF^+$ (50)
$SeCl_2^+$	$SeClF^+$ (40), SeF_2^+ (10)	SeF^+ (40)
Se_2Cl^+	Se_2F^+ (25)	$SeCF_2^+$ (10), $SeCF^+$ (15), $SeClF^+$ (10), SeF_2^+ (15), SeF^+ (20)
$Se_2Cl_2^+$	n/a	Se_2F^+ (10), $SeClF^+$ (20), SeF_2^+ (25), SeF^+ (25)
$TeCl^+$	TeF^+ (100)	$TeCF^+$ (2), TCF_2^+ (2), TeF_2^+ (30), TeF_3^+ (5)
$TeCl_2^+$	$TeClF^+$ (45), TeF_2^+ (20)	TeF^+ (70)
$TeCl_3^+$	$TeCl_2F^+$ (5),	$TeClF^+$ (30), TeF_2^+ (2), TeF^+ (15)

* Maximum relative abundance (%) over the collision energy range of 20 - 90 eV

Species may also be accompained by surface sputtering

† Not any

70 eV yielded results similar to those of $OCNCO^+$ (Figure 23). The reaction was carried out for one hour and a chemical sputtering spectrum was taken using $^{132}Xe^{+\cdot}$ at 70 eV. The chemical sputtering spectrum shown in Figure 24 shows a very minor peak at 28 Th corresponding to CO^+ and a more abundant peak at 44 Th assigned to CS^+. These two peaks are evidence that $OCNCS^+$, like $SiCl_4^{+\cdot}$, can modify a F-SAM surface. Further experimental evidence is being obtained by high resolution TOF-SIMS to help determine the type of modification occurring at the surface.

13. Conclusions

The fundamental principles of low energy polyatomic ion/surface collision phenomena and the instrumentation used to study ion/surface collisions have been described. Surface-induced dissociation can strongly excite ions causing extensive fragmentation and yet still allow fine control over the degree of excitation due to the characteristically narrow internal energy distribution associated with the scattered ions. This excitation is most commonly vibrational, however, a few examples of electronic excitation have been noted. Ion/surface collisions are shown to transfer from translational to vibrational energy amounts on the order of 13 - 20% for various projectiles and particular SAM surfaces. The kinetic energy of the scattered secondary ions is low (< 10% of the translational energy of projectile energy) while the majority of the energy is deposited into the surface. Surface-induced dissociation has been applied to both large and small organic molecules to determine ion structure and to distinguish between isomeric ions. Biological molecules have also been studied using SID, a particular application being peptide sequencing. It was also noted that the nature of the surface has a large effect on SID efficiencies

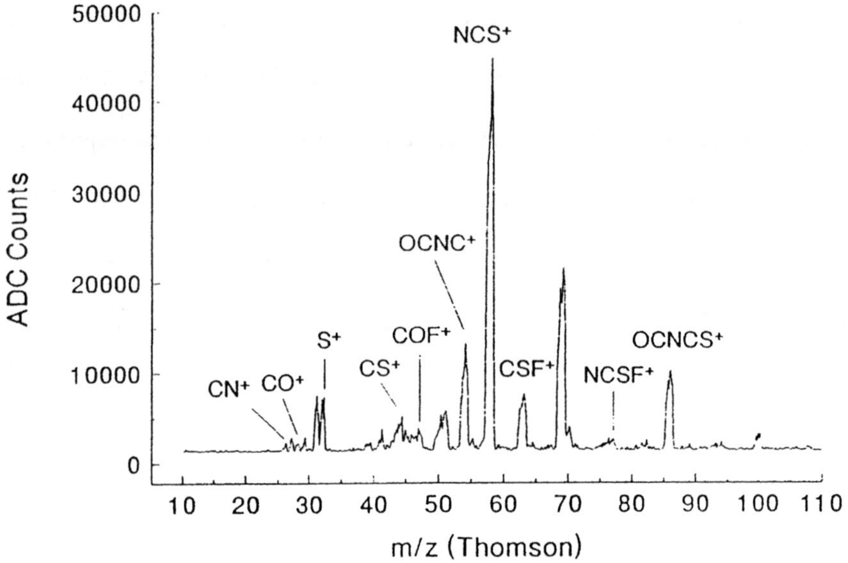

Figure 23: SID spectrum of $OCNCS^+$, generated from the electron impact of ethoxycarbonyl isothiocyanate, upon 70 eV collision with a F-SAM surface.

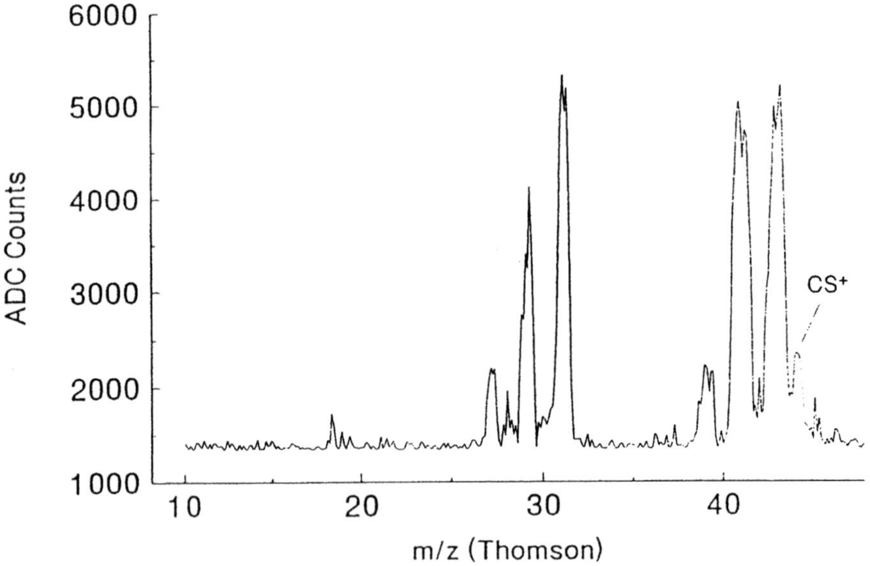

Figure 24: Partial 70 eV $Xe^{+\cdot}$ chemical sputtering spectrum obtained from a F-SAM surface after a 2 hour reaction with $OCNCS^+$ at 70 eV.

(i.e. propensity of a surface to neutralize the projectile) as well as the type of ion/surface reactions and the extent of chemical sputtering.

Two mechanisms have been uncovered for ion/surface reactions, an electron transfer mechanism that predominates for hydrocarbon projectiles and hydrocarbon surfaces and a direct oxidative addition mechanism which applies to metal ion reactions with F-SAM surfaces. Ion/surface reactions have some similarities to many ion/molecule reactions and these reactions are being used to provide insights into the ion/surface reactions. Both exothermic and endothermic reactions occur during ion/surface reactions where the endothermic reactions are driven by the translational energy of the projectile and weak bonds (eg. $Xe\text{-}F^+$) are stabilized by the large excess energy adsorption of the surface.

Selective chemical modification of a F-SAM surface has been demonstrated using appropriate low energy projectile ions. Two examples were given including a dissociative transhalogenation reaction and the reaction of $OCNCS^+$ with an F-SAM surface. Projectiles which react with specific atoms or groups of atoms on a surface are used to probe the outermost monolayer of the surface, *viz.* as a new method of surface analysis. Chemical sputtering, a charge exchange process between the projectile and surface, is also used to probe the outermost monolayer of surfaces by ejection of surface groups as ions. Chemical sputtering and TOF-SIMS have been used to monitor the modification of surfaces by reactive ion beams. This technique could advance materials science by offering the ability to chemically write on a surface, both on a large and small scale, producing new covalently bound groups in the uppermost layers of a surface.

Acknowledgements

We acknowledge the support by NSF (CHE-9223791) and the contributions of collaborators and colleagues cited herein.

References

1. Cooks, R. G.; Ast, T.; Mabud, M. A. *Int. J. Mass Spectrom. Ion Processes* **1990,** *100*, 209-265.

2. Mabud, M. A.; Dekrey, M. J.; Cooks, R. G. *Int. J. Mass Spectrom. Ion Processes* **1985**, *67*, 285-294.

3. Morris, M. R.; Riederer, D. E., Jr.; Winger, B. E.; Cooks, R. G.; Ast, T.; Chidsey, C. E. D. *Int. J. Mass Spectrom. Ion Processes* **1992,** *122*, 181-217.

4. Wysocki, V. H.; Jones, J. L.; Ding, J.-M. *J. Am. Chem Soc.* **1991,** *113*, 8969-8970.

5. Cooks, R. G.; Ast, T.; Pradeep, T.; Wysocki, V. *Acc. Chem. Res.* **1994,** *27*, 316-323.

6. Pradeep, T.; Miller, S. A.; Cooks, R. G. *J. Am. Soc. Mass Spectrom.* **1993,** *4*, 769-773.

7. Kane, T. E.; Somogyi, À.; Wysocki, V. H. *J. Mass Spectrom.*, submitted.

8. Pradeep, T.; Riederer, D. E., Jr.; Ast, T.; Cooks, R. G. *Rapid Commun. Mass Spectrom.* **1993,** *7*, 711-713.

9. Rowe, G. K.; Creager, S. E. *Langmuir* **1991**, *7*, 2307-2312.

10. Callahan, J. H.; Somogyi, À.; Wysocki, V. H. *Rapid Commun. Mass Spectrom.* **1993**, *7*, 693-699.

11. Wainhaus, S. B.; Burroughs, J. A.; Wu, Q.; Hanley, L. *Anal. Chem.* **1994,** *66*, 1038-1043.

12. Rabalais, J. W. *Acc. Chem. Res.* **1994,** *27*, 26-32.

13. Riederer, D. E., Jr.; Cooks, R. G.; Linford, M. R. *J. Mass Spectrom.* **1995,** *30*, 241-246.

14. Williams, E. R.; Jones, G. C., Jr.; Fang, L.; Zare, R. N.; Garrison, B. J.; Brenner, D. W. *J. Am. Chem. Soc.* **1992,** *114*, 3207-3210.

15. Chen, G.; Hoke, S. H., II; Cooks, R. G. *Int. J. Mass Spectrom. Ion Processes* **1994,** *139*, 87-93.

16. Pradeep, T.; Feng, B.; Ast, T.; Patrick, J. S.; Cooks, R. G.; Pachuta, S. J. *J. Am. Soc. Mass Spectrom.* **1995,** *6*, 187-194.

17. Heiland, W. in *Low Energy Ion-Surface Interactions,* Rabalais, J. W., Ed., John Wiley & Sons Ltd: New York, 1994.

18. Cooks, R. G.; Rockwood, A. L. *Rapid Commun. Mass Spectrom.* **1991**, *5*, 93.

19. Cooks, R. G., ed.; *Collision Spectroscopy*, Plenum Press: New York, 1978.

20. Pachuta, S. J.; Kenttämaa H. I.; Sack, T. M.; Cerny, R. L.; Tomer, K. B.; Gross, M. L.; Pachuta, R. R.; Cooks, R. G. *J. Am. Chem. Soc.* **1988**, *110*, 657-665.

21. Detter, L. D.; Hand, O. W.; Cooks, R. G.; Walton, R. A. *Mass Spectrom. Rev.* **1988**, *7*, 465.

22. Pradeep, T.; Riederer, D. E., Jr.; Hoke, S. H., II.; Ast, T.; Cooks, R. G.; Linford, M. R. *J. Am. Chem. Soc.* **1994**, *116*, 8658-8665.

23. Pradeep, T.; Patrick, J. S.; Feng, B.; Miller, S. A.; Ast, T.; Cooks, R. G. *J. Phys. Chem.* **1995**, *99*, 2941-2945.

24. Dagan, S.; Amirav, A. *J. Am. Soc. Mass Spectrom.* **1993**, *4*, 869-873.

25. Somogyi, À.; Kane, T. E.; Ding, J. -M.; Wysocki, V. H. *J. Am. Chem. Soc.* **1993**, *115*, 5275-5283.

26. All thermochemical data is derived from this reference unless otherwise stated: Lias, S. G.; Bartmess, J. E.; Liebman, J. F.; Holmes, J. L.; Levin, R. D.; Mallard, W. G. *J. Phys. Chem. Ref. Data* **1988**, *17*, Suppl. 1.

27. Miller, S. A.; Riederer, D. E., Jr.; Cooks, R. G.; Cho, W. R.; Lee, H. W.; Kang, H. *J. Phys. Chem.* **1994**, *98*, 245-251.

28. Vincenti, M.; Cooks, R. G. *Org. Mass Spectrom.* **1988**, *23*, 317-326.

29. Benninghoven, A.; Rüdenauer, F. G.; Werner, H.W. *Secondary Ion Mass Spectrometry*, John Wiley & Sons Ltd: New York, 1987.

30. Ada, E. T.; Hanley, L.; Etchin, S.; Melngailis, J. *J. Vac. Sci. Techol. B*, submitted.

31. Wu, Q.; Hanley, L. *J. Phys. Chem.* **1993**, *97*, 2677-2683.

32. Perry, S. S.; Somorjai, G. A. *Anal. Chem.* **1994**, *66*, 403A - 415A.

33. Tarlov, M. J.; Newman, J. G. *Langmuir* **1992**, *8*, 1398.

34. Wysocki, V. H.; Ding, J.-M.; Jones, J. L.; Callahan, J. H.; King, F. L. *J. Am. Soc. Mass Spectrom.* **1992**, *3*, 27-32.

35. Mabud, M. A.; Ast, T.; Verma, S.; Jiang, Y. -X.; Cooks, R. G. *J. Am. Chem. Soc.* **1987**, *109*, 7597.

36. Mabud, M. A.; Ast, T.; Cooks, R. G. *Org. Mass Spectrom.* **1987**, *22*, 418-420.

37. Cooks, R. G.; Mabud, M. A.; Horning, S. R.; Jiang, X.-Y.; Paradisi, C.; Traldi, P. *J. Am. Chem. Soc.* **1989,** *111*, 859-865.

38. Hayward, M. J.; Mabud, M. A.; Cooks, R. G. *J. Am. Chem. Soc.* **1988,** *110*, 1343.

39. Jones, J. L.; Dongré, A. R.; Somogyi, À.; Wysocki, V. H. *J. Am. Chem. Soc.*, in press.

40. McCormack, A. L.; Somogyi, À.; Dongre, A. R.; Wysocki, V. H. *Anal. Chem.* **1993,** *65*, 2859-2872.

41. Durkin, D. A.; Schey, K. L. *Proceedings of the 43rd ASMS Conference on Mass Spectrometry and Allied Topics*, May 21 - 26, 1995, Atlanta, Georgia, 1332.

42. McCormack, A. L.; Jones, J. L.; Wysocki, V. H. *J. Am. Soc. Mass Spectrom.* **1992,** *115*, 859-862.

43. Wysocki, V. H.; Kenttämaa, H. I.; Cooks, R. G. *Int. J. Mass Spectrom. Ion Processes* **1987**, *75*, 181-208.

44. Schey, K. L.; Kenttämaa, H. I.; Wysocki, V. H.; Cooks, R. G. *Int. J. Mass Spectrom. Ion Processes* **1989,** *90*, 71.

45. Levsen, K. *Fundamental Aspects of Organic Mass Spectrometry*, Weinheim: New York, 1978; Vol. 4, Chapters 1,2.

46. Cooks, R.G.; Ast, T.; Kralj, B.; Kramer, V.; Žigon, D. *J. Am. Soc. mass Spectrom.* **1990,** *1*, 16 - 27.

47. Vékey, K.; Somogyi, À.; Wysocki, V. H. *J. Mass Spectrom.* **1995,** *30*, 212 - 217.

48. Jones, J. L.; Dongré, A.; Nair, H.; Somogyi, Á; Wysocki, V. H. *Proceedings of the 43rd ASMS Conference on Mass Spectrometry and Allied Topics*, May 21 - 26, 1995, Atlanta, Georgia, 1333.

49. Busmann, H.-G.; Lill, T.; Reif, B.; Hertel, I. V. *Surf. Sci.* **1992,** *272*, 146-153.

50. Busmann, H.-G.; Lill, T.; Reif, B.; Hertel, I. V.; Maguire, H. G. *J. Chem. Phys.* **1993,** *98*, 7574-7580.

51. Kane, T. E.; Somogyi, Á; Wysocki, V. H. *Org. Mass Spectrom.* **1993,** *28*, 1665.

52. Beck, R. D.; St. John, P. M.; Alvarez, M. M.; Diederich, R. L.; Whetten, R. L. *J. Phys. Chem.* **1991,** *95*, 8402.

53. Lill, Th.; Busmann, H.-G.; Lacher, F.; Hertel, I. V. *Chem. Phys.* **1995,** *193*, 199-210.

54. Burroughs, J. A.; Wainhaus, S. B.; Hanley, L. *J. Chem. Phys.* **1995**, *103*, 6706-6715.

55. Franklin, R. L.; Zhong, W.; Somogyi, À; Wysocki, V. H. *Proceedings of the 43rd ASMS Conference on Mass Spectrometry and Allied Topics*, May 21-26, 1995, Atlanta, Georgia, 418.

56. Martin, J. S.; Greeley, J. N.; Morris, J. R.; Feranchak, B. T.; Jacobs, D. C. *J. Chem. Phys.* **1994,** *100*, 6791-6812.

57. Martin, J. S.; Greeley, J. N.; Morris, J. R.; Jacobs, D. C. *J. Chem. Phys.* **1992,** *97*, 9476-9479.

58. Ast, T.; Riederer, D. E., Jr.; Miller, S. A.; Morris, M.; Cooks, R. G. *Org. Mass Spectrom.* **1993,** *28*, 1021-1033.

59. Burroughs, J. A.; Wainhaus, S. B.; Hanley, L. *J. Phys. Chem.* **1994,** *98*, 10913-10919.

60. Riederer, D. E., Jr.; Miller, S. A.; Ast, T.; Cooks, R. G. *J. Am. Soc. Mass Spectrom.* **1993,** *4*, 938-942.

61. Hayward, M. J.; Park, F. D. S.; Manzella, L. M.; Bernasek, S. L. *Proceedings of the 43rd ASMS Conference on Mass Spectrometry and Allied Topics*, May 21-26, 1995, Atlanta, Georgia, 1335.

62. Riederer, D. E., Jr.; Cooks, R. G.; Linford, M. R. *J. Mass Spectrom.* **1995,** *30*, 241-246.

63. Cooks, R. G. et al., unpublished.

64. Pradeep, T.; Ast, T.; Cooks, R. G.; Feng, B. *J. Phys. Chem.* **1994,** *98*, 9301-9311.

65. Coburn, J. W. *Pure Appl. Chem.* **1992**, *64*, 709-713.

66. Scoles, G., Ed. *Atomic and Molecular Beam Methods,* Oxford University Press:Oxford, 1988.

67. Wilson, R.G.; Brewer, G. R. *Ion Beams with Application to Ion Implantation*, Wiley:New York, 1973.

68. Hu, H.-K.; Schultz, J. A.; Rabalais, J. W. *J. Phys. Chem.* **1982,** *86*, 3364-3367.

69. Ast, T.; Mabud, M. D.; Cooks, R. G. *Int. J. Mass Spectrom. Ion Processes* **1988**, *82*, 131.

70. Pradeep, T.; Cooks, R. G. *Int. J. Mass Spectrom. Ion Processes* **1994,** *135*, 243-247.

71. Durkin, D. A.; Thornburg, K. R.; Schey, K. L. *J. Am. Soc. Mass Spectrom.* **1995,** *6,* 257-263.

72. Li, G.; Duhr, A.; Wollnik, H. *J. Am. Soc. Mass Spectrom.* **1992,** *3,* 487-492.

73. Aberth, W. *Anal. Chem.* **1990,** *62,* 609-611.

74. Bier, M. E.; Schwartz, J. C.; Schey, K. L.; Cooks, R. G. *Int. J. Mass Spectrom. Ion Processes* **1990,** *103,* 1-19.

75. Theobold, F.; Budzikiewicz, H. *Proceedings of the 39th ASMS Conference on Mass Spectrometry and Allied Topics,* 1993, 835.

76. Despeyroux, D.; Wright, A. D.; Jennings, K. R.; Evans, S.; Riddoch, A. *Int. J. Mass Spectrom. Ion Processes* **1992,** *122,* 133-141.

77. Bier, M. E.; Amy, J. W.; Cooks, R. G.; Syka, J. E. P.; Ceja, P.; Stafford, G. *Int. J. Mass Spectrom. Ion Processes* **1987,** *77,* 31-47.

78. Schey, K. L.; Cooks, R. G.; Kraft, A.; Grix, R.; Wollnik, H. *Int. J. Mass Spectrom. Ion Processes* **1989,** *94,* 1-14.

79. Beck, R. D.; St. John, P.; Alvarez, M. M.; Diedrich, F.; Whetten, R. L. *J. Phys. Chem.* **1991,** *95,* 8402.

80. Winger, B. E.; Laue, H.-J.; Horning, S. R., Julian, R. K., Jr.; Lammert, S. A.; Riederer, D. E., Jr.; Cooks, R. G. *Rev. Sci. Instrum.* **1992,** *63,* 5613-5625.

81. Wu, Q.; Hanley, L. *J. Phys. Chem.* **1993,** *97,* 2677-2685.

82. Park, K. H.; Kim, B. C.; Kang, H. *J. Chem. Phys.* **1992,** *97,* 2742.

83. Lammert, S. A.; Cooks, R. G. *J. Am. Soc. Mass Spectrom.* **1991,** *2,* 487-491.

84. Ijames, C. F.; Wilkins, C. L. *Anal. Chem.* **1990,** *62,* 1295.

85. Williams, E. R.; Henry, K. D.; McLafferty, F. W.; Shabanowitz, J.; Hunt, D. F. J. *J. Am. Soc. mass Spectrom.* **1990,** *1,* 413.

86. Chorush, R. A.; Little, D. P.; Beu, S. C.; Wood, T. D.; McLafferty, F. W. *Anal. Chem.* **1995,** *67,* 1042-1046.

87. Yeretzian, C.; Beck, R. D.; Whetten, R. L. *Int. J. Mass Spectrom. Ion Processes* **1994,** *135,* 79.

88. St. John, P. M.; Whetten, R. L. *Presentation at the Pittsburgh Conference on Analytical Chemistry and Applied Spectroscopy*, Chicago, Ill., March 1993.

COUPLING OF LIQUID CHROMATOGRAPHY TO MASS SPECTROMETRY: TWENTY-FIVE YEARS OF EXITING DEVELOPMENTS

E.L.ESMANS
University of Antwerp (RUCA -UIA)
Department of Chemistry
Nucleoside Research & Mass spectrometry unit
Groenenborgerlaan 171
B-2020 Antwerp Belgium

1. Introduction

One of the more intriguing analytical methods developed over the past 25 years certainly is the coupling of liquid chromatography to mass spectrometry (LC-MS). The pioneering period of the late '70's was however quite different from the early days of gas chromatography-mass spectrometry (GC-MS): there was no sudden breakthrough of the methodology which was reflected in the limited number of papers in scientific journals . This phenomenon could be explained by the fact that, although LC-MS arose a lot of interest, researchers were a little bit reluctant in marrying two incompatible systems. [1] However the increasing research activities in the area of natural products, the developments in high performance liquid chromatography (HPLC) and other separation techniques and the growing demand for a powerful and universal detection system which would allow on-line identification of polar and

K. R. Jennings (ed.), Fundamentals and Applications of Gas Phase Ion Chemistry, 115–147.

thermolabile compounds , stimulated many researchers to persist in their search for a useful practical LC-MS combination with features comparable to GC-MS.
The characteristics of such an ideal LC-MS combination can be summarized as follows:

- No restriction of solvents to get a maximum of chromatographic performance.
- Utilization of both volatile and non-volatile buffers.
- If necessary capable of handling high flow-rates.
- A choice of ionization techniques for structure identification.
- Positive and negative ion recordings.
- Good sensitivity
- TIC, full scan and multiple ion detection (MID) and eventually multiple reaction monitoring (MRM.) for quantitative measurements.
- Not too much additional peak broadening.
- Capability of transferring polar molecules from the column outlet into the ion source without thermal degradation.

In order to get a better understanding of the scope and limitations of some LC-MS systems it is appropriate to start with an introduction of the principles of some chromatographic techniques which are hyphenated to mass spectrometry.

2. High Pressure Liquid Chromatography (HPLC)

In chromatography in general three parameters are considered as important : the resolution R, the speed of analysis and the column capacity. These factors are not independent of each other so that none of them can be altered without influencing the others.

The resolution is determined by the efficiency (N) , the selectivity (α) and the capacity factor k'. Speed of analysis is a function of column length and particle size (dp) while the capacity is a function of the column and sample size.

Standard HPLC columns in analytical HPLC have internal diameters of 4.6 mm, 2 mm, 1mm. Each type of column is operated at its typical flow-rate. Within this range the best chromatographic performance is obtained.(Table 1)

TABLE 1. HPLC column types : diameters and flow-rates

	I.D.	Inj. Volume	Flow-rate
Conventional	4-6mm	5-20μL	0.5 - 2 mL/min
Microbore	0.5 - 2 mm	0.5-2μL	1-200μL/min
Packed capillary	0.2 - 0.5 mm	60-100 nL	0.5 3 μL/min

From Table 1 it is obvious that a decrease in diameter **d** of the HPLC column leads to much lower flow-rates so that solvent consumption is reduced.

Of course one should also realize that a decrease in column diameter results in a drop in bed volume which signifies that the loadability of the column becomes less (factor 20 going from conventional 4.6 mm ID to 1 mm ID microbore). However since peak volume V_p is proportional to d^2 the peak volume is about 20-times smaller in the microbore column. Therefore the concentration at the maximum of the eluting peak is the same. This is an important in concentration dependent detectors (U.V., electrospray) were a minimum of dilution of the sample inside the column gives an advantageous detection limit. [2]

Separation of analytes can be achieved using different stationary- and mobile phases which permit to realize experimental conditions for optimum selectivity. Straight phase chromatography (adsorption chromatography) is largely done on silica (polar) using as well apolar as polar solvents (hexane, methylene chloride, chloroform, methanol, THF, acetonitrile). Reverse phase chromatography (RPC) is done on a modified silica support covered with e.g. C_8 to C_{18} chains. In this type of chromatography acetonitrile/water or methanol/water is used as mobile phase. Water is often replaced by a buffer system. If ion exchange chromatography is the method of choice aqueous buffers are a prerequisite in order to separate the compounds. Separation of ionic compounds on RP is difficult unless a technique called ion pair chromatography is utilized. In this technique a small amount (0.01 to 0.005 M) of an ion pair reagent [quaternary amines, alkylsulfonates such as sodium dodecylsulfonate (SDS)] is added to the mobile phase. These reagents transform the RP column in a kind of dynamic ion exchange column. [3]

3. Capillary Zone Electrophoresis (CZE)

CZE is most widely used due to its simplicity of operation and versatility. Applications of CZE include the analysis of amino acids, peptides ,ions, enantiomers etc. The technique is simple, gives rapid separations (5 - 50 min.), is highly efficient (10^5 plates/m), is sensitive (10^{-12} to 10^{-21} M), and has a very low buffer consumption (1 - 30 nL/min) [4]

In electrophoresis separation of molecules is obtained by differential migration of solutes in an electric field (V = 10 - 30 KV; E= 100 to 500 V/cm). In high performance capillary electrophoresis the separation of charged analytes is performed in narrow-bore capillaries (I.D. 25 - 200 um, length 10 - 100 cm) which are filled with a buffer. The efficiency is often in excess of 10^5 theoretical plates and is partly due to the plug profile flow of the electro-osmotic flow , an electrophoretic phenomenon generating a bulk flow inside the capillary.

Within this electric field charged molecules will migrate towards the opposite electrode: the electrophoretic flow (EPF). Their velocity is expressed by the following equations:

$$\vec{\upsilon} = \mathrm{EPF} \cdot \vec{E} \qquad (1)$$

$$\mathrm{EPF} = \frac{\varepsilon\zeta_{ion}}{6\pi\eta} \qquad (2)$$

ε = Dielectric constant
ζ = Zeta potential ion
η = Viscosity

Because the capillary is filled with a buffer with a defined pH, the fused silica wall inside the capillary is ionized and becomes negatively charged. This process occurs provided the wall inside the capillary is unmodified and becomes significant at pH values > 4. This negatively charged wall is attracting cations from the solution resulting in the formation of a electric double layer characterized by a ζ - potential. Cations forming the double layer are attracted towards the cathode dragging the bulk of the solution along. This results in the so-called electro-osmotic flow (EOF) expressed by the following equation:

$$\mathrm{EOF} = \frac{\varepsilon\zeta_{\mathrm{wall}}}{4\pi\eta} \tag{3}$$

ε = Dielectric constant
ζ = Zeta potential
η = Viscosity

The EOF is larger than the EPF and all molecules will move to the cathode with a different speed as a result of the vector sum of both phenomenon. (Fig. 1) Since this vector sum is different for each species separation is accomplished.

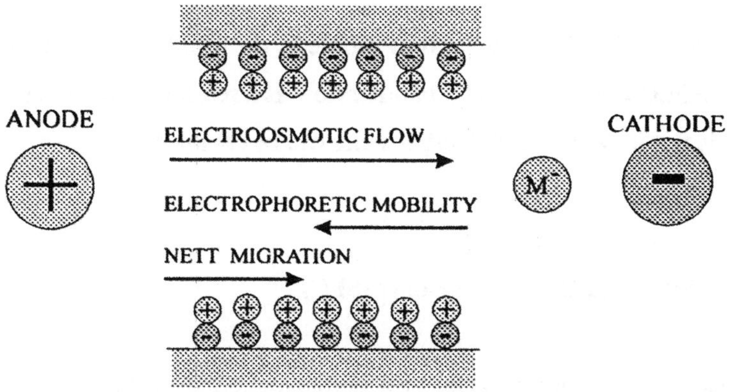

Figure 1. Principle of Capillary Zone Electrophoresis

Separation of neutral molecules by CZE is a however a problem unless a technique Micellar Electrokinetic Chromatography (MEKC) is used. Here the separation of neutral species is accomplished by the use of surfactants in the running buffer. At concentrations above the critical micelle concentration (8 to 9 mM for SDS) charged micelles are formed. Interaction of these micelles with the neutrals by a partition process different for each set of particular neutrals (different k'-value) causes their separation (Fig.2).

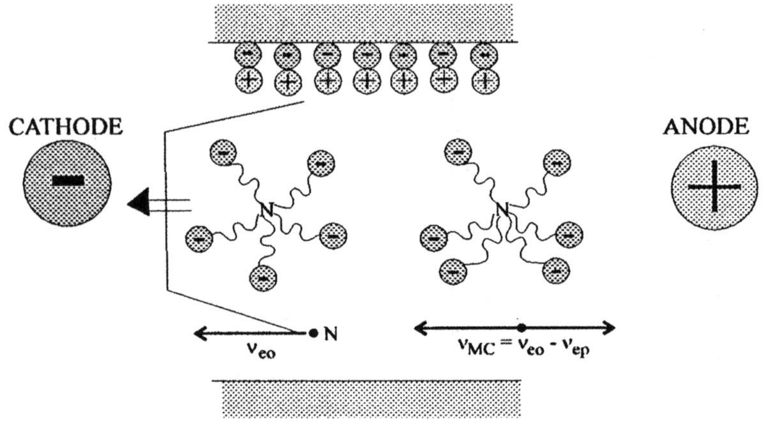

Figure 2. Principle of Micellar Electrokinetic Chromatography

A unique feature of EOF is the flat flow profile in the capillary. This is in contrast to the flow in HPLC which produces a laminar flow as a result of external pumping. As a result of this flat flow profile high efficiencies can be obtained.

4. Supercritical Fluid Chromatography (SFC)

Supercritical fluid chromatography is a chromatographic technique using a mobile phase under supercritical conditions i.e. above its critical temperature and pressure. [5] A supercritical fluid has a density comparable to a liquid, a gas-like viscosity and a diffusion coefficient between gas- and liquid phase (Table 2).

TABLE 2

Phase	Density (mg/cm^3)	Diffusion (cm^2/sec)	Viscosity (mg/cm.sec)
Gas	10^{-3}	0.01 - 1.0	(0.5 - 3) 10^{-4}
SFC	0.2 - 0.9	(3 - 0.1) 10^{-4}	(0.2 - 1) 10^{-3}
Liquid	0.8 - 1.0	(0.5 - 2) 10^{-5}	(0.3 - 2) 10^{-2}

The most popular, safe and practical mobile phase is CO_2 (T_c = 31.3°C ; P_c= 72.9 atm) used with or without an organic modifier as CH_3OH. Carbon dioxide is pumped as a liquid and the pressurized fluid is preheated above its critical temperature before passing into the column via an injection valve. A pressure restrictor is located at the end of the column or the detector to ensure supercritical conditions. If a modifier is necessary the SFC instrumentation becomes more complex since a second pump is needed.

SFC can be done on conventional HPLC columns or on fused silica open tubular capillary columns to which heavily cross-linked bonded stationary phases have been applied. SFC is used to separate thermolabile compounds or compounds who can hardly be handled by GC.

During a SFC run pressure control is necessary (in HPLC flow control !). Provision must also be made for pressure or density programming (at constant T) : slowly increasing the mobile phase density decreases the retention time of the analyte. This is analogous to temperature programming in GC and gradient elution in HPLC. Addition of a modifier also reduces the k'-values and should be used in packed column SFC to elute polar compounds since CO_2 has a polarity most of the time comparable to hexane. We hereby want to emphasize that this effect is due to a modification of the stationary phase (covering of polar silanol groups) and not to an enhancement of the solvating power of the mobile phase.

As to the speed of analysis , packed SFC is faster than capillary SFC although in a capillary SFC column (L = 50 or 100 mm) more theoretical plates can be obtained and therefore can be recommended for the highly efficient separation of complex mixtures.

5. Early approaches to LC-MS

One of the main problems in the coupling of HPLC to MS is the handling of the flow-rates applied in HPLC. Introduction of the total amount of mobile phase into the ion source of a conventional mass spectrometer, even equipped with a pumping system for chemical ionization (C.I.) measurements, is impossible without destroying the vacuum conditions. This is true if standard HPLC and microbore columns are used (Table 1). The pressures allowed in the different compartments of a differentially pumped quadrupole mass spectrometer are summarized in Table 3. The maximum flow-rate which can be handled by a diffusion pump (C_{max} :1300 L/sec) at a source temperature T_s =100°C provided an ion source pressure P_i of 10^{-4} mm Hg is required is shown in Table 4. The diffusion pump is connected to the ion source housing by a tube with diameter d=30 cm and length l=15 cm. [6]

TABLE 3

$A_{nalyzer}$	1.10^{-6}	$<P_a<$	5.10^{-5} mmHg
P_{source}	1.10^{-6}	$<P_s<$	5.10^{-4} mm Hg
$P_{ion\ source}$	1.10^{-6}	$<P_i<$	1 mm Hg

TABLE 4

	H_2O	CH_3OH	CH_3CN
C_{con} (L/sec)	8525	6381	5637
C_{eff} (L/sec)	1127	1079	1056
Q_g (mL/min STP)	7	6.7	6.6
Q_l (μL/min)	5.6	12	15.3

** C_{con}= conductance of the tube; C_{eff}= effective conductance i.e. conductance taken the conductance of the diffusion pump, the connection and the solvent into account; Q_g= maximum amount of gas which can be introduced into the ion source; Q_l= corresponding maximum amount of liquid.*

From Table 4 it is clear that direct introduction of approximately 10 μL/min of mobile phase is allowed into the ion source without destroying the vacuum conditions and that coupling of HPLC columns operating at high flow-rates to MS must be realized by the design of appropriate interface systems.

In view of this early LC-MS experiments were based on the use of a glass capillary [7], a hollow probe [8] and interface systems which had proven to be valuable in GC-MS such as a jet separator [9] and a membrane separator [10]. A complete different approach was introduced by Scott *et al.* [11] utilizing a moving wire as a mechanical transport system to introduce the analyte in the ion source. So most of the interface systems try to cope with the large amounts of mobile phase but minimum attention has been directed towards the tolerance of buffers or ion pair reagents. As we shall see later on modern LC-MS interface systems can handle buffers or salts provided they are volatile: this means that the so often used phosphate buffers are still prohibited if on-line LC-MS experiments are needed.

The most straightforward way to introduce the LC effluent to a mass spectrometer seems at first sight a capillary approach (with or without split depending on the HPLC column used). Mass spectra were acquired under C.I. conditions [12,13]. However clogging of this type of interface due to premature evaporation of the solvent and the occurrence of so-called vacuum-locks were serious disadvantages. They were solved by the development of the Direct Liquid Introduction (DLI) interface by Hewlett-Packard (Palo Alto, California) and Ribermag (Paris , France)(Fig.3). [14,15]

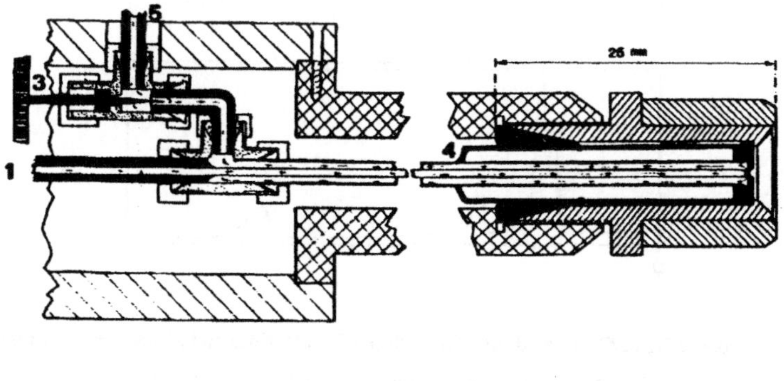

Figure 3. Scheme of a Direct Liquid Introduction Probe:
(1) HPLC-side. (2) diaphragm. (3) needle valve. (4) outer capillary. (5) to waste or U.V.-detector

In this system the tapered capillary is replaced by a diaphragm equipped with a pinhole of a few microns (5 to 10 micron) allowing the direct introduction of approximately

10μL/min if mobile phase straight into the ion source. The probe could be cooled to avoid vacuum-locks. The solvent was evaporated in the ion source , with or without the aid of a disolvation chamber [16,17.] and was ionized by electron impact.. The resulting plasma acted as a c.i.-reagent gas for the analyte molecules eluting from the HPLC column. Although the system was not easy to operate it was able to handle large amounts of H_2O in the effluent, a chromatographic situation often encountered in the HPLC analysis of natural compounds (Fig.4). The DLI approach has been used in combination with 4.6 mm I.D - and microbore columns [18, 19]. Several compounds have been analyzed such as nucleosides [16] , drugs [20], hormones [21] , Pharmaceuticals [22] etc.

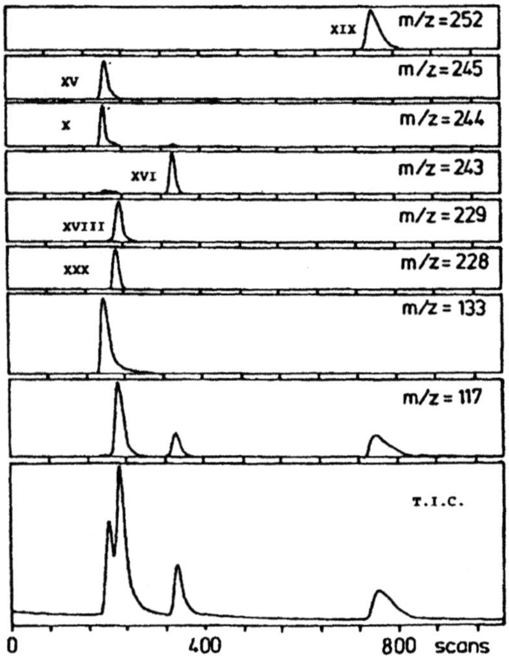

Figure 4. Microbore DLI LC-MS analysis of a mixture of nucleosides and 2'-deoxynucleosides on a RPC_{18} column (50cm x 1mm I.D.). Mobile Phase: 94% $HCOONH_4$ / 6% CH_3OH. Flow-rate: 70μL/min. (Ref. 17).

Due to the pioneering work of Scott [11] a novel transport system was promoted and commercialized by the Finnigan Corporation : the Moving Belt system. This interface is schematically presented in Fig.5

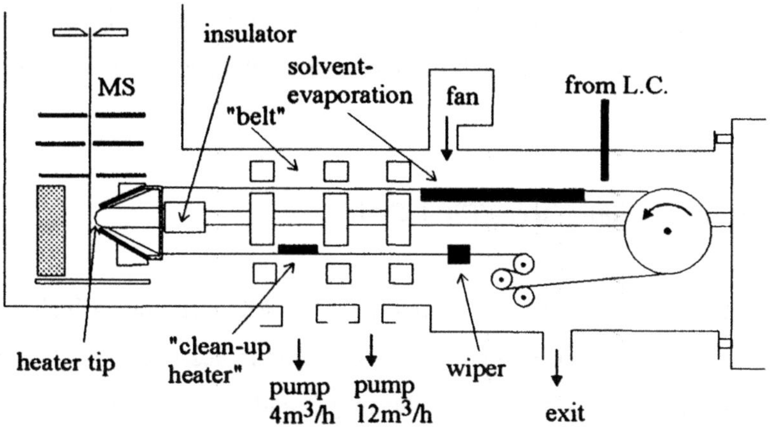

Figure 5. Principle of a Moving Belt Interface

In the first version of this system the eluent from the LC column was deposited on the belt which was moving around at a constant speed. The HPLC band first passed below an infra-red heater , where most of the solvent was evaporated before the belt entered two vacuum locks. In here the remainder of the solvent was pumped away and the compound finally was transported inside the ion source . Here the compounds was flash-evaporated and was ionized by either e.i. or c.i.. Upon its return to its original position the belt, made of Capton™, was cleaned by a second heater. In combination with normal phase chromatography flow-rates of ca. 1 mL/min could be handled however in the case of reverse phase chromatography the presence of high amounts of H_2O lead to droplet formation. In the latter case the flow-rate had to be reduced to 0.1 à 0.3 mL/min. Microbore HPLC , an aerosol deposit device or replacement of the I.R.-heater by a hot plate/turbomolecular pump system have been introduced to cope with this problem [23,24, 25]

This interface has been extensively used for the analysis of alkaloids [26]., pesticides [27], polynuclear aromatics [28.], sugars [29] , etc.. The advantage of this interface is its ability to produce both e.i.- and c.i.-mass spectra.

Although the moving belt system became more or less obsolete it has more recently been demonstrated that the interface is useful in combination with packed SFC [43]. SFC, using CO_2 with polar modifiers can be excellently handled by the system. The mobile phase is deposited on the belt via a spray deposit device (See SFC-MS.)

6. The thermospray interface. (TSP LC-MS)

In 1978 Vestal *et al.* published a paper on a LC-MS technique which they called Thermospray (TSP). The term thermospray was used to describe " the production of a jet of fine liquid or solid particles by heating " and was the first LC-MS interface capable of handling flow-rates up to 1 mL/min [30].

It is now recognized as a practical interface applicable to a wide variety of samples under a broad choice of chromatographic conditions. TSP LC-MS is capable of producing mass spectra of compounds with a molecular mass below 2000 Da and is commercialized by almost every manufacturer of mass spectrometers. Although we may expect the technique to be gradually replaced by Atmospheric Pressure Chemical Ionization (ApCI) valuable TSP LC-MS results still are published

In TSP LC-MS the use of volatile buffers (usually 0.1 M NH_4OAc) is allowed and gradient elution, although not easy to perform, is possible provided a temperature programming unit for the TSP probe is purchased.

A scheme of a TSP interface is presented in Fig.6.

The mobile phase coming from the HPLC column is conducted through a resistively heated capillary tube (I.D. 100 mm) into a heated ion chamber. The temperature of the capillary is adjusted in order to create a kind of superheated mist which is composed of solvent vapor and small, positively and negatively charged, droplets due to the presence of a 0.1 M NH_4OAc solution in the mobile phase. Some of the droplets are positively - others are negatively charged depending on the excess of either NH_4^+ or OAc^- in the droplet. The charged droplets play an important role in ion formation of the analyte ions. The mechanism of ion formation in TSP is not fully understood, but it is a fact that ions can be produced without the presence of an external ionization source. At the moment it is accepted that under " real" TSP conditions ions can directly be ejected from the condensed phase (droplet) into the gaseous phase under the influence of high electrical fields as a results of vaporization and shrinking of the charged droplets. This process is called ion evaporation. However chemical ionization processes, governed by gas phase acidity rules, cannot be entirely excluded.

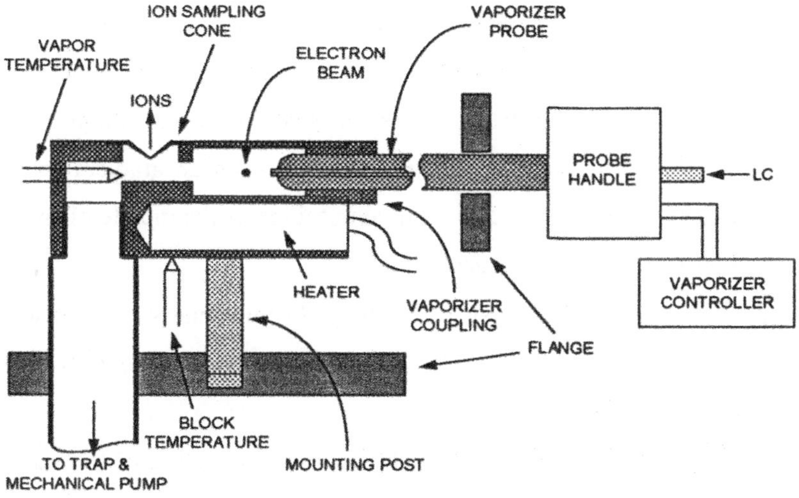

Figure 6. Principle of a modern Thermospray Interface.

Sometimes chromatographic conditions do not allow the use of NH_4OAC solutions. In that case an external ionization source (usually a discharge electrode) is needed. This electrode produces a c.i. plasma which ionizes the sample. Sometimes this technique is referred to as Plasmaspray™ .

Excess of solvent in the ion chamber is pumped away by a rotatory pump (ca. 300 L/min) leaving a pressure of approximately 10 Torr. Analyte ions are introduced in the quadrupole region through a sampling orifice.

In the case of "real" TSP , gradient elution is possible however from the ionization process described above it is obvious that during solvent programming the TSP probe temperature should be changed in order to maintain optimum vaporization conditions. From a practical point of view this is not easy to realize even if a probe temperature programming unit is available.

Under (+)- or (-)TSP conditions intense protonated molecules $[MH]^+$ or $[M\text{-}H]^-$ ions may be observed and fragment ions may be non-existing. This is however not always the case and fragment ions can be observed. This behavior is not only compound dependent but is also extremely influenced by the setting and control of the TSP probe. In many cases the mass spectrometer is tuned by optimalisation of the cluster ions produced by the mobile phase. These temperature settings may not be optimal for the analytes. If maximum sensitivity is required it may be beneficial to tune on a solution of the compound itself. The latter procedure is of course not unique for TSP but applies to all LC-MS techniques. In the case of a linear gradient in HPLC the instrument is tuned at the initial and final composition of the mobile phase. The TSP probe temperature is programmed accordingly.

Thermospray is used for the analysis of a variety of compounds such as pesticides [31], nucleosides [32], DNA adducts [33], drugs [34], glycosides [35] etc., also quantitative measurements are possible.

The majority of papers on TSP LC-MS deal with the coupling of the interface with conventional HPLC columns operating under reverse phase conditions. However the TSP interface has also been coupled to packed column SFC (see SFC-MS).

For a review on the TSP LC-MS topic we refer to P. Arpino [36]

7. The Particle Beam Interface (PB LC-MS)

The PB -interface is derived from the MAGIC interface (Monodisperse Aerosol Generation Interface for Chromatography) developed by Willoughby *et al.* [37].

The technique is based upon the nebulisation of the mobile phase (0.1 - 1 mL/min) with He-gas (ca. 40 psi) in a desolvation chamber held at ambient temperature and pressure (Fig.7).

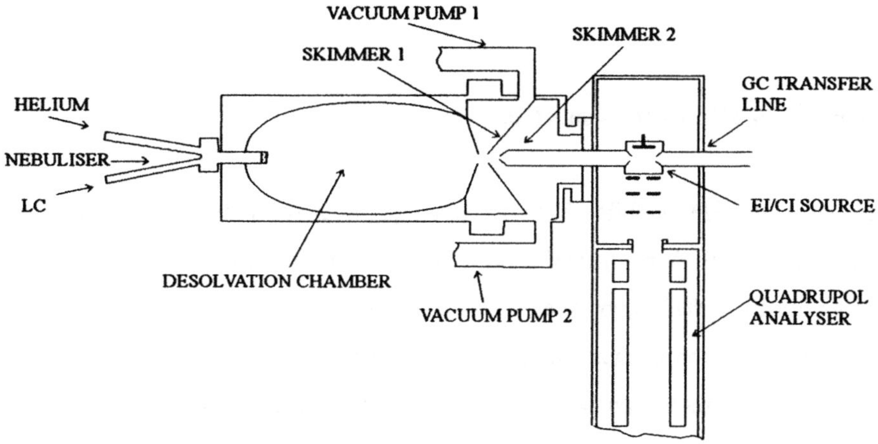

Figure 7. Principle of a Particle Beam Interface.

Both solvent and analyte molecules enter the first low pressure region of the momentum separator , which is pumped by a rotary pump. Through the nozzle a supersonic jet is created which has the shape of a cone and which is enriched in the center by the heavier analyte molecules. These molecules preferentially enter the second stage of the momentum separator through a skimmer - nozzle system and are finally transferred into the ion source. When they collide with the heated source walls they evaporate and can be ionized either under e.i. - or c.i.-conditions using the classical reagent gasses.

The interface is usually coupled to HPLC (4.6 mm I.D - and microbore columns).

PB LC-MS analysis is limited to compounds which hardly can be analyzed by GC-MS and is frequently used in the area of environmental chemistry for the analysis of pesticides [38], herbicides [39] and polyaromatic hydrocarbons [40]. The interface is not suitable for the analysis of thermolabile and high molecular mass compounds.

The detection limits are in the low-ng region and it has been shown that the addition of e.g. volatile salts can enhance the sensitivity of the system. This phenomenon was explained by a so-called "carrier effect " : addition of salts produce larger/heavier particles which are more efficiently transferred through the interface into the ion source. In a recent publication it is stated that the particle size distribution is not the major factor influencing transmission efficiency [41]

8. Coupling of SFC to mass spectrometry

Since the by the SFC mobile phase produced gas-flow rates are high and because it is required to maintain supercritical conditions at the chromatographic end of the system, coupling of SFC to MS is not so easy as it may seem.

In the number of approaches which have been developed for interfacing SFC wit MS two types of interfaces can be found: these interface systems in which the eluant is eliminated and those interface systems acting as direct coupling devices.

The interfaces which eliminate the mobile phase have the advantage of freedom of choice of ionization method (e.i.; c.i.). The drawbacks are poor sensitivity and thermal decomposition. This is e.g. the case if SFC is coupled to a PB interface [42].

Packed SFC can be perfectly combined with the moving belt interface using a heated spray deposit device as has been demonstrated by D.Games *et al* [43]. This type of interface is capable of handling the high flow-rates and its limitations are most of the time found in the thermal decomposition of too thermolabile compounds.

Packed SFC has also been coupled to the TSP interface [44]. In this technique the supercritical fluid is completely vaporized while flowing through the heated capillary tube. The result of this process is the creation of a supersonic jet of vapor consisting of a mist of solution droplets and fine solute particles. If SFC is coupled to the TSP interface one should realize that only the "filament on" mode can be used.

As a result of the development of capillary SFC, direct fluid introduction (DFI) came available. In this approach the the SFC fluid generally expands through a heated restrictor and the expanding jet is directly introduced in the ion source (Fig.8).

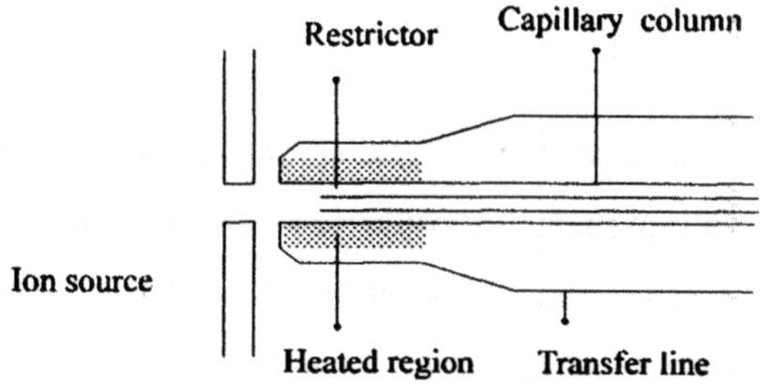

Figure 8. Principle of capillary SFC-MS.

The interface generally exists of a heated transfer line with the restrictor mounted at the end. Approximately 1-5 mL/min of gas is introduced into the ion source. Since in DFI interfaces a large volume of pure CO_2 is introduced the pressure in the ion source should be controlled since a too high pressure in the ion source may adversely effect the e.i. ionization process and the sensitivity often required for trace analysis is not reached (detection limit 10 - 100 ppm). At "higher " pressures CO_2 molecules will be preferentially ionized and will act as as reagent molecules for charge exchange with the analyte molecules [45]. Fortunately, the latter process produces e.i.-like spectra and the resulting mass spectra can be library searched. In the event no e.i.-data are needed, c.i.-spectra can be recorded by introducing standard c.i. reagent gasses such as NH3, methane and isobutane.

For a recent short review on the SFC-MS we refer to Ref.46.

9. LC-MS techniques at Atmospheric Pressure: Atmospheric Pressure/Chemical ionization (APcI) and Electrospray (ES).

Amongst all commercially available LC-MS systems such as Thermospray (TSP), Particle Beam (PB), dynamic FAB, Electrospray (ES) and Atmospheric Pressure Chemical Ionization (APcI), the most popular LC-MS interface systems are at the moment those in which the ion source of the mass spectrometer is operated at atmospheric pressure i.e. ApCI and ES.

9.1 THE DIFFERENT API-INTERFACES

API-interface systems have as a common characteristic feature that ionization of sample molecules is realized under atmospheric pressure conditions.
In summary three steps can be distinguished in API LC-MS :

1. Introduction of sample in solution into the API-ion source via a capillary.
2. Ionization of the analyte and formation of ion/solvent clusters.
3. Declustering of sample ions and extraction of the ions into the mass analyzer.

More details can be found in two reviews written by A.Bruins [47,48].

9.2 ATMOSPHERIC PRESSURE CHEMICAL IONIZATION (APcI)

In the APcI-interface (Fig.9) the eluate flows through a central micro-tube and nebulization is accomplished with the aid of a nebulizing gas (N_2; flow-rate : 400 mL/min), which is coaxially introduced and by the supply of heat (temperature around 400°-600°C). In this way a vapor of solvent and analyte molecules is created in the ion source. By means of a discharge electrode a c.i. plasma is formed of solvent derived ions (for example in the case of methanol/H_2O-mixtures $CH_3OH_2^+$ and $CH_3OH_2^+(H_2O)_n(CH_3OH)_m$-clusters). These clusters will react with a basic site in the analyte molecule producing a $[MH]^+$/solvent-cluster. If a proton from the analyte is transferred to the plasma $[M-H]^-$/solvent clusters are produced. The former c.i.-processes are extremely efficient at atmospheric pressure due to the high collision frequency between molecules.
Next to the proton transfer reactions described above also charge transfer reactions may occur:

$$R^+ + A \rightarrow A^+ + R \qquad \text{A= Analyte molecule}$$

$$R^- + A \rightarrow A^- + R \qquad \text{R = Reactant ion}$$

The reactant ions for charge transfer in ambient air are mainly O_2^+, N_2^+ and to a lesser degree NO^+. These ions react rapidly with H_2O to form proton hydrates. Therefore the sensitivity of the ApCI system is approximately three orders of a magnitude less for charge transfer reactions than for proton transfer. Once the analyte ions are formed they are transferred to the analyzer section.

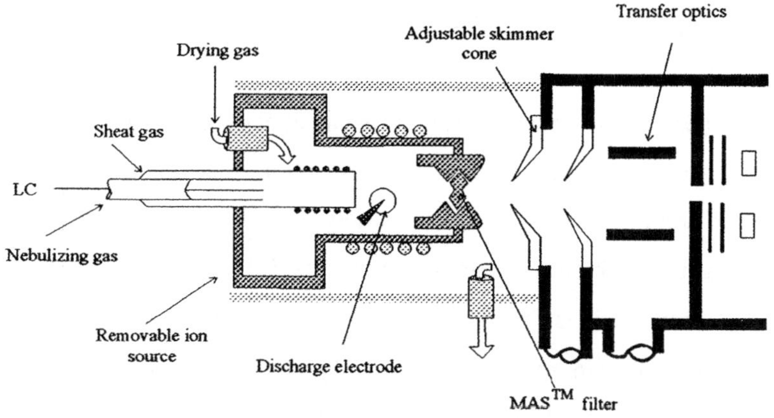

Figure 9. Principle of a ApCI-source

A constant stream of dry nitrogen gas (curtain gas, bath gas) helps to strip off most of the solvent molecules and prevents at the same time the passage of the majority of neutrals into the mass analyzer region. Once in the intermediate pressure zone (ca. 1 Torr) ions are accelerated in a electric field (cone voltage). During flight, ion/solvent clusters will fall apart as a result of collisions with N_2 and $[MH]^+$-ions are focused in the mass analyzer. If the cone voltage is set at higher value, fragmentation of molecular ion species via CID-like mechanisms is observed. This is an important feature since it allows the production of fragment ions out of charged molecules such as $[MH]^+$ or $[M\text{-}H]^-$ which can be used for structure elucidation. Such techniques are also possible but on a more sophisticated level in MS/MS-systems.

Advantages of the APcI-interface are :

1. Coupling with 4.6 mm ID columns is possible without any problems since flow-rates up to 2 ml/min can be handled.

2. Gradient elution : in contrast to TSP temperature tracking during elution is not necessary.
3. Volatile salts such as ammonium acetate do not present a problem at concentrations up to 0.1 M.
4. Compared to Thermospray, APcI is more sensitive and certainly more user friendly.

9.3 ELECTROSPRAY IONIZATION

Electrospray (ES)- and Ionspray (IS or pneumatically assisted electrospray) mass spectrometry have evolved as powerful ionization techniques for the structure determination and quantitation of a wide variety of polar compounds, both of low and high molecular mass which are ionic in solution [49,50)].

Electrospray ionization is however not a new phenomenon and was already investigated some 20 years ago by Dole *et al.*[51] Its revival was due to the work of the group of Fenn [52,53] who demonstrated the analytical potential of the technique.

Before discussing the ionization mechanism somewhat more in detail we wish to emphasize that both ES and IS are governed by the same physical phenomena . A difference is found in the coaxial N_2 gas flow applied in the case of ion spray to facilitate droplet formation . This allowed the introduction of higher amounts of solvent [54]. Compared to "real" electrospray (1 - 10 µL/min) the ionspray probe tolerates 20 to 1000 µL/min.

In both ES and IS an ion containing solution is sprayed via a narrow bore capillary tube (50- 200 µm I.D.)in the ion source which is at atmospheric pressure and a temperature of ca. 80°C. Sample introduction can be realized either by flow-injection (FIA) or coupling with a chromatographic system : HPLC or CZE [55,56].

The ES-probe can be coupled to a 4.6 mm I.D. HPLC column operated at optimum flow-rate (Fig.10). Although it is claimed that the ES probes can handle such high flow-rates we find that it is more elegant to introduce 80 to 150 µL/min using a post-column splitter. In the case of the coupling with microbore HPLC the total effluent can be introduced. This is of course also the case for packed capillary ES LC-MS and

ES LC-MS using a column operating at nannoliter flow-rates (micro LC ES MS). In the case of the coupling with CE a coaxial make-up flow is added in order to obtain stable spray conditions.

In order to produce ions the stainless steel capillary tube is polarized to a high voltage (2-3 KV). The sign (+ or -) of the voltage determines the charge on the droplets. For positive ion formation a positive voltage and for negative ion formation a negative voltage is applied. On their way to the mass analyzer (quadrupole, magnetic sector) the droplets shrink and produce solute ions which are finally stripped from remaining solvent molecules by collision with the N_2 bath gas.

While the exact mechanism of ion formation under ES conditions still is a matter of debate we can state that the following four major processes play an important role in the production of solute ions [57.]:

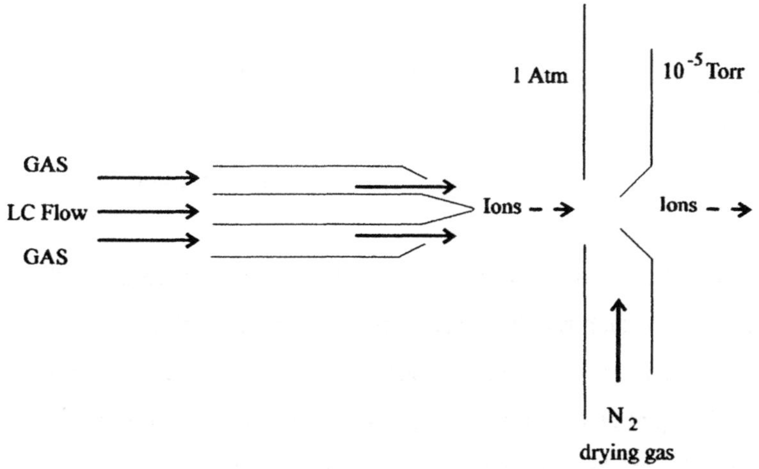

Figure 10. Principle of a Pneumatically assisted Electrospray Source.

1. The production of charged droplets from an electrolyte dissolved in the mobile phase.
2. The shrinkage of these charged droplets by solvent evaporation and droplet disintegration by Raleigh break -up.
3. Production of gas phase ions straight from the condensed phase
4. Modification of these gas-phase ions by declustering and solvent stripping.

The production of charged droplets at the ES capillary is induced by the high electrical field at the tip of the capillary (ca. 10^6V/cm) at a distance *d* from the counter electrode and is influenced by several other parameters as the viscosity of the solvent and the radius of the capillary:

$$E_c = {}^{2Vc}/_{r_c} \ln\left({}^{4d}/_{r_c}\right) \qquad (4)$$

E_c = electrical field at capillary tip in absence of solvent
r_c = radius of capillary
d = distance from counter electrode

At sufficient high voltage, the onset voltage V_{onset}, an unstable Taylor cone is produced from which charged droplets are emitted:

$$V_{onset} = 2.10^{-5}(\gamma\, r_c)^{1/2} \ln(4d/r_c) \qquad (5)$$

γ = surface tension

Solvent evaporation, during flight, reduces the size of the droplet till they become unstable as the Coulomb repulsion surpasses the surface tension. The droplet explodes by uneven fission i.e. the droplet does not split evenly into smaller droplets of approximately equal charge and mass but much smaller offspring droplets are created carrying ca. 2% of the original mass but 15% of the charge. Once these small (approximately 8 nm) highly charged droplets are created preformed solute ions can be ejected from these very small droplets into the gas phase this according to the Iribarne -Thomson theory of the ion evaporation model at atmospheric pressure (IEM). Once these gaseous solute ions are formed they are declustered (bath gas) and sampled in the analyzer section.

Applications of ES MS are found in a wide range of polar compounds of both low and high molecular mass. In the area of phytochemistry we have been looking at the quantitative determination of Cytokinins (plant hormones) from plant tissue [58].

In fig.11 the analysis of a phenylglycidylether/2'-deoxyadenosine adduct (PGE/dAdo) mixture is shown under ES LC-MS conditions on a RP Select B column (4mm I.D x 12.5 cm) , using CH_3OH/ 0.01M NH_4OAc at a flow-rate of 0.8ml/min (post-column split 1/20) [59].

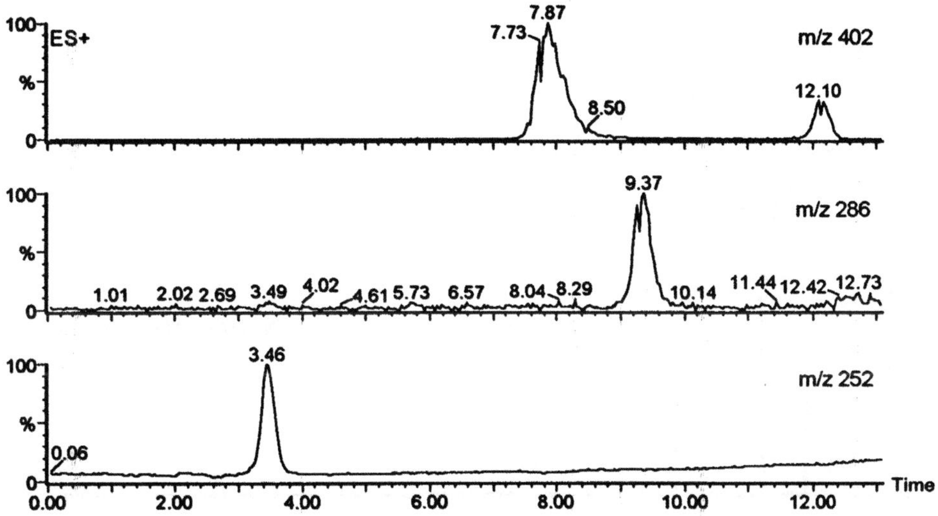

Figure 11. (+) ES LC-MS analysis of a mixture of PGE/dAdo; m/z = 252 (unreacted dAdo); m/z = 402 (N1- and N_6- PGE/dAdo); m/z = 286 (PGE/Adenine). (Ref. 59)

Electrospray is a very soft ionization technique as illustrated by the (+) ES mass spectrum of N1-PGE/dAdo (Fig.12.). Structural information is found in the (+) ES MS/MS spectrum.

In order to get any structural information one has to create fragment ions. This can be realized by either tandem mass spectrometry or by cone voltage fragmentation. In the latter technique the voltage of the sampling orifice is set to a higher value which induces CID like processes producing fragment ions.

Although some very nice data can be obtained for many polar molecules the most striking feature of electrospray mass spectrometry was its ability to record mass spectra of intact high molecular mass compounds such as peptides, proteins [60] and oligonucleotides [61,62] with a conventional mass spectrometric system and the subsequent possibility to calculate the molecular mass with high precision from the observed data. The observation of ions with a m/z-value which fall within the mass range of a conventional mass analyzer such as a H-sector or a quadrupole is explained by the formation of multiply-charged ions which have an m/z value falling within the scan range of such analyzers.

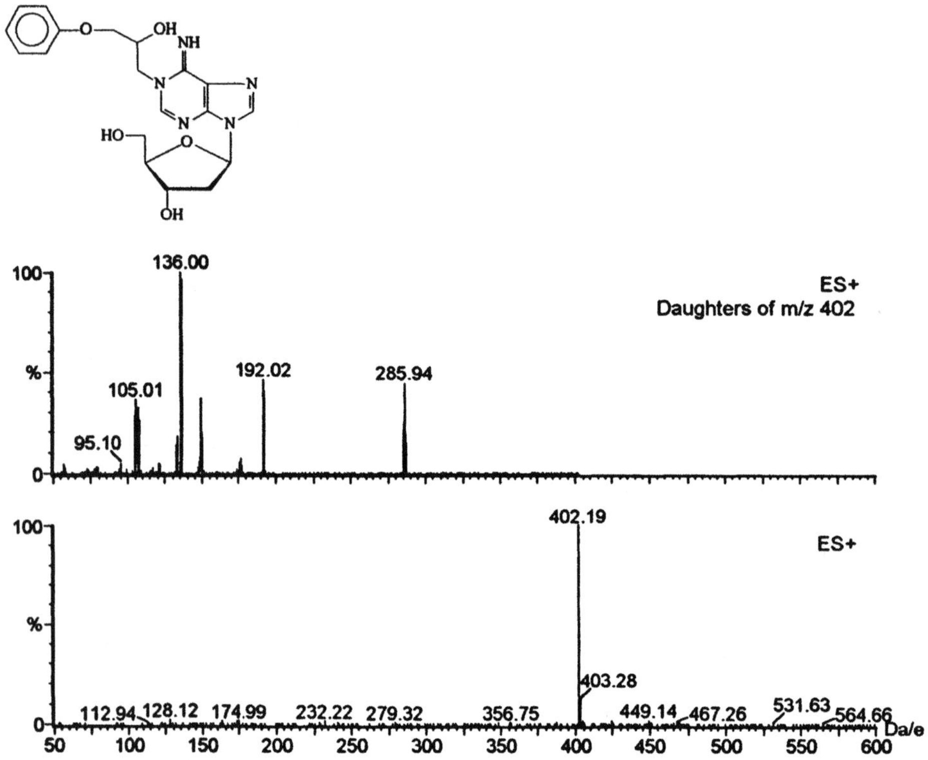

Figure 12. (+) ES MS and the low-energy (+) ES MS/MS data of N_1-PGE/dAdo. (MM: 401).

Usually a series of m/z values is detected which all differ one charge (in (+) ES = H^+). To illustrate this, the (+) ES spectrum of Myoglobin is shown in Fig.13. The number of charges is mentioned above each signal.

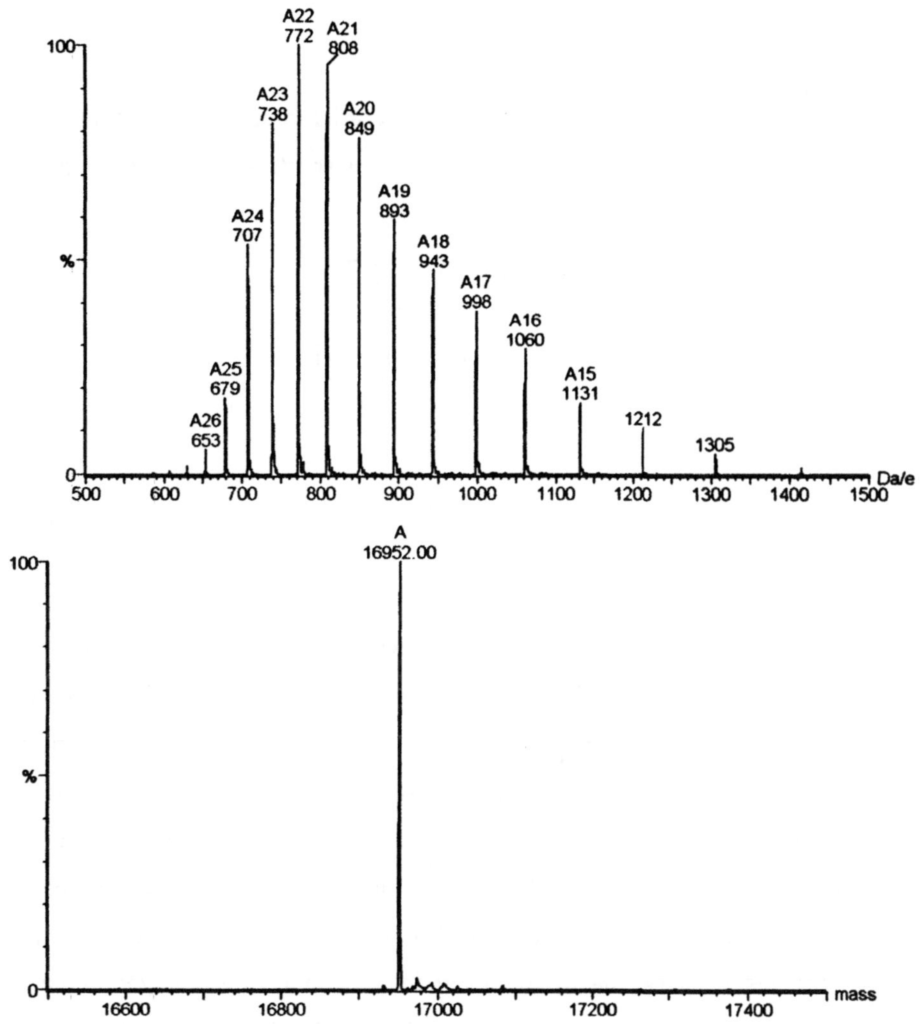

Figure 13. (+) ES mass spectrum of Myoglobin and its calculated molecular mass.

In order to calculate the molecular mass (M_r) from these data the following formula are used and Mr is presented as a single peak on a true M_r scale. The mass spectrometer measures the mass to charge ratio (m/z) of each peak :

$$m = \frac{M_r + nH}{n} \tag{6}$$

From this equation follows that:

$$M_r = n(m - H). \qquad (7)$$

Thus M_r can be calculated provided n is known. In order to do so, one should realize that any two consecutive peaks m_1 and m_2 in the mass spectrum differ by one proton. Therefore m_1 and m_2 can be expressed as :

$$m_2 = \frac{M_r + nH}{n} \qquad (8)$$

$$m_1 = \frac{M_r + (n+1)H}{n+1} \qquad (9)$$

If these equations are solved,

$$n = \frac{m_1 - H}{m_2 - m_1} \qquad (10)$$

The question is how these multiply charged species are formed and whether any explanations can be offered for the number of charges observed. [63]. In the case of peptides and proteins it has been suggested that the maximum number of charges is related to the number of basic residues present. However this explanation is incomplete : the maximum number of charges on an ES ion is related to the total number of polar groups or sites with the constraint that after the ion has left the droplet, the bonding energy of the charge with the a site must be equal or exceed the electrostatic repulsion energy due to Coulombic interaction by other charges on the ion. According to Fenn [63] at low pH a positively charged droplet will have an excess of H^+ on the surface. A protein IN the droplet will be protonated and each charge will be compensated by an anion. This results in a neutral molecule in the bulk solution. However , if this protein-"ion pair" approaches the droplet surface, the anion can be neutralized by a e.g. a surface proton or an unprotonated basic site can be protonated at the surface . This processes leave a net positive charge on the protein. If this is true the number of protons attached on the proton must be dependent on the surface charge density of the droplet.

Since ES MS is a very soft ionization technique which directly can be applied to aqueous solutions the methodology is also investigated to what extend it will allows the study of non-covalent interactions between molecules (guest-host complexes, protein - ligand etc.) [64].

10. Capillary Electrophoresis-Mass Spectrometry (CE-MS)

As already mentioned in the CE section, is electrophoresis the migration of charged particles or ions in solution due to the applied electric field. In order to couple CE with MS an interface must be present and up to know most of the interfaces used in CE-MS are interfaces used in combination with LC such as the ES interface and the dynamic FAB probe . Since the typical flow-rates required in these interface systems are 1- 50 μL/min and because the flow-rates in CE are only a few nL/min to 100 nL/min a make -up flow is required for stable CE-MS operating conditions.

As it is the case with other LC-MS techniques, non-volatile buffers or salts should be replaced by volatile ones in the mobile face even though the flow-rates are very low. This will prevent contamination of the ion source on a long-term basis and will ensure optimal MS detection.

Typical CZE-MS problems are: related to the very high efficiencies which can be achieved which results in very narrow bands which in turn require fast scanning for an adequate characterization of the compound . This, in combination with the small injection volumes, has its implications on the detection limits In order to cope with these problems many CE-MS recordings are done under single ion recording (SIR) conditions. Eventually special CE “tricks” are used which allow sample preconcentration prior to CE-MS analysis [sample stacking and isotachophoresis (ITP)]. For these techniques we refer to the specialized literature [65].

If CZE-MS is performed by using a CF-FAB probe a continuous flow of liquid is introduced through a 50-75 μm I.D. fused silica capillary positioned in the CF-FAB probe. The latter probe is usually introduced in the ion source of a magnetic sector instrument which is a t high potential (ca. 8 KV). As it is required for FAB ionization the mobile phase should contain a few % of glycerol (FAB matrix) . The glycerol can be introduced coaxially as first described by Moseley *et al.* [66.]. This is an elegant approach because it allows the independent optimalisation of the chromatographic conditions. One should also realize that the CZE capillary is positioned in the high-vacuum source which creates a vacuum induced flow of several μL/min. This results in a reduction of the chromatographic efficiency.

If CZE is coupled to ES or pneumatically assisted ES (Ionspray™) again a coaxial make-up flow is needed, not only to establish stable spray conditions but also to enable electric contact between the CZE-buffer and the ES-probe. Since in this system the CZE capillary is inserted in an ion source operating at atmospheric pressure no vacuum induced flow is observed . In the case of CZE ES-MS probe is shown in Fig.14.

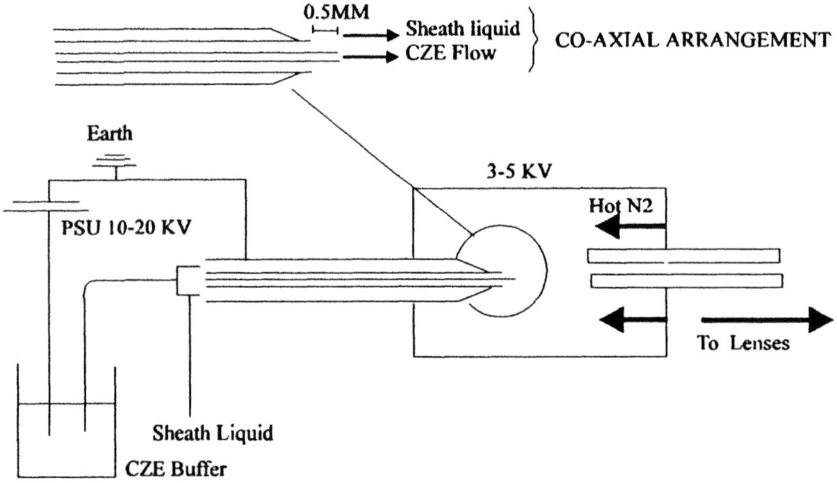

Figure 14. Principle of a CZE-MS interface.

Flow-rates for nebulisation- and bath gas are adjusted to lower values compared to the flow-rates implemented during LC ES MS.

CE-MS has been used for the analysis of a wide variety of compounds such as peptides, proteins, DNA-adducts, sulfonamides, antibiotics, oligonucleotides, etc. For a more detailed introduction to the subject see ref.[66] and[67].

Acknowledgements.

We wish to thank the National Fund for Scientific Research and the Belgian Lotto for financial support. (Project 2.0133.94.) and N.Verhaert for the lay-out.

References.

1. Arpino, P.J. (1982) On line Liquid Chromatography/Mass Spectrometry ? An odd couple!, *Trends in Anal.Chem.* **51**, 154-158.
2. Novotny M. (1984) Liquid Chromatography Columns of Capillary Dimensions, *J.Chromatogr. Lib.* **28**, 194-300.
3. Practical High Performance Liquid Chromatography (1976) Simpson C.F. (ed.)., Heyden and Son Ltd. London.
4. Ewing, G.A, Wallingford, R.A. and Olefirowicz, T.M. (1989) Capillary Electrophoresis, *Anal.Chem.* **61**, 292A-303A.
5. Smith R.M. (1988) Supercritical Fluid Chromatography, Royal Society of Chemistry, Chromatography Monographs, Loughborough, U.K.
6. Arpino, P.J. and Guichon, G. (1979) Optimization of the instrumental parameters of a combined Chromatograph-Mass Spectrometer coupled by an interface for Direct Liquid Introduction, *J.Chromatogr.* **185**,529-547.
7. Arpino, P.J.,Baldwin M.A. and McLafferty F.W. (1974) Liquid Chromatography-Mass Spectrometry II.:Continuous monitoring, *Biomed.Mass Spectrom.* **1**, 80-82.
8. Lovins, R.E., Ellis, S.R.,Tolbert G.D. and McKinney, C.R., Liquid Chromatography-Mass spectrometry: Coupling of a Liquid Chromatograph to a Mass Spectrometer, *Anal.Chem.* **45**, 1553-1556.
9. Takeuchi,T.,Hirata, Y. and Okumura Y. (1978) On-line coupling of Microliquid Chromatography and Mass Spectrometry through a Jet Separator, *Anal.Chem.* **50**, 659-660.
10. Jones, P.R. and Yang, S.K.. (1975) A Liquid Chromatograph/Mass Spectrometer Interface, *Anal.Chem.* **47**, 1000-1008.
11. Scott, R.P.W., Scott, C.G., Monroe, M. and Hess Jr.,J. (1974) Interface for on-line Liquid Chromatograph-Mass Spectroscopy Analysis, *J.Chromatogr.* **99**, 395-405.
12. Henion, J.D. (1978) Drug analysis by continuously monitored liquid chromatograph/mass spectrometry with a Quadrupole Mass Spectrometer, *Anal.Chem.* **50**, 1687-1693.
13. Bruins, A.P. and Drenth, B.F.H. (1983) Experiments with the combination of Micro Liquid Chromatography and Chemical Ionisation Quadrupole Mass Spectrometer using a Capillary Interface for Direct Liquid Introduction, *Int.J. Mass Spectrom. & Ion Phys.* **46**, 213-216.

14. Arpino, P.J. and Guichon G. (1979) LC-MS coupling, *Anal.Chem.* **51**, 682A-700A.
15. McFadden, W. (1980) Liquid chromatography/Mass spectrometry: Systems and Applications, *J.Chromatogr. Sci.* **18**, 97-113.
16. Esmans, E.L., Luyten Y. and Alderweireldt, F.C., LC-MS of Nucleosides using a DLI-probe (1983) *Biomed. Mass Spectrom.* **6**, 347 351.
17. Esmans E.L., Geboes, P , Luyten Y. and Alderweireldt, F.C. (1985) Direct Liquid Introduction LC-MS Microbore Experiments for the Analysis of Nucleoside Material present in Human Urine, *Biomed.Mass Spectrom.* **12**, 241 245.
18. Henion, J.D. and Wachs, T. (1981) Micro Liquid Chromatography/Mass Spectrometry Diaphragm Probe Interface, *Anal.Chem.* **53**, 1963-1965.
19. Arpino, P.J. and Beaugrand C. (1985) Optimization of the instrumental parameters of a combined Liquid Chromatograph/Mass Spectrometer coupled by an Interface for Direct Liquid Introduction, *Int.J.Mass Spectrom. & Ion Process.* **64**, 275-298.
20. Henion , J.D. and Maylin G.A (1980) Drug Analysis by Liquid Introduction Micro Liquid Chromatography Mass Spectrometry, *Biomed.Mass Spectrom.* **7**, 115-121.
21. Mauchamp,B., Lafont, R. and Krien, P. (1981) Analysis of Juvenile Hormones by High Performance Liquid Chromatography coupled with Mass Spectrometry, *Dev.Endocrinol.* **15**, 21-31.
22. Kenyon, C.N., Melera, A. and Erni, F. (1981) Utilization of Direct Liquid Inlet LC/MS in studies of Pharmacological and Toxicological Importance, *J.Anal.Toxicol.* **5**, 216-230.
23. Dobberstein, P., Korte, E., Meyerhoff, G. and Pesh, R. (1983) Investigation of an LC-MS interface for E.I-C.I and FAB ionisation, *Int.J.Mass Spectrom. & Ion Phys.* **46**, 185-188.
24. Games, D.E., McDowell, M.A., Levsen, K., Schafer, K.H., Dobberstein, P. and Gower, J.L. (1984) A Comparison of Moving Belt interfaces for Liquid Chromatography/Mass Spectrometry, *Biomed.Mass Spectrom.* **11**,87-95.
25. Hayes, M.J., Lankmayer, E.P., Vouros, P.,Karger, B.L. and McGuire, J.M. (1983) Moving Belt interface with Spray Deposition for Liquid Chromatography/Mass Spectrometry, *Anal.Chem.* **55**, 1745-1752.
26. Games, D.E. (1978) Soft Ionisation mass spectral methods for Liquid Analysis,. *Chem. Phys Lipids* **21**, 389-402.
27. Karger, B.L., Kirby, D.P., Vouros, P.,Foltz, R.L. and Hidy, B. On line Reverse Phase Chromatography-Mass Spectrometry, *Anal.Chem.* **51**, 2324-2328.
28. Dark, W.A., McFadden, W.H. and Bradford, D.C. (1977) Fractionation of Coal Liquids by HPLC with structural characterization by LC-MS, *J.Chromatogr.Sci.* **15**, 454-464.
29. Games, D.E., Gower, J.L., Lewis, M.G., Pugh, M.E and Rossiter M. (1978) Scope of HPLC/MS and Soft Ionisation Mass Spectrometry, in E.Reid and E.Harwood (eds.) ,*Blood drugs and analytical challenges*, Chichester, London, pp.185-194.
30. Blakley, C.R., McAdams, M.J. and Vestal, M.L. (1978) Crossed-Beam Liquid Chromatograph-Mass Spectrometer Combination, *J.Chromatogr.* **158**,261-276.

31. Nelieu,S., Stobiecki, M., Kahous, L. and Einhorn, J. (1994) Screening and Characterisation of Atrazine metabolites by Liquid Chromatography Tandem Mass Spectrometry, *Rapid Comm. in Mass Spectrom.* **8**, 945-952.

32. Cong, H.N., Bertaux, D., Valencia R., Becue, T.,Fournier, T.,Biou, D. and Porquet D (1994) Separation and Characterisation of the main methylated Nucleobases from nuclear, cytoplasmic and poly(A)(+) RNA by High Performance Liquid Chromatography and Mass Spectrometry, *J.Chromatogr. B. Biomed.Appl.* **661**,193-204.

33. Lemière, F., Esmans, E.L., Van Dongen, W., Van den Eeckhout, E. and Van Onckelen, H. Investigation of Thermospray LC-MS for the Analysis of 2'-deoxynucleoside/ Phenylglycidyl ether Adducts (1993). *J.Chromatogr.* **647**, 211-218.

34. Verweij, A.M.A., Hordijk, M.L. and Lipman, P.J.L. (1995) Quantitative Liquid Chromatography Thermospray Tandem Mass Spectrometric Analysis of some Analgesics and Tranquilizers of Methadone, Butyrophenone or Diphenylbutyl piperidine groups in whole Blood , *J.Anal. Toxicol.* **19**, 65-68.

35. Ducrey, B., Wolfender, J.L., Marston, A. and Hostettmann, K. (1995) Analysis of Flavanolglycosides of thirteen Epilobium species (Ornagraceae) by LV/UV and Thermospray LC-MS., *Phytochem.* **38**,129-137.

36. Arpino,P. (1990) Combined Liquid Chornatography Mass Spectrometry; PartII Techniques and mechanisms of Thermospray, *Mass Spectrom. Rev* **.9**, 631-669.

37. Willoughby, R.C. and Browner, R.F. (1984) Monodisperse Aerosol Generation Interface for Combining Liquid Chromatography with Mass spectrometry, *Anal.Chem.* **56**, 2625-2631.

38. Cappiello, A. and Famiglini, G. (1995) Analysis of thermally unstable Compounds by a Liquid Chromatograph-Mass Spectrometer Particle Beam Interface with a modified Ion Source., *Anal.Chem.* **67**, 412-419.

39. Kambhampti, I., Roinestad, K.S., Hartman, T.G., Rosen, J.D.,Fukuda, E.K., Lippincott, R.L. and Rosen, R.T. (1994) Determination of Diquat and Paraquat in water using High Performance Liquid Chromatography with confirmation by Liquid Chromatography-Particle Beam Mass Spectrometry., *J.Chromatogr.* **688**, 67-73.

40. Mao, J., Pacheco, C.R., Traficante, D.D. and Rosen, W. (1995) Identification and Characterisation of nitrogen compounds in Brazilian Diesel Oil by Particle Beam LC-MS., *Fuel* **74**, 880-887.

41. Wilkes, J.G., Zarrin,F.,Lay, J.O. and Vestal, M.L. (1995) Particle size Distribution is not the major factor explaining variable Analyte Transmission efficiency in Liquid Chromatography-Particle Beam Mass Spectrometry, *Rapid comm. in Mass Spectrom.* **9**, 133-137.

42. Edlund, P.O. and Henion, J.D.(1989) Packed column Supercritical Fluid-Mass Spectrometry via a two-stage momentum separator, *J.Chrom.Sci.* **27**, 274-282.

43. Berry, A.J., Games D.E. and Perkins J.R. (1986) Supercritical Fluid Chromatographic and Supercritical Fluid Chromatographic-Mass spectrometric studies of some Polar compounds, *J.Chromatogr.***363**, 147-158.

44. Berry, A.J., Games, D.E., Mylchreest, I.C., Perkins, J.R. and Pleasance, S. (1988) Packed Column Supercritical Fluid Chromatography/Mass Spectrometry using a Thermospray source in the filament on mode, *Biomed. Mass Spectrom.* **15**, 105-110.

45. Lee, E.D., Hsu, S.H. and Henion J.D. (1988) Electron Ionization like Mass Spectra by Capillary Supercritical Fluid Chromatography/ Charge Exchange Mass Spectrometry, *Anal.Chem.* **60**, 1990 1996.
46. Mertens, M.A.A., Leclercq, P.A. and Janssen, H.G.M (1995) Possibilities and Limitations of Interfaces combined with SFC-MS, *LC-GC International* **8**, 328-332..
47. Bruins, A.P., (1994) Atmospheric Pressure Ionization Mass Spectrometry. 1. Instrumentation and Ionization techniques,*Trends in Anal.Chem.* **13**, 37- 41.
48. Bruins, A.P. (1994) Atmospheric Pressure Ionization Mass Spectrometry. 2. Applications in Pharmacy, Biochemistry and General Chemistry, *Trends in Anal.Chem.* **13**, 81- 90.
49. Fenn, J.B., Mann, M., Meng, C.K., Wong, S.F. and Whitehouse, C.M. (1990).Electrospray Ionization: Principles and Practice. *Mass Spectrom.Rev.* **9**, 37-70.
50. Hamdan, M. and Curcuruto, D (1991) Development of the Electrospray Ionization Technique.*Int.J.Mass Spectrom. & Ion Process.* **108**, 93-113.
51. Dole, M., Mack, L.L., Hines, R.L., Mobley, R.C., Furguson, L.D. and Alice, M.B. (1968) Molecular Beams of Microions,*J.Chem.phys.* **49**, 2240-2249.
52. Fenn , J.B., Mann, M., Meng, C.K. and Whitehouse C.M. (1989) Electrospray Ionization for Mass Spectrometry of Large Biomolecules, *Science* **246**, 64-71.
53. Fenn, J.B. and Mann, M. (1992) Electrospray Mass Spectrometry: Principles and Methods in *Mass Spectrometry. Clinical and Biomedical Applications*, Plenum Press, New-York, London , pp. 1-35.
54. Bruins, A.P., Weidolf, L.O.G. and Henion, J.D.(1987). Determination of Sulfonated Azo Dyes by Liquid Chromatography/Atmospheric Pressure Ionization Mass Spectrometry, *Anal.Chem.* **59**, 2647-2652.
55. Janning, P., Schrader, W. and Lindscheid, M. (1994) A new Mass Spectrometric Approach tot detect modifications in DNA. *Rapid Comm. in Mass Spectrom.* **12**, 1035 -1040 .
56. Hopfgartner, G., Wachs, T., Bean, K. and Henion , J. (1993) High-Flow Ion Spray Chromatography/Mass Spectrometry, *Anal.Chem.* **65**, 439-446.
57. Kebarle,P. and Tang,L. (1993) From Ions in Solution to Ions in the Gas Phase. *Anal.Chem.* **65**, 972A-986A.
58. Prinsen, E., Redig ,P.,Van Dongen, W., Esmans , E.L. and Van Onckelen, H.(1995) Quantitative Analysis of Cytokinins by Electrospray Tandem Mass Spectrometry, *Rapid Comm. in Mass Spectrom.*, in press.
59. Lemière, F., Joos, P., Vanhoutte, K;, Esmans, E.L., De Groot, A., Claeys, M. and Van den Eeckhout, E. (1995) Phenylglycidylether Adducts of dCyd and dAdo: Stability in Solution and Structure Analysis by Electrospray MSMS., submitted for publication.
60. Metzger, J.W. and Eckershorn (1994) Electrospray Mass Spectrometry, in *Microcharacterization of Proteins*, Verlag Chemie GmbH, Weinheim, New-York, pp 167-188.
61. Stults, J.T. and Marsters, J.C. (1991) Improved electrospray of synthetic Oligonucleotides. *Rapid Comm. in Mass Spectrom.* **5**, 359-363.
62. Potier, N., Van Dorsselaer, A., Cordier, Y., Rock, O. and Bisschoff, R. (1994) The Negative Electrospray Ionization Mass Spectra of Synthethic and Chemically modified Oligonucleotides.*Nucl.Acid Res.* **19**, 3895 -3903.

63. Fenn, J.B. (1993). Ion Formation from Charged Droplets: The Role of Geometry, Energy and Time, *J.Amer.Soc.Mass Spectrom.* **4**, 524- 535.

64. Smith, R.D. and Light-Wahl, K.J. (1993) The Observation of Non-Covalent Interaction in Solution by Electrospray Ionization Mass Spectrometry: Promise, Pitfalls and Prognosis. *Biol.Mass Spectrom.* **22**, 493-501.

65. An Introduction High Performance Capillary Electrophoresis (1992) Heiger, D.N. Ed. Hewlett Packard GmbH, Waldbronn , Germany.

66. Moseley, M.A., Deterding, L.J., Tomer, K.B. and Jorgenson, J.W. (1989) Coupling of Capillary Zone Electrophoresis and Capillary Liquid Chromatography with Coaxial Continuous - Flow Fast Atom Bombardment Tandem Sector Mass Spectrometry. *J.Chromatogr.* **480**, 197-209.

67. Niessen, W.M.A., Tjaden, U.R. and van der Greef, J. (1993) Capillary Electrophoresis-Mass Spectrometry. *J.Chromatogr.* **636**, 3 -19.

68. Smith, R.D., Wahl, J.H., Goodlett, D.R. and Hofstadler, St.A. (1993) Capillary Electrophoresis/ Mass Spectrometry, *Anal.Chem.* **65**, 574A -584A.

MASS SPECTROMETRY IN THE STRUCTURE DETERMINATION OF LONG-CHAIN FATTY ACID DERIVATIVES

MAGDA CLAEYS, LIBERATA NIZIGIYIMANA, and
HILDE VAN DEN HEUVEL
University of Antwerp (UIA)
Department of Pharmaceutical Sciences
Universiteitsplein 1
B-2610 Antwerp, Belgium

ABSTRACT. Methodology based on the combined use of fast atom bombardment, high-energy collision-induced dissociation and B/E linked scanning has been applied to the structure determination of fatty acid derivatives including long-chain alkenyl salicylic acids, acylcarnitines and monoacylglycerides. Some mechanistic aspects of charge-remote fragmentation reactions occurring under high-energy collision-induced dissociation conditions in ions containing a stable charge center are discussed. Results obtained with mono- and 1,4-*bis*-unsaturated compounds indicate that a radical mechanism involving an initial C-H bond cleavage followed by radical-induced C-C cleavage reactions should be considered.

1. Introduction

High-energy collision-induced dissociation (CID) in combination with tandem mass spectrometric techniques provides an entirely instrumental approach to the structural characterisation of fatty acyl chains in biomolecules. The potential of this methodology has been recognised by Gross and coworkers; this research group discovered in 1983 [1] that charge-remote fragmentations occur in closed-shell ions, containing a stable charge center, at sites that are removed from the charge. Since then, a great number of studies have dealt with fundamental aspects of charge-remote fragmentation reactions and analytical applications to biomolecules (for reviews, see [2, 3]). The availability of mass spectrometric methods for the detailed characterisation of long-chain fatty acyl parts is very relevant in the structure elucidation of unknown

K. R. Jennings (ed.), Fundamentals and Applications of Gas Phase Ion Chemistry, 149–163.

biomolecules because NMR techniques are not very suitable for that purpose. In addition, the methodology mentioned above is sensitive and only requires a sample amount of a few micrograms or even lower if photodiode array detection technology is used [4, 5].

In this review, we will present some analytical applications of the combined use of fast atom bombardment (FAB), high-energy CID and B/E linked scanning to the structure characterisation of long-chain alkenyl salicylic acids, acylcarnitines and monoacylglycerides [6-8]. More specifically, we will illustrate how this methodology can be used to determine the position of a double bond in mono-unsaturated compounds or to confirm the presence of a *n*-alkyl chain in saturated fatty acid derivatives. In addition, we will also discuss some mechanistic aspects of charge-remote fragmentation reactions; detailed interpretation of data obtained for mono-unsaturated and 1,4-*bis*-unsaturated compounds indicate that the mechanisms proposed by Contado *et al.* [9] and by Wysocki and Ross [10] do not enable us to rationalise all abundant product ions in their high-energy CID spectra and that an alternative mechanism involving an initial cleavage of a C-H bond followed by radical-induced reactions should be considered [11].

2. Mass spectrometric analyses

The mass spectrometric data were obtained on a Fisons VG70SEQ hybrid instrument (EBqQ configuration) which is equipped with a caesium ion source for primary particle bombardment. The acronym FAB will be used in this review to refer to caesium ion bombardment. Glycerol, *m*-nitrobenzylalcohol or *m*-nitrobenzylalcohol saturated with LiI were employed as liquid matrices. High-energy CID spectra of protonated molecules $[M + H]^+$, deprotonated molecules $[M - H]^-$, lithiated molecules $[M + Li]^+$ or dilithiated molecule species $[M - H + 2Li]^+$ were obtained by performing collisions with helium in the first field-free region at a collision energy of 8 keV and 50% beam attenuation, and B/E linked scanning. The low-energy CID spectrum of $[M - 2H + 2Li]^{+\cdot}$ ions of 6-(12'Z-nonadecenoyl)-salicylic acid was taken by activating the ions in the quadrupole collision cell using argon at a collision energy of 50 eV and scanning the quadrupole analyser. The argon pressure measured with an ionisation gauge outside the cell was $2*10^{-6}$ mbar. The product ion spectra were acquired in the continuum mode by accumulating 10 scans.

3. Characterisation of a *n*-alkyl chain in saturated fatty acid derivatives

The typical appearance of high-energy CID spectra of saturated fatty acid derivatives is illustrated with spectra obtained for $[M + H]^+$ ions of palmitoylcarnitine (Fig. 1) and $[M + Li]^+$ ions of 1-monopalmitoylglyceride (Fig. 2). Both spectra show very similar features including the presence of a homologous ion series corresponding to formal losses of elements of alkanes (C_nH_{n+2}, where n = 1-14). Most ions of this homologous ion series have a comparable abundance which is characteristic of a *n*-alkyl chain. In the case of branching or the presence of a double bond a gap would be detected. It is pointed out that for 1-monopalmitoylglyceride this homologous ion series is not produced upon high-energy of $[M + H]^+$ ions. These results are in agreement with those of a previous study by Adams and Gross [12] demonstrating that charge-remote fragmentation in a long-chain fatty acyl chain is favoured in closed-shell ions containing a stable charge center such

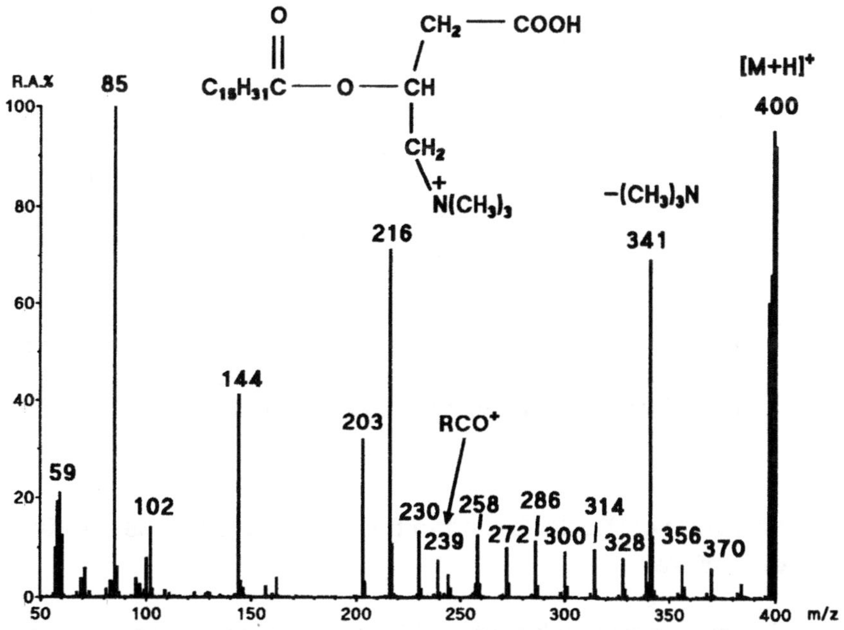

Figure 1. High-energy CID spectrum obtained for protonated palmitoylcarnitine. $[M + H]^+$ ions were generated by FAB using glycerol as liquid matrix

as in dilithiated molecule species $[M - H + 2Li]^+$ of fatty acids. In the case of protonated palmitoylcarnitine the charge center is not too stable which is evident from the occurrence of significant charge-proximate fragmentation resulting, for example, in the loss of $(CH_3)_3N$ (59 u) and yielding the abundant ions at *m*/*z* 341.

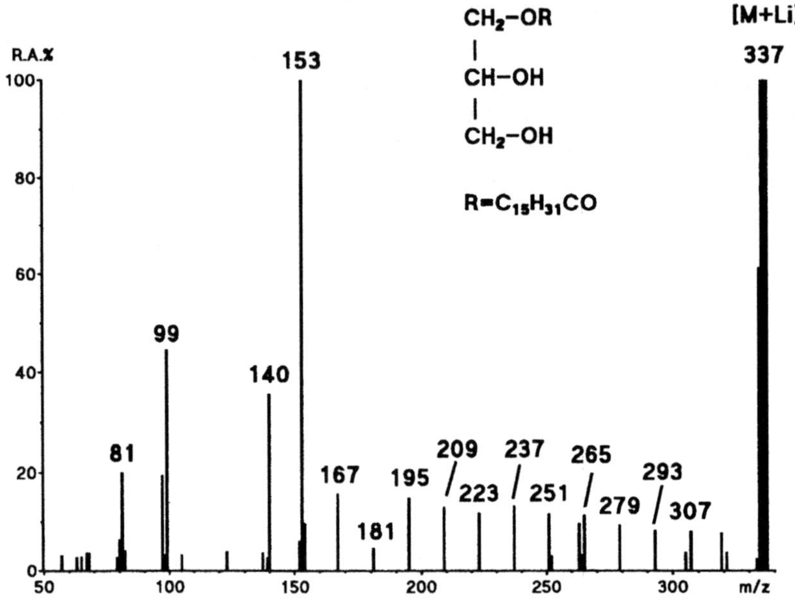

Figure 2. High-energy CID spectrum obtained for lithiated 1-mono-palmitoyl-glyceride. $[M + Li]^+$ ions were generated by FAB using *m*-nitro-benzylalcohol saturated with LiI as liquid matrix

4. Determination of the double bond position in mono-unsaturated fatty acid derivatives

The typical appearance of high-energy CID spectra of mono-unsaturated fatty acid derivatives is illustrated with the product ion spectrum obtained for $[M + Li]^+$ ions of 1-mono-oleoylglyceride (Fig. 3). The charge-remote product ions show a specific pattern; the formation of abundant structure-informative ions involves allylic and homo-allylic cleavages as indicated in Figure 3. A gap

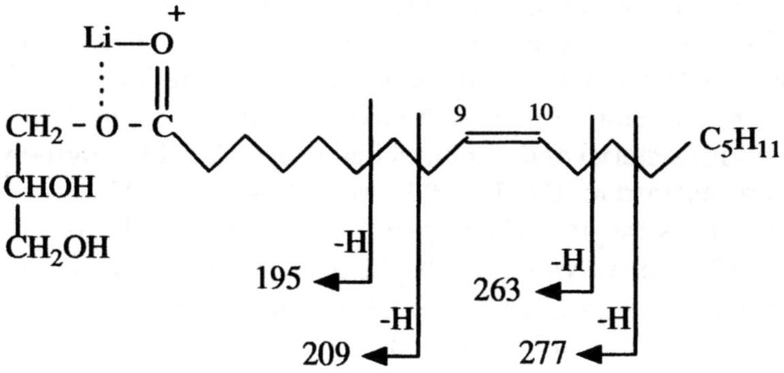

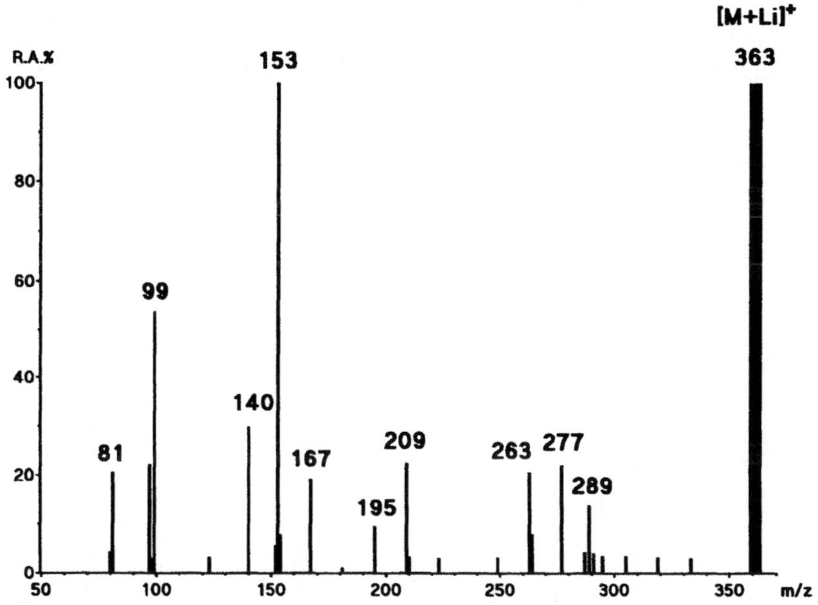

Figure 3. High-energy CID spectrum obtained for lithiated 1-mono-oleoylglyceride. $[M + Li]^+$ ions were generated by FAB using *m*-nitrobenzylalcohol as liquid matrix

of 54 u is present between the ions at *m/z* 209 and *m/z* 263 which is consistent with the presence of a double bond at the C(9) position of the oleoyl chain.

Using this methodology we have determined the double bond position in long-chain 6-alkenyl salicylic acids containing $C_{17:1}$, $C_{19:1}$ and $C_{21:1}$ side chains [6]. These compounds were isolated from the leaves and stems of *Spondias mombin* using bio-assay guided isolation and were found to have interesting antibacterial and molluscicidal properties [13]. High-energy CID spectra were obtained for $[M - H + 2Li]^+$ as well as for $[M - H]^-$ precursor ions. Figure 4 illustrates the product ion spectrum obtained for dilithiated molecule ions of 6-(12'Z-nonadecenoyl)-salicylic acid containing a $C_{19:1}$ side chain. Interpretation of the product ion spectrum enables us to conclude that the double bond is at the ω7 position as found in vaccenic acid, a naturally occurring $C_{18:1}$ fatty acid. A gap of 54 u is present between the ions at *m/z* 289 and *m/z* 343 whose formation involves allylic cleavages. The spectrum also reveals that the charge-remote product ion profile is distorted in the high-mass region due to charge-proximate loss of CO_2 (*m/z* 371). Following the same approach the double bond in the other homologues containing $C_{17:1}$ and $C_{21:1}$ side chains could be localised at the ω7 and ω6 positions respectively.

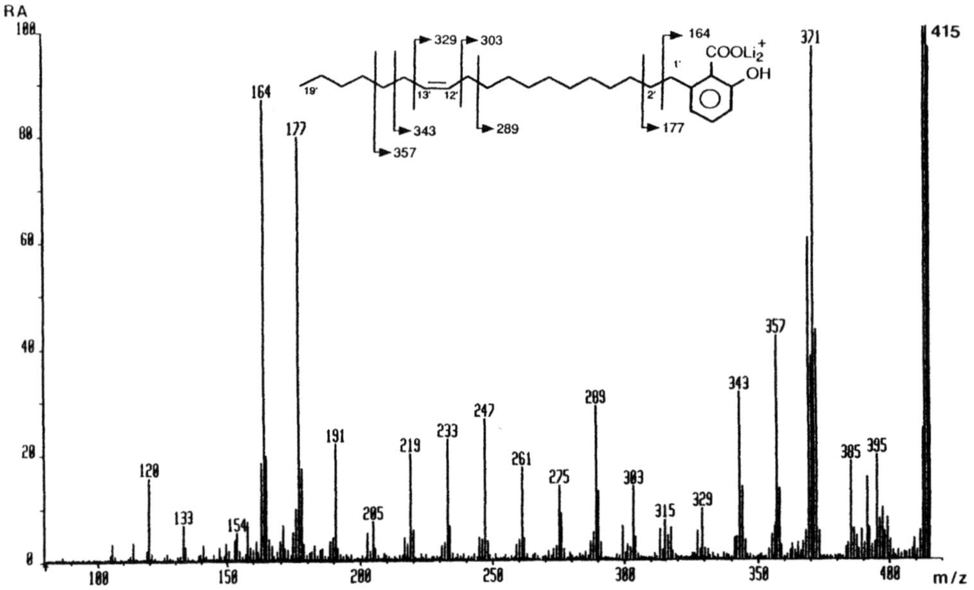

Figure 4. High-energy CID spectrum obtained for $[M - H + 2Li]^+$ ions (*m/z* 415) of 6-(12'Z-nonadecenoyl)-salicylic acid which were generated by FAB using *m*-nitrobenzylalcohol saturated with LiI as liquid matrix

Compared to [M - H]$^-$ ions [M - H + 2Li]$^+$ ions were found to be superior precursor ions for determining the position of the double bond in 6-alkenyl salicylic acids, findings which are similar to those obtained for fatty acids by Crockett *et al.* [14]. In the case of [M - H]$^-$ precursor ions, a very strong loss of CO_2 (44 u) is observed and in addition to the regular series of charge-remote product ions a second ion series is detected corresponding to additional loss of CO_2. Figure 5 shows the high-energy CID spectrum obtained for [M - H]$^-$ ions of 6-(10'Z-heptadecenoyl)-salicylic acid demonstrating these features. It is clear from these results that determination of the double bond is not at all straightforward.

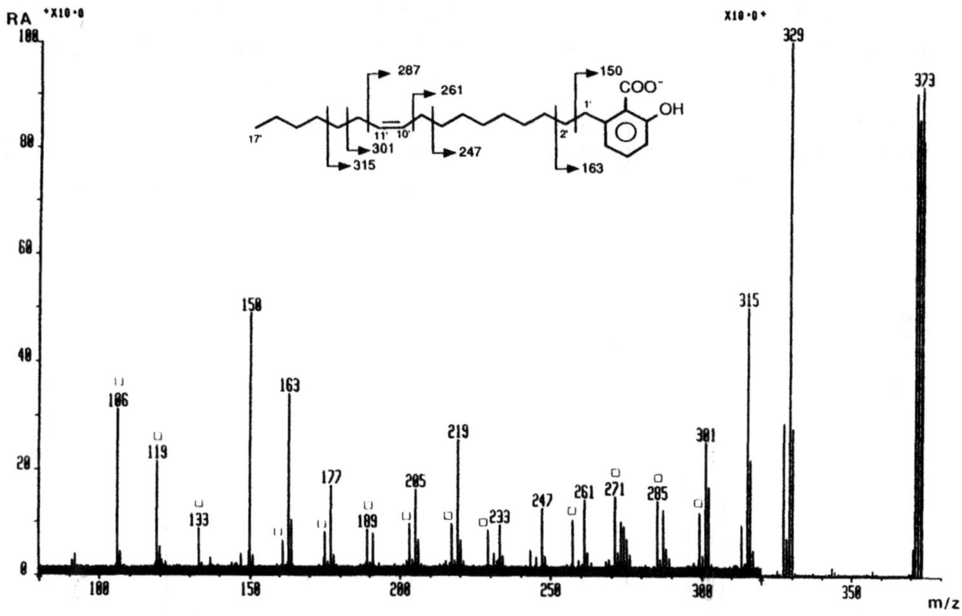

Figure 5. High-energy CID spectrum obtained for [M - H]$^-$ ions (*m/z* 373) of 6-(10'Z-heptadecenoyl)-salicylic acid. The second series of charge-remote product ions corresponding to additional loss of CO_2 is indicated with □

5. Mechanistic aspects of charge-remote fragmentation reactions

5.1. MONO- AND 1,4-*BIS*-UNSATURATED FATTY ACID DERIVATIVES

The formation of abundant remote-charge product ions which are useful for determining the position of a double bond in mono-unsaturated fatty acid derivatives can be rationalised by an initial C-H cleavage followed by radical-induced reactions. Scheme 1 gives plausible fragmentation routes for the formation of the ions at *m/z* 289, *m/z* 357, *m/z* 275 and *m/z* 343, which involve allylic and homo-allylic bond cleavages, from $[M - H + 2Li]^+$ precursor ions of 6-(12'Z-nonadecenoyl)-salicylic acid (Fig. 4). Mechanisms previously proposed by Contado *et al.* [9] for remote-charge fragmentation in mono-unsaturated fatty acids include retro-ene and 1,4-conjugate reactions, and allylic C-C cleavages. In an effort to support our proposal in which initial C-H cleavage is the key step we have examined the low-energy CID behaviour of ions formed by loss of a H· radical [11]. Figure 6 illustrates the low-energy CID spectrum of $[M - 2H + 2Li]^{+\cdot}$ ions of 6-(12'Z-nonadecenoyl)-salicylic acid

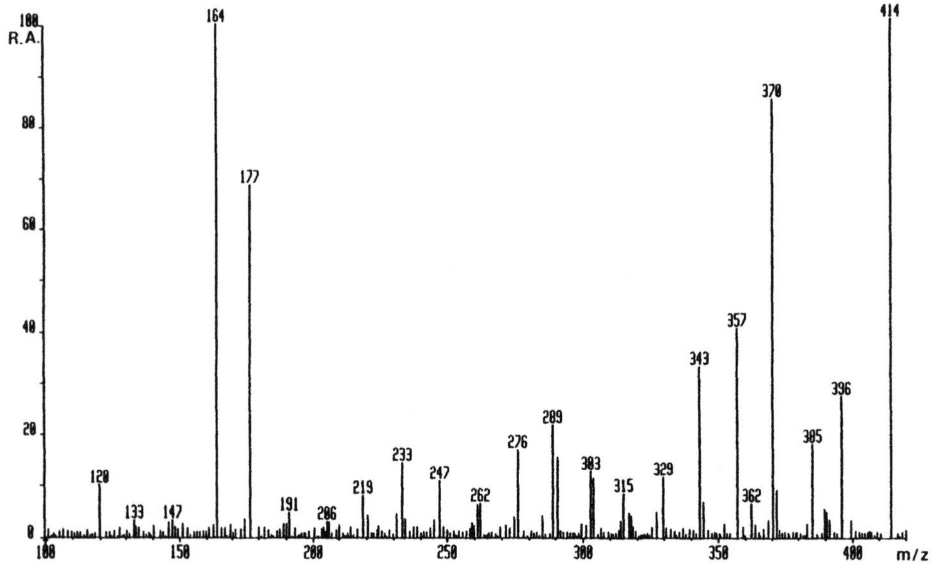

Figure 6. Low-energy CID spectrum obtained for $[M - 2H + 2Li]^{+\cdot}$ ions (*m/z* 414) of 6-(12'Z-nonadecenoyl)-salicylic acid formed during FAB

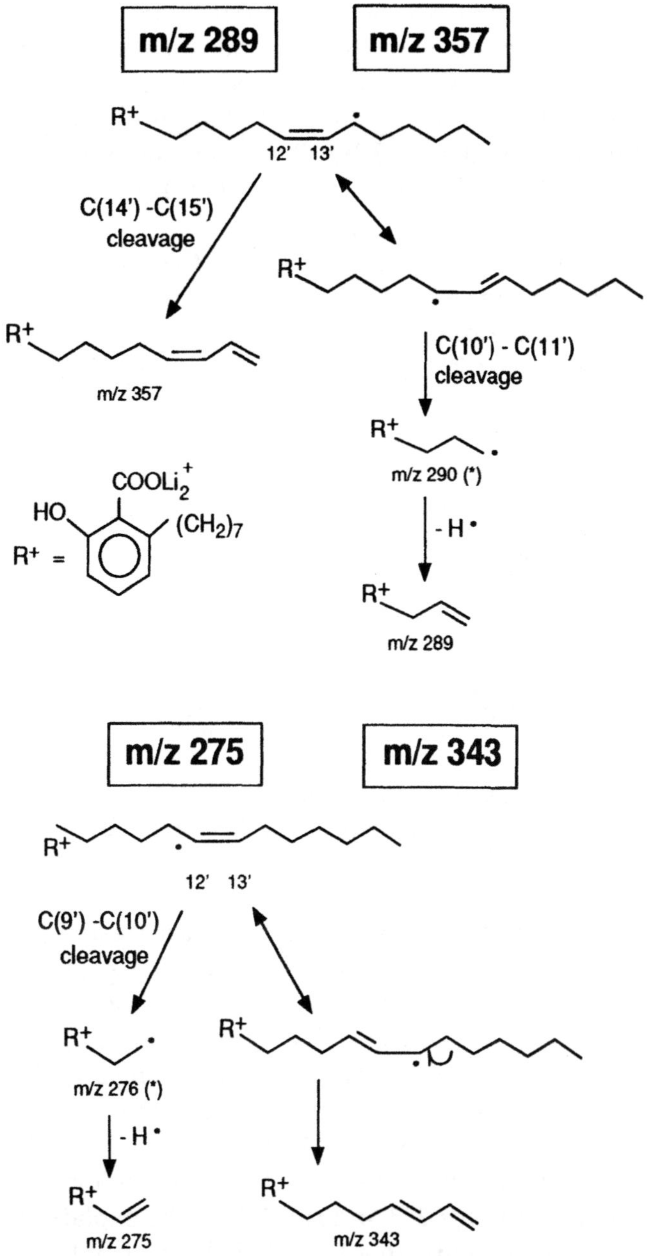

Scheme 1. Fragmentation mechanisms proposed for the formation of the ions at *m/z* 289, *m/z* 357, *m/z* 275 and *m/z* 343 upon high-energy CID of $[M - H + 2Li]^+$ ions of 6-(12'Z-nonadecenoyl)-salicylic acid [11]

which are produced under FAB conditions. It is worth noting that these ions do serve as precursor ions for all the product ions formed by high-energy CID of $[M - H + 2Li]^+$ ions (Fig. 4) including the odd-electron ions at *m/z* 164.

It is pointed out that the pattern of charge-remote product ions is remarkably similar to that found by EI of 3-picolinyl ester derivatives of mono-unsaturated fatty acids reported by Harvey [15]. Upon EI of these fatty acid derivatives, the ionised 3-picolinyl group abstracts H· radicals, predominantly at allylic positions (but also at other positions) in the mono-unsaturated chain, giving rise to radical-induced reactions. Scheme 2 shows the fragmentation mechanisms proposed by Harvey [15] for the formation of the abundant ions at *m/z* 274 and *m/z* 288 formed by EI of the 3-picolinyl derivative of oleic acid.

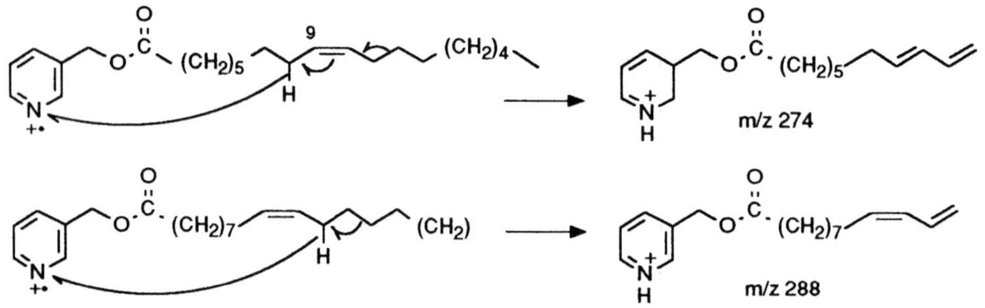

Scheme 2. Fragmentation mechanisms proposed by Harvey [15] rationalising the abundant ions at *m/z* 274 and *m/z* 288 formed by EI of the 3-picolinyl derivative of oleic acid

A 1,4-*bis*-unsaturated system is a key structural feature of the essential fatty acids linoleic ($C_{18:2}$) and arachidonic acid ($C_{20:4}$) which serve as precursors of a whole array of biologically important hormones including the prostaglandins and the leukotrienes [16]. The typical appearance of high-energy CID spectra of fatty acid derivatives containing a 1,4-*bis*-unsaturated system is illustrated with the product ion spectrum obtained for $[M - H + 2Li]^+$ molecule species of 6-(8'Z,11'Z-heptadecadienoyl)-salicylic acid which contains a ω6 1,4-*bis*-unsaturation as found in linoleic acid (Fig. 7). The abundant product ions at *m/z* 233 and *m/z* 327 can be explained in a straight-forward manner starting with an allylic C-H cleavage at the C(10') position followed by isomerisation and radical-induced cleavages (Scheme 3) [11].

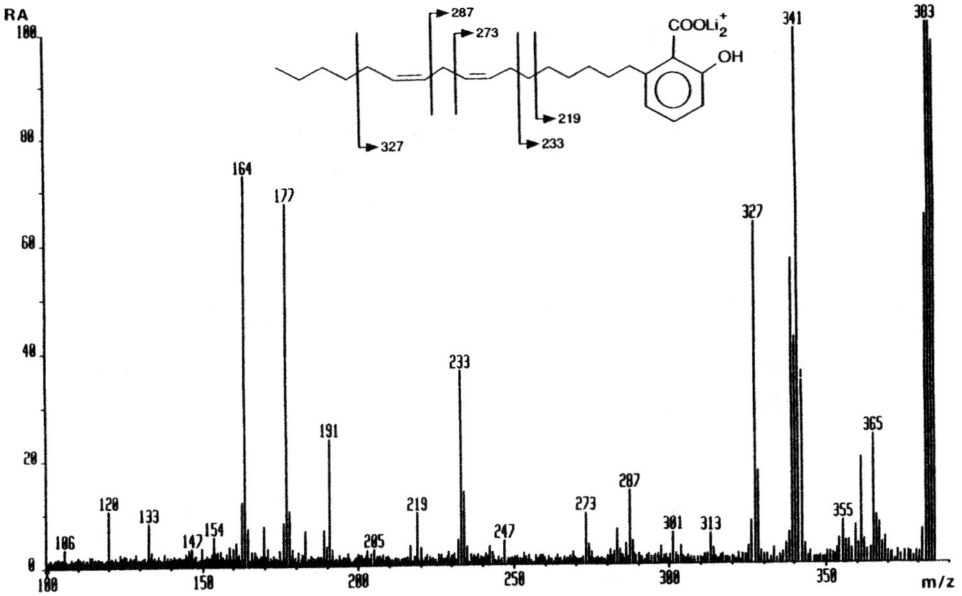

Figure 7. High-energy CID spectrum obtained for the [M - H + 2Li]$^+$ (*m*/*z* 383)molecule species of 6-(8'Z,11'Z-heptadecadienoyl)-salicylic acid

5.2. SATURATED FATTY ACID DERIVATIVES

The mechanism of charge-remote fragmentation reactions in long-chain satured fatty acid ions containing a stable charge center has been studied in detail by Jensen *et al.* [17] and has been proposed to be a 1,4-H_2 elimination reaction (Scheme 4). This reaction results in formal losses of elements of alkanes and the formation of a homologous series of terminally unsaturated ions, and has been supported by deuterium labelling data. However, it has been pointed out by Wysocki et al. [18] and Wysocki and Ross [10] that several charge-remote dissociation pathways cannot be explained by this cyclic mechanism, such as the loss of methane, the formation of odd- and even-electron product ions separated by 1 u, and the enhancement of certain dissociation products (e.g., allylic cleavage product ions) when unsaturation is introduced into a molecule. Wysocki and Ross [10] have proposed an alternative radical mechanism which involves homolytic C-C cleavage followed by radical-induced C-C cleavage based on their observation that the

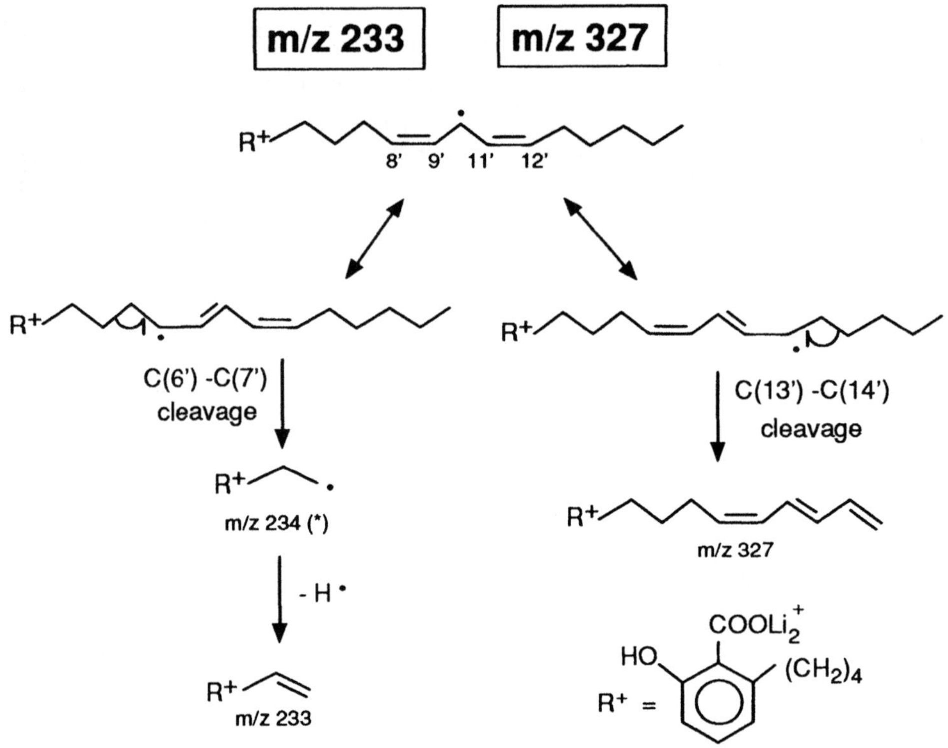

Scheme 3. Fragmentation mechanisms rationalising the abundant product ions at *m/z* 233 and *m/z* 327 formed during high-energy CID of $[M - H + 2Li]^+$ molecule species of 6-(8'Z,11'Z-heptadecadienoyl)-salicylic acid

products of the homolytic C-C cleavage reaction are more abundant when cooler ions are subjected to CID (Scheme 5). It is also worth mentioning that Cordero and Wesdemiotis [19] have characterised the neutral products formed upon charge-remote fragmentation of fatty acid ions using neutralisation-reionisation mass spectrometry. Because they detected alkenes and not alkyl radicals, they concluded that the 1,4-H_2 elimination reaction is the most likely mechanism.

A radical mechanism starting with homolytic C-H cleavage at different positions in a saturated fatty acid chain followed by radical-induced C-C cleavages provides a straightforward mechanism to rationalise the formation of the homologous series of terminally unsaturated ions. This mechanism is supported by the experimental finding that ions formed by initial H· radical loss do serve as precursor ions for all the terminally unsaturated product ions formed upon high-energy CID of ions containing a stable charge center (*vide supra*, Fig. 6) and by results obtained by Wysocki and Ross [10] who demonstrated that the products of the homolytic C-C cleavage reaction are more abundant when cooler ions are subjected to CID. In addition, this mechanism also enables us to rationalise the formation of the abundant odd-electron ions which terminate the homologous series of unsaturated product ions and the decreased abundance of the product ions corresponding to cleavage of the C(5)-C(6) bond [8, 11]. In the neutralisation-reionisation experiments performed by Cordero and Wesdemiotis [19] to characterise the neutral products formed upon charge-remote fragmentation of fatty acid ions, however, no alkyl radicals could be detected. This finding may not be an argument to disregard radical mechanisms.

It is also pointed out that the pattern of charge-remote product ions in saturated fatty acid ions containing a stable charge center is remarkably similar to that found by Harvey [15] for 3-picolinyl esters of saturated fatty acids under EI conditions, for which the formation of the homologous ion series has been explained by assuming hydrogen abstraction from the chain by the nitrogen of the pyridine ring and cleavage at the resulting radical site.

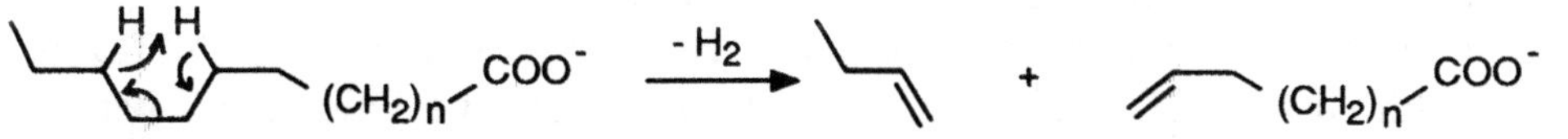

Scheme 4. Mechanism proposed by Jensen et al. [17] for the formation of terminally unsaturated ions upon high-energy CID of fatty acid carboxylate ions

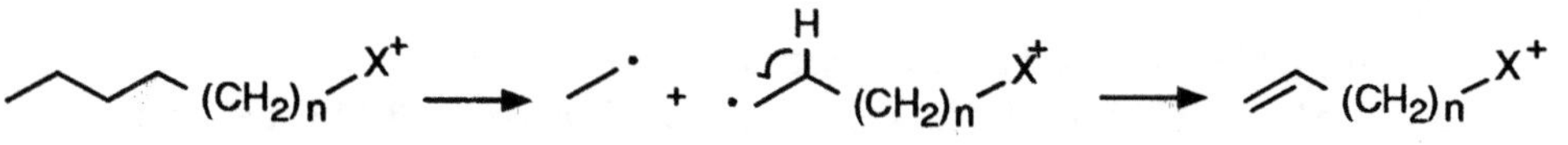

Scheme 5. Alternative radical mechanism proposed by Wysocki and Ross [10] to rationalise the formation of terminally unsaturated ions in long-chain functionalised alkanes

6. Conclusions

It should be clear from the data presented and discussed in this review that high-energy CID in combination with tandem mass spectrometry is a powerful instrumental method for characterising fatty acyl chains and that a stable charge center in the precursor ions is critical for observing charge-remote fragmentation reactions.

With respect to mechanistic aspects of charge-remote fragmentation reactions occurring in fatty acyl chains we have presented arguments in favour of a radical mechanism starting with a homolytic C-H bond cleavage at different positions in the acyl chain followed by radical-induced C-C cleavage reactions. This mechanism enables us to rationalise the CID behaviour of ions of saturated, mono-unsaturated as well as 1,4-*bis*-unsaturated fatty acid derivatives containing a stable charge center and is analogous to the radical mechanism proposed in an earlier EI study by Harvey [15] on 3-picolinyl ester derivatives of fatty acids.

7. Acknowledgements

We are grateful to the Belgian National Fund for Scientific Research (NFRS) for continuous financial support. M. Claeys is indebted to the NFRS as a research director. L. Nizigiyimana acknowledges the Belgian Ministry of Foreign Affairs (ABOS) for a doctoral research fellowship.

8. References

1) Tomer, K. B., Crow, F. W. and Gross, M. L. (1983) *J. Am. Chem. Soc.* **105**, 5487-5488.
2) Adams, J. (1990) *Mass Spectrom. Rev.* **9**, 141-186.
3) Gross, M. L. (1992) *Int. J. Mass Spectrom. Ion Processes* **118/119**, 137-165.
4) Evans, S., Buchanan, R., Hoffman, A., Mellon, F. A., Price, K. R., Walls, F. C., Hall, S., Burlingame, A. L., Chen, S. and Derrick, P. J. (1993) *Org. Mass Spectrom.* **28**, 289-290.
5) Chen, S., Derrick, P. J., Mellon, F. A. and Price, K. R. (1994) *Anal. Biochem.* **218**, 157-169.
6) Claeys, M., Van den Heuvel, H., Claereboudt, J., Corthout, J., Pieters, L. and Vlietinck, A. J. (1993) *Biol. Mass Spectrom.* **22**, 647-653.

7) Van Bocxlaer, J. F., Claeys, M., Van den Heuvel, H. and De Leenheer, A. P. (1995) J. *Mass Spectrom.* **30,** 69-80.

8) Nizigiyimana, L., Van den Heuvel, H. and Claeys, M. (1995) *J. Mass Spectrom.* in press.

9) Contado, M. J., Adams, J., Jensen, N. J. and Gross, M. L. (1991) *J. Am. Soc. Mass Spectrom.* **2,** 180-183.

10) Wysocki, V. H. and Ross, M. M. (1991) *Int. J. Mass Spectrom. Ion Processes* **104,** 179-211.

11) Claeys, M. and Van den Heuvel, H. (1994) *Biol. Mass Spectrom.* **23,** 20-26.

12) Adams, J. and Gross, M. L. (1987) *Anal. Chem.* **59,** 1576-1582.

13) Corthout, J., Pieters, L., Claeys, M., Geerts, S., Vanden Berghe, D. and Vlietinck, A. (1994) *Planta Medica* **60,** 460-463.

14) Crockett, J. S., Gross, M. L., Christie, W. W. and Holman, R. T. (1990) *J. Am. Soc. Mass Spectrom.* **1,** 183-191.

15) Harvey, D. J. (1982) *Biomed. Mass Spectrom.* **9,** 33-38.

16) Claeys, M. (1989) *Mass Spectrom. Rev.* **8,** 1-34.

17) Jensen, N. J., Tomer, K. B. and Gross, M. L. (1985) *J. Am. Chem. Soc.* **107,** 1863-1868.

18) Wysocki, V. H., Ross, M. M., Horning, S. R. and Cooks, R. G. (1988) *Rapid Commun. Mass Spectrom.* **2,** 214-216.

19) Cordero, M. M. and Wesdemiotis, C. (1994) *Anal. Chem.* **66,** 861-866.

MASS SPECTROMETRY IN STUDIES OF FLAVONOID GLYCOSIDES

MAGDA CLAEYS
University of Antwerp (UIA)
Department of Pharmaceutical Sciences
Universiteitsplein 1
B-2610 Antwerp, Belgium

ABSTRACT. Methodology based on the combined use of fast atom bombardment, low- and high-energy collision-induced dissociation and tandem mass spectrometry has been developed for the structure determination of flavonoid glycosides, which are a large group of plant glycoconjugates. Using mass spectrometric methods, it is possible to make a differentiation between *O*- and *C*-glycosides, characterise the aglycone and carbohydrate parts, and, in some cases, also to determine the glycosylation position.

1. Introduction

Mass spectrometric methodologies play a key role in the structure determination of glycoconjugates. The structure elucidation of glycoconjugates has many aspects and involves characterisation of both the carbohydrate and aglycone parts. The discovery of fast atom bombardment (FAB) by Barber and coworkers [1] has been of particular significance for the direct mass spectrometric determination of polar biomolecules and has greatly facilitated their molecular weight determination. The combination of FAB with collision-induced dissociation (CID) and tandem mass spectrometric techniques has given this ionisation technique a new dimension by providing structural information and has enabled the characterisation of underivatised flavonoid glycosides, saponins and glycoalkaloids [2-13].

In this review, the mass spectrometric structure analysis of flavonoid glycosides will be covered. During the past years we have been dealing with the mass spectrometric structure determination of a large variety of polyphenolic secondary plant metabolites, which have been isolated from

K. R. Jennings (ed.), Fundamentals and Applications of Gas Phase Ion Chemistry, 165–179.

common phytochemical preparations (e.g., *Passiflora incarnata*) or plant extracts still widely used in African traditional medicine. In many cases, bioassay guided fractionation in which the antiviral, antifungal, antibacterial, anti-inflammatory or immuno-stimulating activity has been monitored, has led to the isolation of biologically active flavonoids, flavonoid glycosides and proanthocyanidins. These polyphenolic secondary metabolites are of biological interest not only because they show chemotherapeutic activities but also because they are useful markers for chemotaxonomy. The experience that flavonoid glycosides have been regularly encountered has prompted us to work out suitable mass spectrometric methods based on FAB in combination with CID and tandem mass spectrometric techniques [8-11].

Methodologies have been developed for the structure characterisation of flavonoid glycosides containing aglycone parts of the flavone or flavonol type. With respect to the characterisation of flavonoid *O*-glycosides, mass spectrometric methods involving low-energy CID yield information on: (i) the carbohydrate sequence; (ii) the aglycone part; and, (iii) the glycosylation position. The characterisation of flavonoid *C*-glycosides is addressed by high-energy CID, which enables the differentiation between isomeric mono-6-*C*- and 8-*C*-glycosides, and isomeric di-6,8-*C*-glycosides. The structure elucidation of flavone *O,C*-diglycosides is accomplished by resorting to low- and high-energy CID; low-energy CID allows one to determine the type of the terminal *O*-linked carbohydrate residue, whereas high-energy CID enables to identify the type and the linkage position of the *C*-linked carbohydrate part.

2. Mass spectrometric analyses

The mass spectrometric data were obtained on a Fisons VG70SEQ hybrid instrument (EBqQ configuration), which is equipped with a caesium ion source for primary particle bombardment. The acronym FAB will be used in this review to refer to caesium ion bombardment. The positive ion mode has mainly been used based on practical considerations: (i) the sensitivity in the positive ion mode is generally better than in the negative mode; and (ii) the fragmentation of positive ions can more readily be explained than that of negative ions. High-energy CID spectra of protonated flavonoid glycosides were obtained by performing collisions with helium in the first field-free region at a collision energy of 8 keV and 50% beam attenuation, and B/E linked scanning. For recording low-energy CID spectra, precursor ions were mass-selected with a resolution of approximately 1000 using the double-focussing electric/magnetic sector. CID was performed in the quadrupole collision cell with argon at a pressure of approximately 10^{-5} mbar (measured in the quadrupole housing) and using a collision energy of 50 eV. The product ion spectra were obtained employing unit resolution of the quadrupole analyser.

Glycerol was employed as liquid matrix for the analysis of flavonoid glycosides, while *m*-nitrobenzylalcohol was used for the analysis of methylated flavonoid glycosides.

3. Types of flavonoid glycosides studied

In this review we will focus on flavonoid glycosides having a flavone or flavonol aglycone part because compounds containing these aglycones have most regularly been received in our laboratory. Structures of common flavones and flavonols are given in Figure 1. Flavonoid glycosides containing a flavonol part occur as *O*-glycosides, whereas flavone glycosides can either correspond to *O*- or *C*-glycosides or be of a mixed *O,C*-diglycoside nature having a *C*-linked disaccharidic unit. The most common glycosylation positions are indicated in Figure 1 with an arrow.

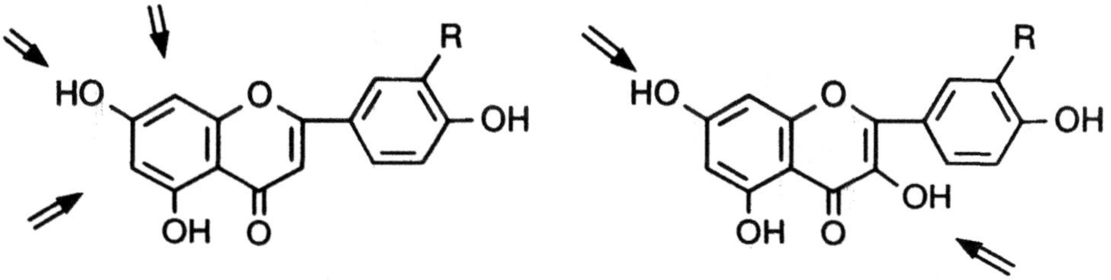

flavone
apigenin : R = H
luteolin : R = OH

flavonol
kaempferol : R = H
quercitine : R = OH

Figure 1. Structures of flavonoid glycosides

4. Differentiation between *O*-glycosides, *C*-glycosides and *O,C*-diglycosides

A differentiation between *O*-, *C*- and *O,C*-diglycosides can be made by examining first-order FAB spectra or low-energy CID spectra [11]. The protonated *O*-diglycosides give rise to Y_1^+ and Y_0^+ ions, which are formed by rearrangement reactions at the interglycosidic bonds. The carbohydrate ion nomenclature, which was introduced by Domon and Costello [12], is given in Scheme 1. Deuterium labelling experiments indicated that hydroxyl hydrogen atoms are involved in the formation of Y_n^+ ions. In the case of *O,C*-diglycosides only Y_1^+ ions, formed by fragmentation in the *O*-linked sugar residue, are detected, whereas in the case of *C*-glycosides only $[M + H]^+$ ions are observed.

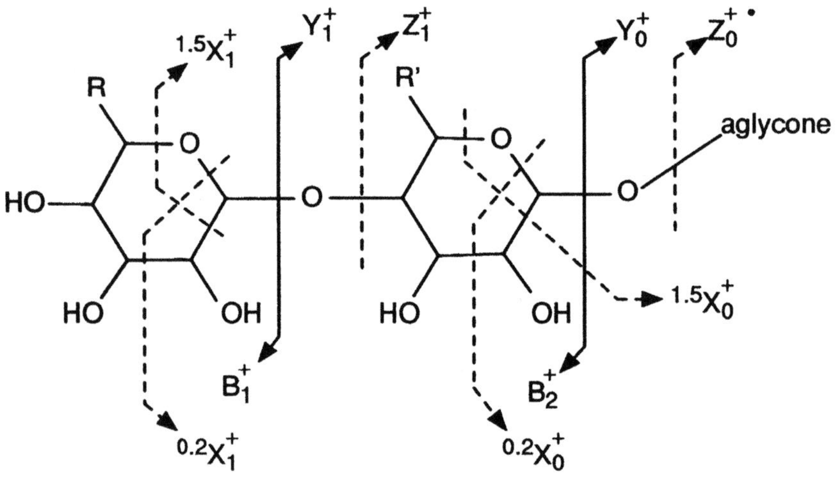

Scheme 1. Carbohydrate ion nomenclature according to Domon and Costello [14]

5. Characterisation of flavonoid *O*-glycosides

A selection of mass spectral data obtained on samples of:

(i) kaempferol 3-*O*-rutinoside, isolated from the leaves of *Morinda morindoides* [15]; and,

(ii) apigenin 7-*O*-neohesperidoside, isolated from the leaves of *Acacia siberiana* (Niyonzima G., unpublished results);

will be discussed in order to demonstrate how mass spectrometric methods can be used to obtain information on:

(i) the carbohydrate sequence;
(ii) the aglycone part; and,
(iii) the glycosylation position.

Figure 2 illustrates the low-energy CID spectrum obtained on $[M + H]^+$ ions (m/z 595) of the sample containing kaempferol 3-*O*-rutinoside. The Y_n^+ ions at m/z 449 (Y_1^+) and m/z 287 (Y_0^+) are consistent with the rutinoside sequence (6-*O*-rhamnosyl-glucosyl); however, the formation of Y_1^+ ions at m/z 433 indicates that the sample also contains an isomer with a terminal hexose residue.

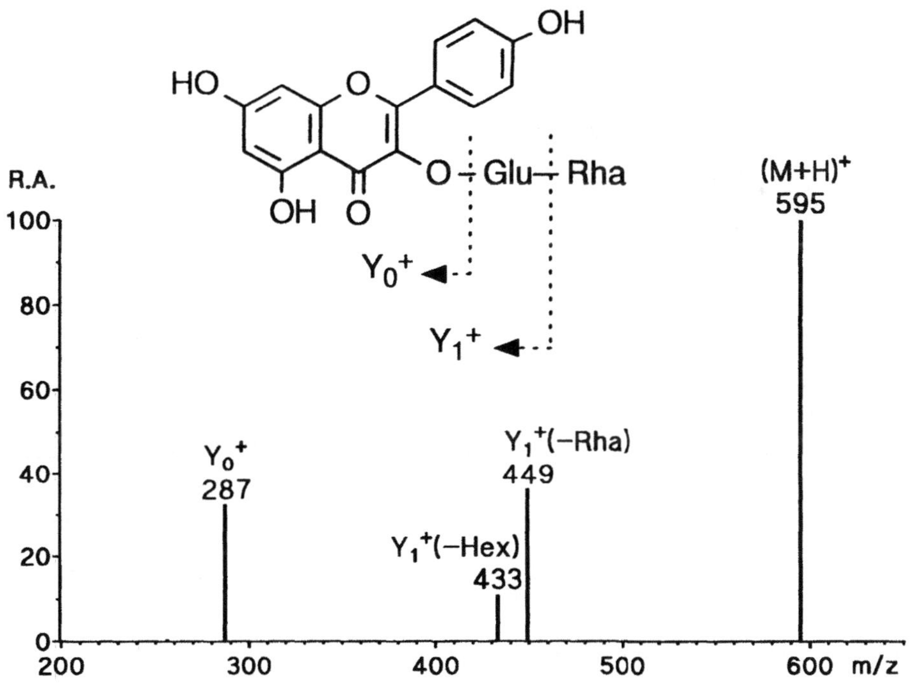

Figure 2. Low-energy CID spectrum obtained on $[M + H]^+$ ions (m/z 595) of a sample containing kaempferol 3-*O*-rutinoside

Figure 3A illustrates the low-energy CID spectrum obtained on the Y_0^+ ions (*m/z* 271) of the sample characterised as apigenin 7-*O*-neohesperidoside. The abundant product ions at *m/z* 119 ($B_{1,3}$) and *m/z* 153 ($A_{1,3}$) are rationalised by Retro Diels Alder fragmentations in the C ring and are typical of the apigenin aglycone part (Scheme 2). With respect to the ion nomenclature used to denote fragment ions typical of the aglycone part, a more systematic ion nomenclature is proposed than that followed in former EI studies (for reviews, see references 16 and 17); in this ion nomenclature, the capital letters refer to the rings which are retained, whereas the bonds broken in the C ring are indicated with subscripts. In a similar way, low-energy CID spectra are also useful for characterising the aglycone part of flavonol glycosides; protonated quercitine and kaempferol which are formed during FAB of corresponding glycosides yield diagnostic product ions of the $A_{1,3}$, $A_{2,6}$ and $B_{2,6}$ types.

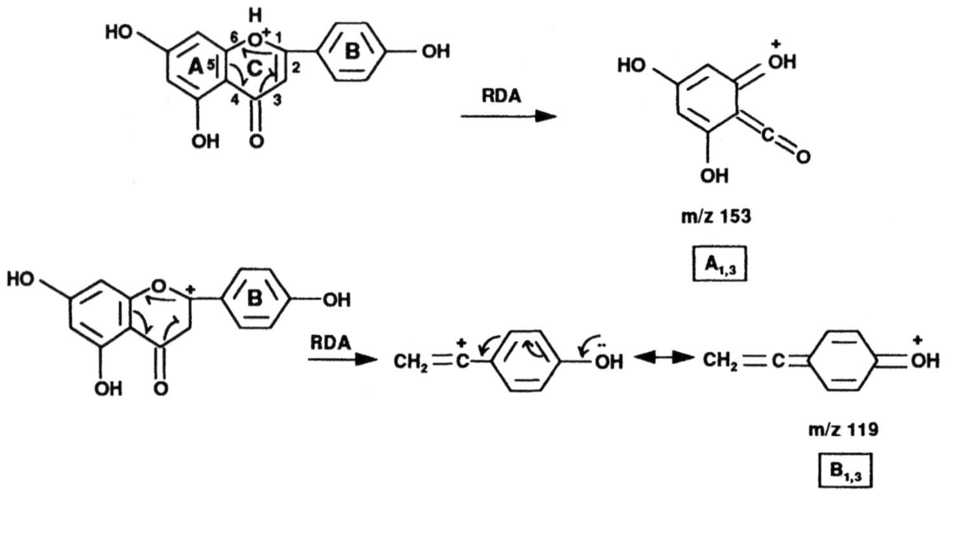

Scheme 2. Fragmentation of protonated apigenin

With the aim of obtaining information on the glycosylation position, we have worked out a simple method based on methylation using diazomethane and re-examination of the Y_0^+ ions formed by fragmentation of the protonated flavone *O*-glycoside methyl ether derivative by low-energy CID. Figure 3B shows the low-energy CID spectrum obtained on the Y_0^+ ions (*m/z* 299) of the

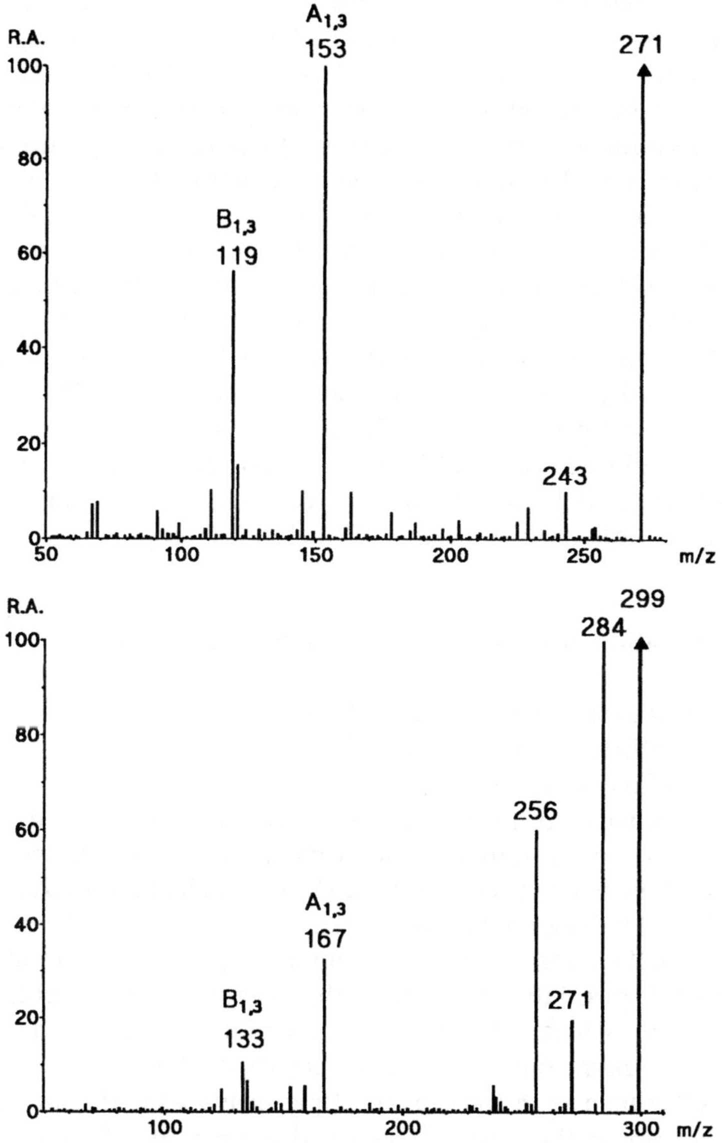

Figure 3. Low-energy CID spectra obtained on Y_0^+ ions of a sample identified as apigenin 7-*O*-neohesperidoside: (top) without (*m*/*z* 271) and (bottom) with (*m*/*z* 299) methylation

same sample of apigenin 7-*O*-neohesperidoside following methylation with diazomethane. The $B_{1,3}$ ions (*m/z* 133) have shifted by 14 u, indicating that the 4'-hydroxyl group of the B ring was free. The $A_{1,3}$ ions (*m/z* 167) also show a mass increase of 14 u which is consistent with glycosylation in the A ring. Taking into account the lower reactivity of the 5-hydroxyl group towards glycosylation, it is logical to propose glycosylation at the 7-*O* position. It is also pointed out that the Y_0^+ ions fragment by loss of a methyl radical; this rather uncommon transition from even-electron to odd-electron ions was found to be characteristic of phenolic methyl ether groups. The structure elucidation of this sample, identified as apigenin 7-*O*-neohesperidoside, was completed using NMR methods; more specifically, NMR methods allowed to characterise: (i) the inner sugar residue as glucose; (ii) the anomeric interglycosidic linkages as α and β; and, (iii) the attachment between rhamnose and glucose as 1→2. It may be feasible, however, to obtain information on the carbohydrate linkages by derivatisation into methyl ethers and resorting to high- or low-energy CID methods [16,17].

6. Characterisation of flavone mono- and di-*C*-glycosides

A selection of mass spectral data obtained on samples of:

(i) orientin and iso-orientin; and,

(ii) schaftoside and isoschaftoside,

which were isolated from extracts of *Passiflora incarnata*, used as a phytomedicine for minor tranquilising purposes, will be discussed in order to demonstrate how mass spectrometric methods can be used to differentiate between 6-C- and 8-C-monoglycosides, and isomeric 6,8-*C*-diglycosides. High-energy CID was used to fragment protonated mono- and di-C-glycosides. Figure 4 illustrates the high-energy CID spectra obtained for the $[M + H]^+$ ions (*m/z* 449) of orientin, a mono-8-*C*-glucoside, and its 6-C-isomer, iso-orientin. Ions are present at *m/z* 329 ($^{0,2}X^+$), *m/z* 315 ($^{1,5}X^+$) and *m/z* 299 ($^{0,1}X^+$), which are formed by multiple cleavages in the carbohydrate unit and characterise the C-linked carbohydrate part as a hexosyl residue. Detailed comparison of the product ion spectra reveals that the 6-C and 8-C glucosides can be distinguished on the basis of abundance differences observed for the $[M + H - (3 \times H_2O)]^+$ ions (*m/z* 395) and $^{0,4}X_{-2w}^+$ ions (*m/z* 353).

Schaftoside and isoschaftoside are a pair of isomeric di-6,8-*C*-glycosides, which have arabinose and glucose units C-linked to apigenin. Characterisation of the positions of the arabinose and glucose substituents in these isomers is particularly significant because these structural features are difficult to infer by NMR techniques. Figure 5 illustrates the high-energy CID spectra obtained for the $[M + H]^+$ ions (*m/z* 565) of schaftoside and isoschaftoside. Differentiation between the two isomers can be made based on the more pronounced fragmentation occurring in the 6-C-linked carbohydrate part. A series of product ions is formed which originate from fragmentation in the 6-C carbohydrate substituent (${}^{0,4}X_{6\text{-}2w}{}^+$, ${}^{0,2}X_6{}^+$, ${}^{1,5}X_6{}^+$ and ${}^{0,1}X_6{}^+$); the 8-C carbohydrate unit was found to lead to another group of diagnostic product ions (${}^{0,2}X_{8\text{-}w}{}^+$, ${}^{0,2}X_{82w}{}^+$ and ${}^{0,2}X_{8\text{-}3w}{}^+$)(Scheme 3).

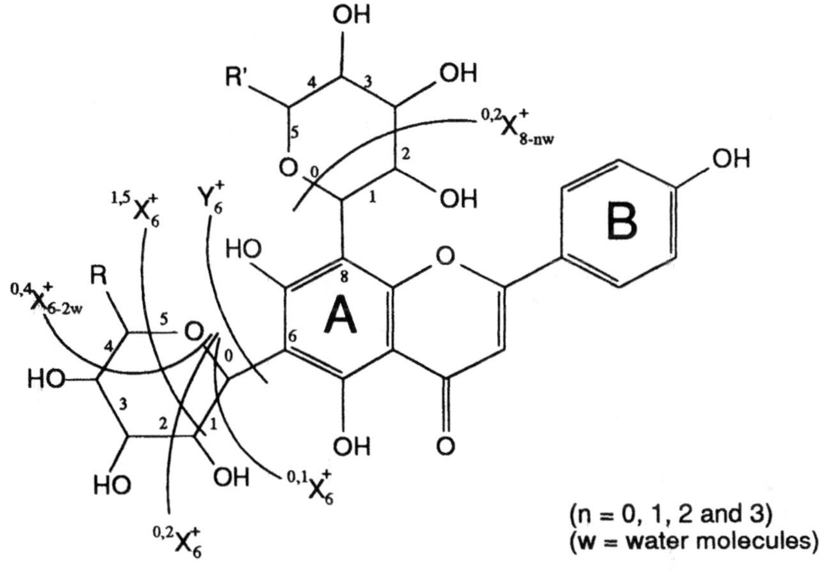

Scheme 3. High-energy CID of protonated schaftoside and isoschaftoside. Schaftoside: R = CH_2OH, R' = H; Isoschaftoside: R = H, R' = CH_2OH

7. Characterisation of flavone *O,C*-diglycosides

The mass spectrometric characterisation of a flavone glycoside of mixed *O,C*-diglycoside nature, having a disaccharidic unit C-linked to a flavone, will be illustrated with spectra obtained on vitexin 2"-*O*-rhamnoside (Roth). The first-order FAB spectrum (not shown) showed $[M + H]^+$ ions (*m/z* 579) and Y_1^+

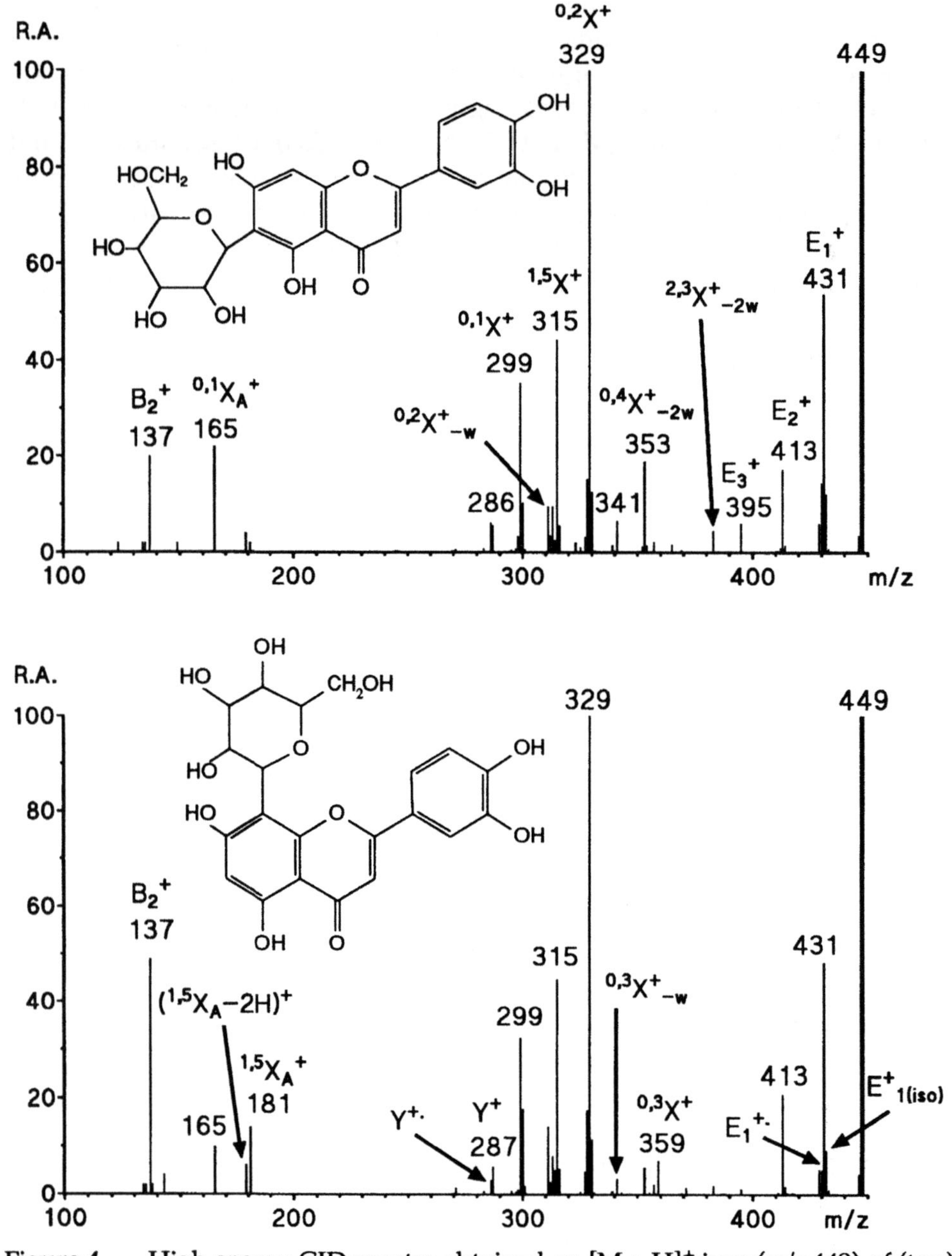

Figure 4. High-energy CID spectra obtained on $[M + H]^+$ ions (m/z 449) of (top) iso-orientin and (bottom) orientin

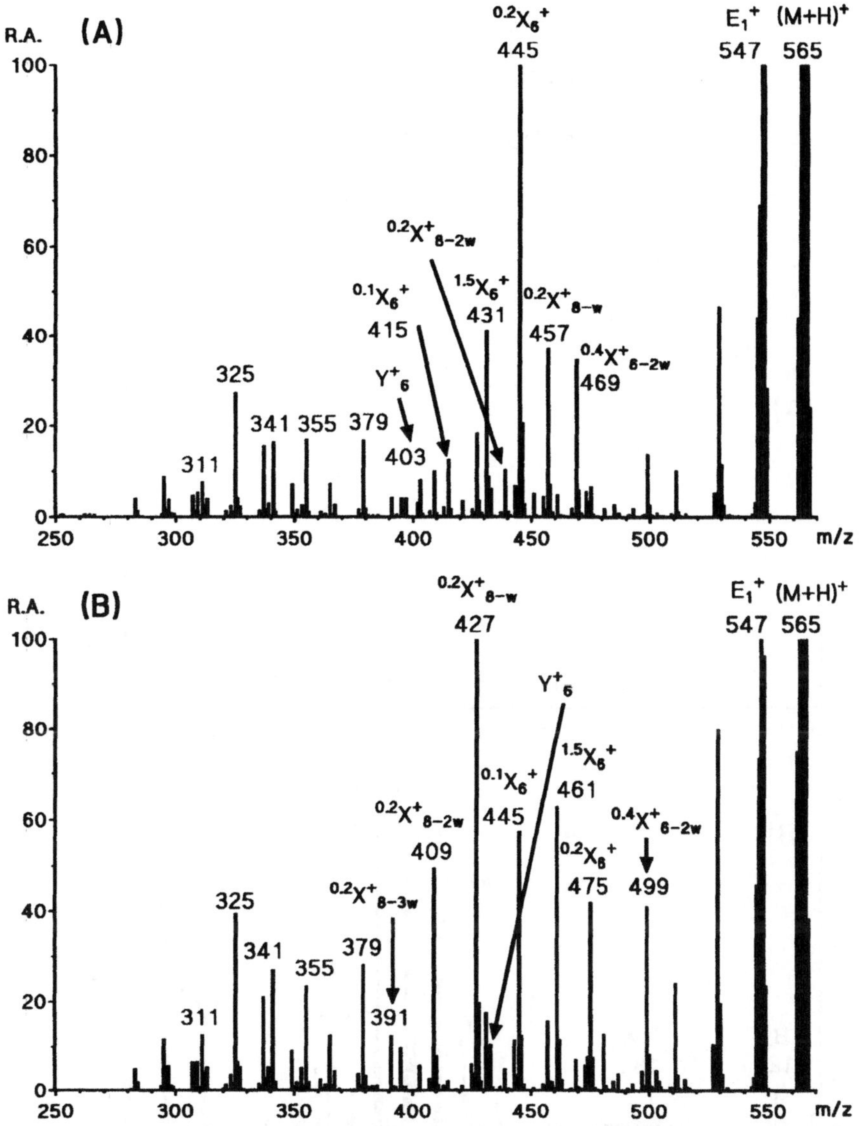

Figure 5. High-energy CID spectra obtained on $[M + H]^+$ ions (m/z 565) of (A) schaftoside and (B) isoschaftoside; for structures, see Scheme 3

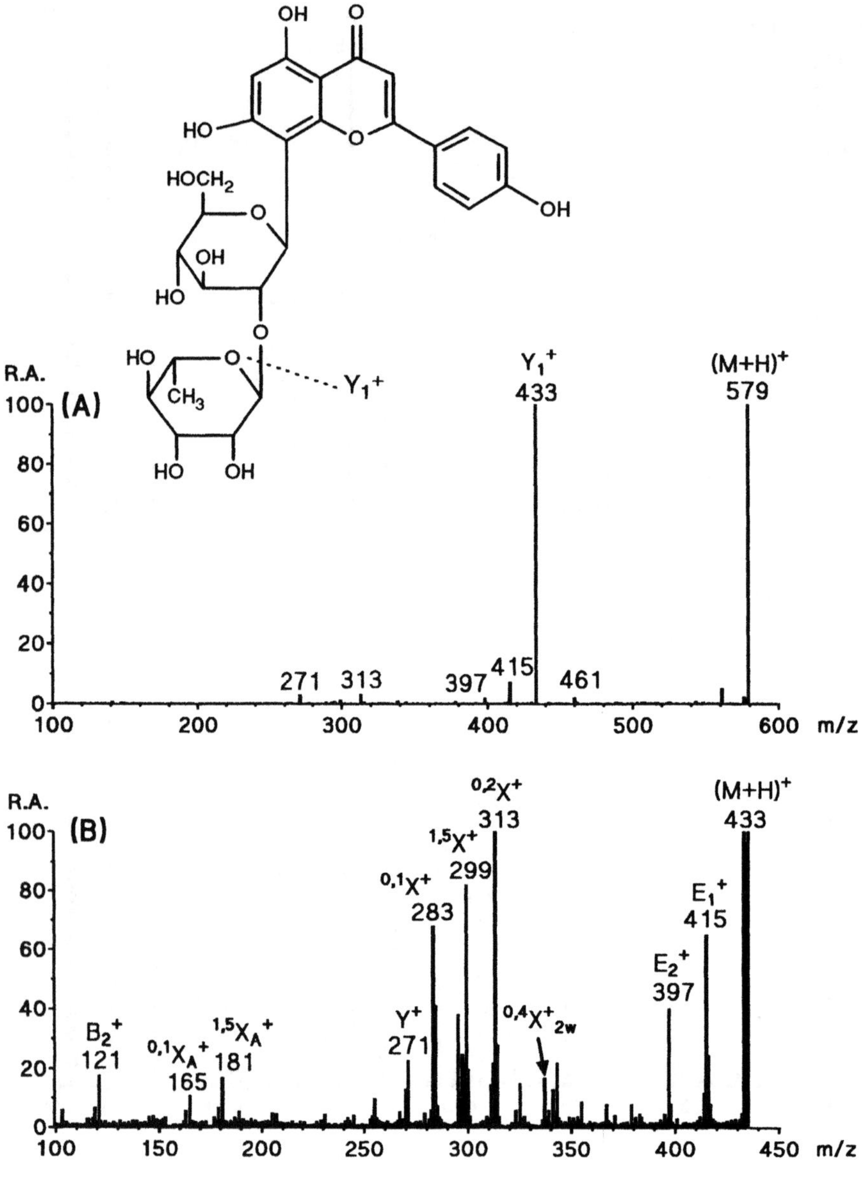

Figure 6. (A) Low-energy CID spectrum obtained on $[M + H]^+$ ions (*m/z* 579) of vitexin 2"-*O*-rhamnoside; (B) High-energy CID spectrum obtained on FAB-generated Y_1^+ ions (*m/z* 433) of vitexin 2"-*O*-rhamnoside

ions (m/z 433), which are formed by fragmentation in the terminal *O*-linked rhamnose residue. Figure 6A illustrates the low-energy CID spectrum obtained for the $[M + H]^+$ ions (m/z 579), which essentially yields the same information as that provided by the first-order FAB spectrum. In order to obtain support for the type and the site of the *C*-linked glycosyl part, a high-energy product ion spectrum was obtained on the Y_1^+ ions (m/z 433) (Fig. 6B). As expected, the mass spectral fragmentation was in agreement with that found for vitexin, which contains a 8-*C*-linked glucosyl unit. Losses of neutral fragments corresponding to 120, 134 and 150 u are consistent with the presence of a *C*-linked hexosyl unit, whereas the low abundance of the $[M + H - (3 \times H_2O)]^+$ ions (m/z 379) and ${}^{0,4}X_{-2w}^+$ ions (m/z 347) points to an 8-*C* linkage.

8. Conclusions

In this chapter, we have focused on methods for the structure determination of flavonoid glycosides, containing aglycones of the flavone and flavonol type. The combination of fast atom bombardment, high- and low-energy collision-induced dissociation and tandem mass spectrometry form the basis of a powerful instrumental method for these plant glycoconjugates. The methodology has been applied to real world samples isolated from medicinal plants. Future prospects should include applications and development of mass spectrometric methods for linkage determination in the carbohydrate part and characterisation of sugar residues.

It should be pointed out that the mass spectrometric methods summarised in this review alone are not sufficient to elucidate the complete chemical structures of unknown flavonoid glycosides; more specifically, NMR spectroscopic and chemical methods are required to determine specific carbohydrate residues, and substitution positions on both carbohydrates and aglycone. However, the integration of modern tandem mass spectrometric techniques with advanced NMR techniques is of significant power and can greatly facilitate the structure elucidation of plant glycoconjugates such as flavonoid glycosides, phenylethanoid glycosides and saponins. Structure elucidation by NMR techniques requires sample amounts which are at least 100 fold higher than for tandem mass spectrometry; thus, in cases where the sample amount is limited tandem mass spectrometry, especially in combination with sensitive photodiode array detection, is indicated to obtain partial chemical information [12,20].

9. Acknowledgements

We are grateful to the Belgian National Fund for Scientific Research (NFRS) for continuous financial support. M. Claeys is indebted to the NFRS as a research director.

10. References

1) Barber, M., Bordoli, R. S., Sedgwick, R. D. and Tyler, A. N. (1981) *J. Chem. Soc. Chem. Commun.* 325-327.
2) de Koster, C. G., Heerma, W., Dijkstra, G. and Niemann, G. J. (1985) *Biomed. Mass Spectrom.* **12**, 596-601.
3) Crow, F. W., Tomer, K. B., Looker, J. H. and Gross, M. L. (1986) *Anal. Biochem.* **155**, 286-307.
4) Becchi, M. and Fraisse, D. (1989) *Biomed. Environm. Mass Spectrom.* **18**, 122-130.
5) Facino, R. M., Carini, M., Traldi, P., Belli, B., Gioia, B. and Arlandini, E. (1987) *Biomed. Environm. Mass Spectrom.* **14**, 187-194.
6) Chen, Y., Chen, N., Li, H., Zhao, F. and Chen, N. (1987) *Biomed. Environm. Mass Spectrom.* **14**, 9-15.
7) Tomer, K. B. and Gross, M. L. (1988) *Biomed. Environm. Mass Spectrom.* **15**, 89-98.
8) Li, Q. M., Van den Heuvel, H., Delorenzo, O., Corthout, J., Pieters, L. A. C., Vlietinck, A. J. and Claeys, M. (1991) *J. Chromatogr. B* **562**, 435-446.
9) Li, Q. M., Van den Heuvel, H., Dillen, L. and Claeys, M. (1992) *Biol. Mass Spectrom.* **21**, 213-221.
10) Li, Q. M., Dillen, L. and Claeys, M. (1992) *Biol. Mass Spectrom.* **21**, 408-410.
11) Li, Q. M. and Claeys, M. (1994) *Biol. Mass Spectrom.* **23**, 406-416.
12) Chen, S., Derrick, P.J., Mellon, F.A. and Price, K.R. (1994) *Anal. Biochem.* **218**, 157-169.
13) Claeys, M., Van den Heuvel, H., Chen, S., Derrick, P. J., Mellon, F. A. and Price, K. R. (1995) *J. Am. Soc. Mass Spectrom.* in press.
14) Domon, B. and Costello, C. (1988) *Glycoconjugate J.* **5**, 397-409.
15) Cimanga, K., De Bruyne, T., Lasure, A., Li, Q. M., Pieters, L., Claeys, M., Vanden Berghe, D., Kambu, K., Tona, L. and Vlietinck, A. (1995) *Phytochemistry*, **38**, 1301-1303.

16) Mabry, T. J. and Markham, K. R. (1975), *The Flavonoids*, in J. B. Harborne, T. J. Mabry and H. Mabry (eds.), Chapman and Hall, London, pp. 78-126.
17) Mabry, T. J. and Ulubelen, A. (1980), *Biochemical Applications of Mass Spectrometry*, in G. R. Waller and O. C. Dermer (eds.), Wiley, New York, pp. 1131-1158.
18) Lemoine, J., Fournet, B., Despeyroux, D., Jennings, K. R., Rosenberg, R. and de Hoffmann, E. (1992) *J. Am. Soc. Mass Spectrom.* **4**, 197-203.
19) Reinhold, V. N. , Reinhold, B. B. and Costello, C. E. (1995) *Anal. Chem.* **67**, 1772-1784.
20) Evans, S., Buchanan, R., Hoffman, A., Mellon, F.A., Price, K.R., Walls, F. C., Hall, S., Burlingame, A. L., Chen, S. and Derrick, P. J. (1993) *Org. Mass Spectrom.* **28**, 289-290.

CARBOHYDRATE MASS SPECTROMETRY

A Characterization of Sequence, Linkage and Branching Using ES-MS and CID

V. N. Reinhold and B. B. Reinhold
Mass Spectrometry Resource
Boston University Medical School
80 East Concord Street
Boston, MA 02118

1. Introduction

As biological researchers try to bring chemical reasoning to physiological function, carbohydrates and their conjugates are often found as crucial participants (1, 2). The evidence for this involvement has developed through the use of carbohydrase inhibitors (blocking glycosylation), deglycosylating enzymes (hydrolyze specific carbohydrate residues), protein site-directed mutagenicity, (avoiding glycosylation by altering a required protein consensus sequence), through the application of an improving set of physical techniques for characterization, including nuclear magnetic resonance, crystallography, and mass spectrometry. From structure-function studies no single, universal role for protein glycosylation has emerged. However, a common theme centers on the modulation of protein structure, in some cases to direct folding, hence intracellular trafficking, in other cases rendering the glycoprotein resistant to proteolysis. In other roles the carbohydrates initiate leukocyte adhesion in the inflammation process and serve as cell surface receptors for bacterial invasion. Interestingly, changes in glycosylation patterns are indicators of untoward function and disease states, including alcoholism, rheumatoid arthritis and cancer. Protein glycosylation, and temporal changes to these structures, both *in vivo* and *in vitro,* appear to be highly sensitive results of alterations in cellular environment. This biological feature is reflected in batch production of glycoprotein therapeutics where subtle alterations in culture conditions induce numerous changes in the glycans produced.

The study of carbohydrate-related function, frequently referred to as glycobiology (3), is a fast developing discipline which seems to have a relationship to every aspect of biology. Few disciplines, however, would benefit more from a detailed understanding of carbohydrate structure than that of immunology. Adhesion molecules, which function through carbohydrate epitopes, are seen as dynamic contributors to embryonic development, wound healing, and inflammation (4-6). Numerous parasites, through molecular mimicry, have capitalized on these intricate functional

K. R. Jennings (ed.), Fundamentals and Applications of Gas Phase Ion Chemistry, 181–200.

roles to subvert immune surveillance, invade, and thrive within cells (7). For example, the oligosaccharide lacto-*N*-fucopentaose III (LNFP-III), from the egg surface of the parasite *Schistosoma mansoni,* diverts the immune system into an ineffective antibody response by interfering with T-helper cell commitment (8). This allows for proliferation of the organism within the host cell and suggests a paradigm for other parasites (9).

From a structural standpoint oligosaccharides are ideally suited for events requiring molecular specificity. Their polyhedric character provides numerous functional groups for modification and interaction, while monomers within an oligomer can participate in multiple linkage and branching patterns, creating a vast array of structural polymers from a limited monomer set. Thus, the idea of sequencing, as understood with linear biopolymers of amino and nucleic acids, must be expanded for the variably linked and multiply branched topologies observed in carbohydrate polymers. Microheterogeneity, often a consequence of variable antennae termination, adds further complexity. This diversity of structural detail and the reproducibility of its biosynthesis in the absence of a defined template message represents one of the many open questions in glycobiology.

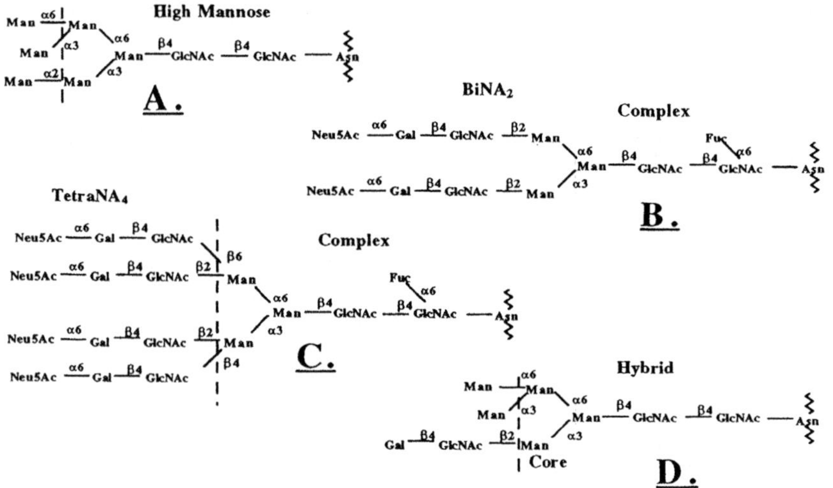

Scheme 1

Considering this potential chemical diversity, there is some solace in the recognition that evolution has created carbohydrate polymers with a comparatively small number of structural motifs, with minor (albeit important) deviations from these patterns. Thus, structural elucidation often reduces to a classification problem and a number of chromatographic and enzymatic techniques have appeared for this purpose. As an example, when proteins are glycosylated, glycans are linked to asparagine through the amide nitrogen, giving rise to the definition of N-linked glycans, or at serine and threonine residues through an oxygen, (i.e., O-linked glycans). The proximal carbohydrate moiety initiating a glycan branch is an amino sugar, (2-acetamidoglucose for the N-linked and 2-acetamidogalactose for

O-linked), although neutral sugar residues have been found (10, 11). Three basic motifs are observed for the N-linked glycans (high mannose, complex, hybrid), and each glycotype comprises numerous homologs (e.g., Man_n, $Neu5Ac_n$) and analogs (e.g., fucosylated, bisected). All N-linked glycans share the same $Man_3GlcNAc_2$ core. The heterogeneity often associated with N-linked glycans is displayed in the degree of chain extension, referred to as a glycoform distribution. With all this potential diversity, glycosylation sites show remarkable fidelity in glycotype and distribution when grown under comparable conditions. Representative glycotypes are presented in Scheme 1. The O-linked glycans lack comparable motifs, but do display heterogeneity and a greater diversity in polymerization (Scheme 2). O-linked glycans are major constituents of mucins where their physical properties dominate the attributes of the molecule as a whole.

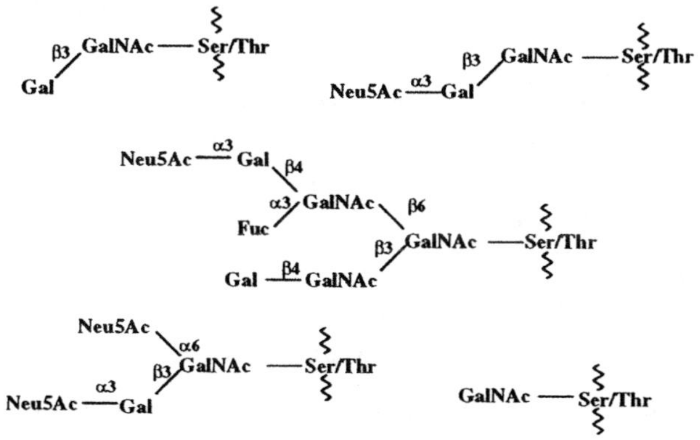

Scheme 2

Glycosylated lipids (glycolipids, glycosphingolipids) represent another major group of glycoconjugates assembled to anchor the hydrophilic carbohydrate residues to a membrane surface. These structures utilize a ceramide moiety for the lipophilic anchor, which also contributes alkane microheterogeneity (unsaturation, hydroxylation, chain length variations) (Scheme 3). Studies of microbial attachment, penetration and tissue invasion (12), cellular differentiation (13), and tumor-related or aberrant glycosylation (14), have directly implicated these structures.

A precise understanding of the functional roles played by glycosylation is frequently compromised by the need to understand its structural diversity, and as in many areas of biological science, understanding will require the pursuit of greater molecular detail. This chapter outlines one structural strategy to better understand protein glycans and glycolipids using electrospray ionization (ES-MS) and collision induced dissociation (CID). In the applications described the carbohydrates are chemically modified (methylated) prior to analysis. For N-linked glycans the carbohydrate moieties are released before methylation. These products extract well into organic solvents facilitating the removal of unwanted biological debris and salts. Organic phase extraction decreases the ionic background in the infused solutions augmenting the high sensitivity of the ES-MS of the methylated sugars. Along with this increased sensitivity, derivatization also converts very dissimilar molecules into a homogeneous family for extraction and ionization. For example, glycans consisting of acidic (Neu5Ac, uronic acid), basic (GlcNAc, GalNAc), and neutral residues (Glc, Gal, Man) extract easily and show similar ionization efficiencies, a feature fundamental to quantitative analysis by ES-MS. From these molecular weight profiles, structural motifs can be proposed and compositions projected. At a first level of structural inquiry profile alterations in conjunction with biological function have provided an effective technique to assess structure-function relationships. An understanding of linkage and branching detail, however, requires CID and tandem mass spectrometry. In this more detailed approach, spectral identification relies on piecing together the glycosidic fragments for sequence and branching information, while a series of cross-ring cleavages provide considerable linkage detail.

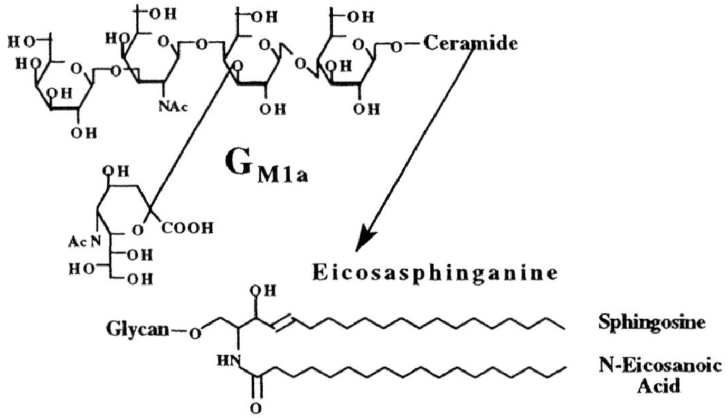

Scheme 3

No single methodology exists, nor can be foreseen, that will resolve the manifold structural components inherent in an oligosaccharide sample, however, sensitivity and working with complex mixtures are well-known strengths of mass spectrometry most important to biological studies. Degradative procedures and determining residue and linkage compositions on isolated fractions remain important elements in the solution of an unknown. Structural details such as the anomeric and monomeric configurations are difficult to obtain by MS. In some cases, these problems can be resolved by enzymatic procedures, or given adequate sample amounts, they are often resolved by NMR. Thus, a comprehensive carbohydrate characterization will still require multiple techniques providing orthogonal information. However, the structural motifs established during carbohydrate biosynthesis have eased the burden of characterization, and in practice, identification often resorts to probing a postulated structure with alternative analytical techniques and chemical derivatives. With these considerations mass spectrometry in combination with on-line purification and CID will continue to have a major role in resolving a specific carbohydrate sequence and in contributing to a better understanding of the difficult structural questions in glycobiology.

2. Methods

2.1 N-GLYCOSIDASE F DIGESTION

Glycoproteins (50 - 75 μg) was dissolved in 0.1 ml of 20 mM sodium phosphate pH 7.2 and digested with 0.025U/μg of N-Glycosidase-F (Boehringer Mannheim) for 16 hr at 37°C. Complete glycan release was aided with 0.1% SDS (Bio-Rad) and 0.7% Nonidet P-40 (Sigma).

2.2 ISOLATION OF N-, AND O-LINKED GLYCANS

Protein and detergent was removed after incubation by elution of the released glycans from a C18 SEP-PAK® cartridge. The cartridge (Waters #20515) was attached to a 5 ml syringe (Hamilton #1005), and rinsed with 7-8 ml of acetonitrile followed by 7-8 ml of 95%H_2O/5% AcCN. The rinse was assisted by the use of gentle air pressure. Digests were loaded onto the cartridge and the N-linked glycans eluted with 4 ml of 95% H_2O/5% AcCN. The protein and O-linked glycoproteins (if present) are then eluted with 70% AcCN/H_2O. The volume of the O-linked fraction was first reduced on a sand bath with a N_2 stream, then both fractions were concentrated on a Speed Vac (Savant). The N-linked sample is transferred to a 1.5 ml glass Reactivial, dried, and vacuum desiccated prior to permethylation. The O-linked glycans were released by beta-elimination as described below.

2.3 ISOLATION OF O-LINKED GLYCANS

The fraction containing the O-linked glycoprotein was dissolved in 100 μl of 2.0 M sodium borohydride in 0.1 N sodium hydroxide, and reacted for 3 days @ 37°C. The reaction is stopped with 30% Acetic Acid and taken to dryness, with several additions of methanol, on the Speed Vac. The residue was redissolved in H_2O and loaded onto a C18 SEP-PAK cartridge. The O-linked glycans were eluted with 4 ml of 95% H_2O/AcCN, and prepared for permethylation.

2.4 PERMETHYLATION OF RELEASED GLYCANS

Vacuum desiccated samples were dissolved in 100 μl of a NaOH/DMSO suspension, prepared by vortexing DMSO and powdered sodium hydroxide. After 1 hour at RT, 30 μl of methyl iodide was added and the solution set for 1 hour at RT with occasional vortexing. Glycans were partitioned by adding 100 μl of chloroform and the suspension back extracted a half dozen times with approx. 0.5 ml of H_2O and the chloroform layer taken to dryness. Permethylation was repeated. All samples were stored at -20 °C.

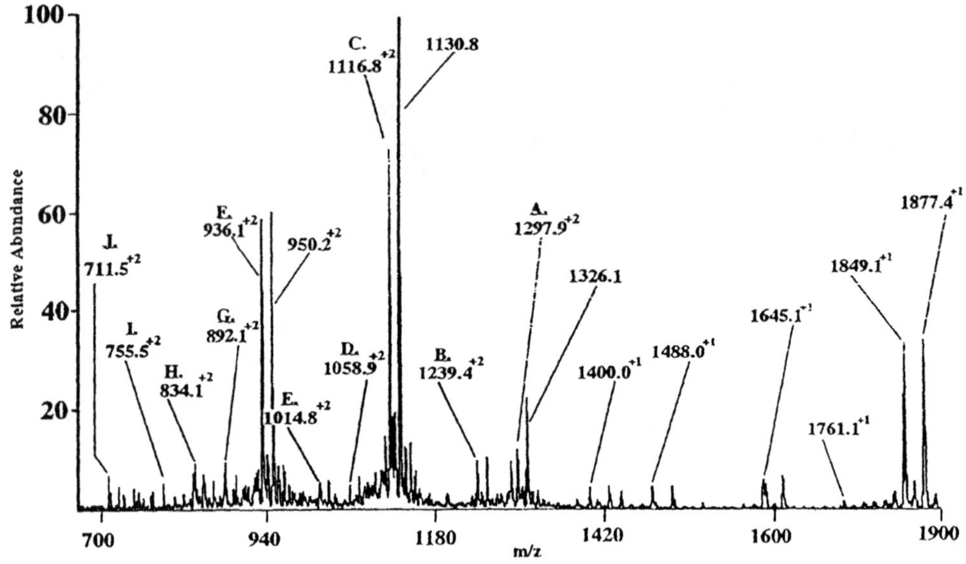

Figure 1. ES-MS profile of a bovine brain glycosphingolipid extract following methylation. Structures A-J; see Table 1.

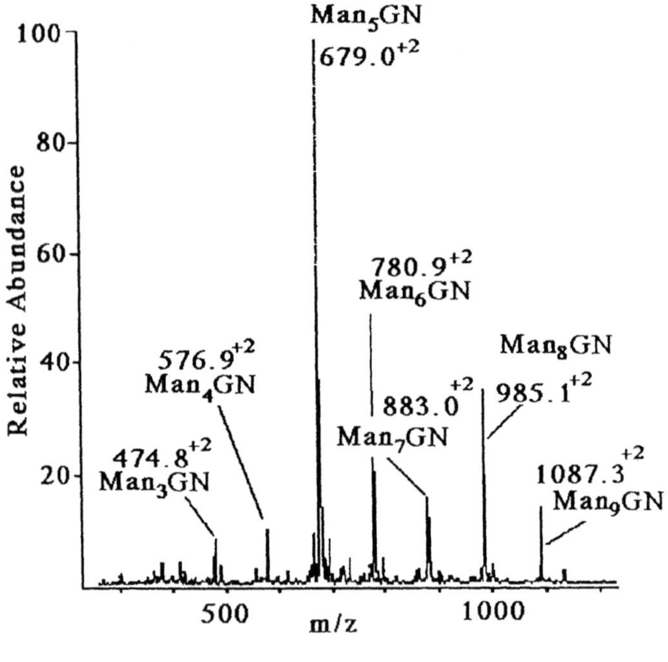

Figure 2. ES-MS profile of methylated high-mannose glycans released from ribonuclease B with Endo-H. All ions are doubly charged.

2.5 ELECTROSPRAY IONIZATION MASS SPECTROMETRY

The instrument used in this study was a Finnigan-MAT TSQ-700 (Finnigan-MAT Corp., San Jose, CA) equipped with an electrospray ion source. Methylated glycans were dissolved in water-methanol solutions (4:6 v/v) containing 0.25 mM NaOH (which provided Na^+ cations for adduction). Solutions were injected at a rate of 0.75 µL/min directly into the electrospray chamber through a stainless steel hypodermic needle. The voltage difference between the needle tip and the source electrode was -4.0 kV. A unique feature of ES is the generation of multiply charged ions from a single molecular species. Unlike FAB-MS which has difficulty in obtaining spectra of high molecular weight glycans, multiple charging brings the instrumentally measured mass (m/z) of larger glycans to an easily determined lower value. ES data was analyzed directly or deconvoluted. This was accomplished by pattern recognition using a relative, or level-2, entropy algorithm (15). Treatment of the data in this manner suppressed artifacts associated with alternative deconvolution algorithms.

3. Profiling By ES-MS.

Mass spectrometry of carbohydrate samples has been largely based on ballistic ionization strategies, e.g., fast atom bombardment (FAB), or secondary ion (SI), which exhibit poor ionization efficiencies and allow considerable ion source fragmentation. For the purposes of carbohydrate analysis, ES and MALDI are the most effective and generally applicable techniques currently available for transforming these molecules into gas-phase ions. Electrospray makes use of high electric fields to aerosolize solutions, creating a fine mist in which each droplet carries an excess of charge on its surface. A bath gas at ambient pressures evaporates neutral solvent from the droplet surface, reducing its size and hence increasing its surface charge density. Eventually electrostatic repulsion disrupts the liquid surface, leading to the ejection of smaller charged droplets and ultimately, gas-phase ions. Ions are produced at high efficiency and with minimal internal energy to provide the "softest" ion desorption technique currently available (16).

3.1 GLYCOSPHINGOLIPIDS (GSLs)

The advantages of derivatization and ES profiling can be demonstrated with a glycolipid sample. Methylation of a ganglioside extract from brain tissue, followed by extraction and direct ES-MS, provided a molecular weight profile of several GSL components, (Fig. 1). The low background and high ES ionization efficiency provides a broad dynamic range for signal amplification. Scale expansion clearly identifies the ions G_{D3}, G_{M2}, G_{A1}, and G_{M3}, (G. - J.,, Table 1). The lower mass range of the spectrum, (m/z

700 - 1400), identifies a series of ten paired ions in the two charge state, five components of which are observed in the single charge state above m/z 1400, (Table 1). The ion pairs are a consequence of heterogeneity in the ceramide moiety (octadecasphinganine, C_{18} and eicosasphinganine, C_{20}, see Scheme 3) that is known to occur in these samples (17). This natural heterogeneity (28 u), and the presence of alternative charge states brings an added measure of confidence in characterization. The charge state z, and detected mass to charge ratio, provide a direct indication of the molecular weight M_r through the relationship, $M_r = z[m/z - 23]$, where 23 is the mass of the adducting sodium cation. Of the ten materials identified in this extract, eight comprise components of the "o", "a" and "b" GSL biosynthetic pathway (18) representing a glycoform distribution analogous to that obtained from glycoprotein glycans.

TABLE 1

Structure[a] (C-18 homolog)		1+ Charge state	2+ Charge state	3+ Charge state
A.	$G_{T1a/b/c}$		1297.9	872.9
B.	$G_{D1a/b}$+HexN		1239.8	834.2
C.	$G_{D1a/b}$		1116.8	752.2
D.	$G_{M1a/b}$+HexN		1058.9	713.7
E.	G_{D2}		1014.8	
F.	$G_{M1a/b}$	1849.1	936.1	
G.	G_{D3}	1761.1	892.1	
H.	G_{M2}	1645.1	834.1	
I.	G_{A1}	1488.0	755.5	
J.	G_{M3}	1400.0	711.5	

[a]All major ions exhibit higher (28 u) ceramide homolog (17). The nomenclature is based on the initial IUB-IUPAC CBN recommendations and recent update: Nomenclature of Glycoproteins, Glycopeptides and Peptidoglycans. *J. Biol. Chem.*, (1987), **262**, 13-18.

3.2 GLYCOPROTEIN GLYCANS

In like manner, a protein's glycan distribution may be profiled following deglycosylation and methylation. As with the glycolipids, these lipophilic products protonate poorly compared to peptides, but large improvements in sensitivity are observed when alkali metal ions are added to the infusion solvent, e.g., 0.25 mM NaOH (19). This analysis of methylated glycans brings together the important features of easy sample cleanup and a single adduct type (Na^+) appropriate for quantitative profiling. Alternative derivatives, e.g., CD_3-, OAc-, Et-, or analysis following periodate chemistry (19) provides a check on composition while ORM contributes to linkage and branching information. Quantitative permethylation of these samples is easy to achieve, but, the basic conditions do cause elimination of O-acyl groups if they are present. Neutral methylation offers an alternative strategy, and this divergent approach has several advantages (11, 20). The chemistry, however, is more difficult and the method does not work when the oligosaccharides contain amino sugars.

3.3 N-LINKED HIGH MANNOSE GLYCANS.

Deglycosylation techniques are well established either by hydrazinolysis or enzymatic release. Hydrazinolysis causes small losses in neuraminic acid, and when *N*-glycolyl groups are present, the required re-N-acetylation converts these residues on neuraminic acid to an *N*-acetyl group, hiding this modification. Moreover, when followed by methylation and electrospray, the overall sensitivity remains significantly diminished. Deglycosylating enzymes, (amidases and endoglycosidases), may suffer from glycan inaccessibility, and if these problems persist, quantitative glycan recovery can be obtained by some combination of proteolysis, denaturization, or reduction and alkylation. As an example, the N-linked glycans from ribonuclease B were released enzymatically, (endo H, splitting between the two GlcNAc of high mannose residues), and their ES-MS profile is presented in Figure 2. The spectrum shows the ions *m/z* $(474.8)^{+2}$, $(576.9)^{+2}$, $(679.0)^{+2}$, $(780.9)^{+2}$, $(883.0)^{+2}$, $(985.1)^{+2}$, and $(1087.3)^{+2}$, with mass intervals of 204/2, establishing a doubly charged glycoform distribution fitting the composition $(Hex)_n$GlcNAc, $n = 3 - 9$. Coupled with enzyme specificity, the composition masses and ion intervals, (including the correct number of methyl groups incorporated following methylation), this spectrum strongly suggests a high mannose motif (A., Scheme 1).

3.4 N-LINKED COMPLEX GLYCANS.

Complex glycans also exhibit a structural motif with a linkage and monomer branching pattern analogous to structure C., (Scheme 1). ES-MS profiles of these materials provides an assessment of antenna extension and branching, (e.g., neuraminic acid capping, fucosylation), heterogeneity frequently encountered with this glycotype. The complex glycans profiled from leukocyte function-associated antigen-3 (LFA-3) provided three homologous ion series of bi-, tri-, and tetra-branched structures with variations in neuraminyl capping (Fig. 3). Most abundant was the ion m/z $(1545.1)^{+3}$, accounted for as a trisodiated tetra-branched and fucosylated glycan, (C., Scheme 1). Homologous ions, deficient one, two, and three neuraminyl residues are observed at *m/z* $(1424.5)^{+3}$, $(1303.7)^{+3}$, and $(1183.1)^{+3}$. These masses and intervals (one-third that of a neuraminic acid residue) indicate a TetraNA$_n$ series in the three-charge state with fucosylation, where $n = 4, 3, 2$, and 1, respectively. A second set of ions in the two charge state defines a biantennary glycan series starting from *m/z* $(929.4)^{+2}$, and incrementing to *m/z* $(1031.4)^{+2}$,, $(1314.1)^{+2}$, and $(1495.6)^{+2}$, for the structures BiNA$_0$-2Gal, BiNA$_0$-Gal, BiNA$_1$, and BiNA$_2$ (B., Scheme 1). The third ion series can be observed for the TriNA$_n$ glycans, (Scheme 4). Again, as in the GSLs spectrum, ions were observed

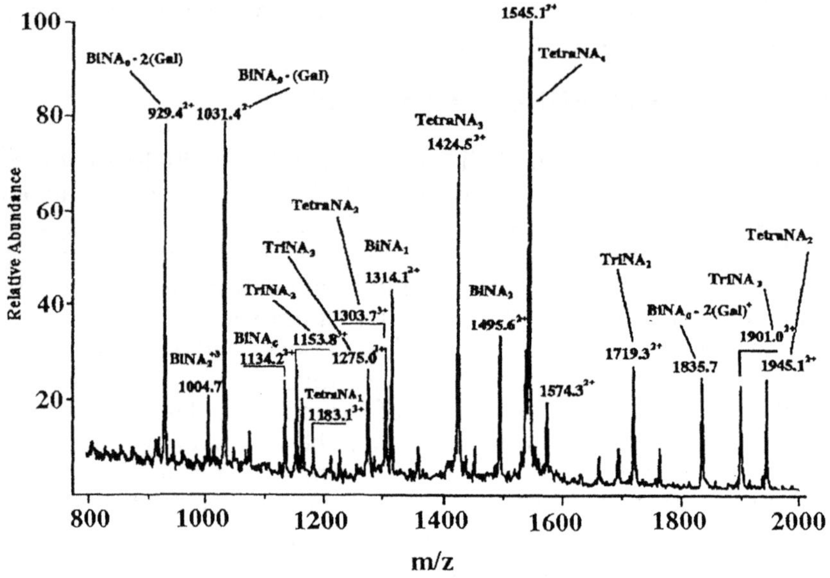

Figure 3. Spectrum of complex glycans enzymatically released from LFA-3, methylated, extracted and ES-MS.

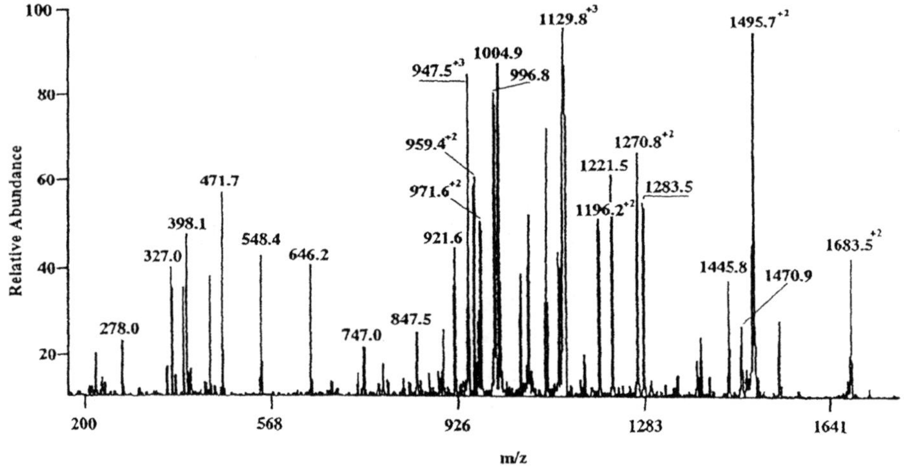

Figure 4. Collision induced dissociation of a mono-fucosylated glycosphingolipid sample extracted from HL-60 cells. Focused parent m/z 1129.8^{+3}

in different charge states which, in combination with series homology, gives added confidence to assignment. The two examples mentioned above represent a profile of all N-linked glycans. If information on site specificity is desired, the protein must be mapped, glycopeptides isolated, and each treated accordingly. These ion profiles and their relationships supply information on glycotype and glycoform distributions. Although the spectra do strongly suggest glycomer structure, there is no direct indication of linkage or branching. These matters should be pursued separately whenever new structures are isolated.

4. Glycan Sequence, Linkage and Branching

For many glycans, release, methylation, and ES-MS profiling is appropriate for characterization. However, carbohydrate samples are replete with isomeric structures and alternative techniques are needed for this identification. Consider the relatively simple high mannose glycans released from bovine ribonuclease, $(Man)_n GlcNAc$, $n = 4 - 9$ (Fig. 2). Although these mannose variations can be observed, an additional level of heterogeneity exists within each value of *n*. This diversity is usually not investigated, but has been studied in detail for ribonuclease B (21). In a similar way the structure $BiNA_1$, has an isomeric glycotype depending on the top or bottom location of the single *N*-acetyl neuraminic acid. Locating terminal residues and branches are difficult problems for all glycans with multiple antennae. Even for fully sialylated glycans multiple isomers have been detected within the same structural motif. As an example, for $TriNA_3$, one might expect four isomers depending on the position of branching at the 4 and 4' mannose residue. Only two, however, have been identified, (A. and B., Scheme 4). The absence of variations in the 2-link glycans (C. and D., Scheme 4) suggests these to be fixed in a biosynthetic precursor before a third antenna is incorporated. The two former structures can be differentiated by oxidation chemistry (see below) before profiling (19).

4.1 OXIDATION, REDUCTION, AND METHYLATION (ORM)

Periodate oxidation provides a selective strategy to probe glycan motifs suggested by ES-MS profiling. This chemistry cleaves the carbon-carbon bonds between hydroxyl groups (23). Since hydroxyl group pairing may be interrupted by linkage, any procedure that measures this change indirectly relates to linkage type. Reduction of the oxidized product increases the mass by 2 Da for each 1→2 and 1→4 linkage, and decreases the mass by 42 Da for 1→6 linked monomers. The 1→3 linked residues are not altered during ORM (Scheme 4). This chemistry does not identify linkage, but occasions an altered molecular weight that reflects linkage composition. The conditions under which ORM occurs are sufficiently mild that other structural features, (acetylation, phosphorylation, sulfation, acetamido, furan ring forms), remain intact. The important contribution of ORM chemistry lies with the fact that an altered structure is easily generated and profiled at considerable sensitivity. Within the context of established motifs, these contrasting profiles corroborate structural elements and distinguish isobaric glycans transparent in the initial profile. As an example, location of

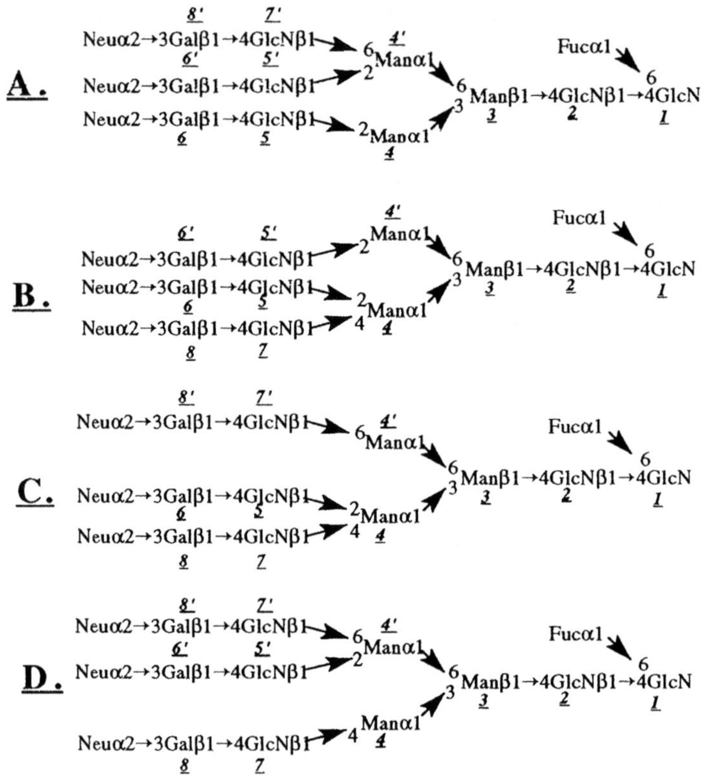

Scheme 4

antennae on the 4. or 4' core mannose in the four triantennary structures (Scheme 4), are isobaric as are the glycan pairs $BiLac_1NA_2$ and $TriNA_2$; $TriLac_1NA_2$ and $TetraNA_2$, (Lac = Gal-GlcNAc). All of these latter glycans, as well as A. and B., (Scheme 4) yield unique products following ORM chemistry. Unfortunately, glycans showing greater substitution and branching are impossible to oxidize, e.g., can not distinguish between 3-O-, and 4-O-linkage positions to amino sugars. To resolve these problem, tandem analyzers and CID have dramatically increased the structural information available by mass spectrometry. Since ionization methods and analyte derivatization greatly influence the fragmentation process, we will discuss only the recent union of electrospray ionization, methylation and low-energy CID. In this focus, prior developments using electron (24), chemical (25) and fast atom bombardment ionization in high-field instruments of methylated and derivatized oligosaccharides (26, 27) and glycolipids (17, 28, 29) have had a major influence.

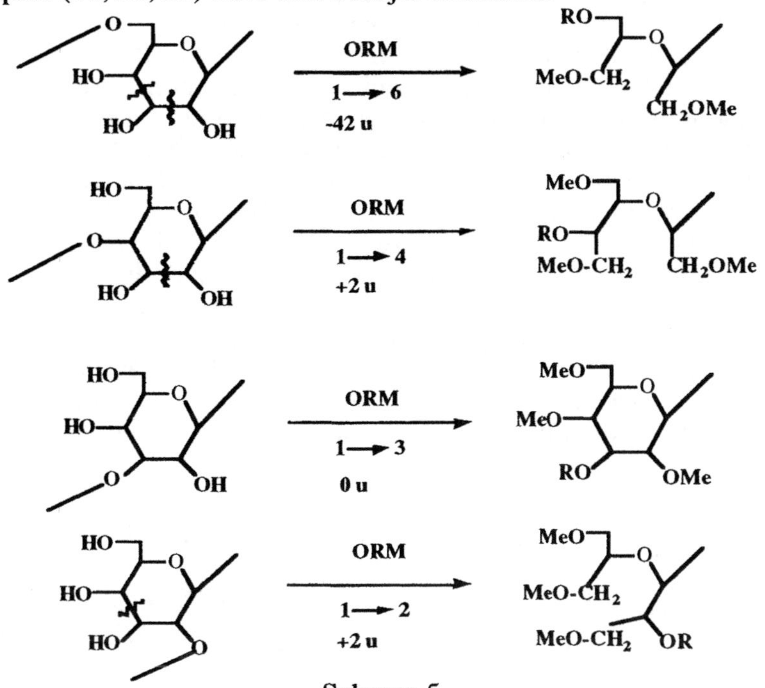

Scheme 5

4.2 COLLISION INDUCED DISSOCIATION (CID).

Monosaccharide residues are connected through glycosidic bonds easily ruptured by low-energy collision. The product ions arise from a C_1-O cleavage with elimination across the proximal C_1-C_2 bond. The leaving oxyanion picks up a proton analogous to a classical four-centered reaction. For each bond ruptured two ion series are generated, comprising the reducing and non-reducing termini (Scheme 6).

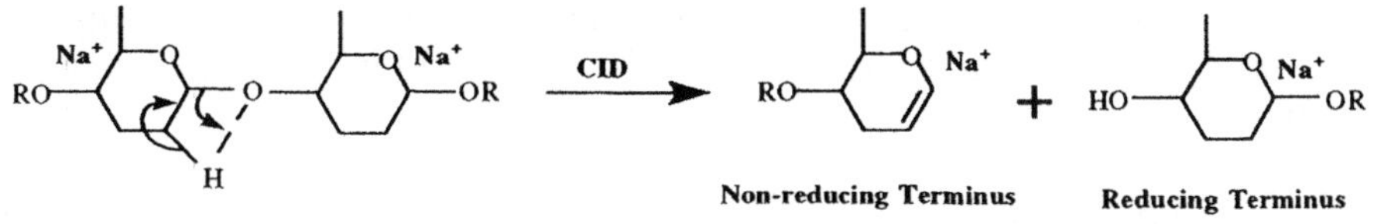

Scheme 6

4.2.1 *Branching and Sequence*

A sequence can be determined from either CID products, reducing or non-reducing. These ions, in turn, may undergo an additional glycosidic cleavage to generate a third ion set of internal fragments, and all can be useful when constructing a sequence. Although there is considerable abundance variation, the three ion series are usually observed in the spectra of most protein glycans, oligosaccharides, and their conjugates. As an example, the three ion types were observed in the CID spectra of the brain GSL (17) and comparable results were observed in fucosylated GSLs involved in cell adhesion. These latter materials, isolated from neutrophils, were shown to be the epitopes recognized by E-selectins. Profiling several active fractions suggested the GSLs were composed of a polylactosamine backbone, a single sialyl group, with one, two, and three fucosyl residues. CID studies of these fractions indicated an unbranched backbone, terminal sialylation, and internal fucosylation, (Neu5Ac-LacN-LacN-LacN-LacN-Hex-Hex-Cer, (LacN = Gal→GlcNAc)). Collision of the monofucosylated sample provided a spectrum Fig. 4) from which the sequence was assigned, (Scheme 7). This, and related structures, were confirmed from their spectral interrelationships, deuterioreduction, deuteriomethylation, and sphingonine heterogeneity, (30). Identification of neuraminyl containing ions in the CID spectra was accomplished by observing ion shifts following a repeat CD_3-methylation step to back exchange the labile neuraminyl methyl ester.

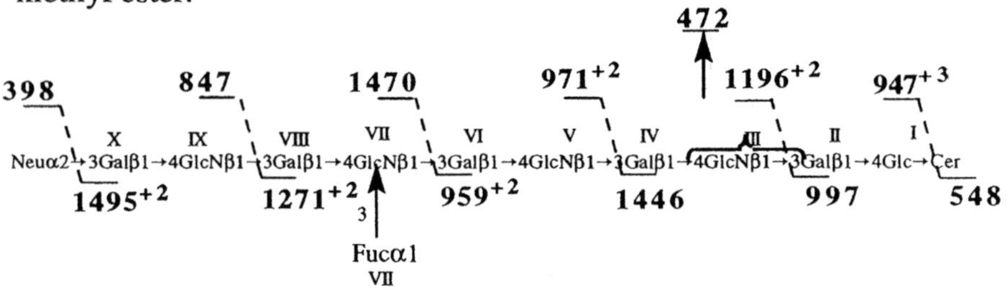

Scheme 7

Protein glycans are often branched and comprehension of this topology requires hydroxyl group blocking. This information in CID spectra resides with ions from the reducing side of a glycosidic rupture that indicate two or more hydroxyl groups exposed during the fragmentation. That is, a reducing-end ion downshifted 14 Da ($-CH_3 + H$) from the normal glycosidic fragment ion indicates the loss of an additional branch. The specific residue(s) involved in branching can be determined by comparing all reducing-end ions and observing at which point(s) in the sequence a shift is initiated.

4.2.2. *Linkage Detail.*

Fragments that define linkage are derived from a double cleavage across the pyranose ring. This rupture results in fragments of the ring attached to both ion series. Non-reducing termini, which differ with linkage position, occur as satellite peaks above their respective glycosidic ions (Scheme 8). Using electron impact ionization on methylated di- and trisaccharides, Chizhov, et.al, (31) first reported these fragments for the three linkage types (1→6,1→4, and 1→2). The observations were confirmed over the next decade by a number of workers (32-34), but all were limited to low molecular weight samples. These observations were extended to oligosaccharides by fast atom bombardment using high energy collisions (26), and more recently by ES-MS using low energy CID (11, 17, 20).

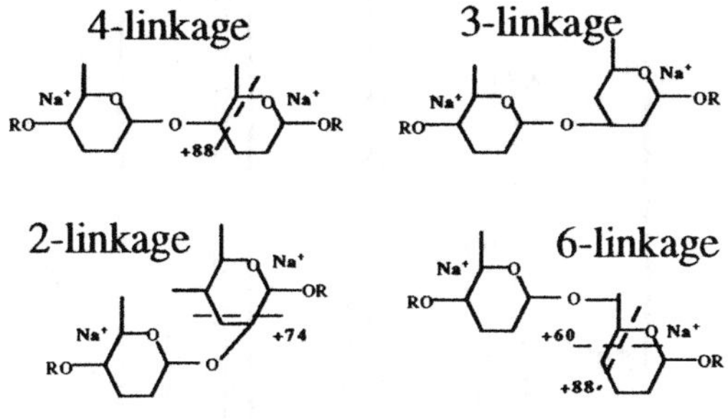

Scheme 8

As summarized in Scheme 8, pyran-ring rupture leaves fragments that vary with linkage position attached to non-reducing termini: from a 1→2 linkage a 74 Da fragment, from a 1→4 linkage, an 88 Da fragment, and from a 1→6 linkage, two fragments at 60 and 88 Da. A 1→3 linkage shows no related fragments. The cross-ring +74, +60, and +88 increments to non-reducing terminal ions are minor fragments, and the importance of the low backgrounds generated by ES is realized here. When acyl, acidic, or basic residues are present, these fragments are absent, and alternative procedures are required to assign inter residue linkages for these samples.

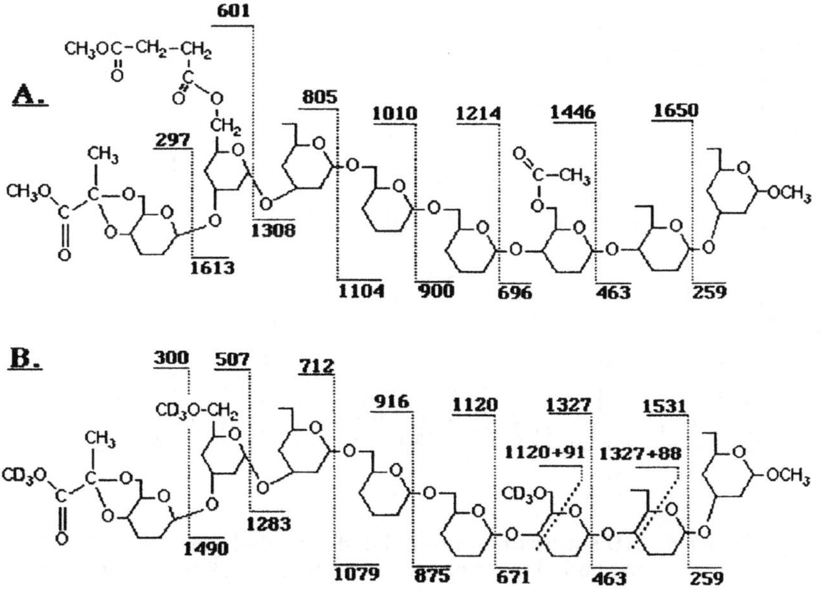

Scheme 9

Such a problem was experienced in assigning the linkage structures for a specifically acylated exopolysaccharide isolated from a bacterial cell wall (20). With this problem, methylation was carried out first under neutral conditions, (acyl groups stable), which identified the oligomer sequence and

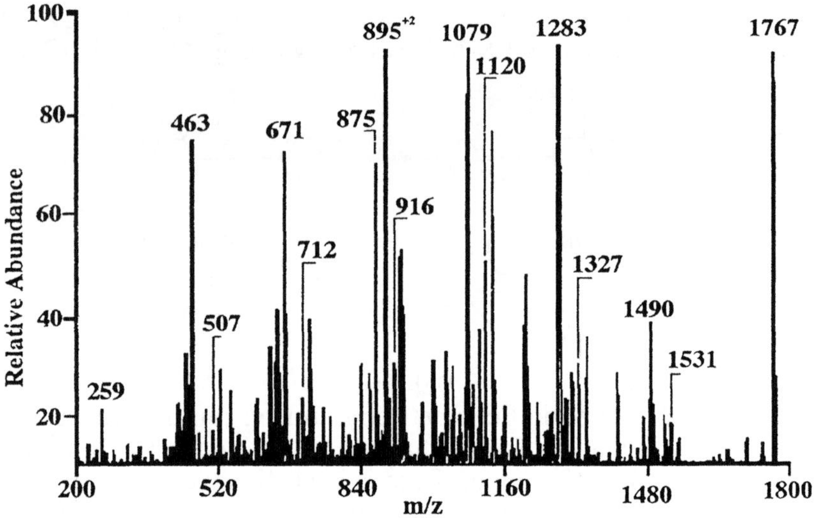

Figure 5. CID spectrum of methylated rhizobial exopolysaccharide following base remethylation with CD_3I. Focused parent m/z 895^{+2}

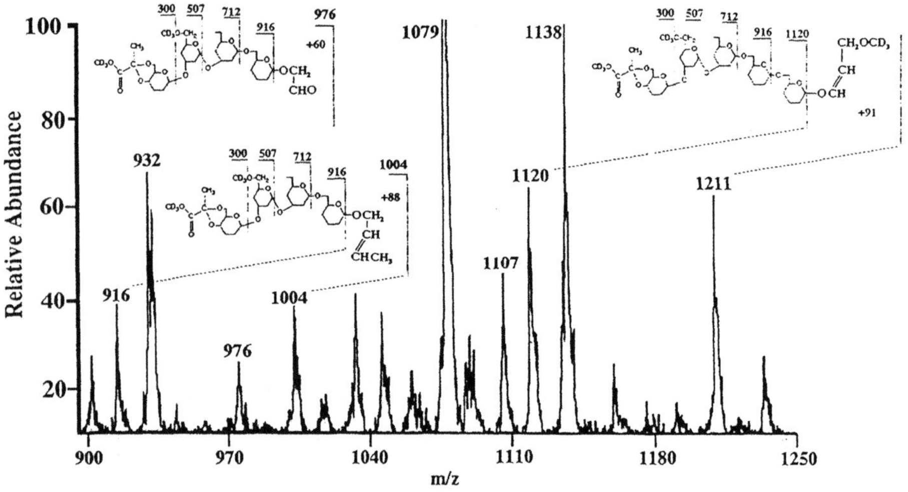

Figure 6. CID spectrum of methylated rhizobial exopolysaccharide. Expanded mass range from Fig. 5. m/z 900 to 1250 to observe 4-linked cross-ring cleavage fragment, (1120 + 91 + 1211).

localized the residues acylated (A., Scheme 9). The sample was then remethylated under basic conditions with CD_3-methyl iodide, (replacing base-labile acyl moieties), which introduced a group that both: identified the former acyl positions and provided a derivative suitable for linkage analysis by CID Fig.5, and B., Scheme 9). Two non-reducing sequence ions in this latter spectrum, (m/z 712 and 916), each showed a pair of satellite fragments indicating them to 1→6 linked. The two fragments for m/z 916 can be observed in Figure 7, m/z 976 (+60 Da) and m/z 1004 (+88 Da). Two other sequence ions (m/z 1120, and 1327) each showed a single satellite fragment (+91 and +88 Da, respectively) indicating each to be 1→4 linked (B., Scheme 9). The increment from 88 Da to 91 Da was expected for the deuterated fragment and defined the former point of acetylation at the 6-position, Fig.6).

5. Summary

Electrospray ionization and collision induced dissociation provide the carbohydrate analyst with two powerful techniques for structural elucidation. Earlier problems of sample ionization, for decades a major obstacle, have now been largely mastered. Ion profiles with the absence of fragmentation and with fidelity in glycoform distribution are available with the potential for quantification in a single analysis. These techniques are being matched with developments in MS analyzers where ion trap technology promises unprecedented gains in sensitivity and resolution. Molecular weight profiling provides a measure of composition to assess glycotype (complex, hybrid, high mannose), and their concentration (glycoform distribution), while detailed studies may proceed on any selected glycan using CID and MS/MS. As we proceed to study glycoconjugates of greater complexity, comprising epitopes of diverse function and vanishing concentration, these techniques can play a significant role.

Acknowledgment

The authors acknowledge gifts of numerous samples and support from Susan Goelz and Werner Meier of Biogen Corp. This study has been supported in part by NSF grants DIR-9002024 and MCB-9400633.

Abbreviations Used

Neu5Ac, N-acetylneuraminic acid; GlcNAc, 2-acetamido-2-deoxy-glucose; GalNAc, 2-acetamido-2-deoxy-galactose; Glc, glucose; Gal, galactose, Man, mannose; ES-MS, electrospray mass spectrometry; MALDI, matrix-assisted laser desorption; FAB, fast atom bombardment; CID, collision induced dissociation; MS/MS, tandem MS; NMR, nuclear magnetic resonance; GSL, glycosphingolipids.

6. References

1. Varki, A. (1993) Biological roles of oligosaccharides: all of the theories are correct, *Glycobiology* **3**, 97-130.
2. Dwek, R. A. (1995) Glycobiology: Towards understanding the function of sugars, *Biochem Soc Trans.* **23**, 1-25.
3. Rademacher, T. W., Parekh, R. B., and Dwek, R. A. (1988) Glycobiology, *Annual Review of Biochemistry* **57**, 785-838.
4. Ratner, S. (1992) Exploring the boundries between lymphocyte traffic and tumor metastasis, *Invasion Metastasis.* **12**, 61-65.
5. Springer, T. A. 1990. Adhesion receptors of the immune system, *Nature* **346**, 425-434.
6. Lasky, L. A. (1992) Selectins: interpreters of cell-specific carbohydrate information during inflammation, *Science* **258**, 964-969.
7. White, J. M. (1993) Integrins as virus receptors, *Current Biology* **3**, 596-599.
8. Velupillai, P., and Harn, D. A. (1994) Oligosaccharide-Specific induction of interleukin-10 production by b220+ cells from Schistosome-Infected mice - a mechanism for regulation of CD4+ T-Cell subsets, *Proc Natl Acad Sci USA.* **91**, 18-22.
9. Clerici, M., and Shearer, G. M. (1994) The Th1-Th2 hypothesis of HIV infection: New insights, *Immunol .Today*, **15**, 575-581.
10. Messner, P., Christian, R., Kolbe, J., Schulz, G., and Sleytr, U. B. (1992) Analysis of a novel linkage unit of O-linked carbohydrates from the crystalline surface layer glycoprotein of Clostridium thermohydrosulfuricum S102-70, *J. Bacteriology* **174**, 2236-2240.
11. Reinhold, B. B., Hauer, C. R., Plummer, T. H., and Reinhold, V. N. (1995) Detailed structural analysis of a novel, specific O-linked glycan from the prokaryote Flavobacterium meningosepticum, *J. Biol. Chem.* **270**, 13197-13203.
12. Karlsson, K. A., Angstrom, J. Bergstrom, J. and Lanne, B. (1992) Microbial Interaction with Animal Cell Surface Carbohydrates, *APMIS*, **100**, 71-83.
13. Zhu, Z., Cheng, L., Tsui, Z., Hakomori, S., and Fenderson, B. A. (1992) Glycosphingolipids of rabbit endometrium and their changes during pregnancy, *J .Reprod. Fertil.* **95**, 813-823.
14. Hakomori, S. (1985) Aberrant glycosylation in cancer cell membranes as focused on glycolipids: overview and perspectives *Cancer Research* **45**, 2405-2414.
15. Reinhold, B. B., and Reinhold, V. N. (1992) Electrospray Ionization Mass Spectrometry: Deconvolution by an Entropy-Based Algorithm, *J. Am. Soc. Mass Spectrometry* **3**, 207-215.
16. Fenn, J. (1990) Electrospray Ionization - Principles and Practice. *Mass Spectrometry Reviews* **9**, 37.
17. Reinhold, B. B., Chan, S.-Y. Chan, S. and Reinhold, V. N. (1994)

Profiling Glycosphingolipid Structural Detail: Periodate Oxidation, Electrospray, Collision-induced Dissociation and Tandem Mass Spectrometry, *Organic Mass Spectrometry* **29**, 736-746.

18. van Echten, G., and Sandhoff, K. (1993) Ganglioside Metabolism, *J Biol. Chem.* **268**, 5341-5344.
19. Linsley, K. B., Chan, S. Y., Chan, S., Reinhold, B. B., Lisi, P. J. and Reinhold, V. N. (1994) Applications of electrospray mass spectrometry to erythropoietin N- and O-linked glycans, *Anal Biochem.* **219**, 207-217.
20. Reinhold, B. B., Chan, S.-Y. L. Reuber, Walker, G. C., and Reinhold, V. N. (1994) Detailed Structural Characterization of the Exopolysaccharide from Rhizobium meliloti, *J. Bact.* **176**, 1997-2002.
21. Fu, D. T., Chen, L., and O'Neil, R. A. (1994) A detailed structural characterization of ribonuclease B oligosaccharides by H-1 NMR spectroscopy and mass spectrometry, *Carbohydr. Res.* **261**, 173-186.
22. Watson, E., Bhide, A. and Vanhalbeek, H. (1994) Structure determination of the intact major sialylated oligosaccharide chains of recombinant human erythropoietin expressed in chinese hamster ovary cells, *Glycobiology* **4**, 227-237.
23. Angel, A. S., and Nilsson, B. (1990) Linkage positions in glycoconjugates by periodate oxidation and fast atom bombardment mass spectrometry. *Methods In Enzymology*, **193**, 587-607
24. Karlsson, K.-A. (1976) Microscale Fingerprinting of Blood-Group Fucolipids by Mass Spectrometry. pp. 97-122. *Glycolipid Methodology*. Edited by L. A. Witting. Champaign, IL: American Oil Chemists Society.
25. Reinhold, V. N. (1987) Direct chemical ionization mass spectrometry of carbohydrates. In: *Complex Carbohydrates*, Part E, ed. Ginsberg, V., vol. **138**, pp 59-84. Academic Press Inc., 525 B Street, Suite 1900, San Diego, CA 92101-4495.
26. Lemoine, J., Fournet, B., Despeyroux, D., Jennings, K. R., Rosenberg, R., and de Hoffmann, E. (1993) Collision-Induced Dissociation of Alkali Metal Cationized and Permethylated Oligosaccharides: Influence of the Collision Energy and of the Collision Gas for the Assignment of Linkage Position, *J. Am.Soc. Mass Spectrom.* **4**, 197-203.
27. Dell, A., Reason, A. J., Khoo, K. H., Panico, M., McDowell, R. A., Morris, H. R. (1994) Mass spectrometry of carbohydrate-containing biopolymers. In Guide to Techniques in Glycobiology, eds. W. J. Lennarz, G. W. Hart, vol. **230**, pp. 108-132. Academic Press Inc., 525 B Street, Suite 1900, San Diego, CA 92101-4495.
28. Costello, C. E., and Vath, J. E. (1990) Tandem mass spectrometry of glycolipids *Methods In Enzymology* **193**, 738-68.
29. Domon, B., Vath, J. E., and Costello, C. E. (1990) Analysis of derivatized ceramides and natural glycosphingolipids by high performance tandem MS, *Anal. Biochem.* **184**, 151-164.
30. Stroud, M. R., Handa, K. Ito, K. Salyan, M. Fang, H. Levery, S. B.

Hakamori, S. Reinhold, B. B. and Reinhold, V. N. (1995) Myeloglycan, a series of E-selectin-binding polylactosaminolipids found in normal human leukocytes and myelocytic leukemia HL60 cells, *Biochem. Biophys. Res. Commun.* **209**, 777-787.

31. Chizhov, O. S., Polyakova, L. A., and Kochetkov, N. K. (1964) Investigation of carbohydrates by mass spectrometry methyl ethers of disaccharides, *Doklady Akademii Nauk SSSR* **158**, 685-688.
32. Egge, H., von Nicolai, H. and Zilliken, F. (1974) *FEBS Lett.* **39**, 341.
33. Moor, J., and Waight, E. S. (1975) *Biomed. Mass Spectrom.* **2**, 36-41.
34. de Jong, Heerma, E. G., Haverkamp, W., and Kammerling, J., J. P. (1979) Electron impact mass spectrometry of permethylated disaccharides, *Biomed Mass Spectrom.* **6**, 72-77.

DIRECT AND INDIRECT ANALYSIS OF SYNTHETIC POLYMERS USING MASS SPECTROMETRY

J H Scrivens, A T Jackson
ICI Research and Technology Centre
PO Box 90
Wilton
Cleveland
England
TS90 8JE

INTRODUCTION

Synthetic polymers play a major part in the commercial and everyday life of everyone in the developed world. The systems employed, which are rarely pure compounds, have been developed over a number of years and have been optimised for a large number of varied applications. In addition to the natural distribution of oligomers that make up the polymeric component the formulation can contain, among other components: residual monomers, low molecular weight oligomers (trapped in the polymer matrix), additives, modifiers, catalyst residues and reactants (e.g. crosslinking agents). The number of individual components present in the formulation can vary from a few hundred to many thousands and few analytical techniques can cope with this complexity especially since the volatility and polarity of the components can vary widely.

The most common approach is to separate the non-polymeric components of the formulation from the native polymer to aid characterisation. Mass spectrometry, either directly, or integrated with separation science approaches such as liquid chromatography (LC), gas chromatography (GC), super-critical fluid chromatography (SFC) and others, is used extensively to characterise these non-polymeric components. The polymeric components are either studied directly using mass spectrometry techniques such as field desorption-mass

K. R. Jennings (ed.), Fundamentals and Applications of Gas Phase Ion Chemistry, 201–234.

spectrometry (FD-MS) or, more recently, matrix assisted laser desorption/ionisation-MS (MALDI-MS) or, more commonly, are treated, either thermally or chemically, to break them down into a molecular weight range which is accessible to the mass spectrometer. This general scheme, for polymer analysis using mass spectrometry, is shown in Figure 1. The use of mass spectrometry in the characterisation of polymeric formulations has recently been reviewed [1].

INDIRECT ANALYSIS

ANALYSIS OF NON-POLYMER COMPONENTS

The vast majority of polymer additives are analysed, subsequent to solvent extraction, by probe analysis using electron impact (EI) or chemical ionisation-MS (CI-MS). The interpretation of the fragments formed, or comparison with library spectra, enables most components to be readily identified. Mass spectrometry is often used in combination with separation science techniques such as GC, LC or SFC to characterise the, often complex, mixtures of formulations used. Recent developments in the manufacture of polymer additives have led to molecules, or combinations of molecules, that are more complicated and generally higher in molecular weight. This has led to the use of soft ionisation methods such as fast atom bombardment (FAB), liquid secondary ion mass spectrometry (LSIMS) and field desorption (FD). The advantages of these techniques are that they extend the molecular weight range accessible by means of mass spectrometry and that they induce little fragmentation of the additives which simplifies the characterisation of the components present. This in itself poses a problem however since, if unknowns are present, structural information is required for identification. This can be resolved by combining soft ionisation techniques with tandem mass spectrometry (MS/MS).

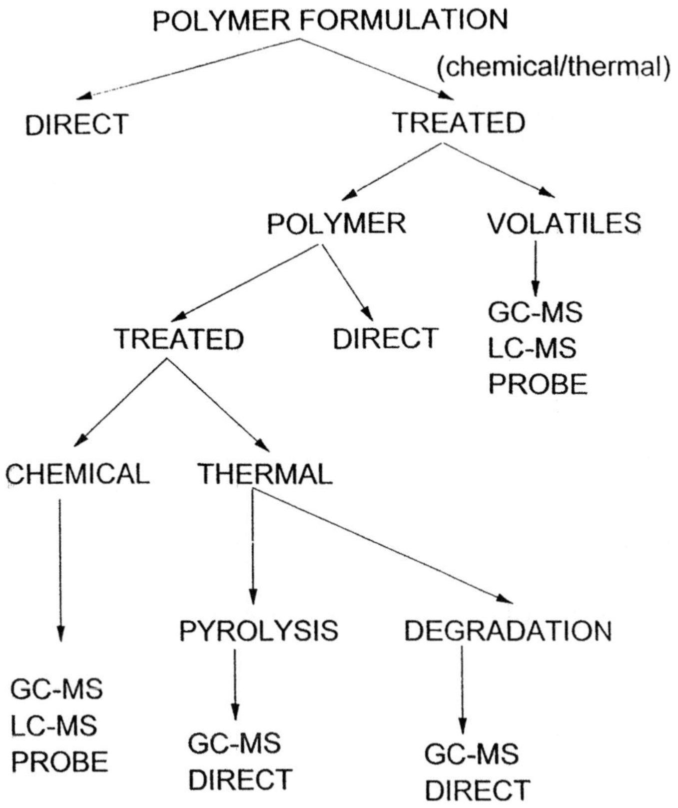

Figure 1. Schematic of Polymer Analysis using Mass Spectrometry

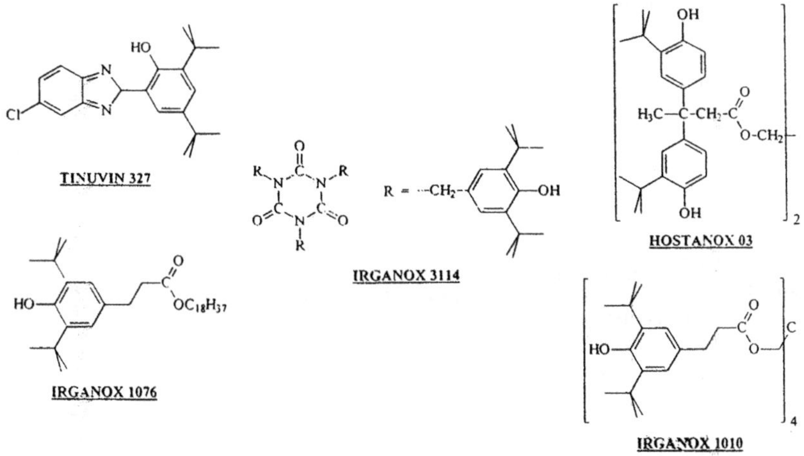

Figure 2. Polymer Additives Studied

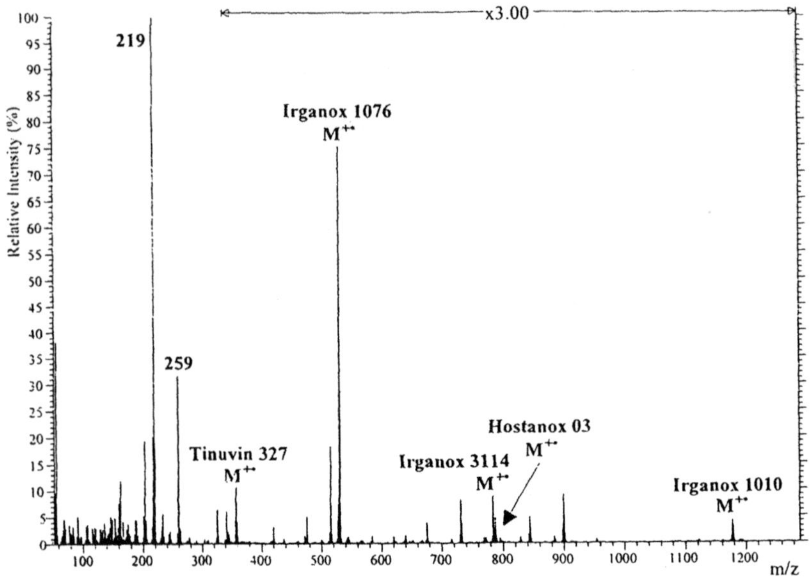

Figure 3. LSIMS-MS Spectrum of Equimolar Mixture of Additives

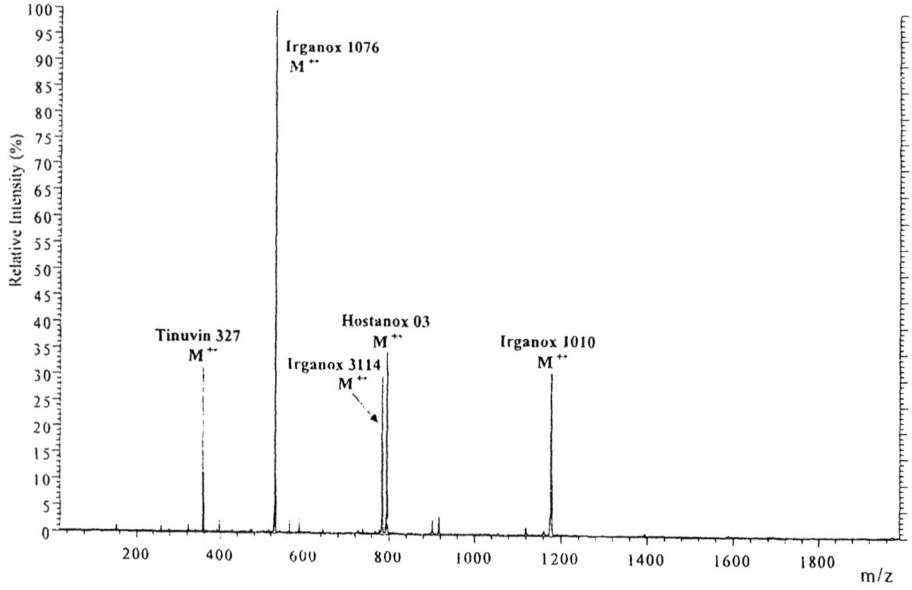

Figure 4. FD-MS Spectrum of Equimolar Mixture of Additives

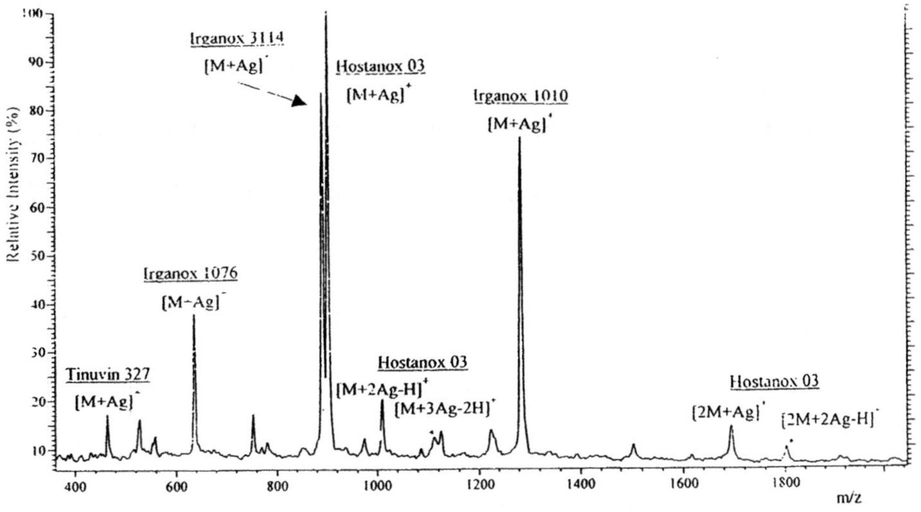

Figure 5. MALDI-MS Spectrum of Mixture of Additives

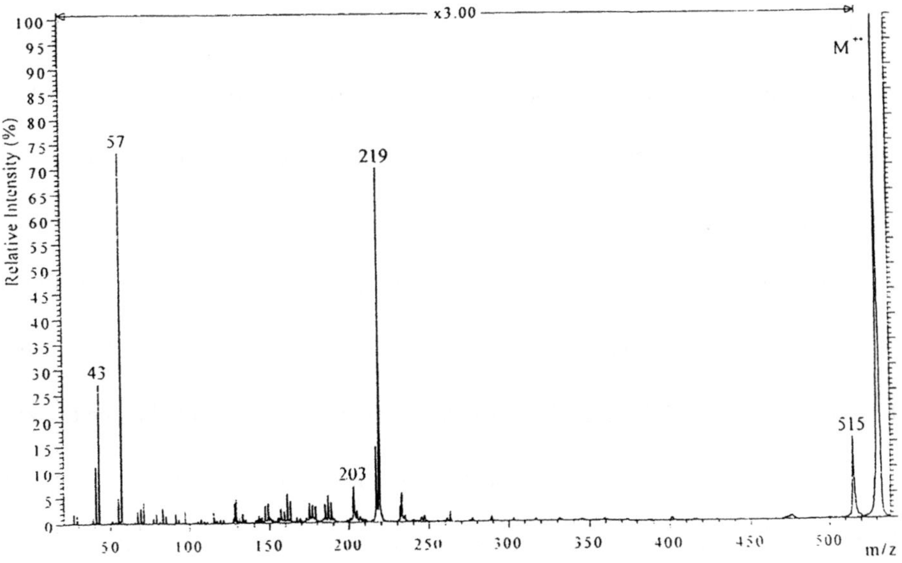

Figure 6. LSIMS-MS/MS Spectrum of Irganox 1076, $M^{+\bullet}$ m/z 530, from an Equimolar Mixture of Additives

Figure 2 shows a number of common polymer additives that have been studied by means of mass spectrometry. These compounds cover a range of compound classes and volatilities. They represent additives which are typical of polymer formulations such as polypropylene and polyethylene. Figures 3 and 4 show the LSIMS and FD spectra of the additive mix. The LSIMS spectrum was obtained with *meta*-nitro-benzyl alcohol (NBA) as the matrix in positive ion mode A high signal-to-noise is observed in the LSIMS spectrum, obtained from 1 nmol of each component placed on the probe tip. There is considerable variation of intensities of component with the matrix employed, which is what might be expected from what is essentially a surface sensitive technique. A significant amount of fragmentation is seen in the spectrum which can be confusing if unknowns are present. The FD spectrum gives a much clearer idea of the components present but at the expense of a much lower signal-to-noise ratio. The FD spectrum also takes a great deal longer to obtain (approximately 60 minutes *versus* 5 minutes). This mixture of additives has also been examined using the more recently introduced technique of matrix assisted laser desorption/ionisation-mass spectrometry (MALDI-MS). In this experiment the sample to be studied is dissolved in a matrix that absorbs typically UV laser energy. The ions that are formed are analysed using, in most cases, a time-of-flight mass spectrometer. A MALDI spectrum of the additive mixture is shown in Figure 5. The matrix used was dithranol and silver trifluoroacetate was used as a cationisation reagent. The additives all form complexes with silver in the presence of the trifluoroacetate salt. The reproducibility of this technique was relatively poor which could be due to variation in sample homogeneity in addition to the intrinsic variation of the experiment. The spectrum is complicated by the various silver adduct ions present. For Hostanox 03, for example, adducts corresponding to $[M+Ag]^+$, $[M+2Ag-H]^+$, $[2M+Ag]^+$ and $[2M+2Ag-H]^+$ were seen. This, together with the relative difficulty involved in obtaining MS/MS data using this technique, makes the use of MALDI for this type of mixture analysis limited.

The use of FD and FAB in the study of polymer additives has been compared [2]. The conclusion reached was that, in FAB, some polymer additives suppressed the ionisation of other species present. This is in agreement with the above results. The use of MS/MS in conjunction with soft ionisation approaches can provide a powerful analytical technique. Figure 6 shows the product ion spectrum of the molecular ion at m/z 530, formed in the LSIMS spectrum of the polymer additive mixture. This can easily be identified as that of Irganox 1076 as fragment ions characteristic of this additive, such as m/z 219 and 57, are observed in the spectrum [3].

THERMAL STUDIES

Introduction

Polymer formulations are often subjected to thermal treatment in order to form fragments in a compatible molecular weight range for study by means of mass spectrometry, . This can range from imparting sufficient thermal energy to melt the polymer, releasing any trapped volatiles that might be present, to providing enough energy to thermally degrade the polymer. If more energy is provided the polymer will break down into small moieties which may be characteristic of the polymer structure. This range of thermal experiments is shown in Figure 7. A number of variables must be carefully controlled in order to obtain reproducible data. These include the sample size and orientation, the atmosphere that the thermal treatment takes place in, the initial and final temperatures and the rate of heating. This thermal treatment can take place outside the mass spectrometer or directly inside the source. When the thermal

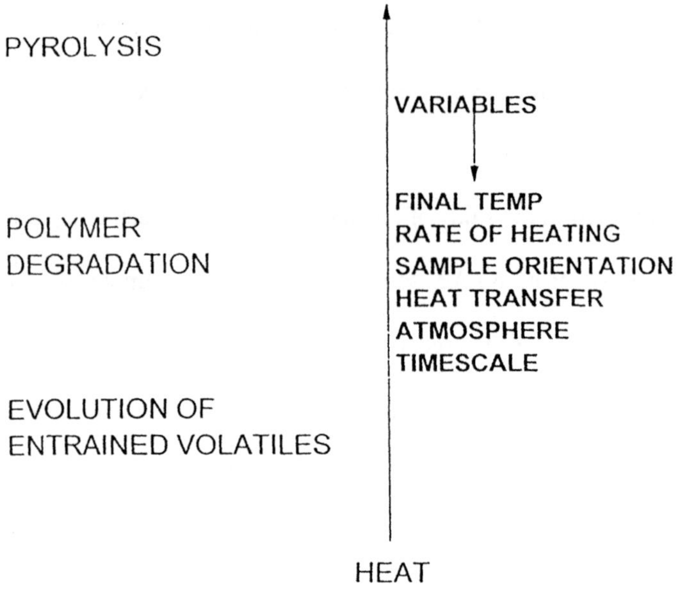

Figure 7. Schematic of the Thermal Treatment of Polymers

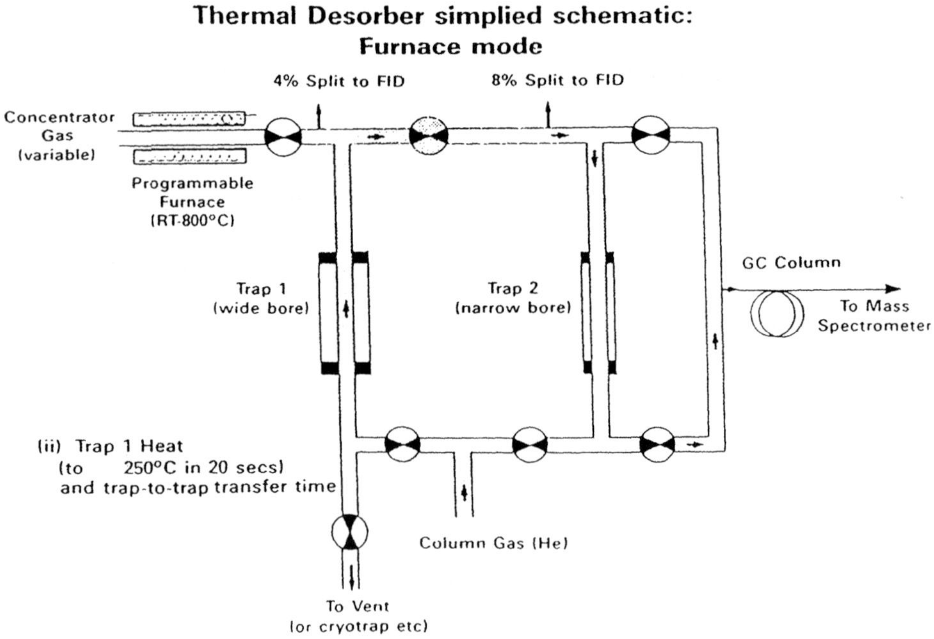

Figure 8. Schematic of the Envirochem 810A Thermal Desorption Unit

treatment is performed outside the mass spectrometer gas chromatography is often used to separate the, often complex, mixtures of compounds that are formed. This gives the advantage of simplifying the characterisation of the complex mixture but introduces the disadvantage of restricting the range of volatilities and polarities of the compounds that may be studied. In addition, the time frame associated with the experiment means that transiently stable, or reactive, species can not be detected.

Thermal desorption gas chromatography-mass spectrometry (TD-GC-MS)

Thermal desorption methods in which the thermal energy imparted may be varied over a wide range, in a controlled manner, can be very useful in the characterisation of polymeric systems. Figure 8 shows a schematic of an Envirochem 810A thermal desorption system coupled to a high temperature furnace unit [4]. A sample of the polymer to be studied is placed in the furnace and the gas that passes over the sample is varied to mimic various experimental conditions. The volatiles generated in the experiment are trapped initially on a wide bore trap containing different materials of varying polarity. The flow is then reversed and the trapped volatiles are carried onto a second trap containing similar materials to provide a narrow band for subsequent injection onto a GC column. The extent of the evolution of volatiles and the amount transferred to the second trap can be monitored via part of the flow which is diverted to a flame ionisation detector (FID). Figure 9 shows the FID trace of the volatiles released from a polyarylsulphone polymer over the range 200-450°C. The trace clearly shows two peaks which correspond to the evolution of entrained volatile material and the onset of thermal degradation respectively. This differentiation of thermal processes can be difficult with other thermal experimental approaches. The reconstructed ion current (RIC) traces of the volatiles evolved from the aromatic polymer are shown in Figure 10. The initial temperature ramp from

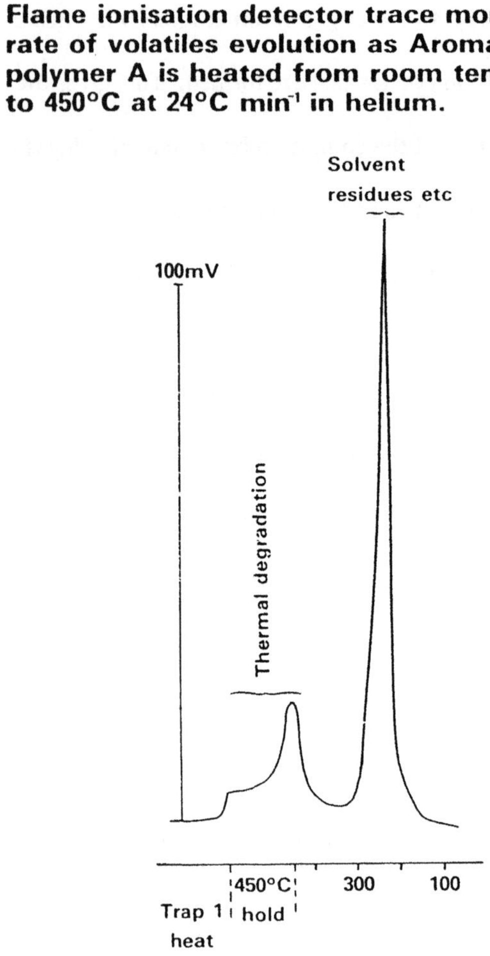

Figure 9. FID Trace of Evolved Volatiles from Poly Aryl Sulphone

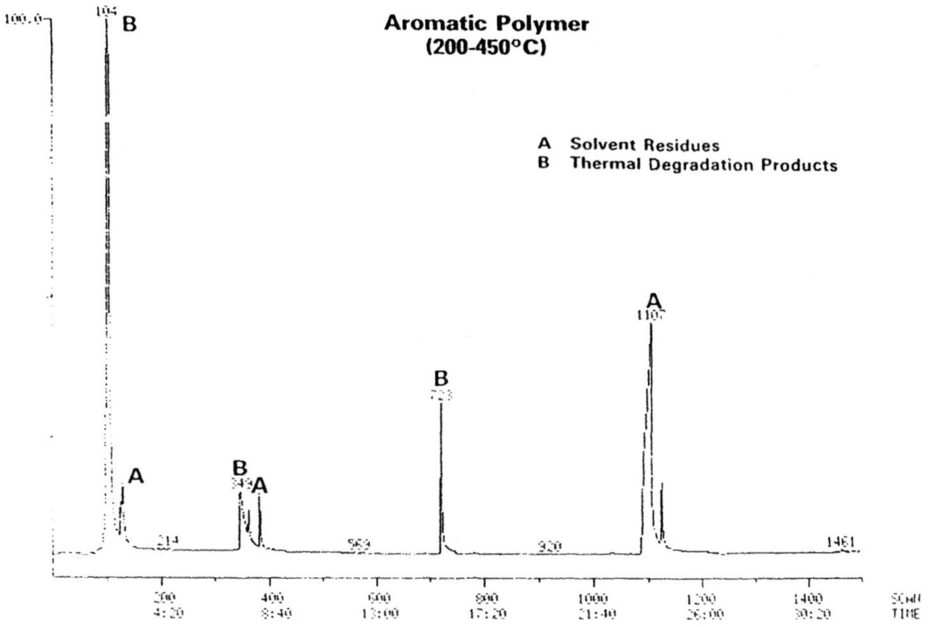

Figure 10(a). RIC Trace of Evolved Volatiles (200-450 °C)

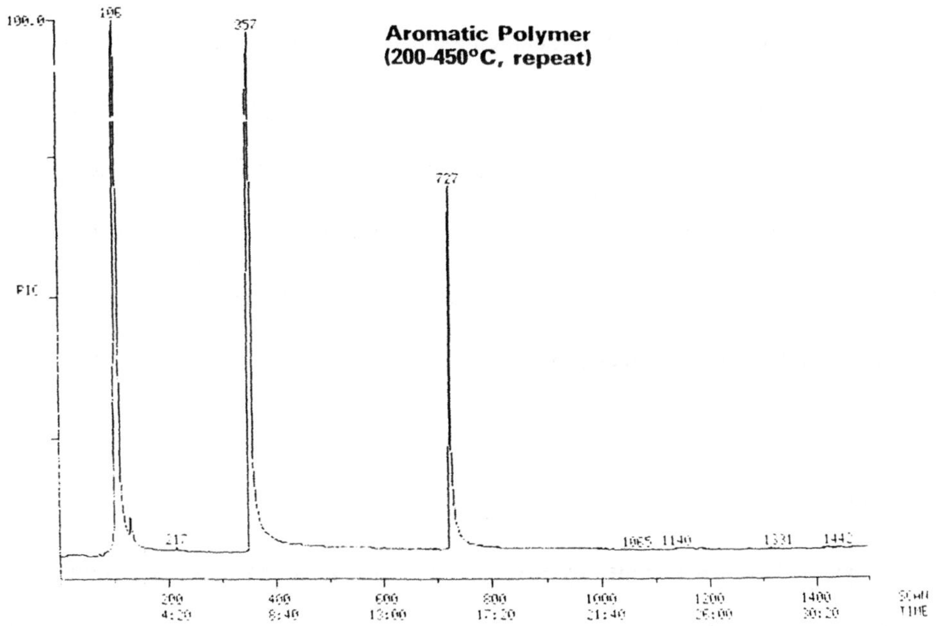

Figure 10(b). RIC Trace of Evolved Volatiles (200-450 °C) (Repeat)

200-450°C shows peaks due to the evolution of trapped solvent volatiles and peaks due to polymer degradation. The repeat run only includes peaks due to polymer degradation.

Figure 11 shows an RIC trace of an acrylic based polymer heated from room temperature to 280°C. In addition to peaks corresponding to the monomer and higher oligomers peaks are also present which are characteristic of the catalyst and chain terminators used. The ability of this experimental approach to allow the atmosphere, in which the thermal treatment takes place, to be varied is very useful since the thermal products vary greatly between inert and oxidising atmospheres. Is has also been used to investigate the effect of wet and dry air on thermal products.

In some cases the entrained volatiles themselves can be used to provide considerable detail regarding the source and history of polymer systems. The TD-GC-MS approach has been used to collect the volatiles evolved from a series of polypropylene formulations [5]. The experiment was stopped just short of degrading the polypropylene itself since the common fragments formed overwhelmed the data collected. The resulting GC-MS data sets were averaged to produce an overall composite mass spectrum of the volatile components. By using principal component analysis (PCA) to extract the data containing the most information it was possible to characterise the systems in terms of method of manufacture, site of manufacture and additive package used.

Direct pyrolysis

Many polymers have been studied by heating them directly inside a mass spectrometry source. The advantage of this approach is that there is little delay between the evolution of pyrolysis products and their analysis. The range of products that can be studied is limited by the ionisation method used and the performance of the mass spectrometer. The atmosphere in which the heating takes place can also not be easily varied.

Figure 12(a) and 12(b) show the total ion current (TIC) and the averaged mass spectrum of an unknown polymer formulation. The spectra were obtained using electron impact (EI) ionisation in a ZAB-T (Micromass, Manchester, U.K.) [6] magnetic tandem mass spectrometer. Heating was from 50-600°C at 50°C per minute. The TIC shows the early evolution of trapped volatiles as the polymer melts together with the later degradation. In contrast to the thermal desorption approach described earlier relatively high molecular weight products are observed. Further information may be obtained from this experiment by utilising the MS/MS capability of the instrument. Figure 13 shows the product ion spectrum of a selected ion at m/z 670. The interpretation of this data, together with other selected product ion spectra, enable the polymer to be identified as a hexanediol, neopentylglycol and adipic acid co-polymer and the fragment at 670 Da to be made up of two hexanediol and one neopentyl glycol units together with three adipic acids. This direct pyrolysis experiment is very commonly used in the identification of unknown polymer systems.

An extension of this method which provides additional data is provided by utilising field ionisation (FI). In this experiment the application of a large electric potential to a surface of high curvature results in an intense localised electric field being produced. The sample is introduced into the mass spectrometry source, as in the above experiment, in the gaseous phase. Ionisation occurs predominately via an electron tunnelling mechanism and very little internal energy is imparted which results in predominately molecular ions being formed. This

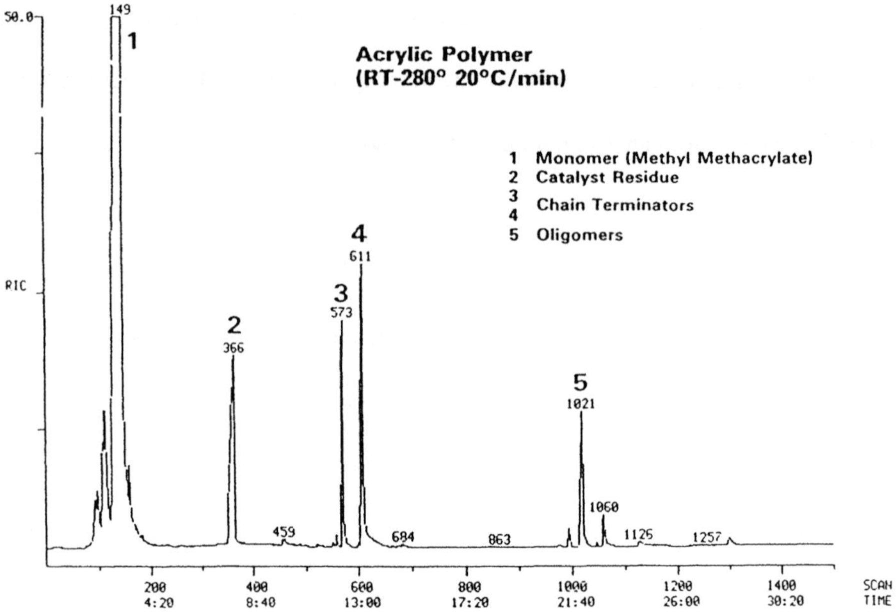

Figure 11. RIC Trace of Evolved Volatiles from an Acrylic Polymer

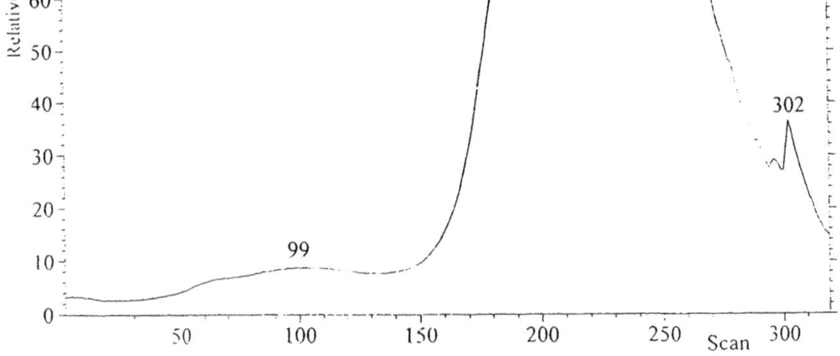

Figure 12(a). Probe TIC of Polymer Formulation

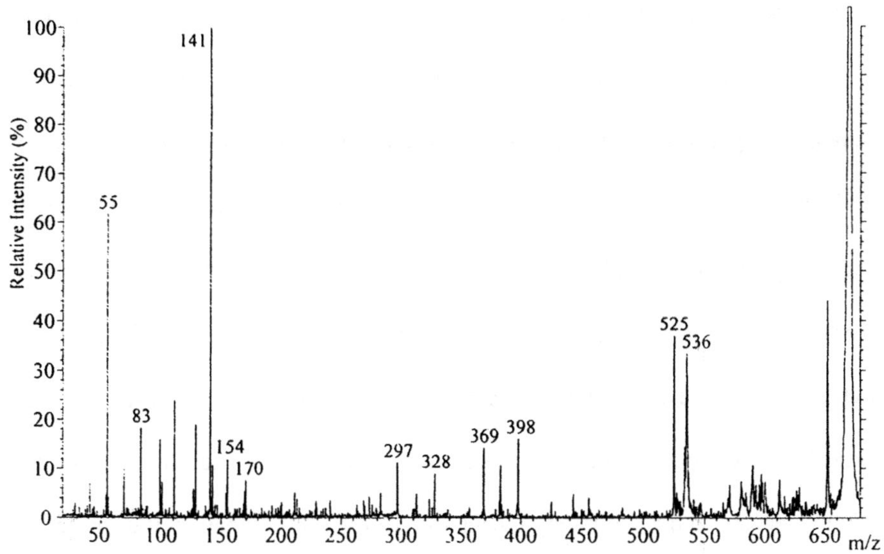

Figure 12(b). Averaged Mass Spectrum of Polymer Formulation

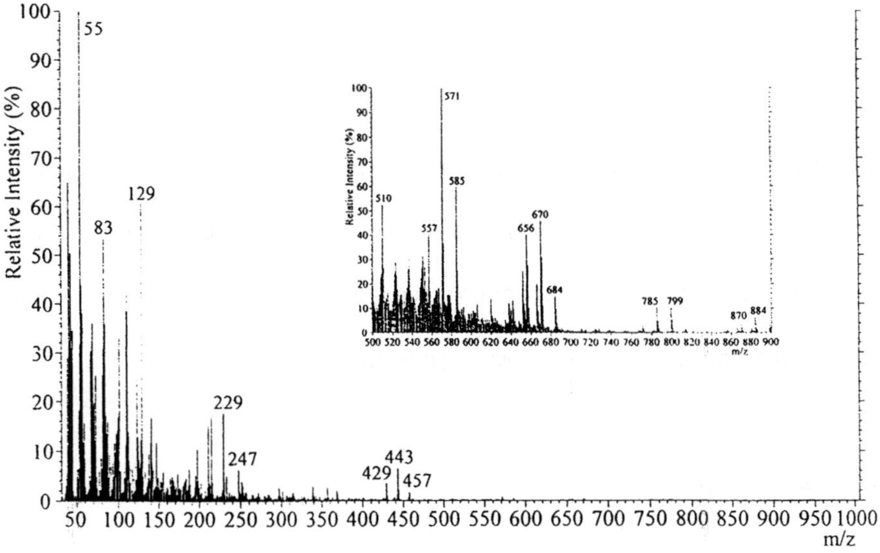

Figure 13. Product Ion Spectrum of m/z 670 from the Polymer Formulation

technique therefore has the advantage that the pyrolysis fragments are introduced directly into the source enabling reactive fragments to be studied. In addition, the very soft nature of the ionisation results in pyrolysis fragments of significantly higher molecular weight being observed. The ionisation method is relatively insensitive however and this, together with the high voltages used, has restricted the experiment to sensitive double focusing magnetic sector instruments of relatively high cost. Figure 14 shows the averaged pyrolysis FI spectrum of a polypropylene polymer. Thermal fragments of > 2000 Da may be observed showing the greatly extended mass range afforded by this approach. Polypropylene is thought to degrade via homolytic cleavage and a backbiting mechanism to form a terminally unsaturated species (Series A). This then undergoes further decomposition to form a diene oligomer (Series E) [7]. These structures are shown below.

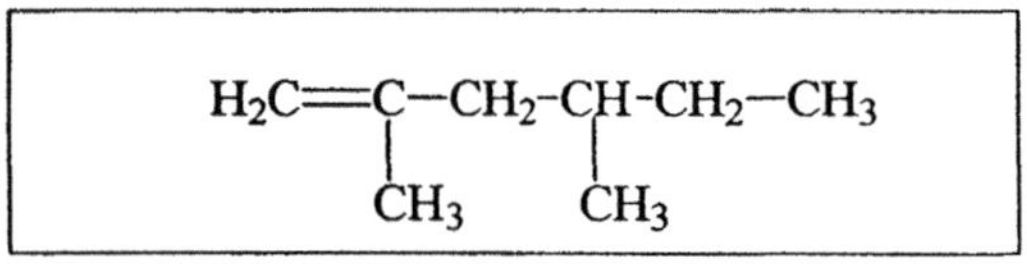

SERIES A

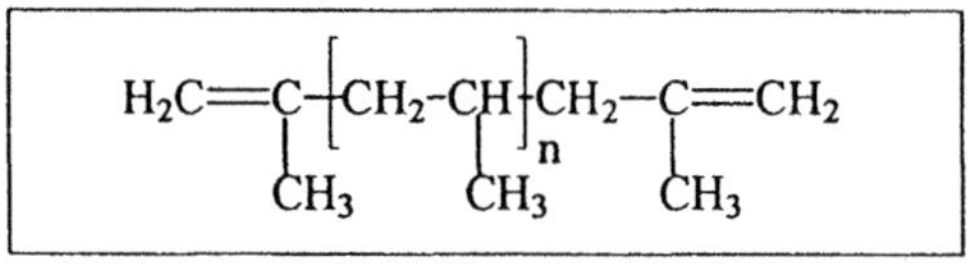

SERIES E

Figure 15(a) shows the low temperature and Figure 15(b) the high temperature region. These ions correspond to series A and series A+E respectively. This approach can be used to study the degradation mechanisms of a number of polymer systems in addition to being a good method for the characterisation of unknown polymer systems.

DIRECT METHODS

Introduction

The direct analysis of polymer systems of more than a few thousand daltons has only just begun to be possible using mass spectrometry. Ionisation methods such as field desorption

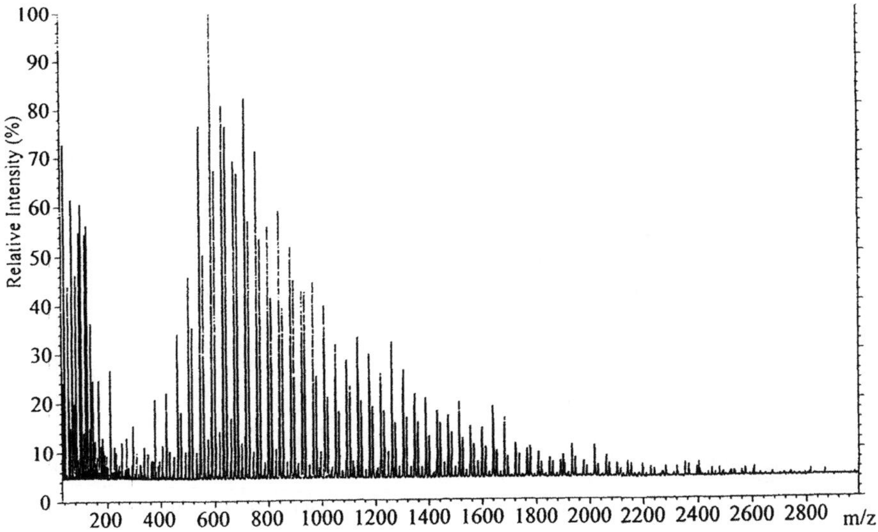

Figure 14. Pyrolysis-FI-MS Spectrum of Polypropylene (Averaged)

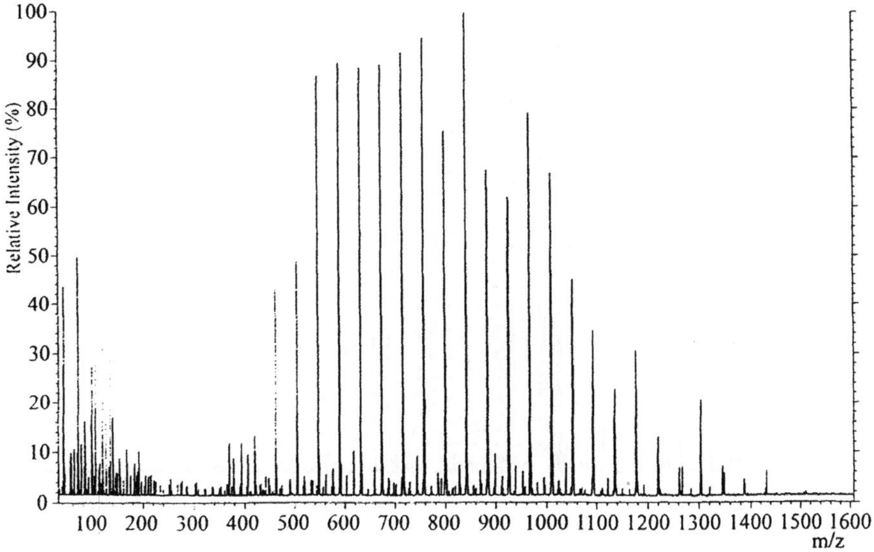

Figure 15(a). Pyrolysis-FI-MS Spectrum of Polypropylene (Series A-Low Temperature Region)

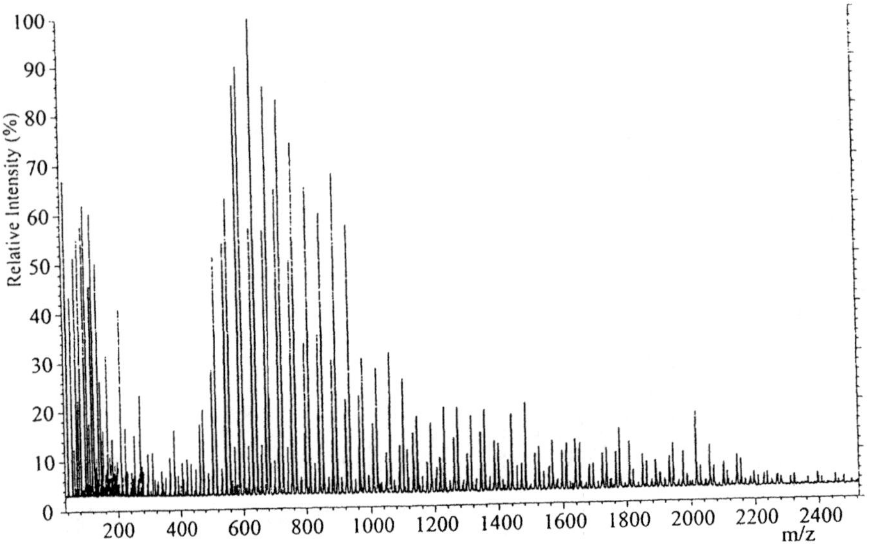

Figure 15(b). Pyrolysis-FI-MS Spectrum of Polypropylene (Series A & E-High Temperature Region)

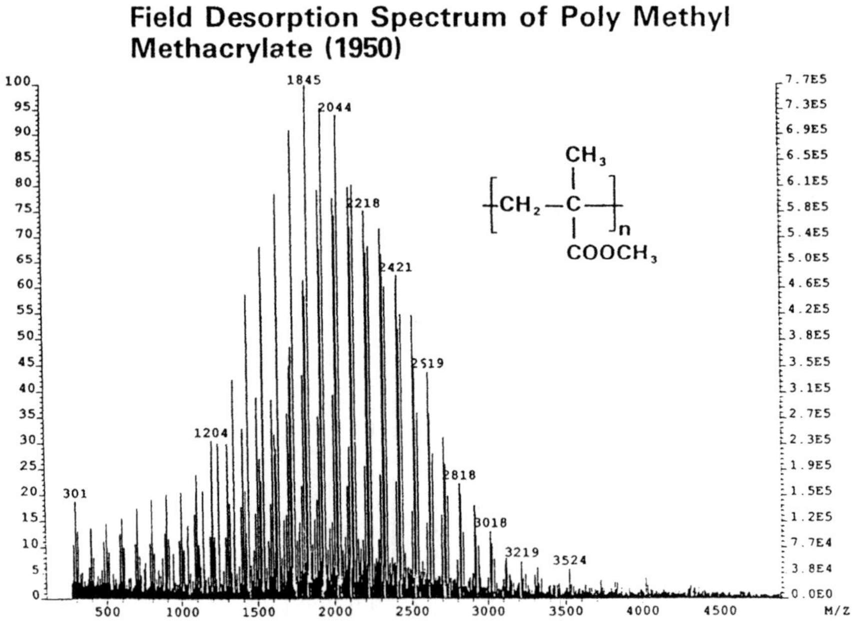

Figure 16. FD-MS Spectrum of Polymethylmethacrylate (PMMA) with a M_p of Approximately 1950 Da

(FD), electrospray ionisation (ESI) and matrix assisted laser desorption/ionisation (MALDI) have led to molecule ions of much larger species being observed. There is now evidence that molecules of >500 kDa may be able to be directly observed. These developments have led to a revaluation of the utility of mass spectrometry for direct polymer analysis and a requirement for mass spectrometrists to engage in a dialogue with the polymer chemists to establish what information is required and which is currently difficult to obtain. The most directly comparable technique to mass spectrometry in the current polymer chemists arsenal is probably size exclusion chromatography (SEC). This technique, often referred to as gel permeation chromatography (GPC), is the most commonly used approach to the determination of molecular weight distributions of polymers. This approach is based on the measurement of polymer shape and therefore calibration can, under certain circumstances, be difficult but nevertheless these measurements are carried out routinely in all polymer characterisation laboratories. A very wide range of polymer molecular weights can be studied (10^2-10^7 Da) and the approach is very widely applicable, capable of automation and requires little interpretation. The resolution is limited however and, if a number of specialist detectors are employed the relative costs can be similar. Achieving comparable results for standard samples is important at this stage of the development of direct mass spectrometric methods.

Field desorption

The use of field desorption in the study of, particularly, non-polar polymeric systems has been in limited use for the last decade [8]. The approach has not received wider popularity due to the relatively difficult nature of the experiment together with the requirement for expensive double focussing mass spectrometers. In addition, the mass range is limited by the current required to be provided to the emitter. When this gets too high thermal decomposition can occur particularly for thermally labile systems. Nevertheless the technique can give useful data for a range of polymer systems. Figure 16 shows an FD spectrum of a polymethylmethacrylate (PMMA) with a M_p of approximately 1950 Da and Figure 17 is a FD spectrum of an alkyl phenol formaldehyde resin. The resin shows the molecular distribution and, in addition, gives the carbon number of the alkyl chains present (C_8-C_{10}). Figure 18 shows the FD spectrum of a polystyrene sample with a M_p of approximately 12500 Da. The resolution is approximately 2000 (FWHM) and baseline resolution of the oligomeric species is achieved. The symmetrical nature of the distribution also shows that, at least from m/z 9500 to 14500 there is little dependance of the detector on mass. This is a time consuming and challenging experiment, however, taking ninety minutes to perform.

Electrospray ionisation

ESI is a desorption ionisation technique in which dilute solutions of sample are sprayed into a gas filled chamber. The solution is forced through a syringe held at high potential (a few kilovolts), so that the resulting aerosol droplets are charged. The droplets break down losing solvent, sample and charge, so that singly, or more often, multiply charged ions are formed. This technique has had a major impact in the study of biological molecules where masses of greater than 100,000 Da may be studied using mass spectrometers of modest mass range (since mass-to-charge is the parameter measured and ESI often produces ions with many charges). The extension of this technique into polymer analysis has not been easy however for a number of reasons.

Synthetic polymers, unlike biopolymers have, in general, few acid or basic functional groups that can be used for ion formation through acid-base equilibria and so formation of multiply

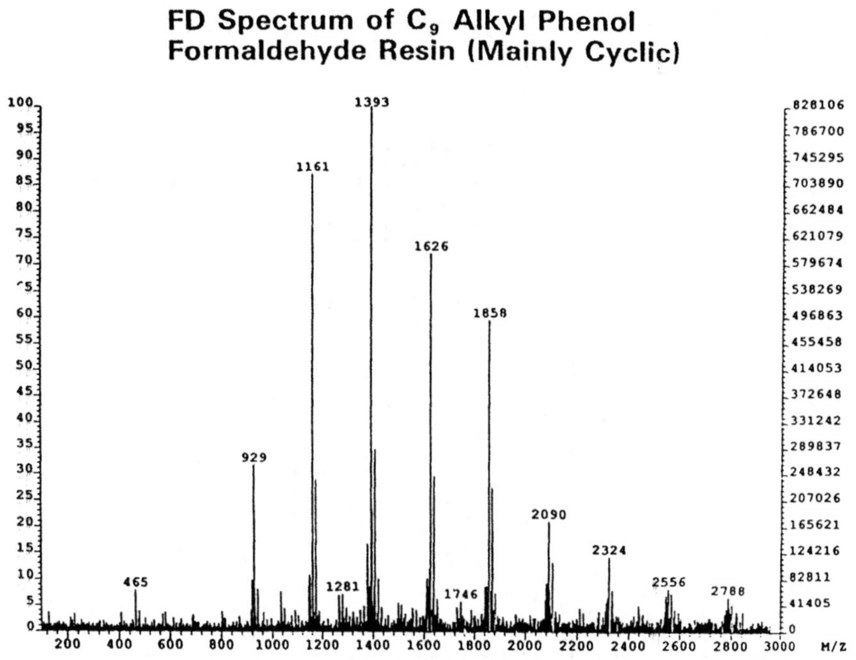

Figure 17. FD-MS Spectrum of Phenol Formaldehyde Resin

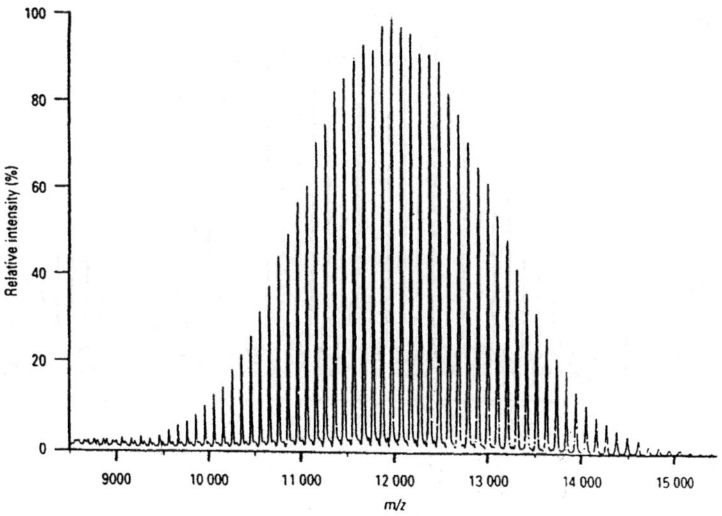

Figure 18. FD-MS Spectrum of Polystyrene with a M_p of Approximately 12500 Da

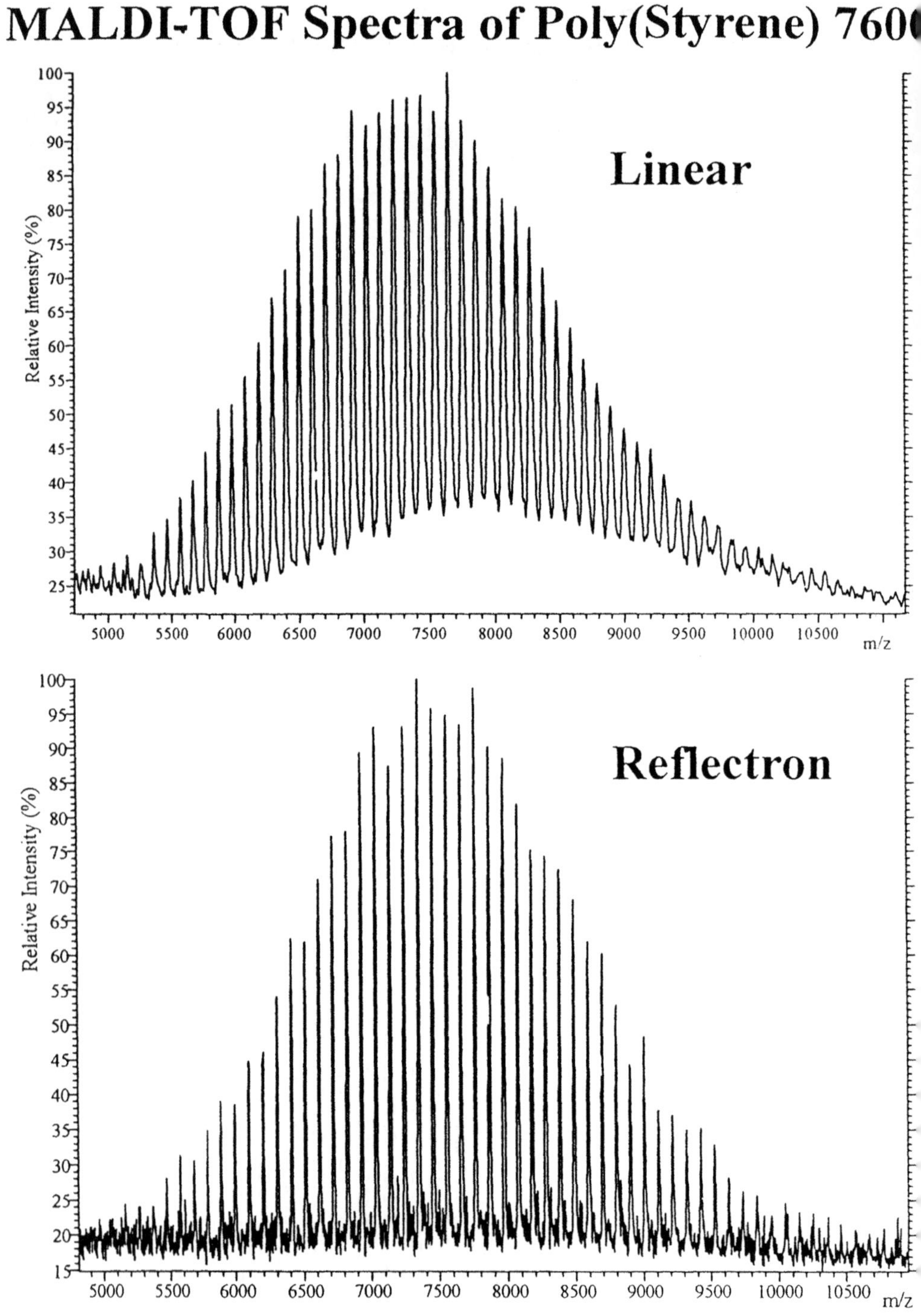

igure 19. MALDI-MS Spectra of Polystyrene with a M_p of Approximately 7600 Da in (a) Linear Mode and (b) Reflectron Mode

charged ions is more difficult. The nature of polymer spectra, with the large numbers of discrete species present, coupled with the range of charged species present makes interpretation of the data, particularly using the quadrupole analysers which are typically used, difficult. In addition the quadrupole analysers do possess a mass dependance and so obtaining good quantitative mass distribution data can be a problem. Some of these difficulties can be overcome by using alternative analyser systems such as magnetic sectors or Fourier transform ion cyclotron resonance (FTICR) instruments which possess higher resolution and greater mass range. ESI has been successfully used for relatively low molecular weight polymers [9] and also in the study of starburst polymers [10].

Matrix assisted laser desorption/ionisation

Laser desorption ionisation has been used with limited success in the study of polymeric systems for some time but major developments in this area followed the introduction of MALDI. MALDI differs from other laser desorption methods in that a UV laser is typically used to promote electronic excitation of the matrix molecule. In addition to the matrix materials found to be useful in biological systems such as dihydroxybenzoic acid and sinapinnic acid, other matrix materials like dithranol, indole acrylic acid and others have been found to give good results for polymers. The ionisation is commonly by means of cation attachment. Laser desorption is normally performed using a nitrogen laser operating at 337 nm with, typically, a 3 ns pulse with a power of 10^7-10^8 Wcm^{-2} and an approximate spot size of approximately 0.2 mm^2. The analyser system employed in the vast majority of the experiments undertaken is a time-of-flight system either of linear design or using a reflectron device to improve the resolution obtained. The large dynamic range requirements imposed on the analysis by the large number of components in a typical system and the high masses that may be present places special requirements on the detection system and modifications to the conventional dual microchannel plate may be neccessary. The instrument may be calibrated using an external two point calibration when a mass accuracy of 0.2% is common. Calibrating using a multi-point calibration using similar compounds and experimental conditions can improve the mass accuracy to approximately 0.02%. Using an internal standard can improve these values but is difficult to perform with polymer samples. Montaudo et al. [11] have made use of the fact that the repeat unit in polymer spectra is often known to internally calibrate their spectra. Use of a higher resolution analyser such as in a FTICR instrument can improve these values down to sub ppm levels.

Figure 19 shows the MALDI spectra of a polystyrene sample with an approximate M_p of 7600 Da obtained using a linear system and using a reflectron in a TofSpec (Micromass, Manchester, U.K.) time-of-flight mass spectrometer. The improvement in resolution in going to a reflectron is approximately a factor of two. The sample was run in dithranol as a matrix with silver trifluoroacetate added. The oligomers are cationised with silver. Figure 20 shows a MALDI spectrum of polystyrene with an approximate M_p of 12800 Da in which a technique known as delayed extraction first shown by Wiley and McLaren in 1955 [12] and applied to MALDI-TOF by Brown et al. [13] has been employed. Using this approach the ions in the source are initially allowed to spread out in a near zero electric field. The distance travelled by the ions is dependent on their initial velocities. At an appropriate time an electric field is applied to the ions. Faster ions, which have travelled further, receive less energy from the applied field than the slower ions. In this way velocity focussing can be achieved and ions of the same mass arrive at the detector at the same time independant of their initial velocities. By correct arrangement of the pulse time pulse intensity and source geometry second order focusing can be achieved. The delay may be varied between 0.5 and 10 μs and the electric field between 500 and 5000 V. This approach significantly improves the resolution and signal-to-noise that can be obtained and results in data that, for the first time, can be compared to that obtained from FD-MS (Figure 18). MALDI-MS has been used successfully to determine the mass of the end group in a selection of polymers. Figures 21(a-d) show the MALDI spectra of a series of polymethylmethacrylate (PMMA) samples with varying end groups as shown below.

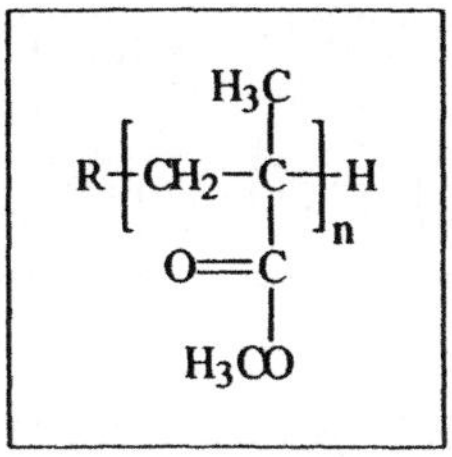

STRUCTURE OF PMMA

Structure of end group (R)	Mass of end group (R)
H	1
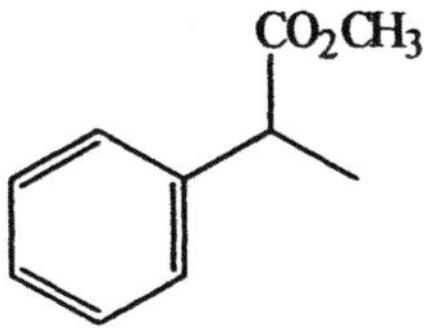	149
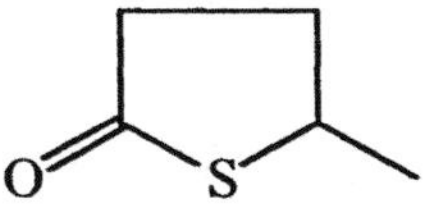	101
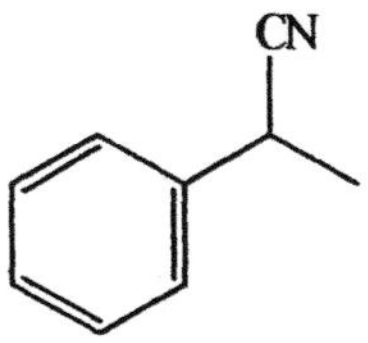	116

STRUCTURES OF END GROUPS

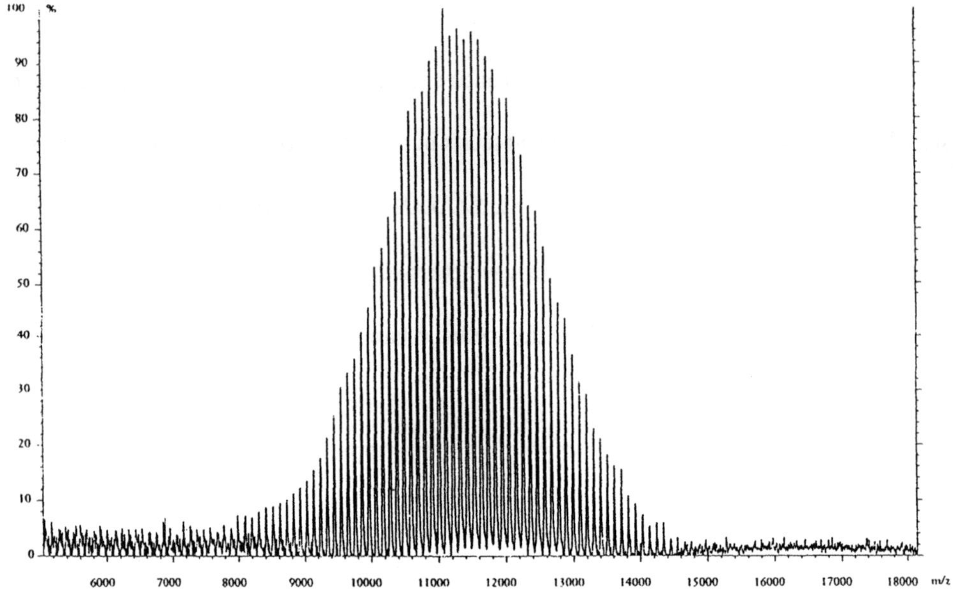

Figure 20. MALDI-MS Spectrum of Polystyrene with a M_p of Approximately 12800 Da in Linear Mode Using Delayed Extraction

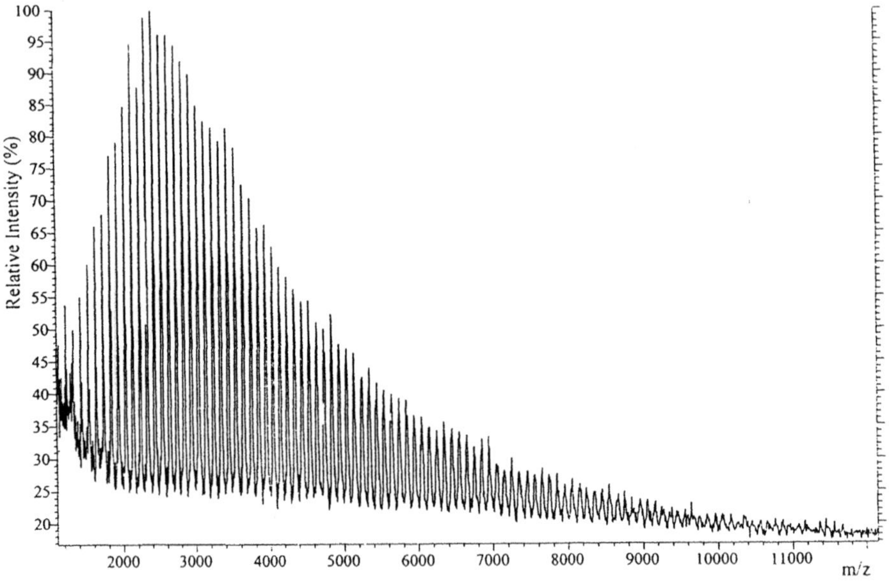

Figure 21(a). MALDI-MS Spectrum of Polymethylmethacrylate (PMMA) (R = 101) in Linear Mode

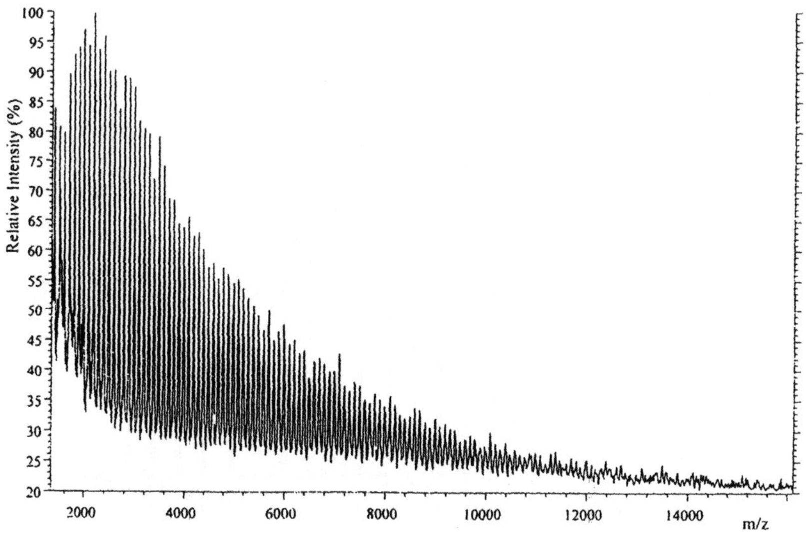

Figure 21(b). MALDI-MS Spectrum of Polymethylmethacrylate (PMMA) (R = 149) in Reflectron Mode

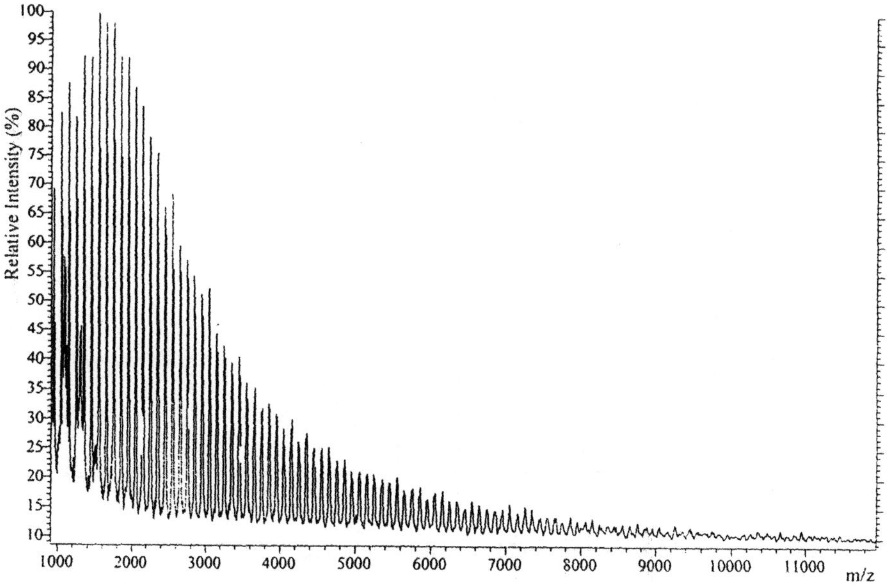

Figure 21(c). MALDI-MS Spectrum of Polymethylmethacrylate (PMMA) (R = 116) in Linear Mode

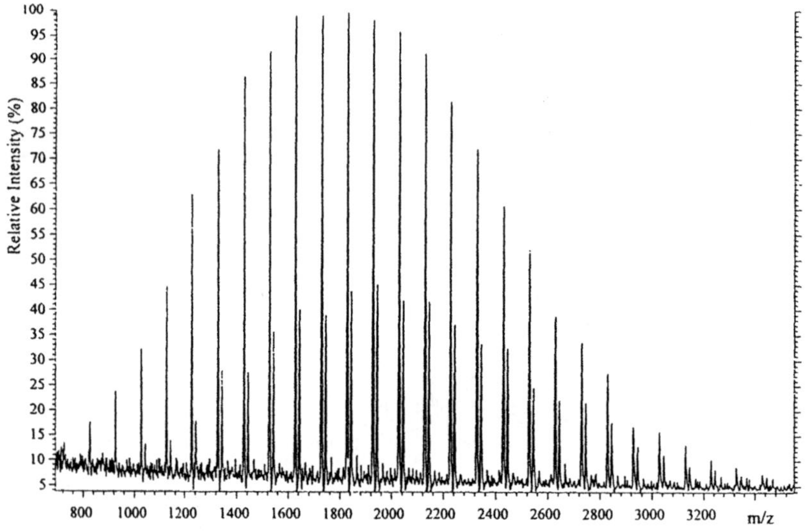

Figure 21(d). MALDI-MS Spectrum of Polymethylmethacrylate (PMMA) with a M_p of Approximately 2010 Da (R = 1) in Reflectron Mode

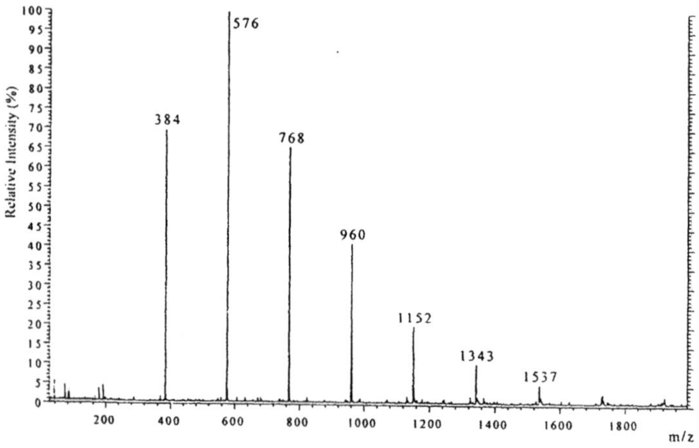

Figure 22. FD-MS Spectrum of Fluorinated Oligomers

The spectra were obtained in a TofSpec instrument using indole acrylic acid as a matrix in a 10:1 matrix:sample ratio. An external multi-point calibration was used enabling the end group to be determined to within one dalton.

End group determination using MS/MS techniques

MALDI and FD have been shown to be good methods for analysing many polymers directly and, when spectra are obtained, end groups can be measured to within one dalton. In some cases, for example in the above PMMA data set, determining the end group to within one dalton is insufficient since two of the end groups differ by the mass of the PMMA monomer. More structural details can often be obtained using MS/MS techniques.

Figure 22 is the FD-MS spectrum of a product that was obtained when 2,3,5,6-tetrafluoro-4-hydroxystyrene was polymerised. The spectrum indicates that oligomeric species have been generated, the 2-mer to the 10-mer, with mass-to-charge ratios which are consistent with either a cyclic structure or a linear structure with an unsaturated end group. Negligible fragmentation is observed in the FD-MS spectrum as a consequence of the low internal energy deposition during the ionisation process. Structural information was obtained from the FD-MS/MS spectrum from the molecular ion of the trimer, m/z 576, which is shown in Figure 23. The low ion currents generated by FD ionisation made MS/MS spectra technically challenging to obtain, but intense ion peaks were observed as a consequence of the combination of the array detector and the high energy band pass of the second mass spectrometer of the ZAB-T. A linear structure with a ketone end group is consistent with the fragmentation pathways observed (Figure 24).

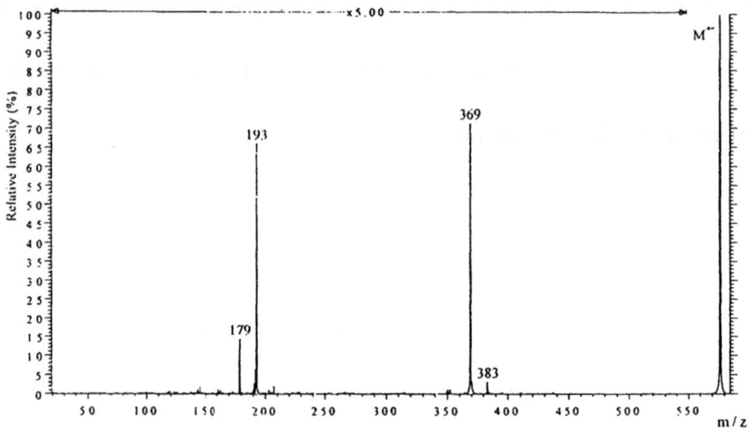

Figure 23. Product Ion Spectrum of a Fluorinated Oligomer, m/z 576

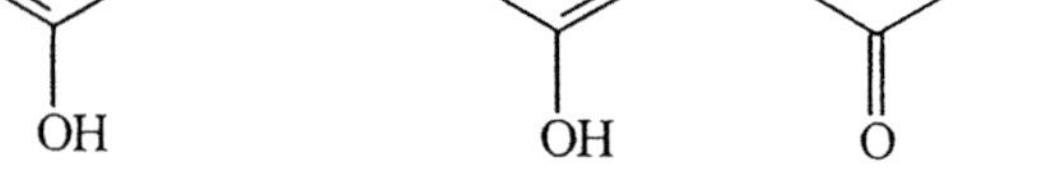

Figure 24. Proposed Fragmentation Scheme for the Fluorinated Oligomer, m/z 576

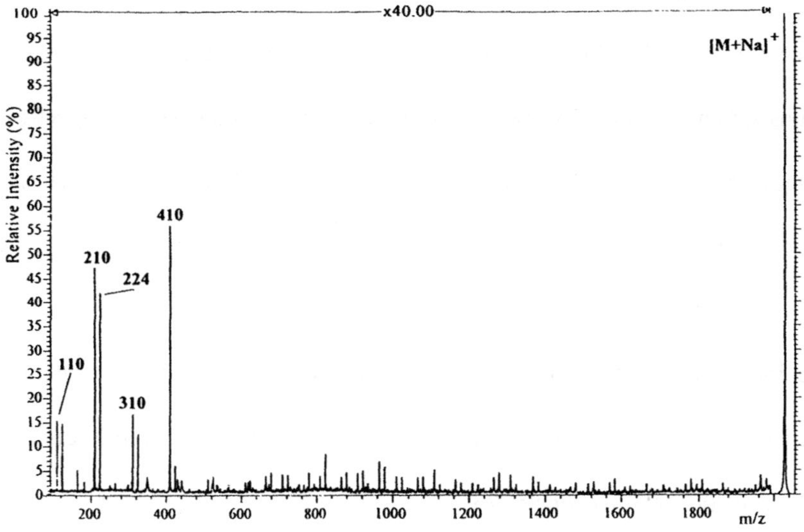

Figure 25. Product Ion Spectrum of Polymethylmethacrylate (PMMA) 2010, $[M+Na]^+$ m/z 2025, Generated by LSIMS Ionisation

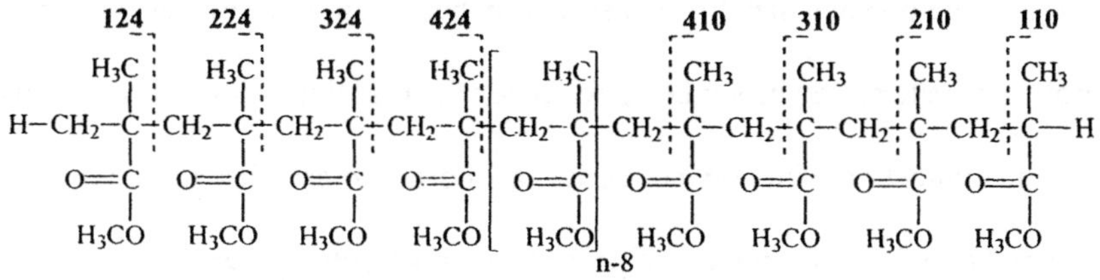

Figure 26. Proposed Fragmentation Scheme for PMMA 2010

Figure 25 is the LSIMS-MS/MS spectrum of m/z 2025, $[M+Na]^+$, from polymethylmethacrylate (PMMA) which is dominated by fragment ions below m/z 500. These fragment ions aid end group identification and were observed in the MS/MS spectra obtained from all oligomers of PMMA studied ranging from the 14-mer to the 25-mer. These fragmentation pathways are shown in Figure 26.

the product ion spectrum of m/z 890, $M^{+\bullet}$, from a low molecular weight polystyrene standard is shown in Figure 27. Two of the fragment ions, at m/z 91 and 161, aid the confirmation of the end groups of the polymer (Figure 28).

FUTURE DEVELOPMENTS

The characterisation of polymer systems using mass spectrometry is a rapidly developing field. The identification of additives and other volatile components using a combination of separation science and mass spectrometry continues to be important, as is the use of thermal methods to probe polymer stability and structure, but recent developments mean that realistic intact polymer systems can be studied directly for the first time. A great deal of work is required, in conjunction with polymer chemists, in order to establish the utility of this approach and to achieve agreement, if possible, with accepted polymer characterisation methods such as size exclusion chromatography (SEC) and nuclear magnetic resonance (NMR) spectroscopy. For matrix asisted laser desorption/ionisation (MALDI), in particular, improvements in resolution, mass range and mass accuracy are needed together with the requirement for the approach to cover the majority of polymer systems. In addition, the capability of performing MS/MS experiments would enhance the data obtained. In all of these areas developments are underway that give cause for optimism that the majority of these

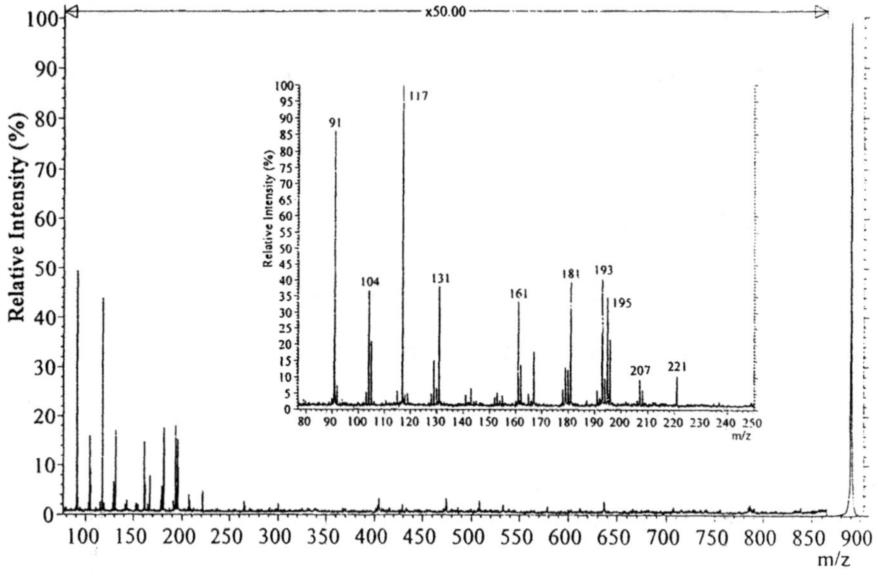

Figure 27. Product Ion Spectrum of Polystyrene 650, $M^{+\cdot}$ m/z 890, Generated by FD Ionisation

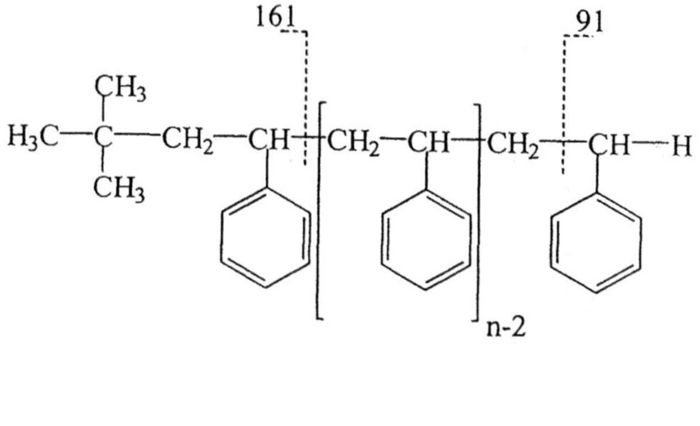

Figure 28 Proposed Fragmentation Scheme for Polystyrene

requirements can, in the future, be realised and it is likely that mass spectrometry will become an increasingly important approach in the armoury of tools for polymer characterisation.

ACKNOWLEDGEMENTS

We would like to thank all those people who contributed to the work reported here. In particular the mass spectrometry team at Wilton: Hilary Yates, Richard Jennings, Mark Redmond, George Paterson and Michael Taylor and from Warwick University: Keith Jennings and Michael Deery.

THEORY OF GAS PHASE ION CHEMISTRY: ENERGETICS, KINETICS AND DYNAMICS OF UNIMOLECULAR AND BIMOLECULAR REACTIONS OF POLYATOMIC IONS

EINAR UGGERUD

Department of Chemistry
University of Oslo
P. O. Box 1033 Blindern
N-0315 Oslo
Norway

1. Introduction

The development in mass spectrometry throughout this century has been formidable. As a result of this the mass spectrometrist now has a range of powerful techniques at hand which helps him tackle the most challenging analytical problems. No matter how advanced the technology becomes, the result of a mass spectrometric investigation is always a mass spectrum. In order to interpret a mass spectrum some basic understanding of the chemical and physical processes which determine the appearance of the spectrum is required. The sub-discipline gas phase ion chemistry has emerged as the result of this and is focused on these matters.

Besides being a supporting discipline for mass spectrometry gas phase chemistry is a discipline in its own right. It has often been pointed out that one great advantage of studying gas phase chemistry (compared to condensed phase chemistry) is that it gives an opportunity to observe reactions in a solvent free environment. Under such conditions the molecules display their pure intrinsic properties, free from interaction with neighbouring solvent molecules. In addition the gas phase situation is much simpler to describe theoretically, because considerably fewer interactions have to be taken into account. One great advantage of studying species which carry electrical charges is that this provides the experimentalist with a unique possibility to control experiments by manipulating the molecules in time and space. This is generally much more difficult to achieve for electrically neutral species.

The purpose of this paper is to outline the basic theory of gas phase ion chemistry. Emphasis will be put on the theoretical description of the chemical reactions which may occur once ions are formed, but not on the ionisation techniques used to produce the ions.

K. R. Jennings (ed.), Fundamentals and Applications of Gas Phase Ion Chemistry, 235–262.

It will be demonstrated how current theories for reaction kinetics and dynamics are derived form first principles, i.e. from the basic laws of physics. In order to arrive at applicable formulae and computational methods a series of approximations must be made. A chain is no stronger than its weakest link, so also for a chain of thoughts. The theories presented here are no longer valid if one of the assumptions made can be proved wrong. It is therefore of utmost importance that the reader is fully aware of this point and that the theoretical formulations should only be applied after critical assessment of the underlying assumptions.

Insight to a chemical reaction can be obtained at different levels. The following hierarchy is useful:

i) Characterisation and identification of products.

ii) Reaction mechanism (the detailed order by which bonds are formed and broken). The mechanism is often inferred from kinetic measurements, consideration of reaction energies, isotope effects, etc. Quantum chemical calculations are often valuable.

iii) Reaction dynamics (the detailed atomic movements during reaction). This is the most detailed level. This requires careful preparation of the reactants (electronic, vibrational, rotational and translational state selection) and determination of the products´ final states. Theoretical modelling is usually of great importance.

In this paper we are mainly concerned about the theoretical description at levels ii) and iii) and the connection between them. Some basic concepts of reaction kinetics and dynamics have to be explained before dealing with the theories themselves. The concept of phase space is of fundamental importance and will introduced after a discussion of the topography of potential energy surfaces, and some basic ideas of chemical reaction dynamics. In the following sections the theoretical description of ion molecule reactions and the statistical theory of unimolecular reactions will be discussed. The last section covers some illustrative and instructive examples taken from the literature. Although numerous equations appear in the text, the approach is intended to be practical. The hope is that this can give the reader the background to start applying theoretical methods in his/her own research.

In gas phase ion chemistry it is only necessary to consider two basic types of elementary reactions, unimolecular and bimolecular. In a unimolecular reaction a polyatomic ion, A^+, decomposes in one elementary step and produces two fragments:

$$A^+ \rightarrow B^+ + C \qquad (1)$$

In order to decompose it is necessary that the reactant has sufficient internal energy to overcome the activation barrier. Many different possibilities exist for preparation of the reactant ion prior to the unimolecular decomposition including collisional activation and photo activation. It may also be formed in a foregoing unimolecular reaction or in an association reaction between an ion and a neutral molecule.

The other reaction type we have to consider is that between an ion and a neutral molecule:

$$B_1^+ + C_1 \rightarrow B_2^+ + C_2 \qquad (2)$$

We call reactions of this type ion/molecule reactions. An ion/molecule reaction always occurs in more than one elementary step, of which only the first, association of B_1^+ and C_1 is truly bimolecular, while subsequent steps usually are unimolecular. Although the two equations above are written for positive ions, they are of course equally valid for negatively charged or multiply charged species.

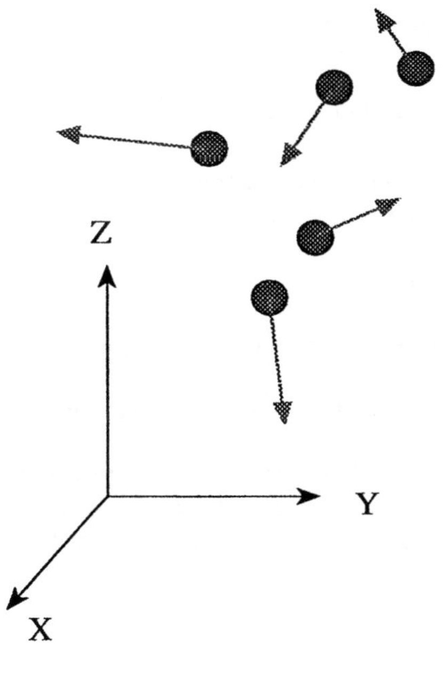

Figure 1. Arbitrary molecular system consisting of N = 5 atoms (only the nuclei are shown). The position of each nucleus, j, is given by its Cartesian position co-ordinates (x_j,y_j,z_j), and Cartesian momentum co-ordinates $(p_{x,j},p_{y,j},p_{z,j})$. As explained in the text this gives rise to a 3N dimensional position vector, $\boldsymbol{q}$, and a 3N dimensional momentum vector, $\boldsymbol{p}$, characterising the system.

2. The Potential Energy Surface (PES)

2.1. SEPARATION OF ELECTRONIC AND NUCLEAR MOTION

A molecule is a mechanical system consisting of N nuclei and n electrons. The position and momentum co-ordinates of the nuclei are given by (see Figure 1)

$$\boldsymbol{q} = \{q_1, q_2, \ldots, q_{3N}\} \tag{3}$$

and

$$\boldsymbol{p} = \{p_1, p_2, \ldots, p_{3N}\} \tag{4}$$

The components of the $\boldsymbol{q}$ vector are the N Cartesian 3-tuples defining the positions of nuclei. The $\boldsymbol{p}$ vector consists of the corresponding momentum co-ordinates. The masses of the nuclei are given by the 3Nx3N diagonal matrix whose elements are given by:

$$m = \mathrm{diag}\{\{m_1,m_1,m_1\},\{m_2,m_2,m_2\}\ldots,\{m_N,m_N,m_N\}\} \tag{5}$$

The position and momentum co-ordinates of the n electrons are given (in non-standard notation) by

$$\boldsymbol{x} = \{x_1, x_2, \ldots, x_{3n}\} \text{ and } \boldsymbol{w} = \{w_1, w_2, \ldots, w_{3n}\} \tag{6}$$

The physical state of the molecule is completely described by the molecule´s wave function, $\Psi(\boldsymbol{x},\boldsymbol{q})$. The total energy is obtained by solving Schrödinger´s equation

$$\mathcal{H}\Psi(\boldsymbol{x},\boldsymbol{q}) = E\Psi(\boldsymbol{x},\boldsymbol{q}) \tag{7}$$

The energy operator (Hamiltonian), $\mathcal{H}$, consists of five parts

$$\mathcal{H}(\boldsymbol{w},\boldsymbol{p},\boldsymbol{x},\boldsymbol{q}) = \mathcal{T}_e(\boldsymbol{w}) + \mathcal{T}_n(\boldsymbol{p}) + \mathcal{V}_{ee}(\boldsymbol{x},\boldsymbol{q}) + \mathcal{V}_{ne}(\boldsymbol{x},\boldsymbol{q}) + \mathcal{V}_{nn}(\boldsymbol{q}) \tag{8}$$

The first two terms are the kinetic energy of the electrons and nuclei. The latter three potential energy terms are due to the forces between electrons (ee), between nuclei and electrons (ne), and between nuclei (nn). The kinetic energy of the nuclei is given by

$$\mathcal{T}_n = \sum_{i=1}^{3N} \frac{p_i^2}{2m_i} \tag{9}$$

In the quantum mechanical formalism $\boldsymbol{p}$ is to be replaced by the expression for the momentum operator. The kinetic energy of the electrons is defined in a completely analogous manner. The key assumption of the Born-Oppenheimer (adiabatic) approximation is that $\mathcal{T}_n$ is small, which means that it can be treated as a perturbation. [1] This means that we can expand the $\boldsymbol{x}$ dependence of the wave function:

$$\Psi = \psi_e(\boldsymbol{x},\boldsymbol{q})\chi_n(\boldsymbol{q}) \tag{10}$$

where the electronic Schrödinger equation is solved by

$$[\mathcal{T}_e + \mathcal{V}_{ee}(\boldsymbol{x},\boldsymbol{q}) + \mathcal{V}_{ne}(\boldsymbol{x},\boldsymbol{q}) + \mathcal{V}_{nn}(\boldsymbol{q})]\psi_e(\boldsymbol{x},\boldsymbol{q}) = \varepsilon(\boldsymbol{q})\psi_e(\boldsymbol{x},\boldsymbol{q}) \tag{11}$$

It can be shown that to a good approximation the following holds true

$$[\mathcal{T}_n + \varepsilon(\boldsymbol{q})]\chi_n(\boldsymbol{q}) = E\chi_n(\boldsymbol{q}) \tag{12}$$

With this procedure the electronic and nuclear parts of the Hamiltonian have been separated. The parametric dependence of the nuclear potential energy upon the position co-ordinates of the nuclei is noticed

$$\mathcal{V} = \varepsilon(\boldsymbol{q}) \tag{13}$$

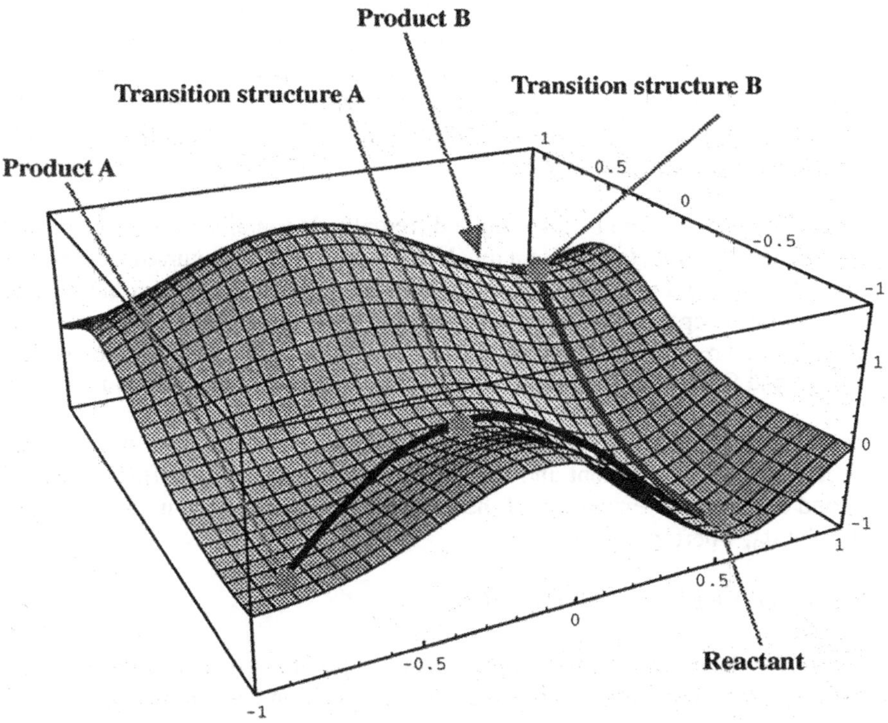

Figure 2. A model potential energy surface showing a reactant (local minimum) which may react via two different routes to give the products A and B (local minima). The two transition structures correspond to saddle points. The indicated lines connecting the points (reactant, TS, product) are the intrinsic reaction co-ordinates (the model PES has been constructed by Prof. H. B. Schlegel, Wayne State University, and is used in agreement with the constructor).

This potential energy function is usually referred to as the potential energy surface (PES). The reason for this is that the potential energy function can be represented as a (3N+1) dimensional hyper surface with respect to the dimensions of what is called configuration space (the space spanned by all possible values of $\boldsymbol{q}$). If the electronic energy and wave function is determined for all possible values of $\boldsymbol{q}$, the motion of the nuclei can be completely described by equation (12). How this is done is the topic of the chapter on reaction dynamics.

2.2. STATIONARY POINTS: SEPARATION OF NUCLEAR CO-ORDINATES

The topography of potential energy surfaces is of fundamental importance in chemical reaction theory. [2] Of particular interest to the chemist is the notion of stationary points. A stable molecule is identified by a local minimum, corresponding to the equilibrium molecular geometry. The molecule resides in a narrow region (potential energy well, or perhaps better a valley bottom of the potential energy surface) of configuration space. During its existence the molecule traverses this local region around the minimum, and is incapable of leaving it, required that the total energy is sufficiently small and that the potential energy well is sufficiently deep. The molecular properties (bond lengths, vibrational frequencies, moments of inertia) are determined from the detailed topography of the potential energy surface in this region.

A reactive molecular system behaves quite differently. It contains excessive amounts of energy and is therefore able to explore a much larger region of configuration space than a stable molecule. During a reaction the molecular system has to pass from the reactant region of configuration space into the product region. This means that a representative point, $\boldsymbol{q}$, has to leave a reactant configuration (one local minimum), pass a region of high potential energy, and finally enter a product configuration (another local minimum).

In order to treat these situations in a general way it is necessary to bring in two topological definitions, the gradient and the Hessian. These definitions follow directly from the second order Taylor expansion of the potential energy function in the vicinity of a reference point $\boldsymbol{q}_0$, where $\boldsymbol{q} = \boldsymbol{q}_0 + \Delta\boldsymbol{q}$

$$\mathcal{V}(\boldsymbol{q}) = \mathcal{V}(\boldsymbol{q}_0) + \boldsymbol{g}^{\mathrm{T}}\,\Delta\boldsymbol{q} + 1/2\,\Delta\boldsymbol{q}^{\mathrm{T}}\,H\,\Delta\boldsymbol{q} \tag{14}$$

The gradient of the potential energy function at any point is given by a vector whose components are the first derivatives with respect to each position co-ordinate, q_i:

$$\boldsymbol{g}(\boldsymbol{q}) = \left\{\frac{\partial\mathcal{V}(\boldsymbol{q})}{\partial q_1},\frac{\partial\mathcal{V}(\boldsymbol{q})}{\partial q_2},\ldots,\frac{\partial\mathcal{V}(\boldsymbol{q})}{\partial q_{3\mathrm{N}}}\right\}^{\mathrm{T}} \tag{15}$$

This means that the gradient is the negative of the force. The Hessian, the second derivative matrix, defines the curvature of the potential energy surface at each point $\boldsymbol{q}$:

$$H(\boldsymbol{q}) = \begin{pmatrix} \frac{\partial^2 V(\boldsymbol{q})}{\partial q_1 \partial q_1} & \frac{\partial^2 V(\boldsymbol{q})}{\partial q_1 \partial q_2} & \cdots & \frac{\partial^2 V(\boldsymbol{q})}{\partial q_1 \partial q_{3N}} \\ \frac{\partial^2 V(\boldsymbol{q})}{\partial q_2 \partial q_1} & \frac{\partial^2 V(\boldsymbol{q})}{\partial q_2 \partial q_2} & \cdots & \frac{\partial^2 V(\boldsymbol{q})}{\partial q_2 \partial q_{3N}} \\ \cdots & \cdots & \cdots & \cdots \\ \frac{\partial^2 V(\boldsymbol{q})}{\partial q_{3N} \partial q_1} & \frac{\partial^2 V(\boldsymbol{q})}{\partial q_{3N} \partial q_2} & \cdots & \frac{\partial^2 V(\boldsymbol{q})}{\partial q_{3N} \partial q_{3N}} \end{pmatrix} \tag{16}$$

A point $\boldsymbol{q}$ where the gradient vanishes, i.e. with $\boldsymbol{g}(\boldsymbol{q}) = \boldsymbol{0}$, is called a stationary point. We are interested in two types of stationary points: minima and transition structures (Figure 2).

If all eigenvalues of H are positive (except those corresponding to overall translation and rotation, which are zero) a stationary point $\boldsymbol{q}$ is a local potential energy minimum. Minima correspond to stable molecules (deep wells) or short lived intermediates (shallow wells).

If all eigenvalues of H are positive, except one which is negative, we term the stationary point a transition structure (TS). From differential calculus the equivalent term a saddle point may be known. A transition structure is a saddle point of first order. Stationary points of this type mark the transition from one minimum structure region into another. In this context we are interested in the exact location of the watershed between the reactant valley and the product valley. Several transition structures may exist for one reaction, but usually we are interested in the one which has lowest potential energy. Transition structures are of key importance in unimolecular reaction theory, as will be evident later. The negative eigenvalue of the Hessian H (with its accompanying imaginary frequency) belongs to a eigenvector which is called the reaction co-ordinate. The exceptional property of this mode of vibration is that the potential energy decreases upon displacement from the stationary point (negative curvature).

During a reaction the $\boldsymbol{q}$ vector traces out a path on the PES, $\boldsymbol{q}(t)$, passing from the reactant region to the product region, determined by the equations of motion. We call a line of this type a reaction trajectory. A reaction trajectory makes many visits into remote regions of the configuration space before it finds its way to the products, so in general it is a long and complex line. Because of this and because all actual reaction trajectories are different (they depend on the initial conditions), it would be practical if we instead could determine one simple line on the PES which describes the reaction progress from reactant to product. In fact a unique line which has some very useful properties can be defined. It is called the intrinsic reaction co-ordinate (IRC). [3] The IRC defines the reaction co-ordinate, not only at the transition structure, but everywhere on the line which is set to pass directly through the reactant, transition structure, and product. Because it follows the gradient extremal downhill from the transition structure in both directions, it depends on the shape of the potential energy surface. The choice of co-ordinate system influences the line. It has been found that use of mass weighed Cartesian co-ordinates is best because this leads to separation of the Hamiltonian in a way which allows for vibrational co-ordinate analysis everywhere on the IRC.

Separation of the three overall translation and three rotation degrees of freedom leaves 3N-6 (3N-5 for linear molecules) so-called internal co-ordinates. Technically this is achieved by projection of the 3Nx3N Hessian into a (3N-6)x(3N-6) subspace by applying the so-called Eckart condition. [4] This procedure is only exactly valid at a stationary point in the sense that the rotational degrees of freedom only have zero eigenvalues of the Hessian there. The most often used approximation to achieve separability of the remaining 3N-6 internal degrees of freedom is to assume that they can be treated as 3N-6 one-dimensional harmonic oscillators. This is equivalent to truncating the Taylor expansion after the second term as we did in Equation (14) above. To obtain the 3N-6 harmonic frequencies of vibration Wilson's FG-matrix method is used. [4] These are in general slightly different from true experimental frequencies due to the second order approximation. In the case of a minimum all frequencies are real, but for a transition structure all are real except for the frequency of vibration of the reaction co-ordinate which is imaginary.

The second order approximation ensures separability of the Hamiltonian giving 3N-6 independent real harmonic oscillators for a minimum, and 3N-7 real plus one imaginary for a transition structure. One valuable trade-off with the definition of the IRC above is that it guarantees that there are 3N-7 independent real frequencies also outside the transition structure. The remaining co-ordinate is the reaction co-ordinate. This means that it is possible to evaluate frequencies of vibration and related properties (densities of states and zero point vibration energy) all the way along the reaction co-ordinate.

A reaction with a potential energy surface which has a well-defined transition structure separating the reactant and product regions is termed a Type II reaction. Many reactions do not have potential energy barriers on the product side, which means that a transition structure is lacking. A reaction of this kind, termed a Type I reaction, therefore proceeds along a potential energy asymptote towards the product side. These two reaction types are illustrated in Figure 3.

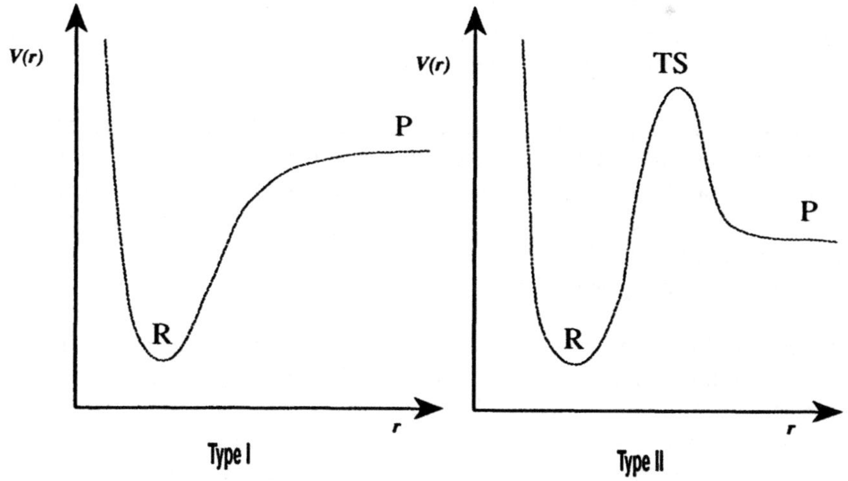

Figure 3. Potential energy functions for Type I and Type II reactions. In a type I reaction there is no barrier for the reverse reaction. Direct heterolytic bond cleavages, e. g. $CH_3OH_2^+ \rightarrow CH_3^+ + H_2O$, often give rise to potentials of this type. A type II reaction is accompanied by a significant barrier for the reverse reaction. Reactions which require simultaneous formation and dissociation of bonds, e. g. $CH_3OH_2^+ \rightarrow CH_2OH^+ + H_2$, often give rise to potentials of this type.

2.3. PRACTICAL CALCULATION OF POTENTIAL ENERGY SURFACES

There are two principally different approaches which are used to obtain potential energy surfaces. The potential may be obtained by a suitable parametrisation based on experimental quantities (bond energies, bond distances, vibrational frequencies etc.), and is then called empirical. Alternatively, and usually better, the potential is calculated *ab initio*. The great advantage of *ab initio* potentials is that they in principle are globally valid, while empirical potentials may be very accurate around minima, but have to rely on numerous assumptions in regions of the potential energy surface where experimental numbers are missing. Due to the simple parametric dependence, empirical potentials can be evaluated quickly for any point $\boldsymbol{q}$. The *ab initio* method is often very time consuming, and it is tedious and often prohibitive to calculate $\mathcal{V}(\boldsymbol{q})$ for the number of points necessary to give a sufficiently precise description. For this reason a common procedure is to calculate $\mathcal{V}(\boldsymbol{q})$ for a limited number of points and to use a parametrisation to interpolate between them.

The quality of the wave function is a point of great concern in *ab initio* calculations. Because bonds are broken and formed during a reaction, inclusion of a method for proper handling of electron correlation is required. Various *ab initio* schemes are available and can be used depending on the desired chemical accuracy. For most applications a medium sized basis set (e.g. 6-31G(d,p)) used in conjunction with a method for electron correlation (Møller-Plesset perturbation theory, configuration interaction, complete active space) turns out to be sufficiently accurate. Recently, density functional theory based methods have spread in use and may often be useful. Use of semi-empirical methods appears to be in decline these days, except for large systems.

From *ab initio* methods potential energies, gradients and Hessians are readily available. The first and simplest step to be taken in describing a reaction is to locate the geometries of the minima; reactants (q_{reac}) and products (q_{prod}). This is usually achieved by geometry optimization starting from an experimentally known or guessed structure. Once the optimum structures are obtained, the potential energy, vibrational frequencies, and moments of inertia are easily determined.

The next step is to locate the transition structure if the reaction is of type II. This is often difficult, and requires some experience and skill. The simplest approach is to interpolate between the geometries q_{reac} and products q_{prod}, and use this as the starting point for the geometry optimization procedure. This approach often fails because the guessed structure lies outside the region of the potential energy surface which contains the actual transition structure (q_{ts}). In such cases more elaborate methods exist. This author has had good experience with a method called linear synchronous transit. [5] An extended version, quadratic synchronous transit, appears promising, but is new to the author. If these and other systematic techniques also fail, the worker has to rely on trial-and-error. In such cases experience with similar cases turns out to be valuable in picking a good starting geometry by hand.

When a stationary point is located the Hessian has to be checked to see if it has the correct number of real and imaginary eigenvalues for a transition structure, and the reaction co-ordinate must be checked to see that its displacement vector has the expected features for the chemical transformation under investigation. Sometimes erroneous transition structures are reported because these simple checks have not been done. Once a candidate q_{ts} is located in this manner the IRC should be calculated to assure the correct correspondence between q_{ts}, q_{reac} and q_{prod}. A thorough search of the PES for other possible transition structures for the reaction must also be done to make sure that passage through the found transition structure represents the minimum energy path.

For a type I reaction, a transition structure will not be found. In this case a procedure for following the reaction path must be invoked to ensure that the potential energy increases monotonically from reactants to products. [6]

Once the points q_{ts}, q_{reac} and q_{prod} have been located, the potential energies $\mathcal{V}(q_{ts})$, $\mathcal{V}(q_{reac})$ and $\mathcal{V}(q_{prod})$, and the corresponding vibrational frequencies are known. The zero point vibrational energies, E_{zpv}, are obtained from the calculated vibrational frequencies after scaling. The reaction energy is obtained as $\Delta E = \mathcal{V}(q_{prod}) + E_{zpv}(q_{prod}) - (\mathcal{V}(q_{reac}) + E_{zpv}(q_{reac}))$. For a type II reaction the reaction's critical energy (this definition is more precise than the more often used term activation energy) is given by $E_0 = \mathcal{V}(q_{ts}) + E_{zpv}(q_{ts}) - (\mathcal{V}(q_{reac}) + E_{zpv}(q_{reac}))$. For a type I reaction $E_0 = \Delta E$. Once these relative energies have been calculated they should be compared with available experimental data. If the agreement turns out to be unsatisfactory, this probably means that the wave function is ill-behaved, and a better should be used.

Nowadays, *ab initio* program packages are available for computers covering the whole range from personal computers to main frame super computers. The most widely used *ab initio* program suites includes GAUSSIAN 92 (and GAUSSIAN 94) [7], HONDO [8], GAMESS [9]. The package MOPAC [10] includes most semi-empirical MO methods.

3. Chemical Reaction Dynamics

3.1. HAMILTON'S EQUATIONS

The time evolution of a chemical system is generally described by Schrödinger's time dependent equation. Various approximate methods exist, the most basic of which includes electron-nuclear coupling (thereby disregarding the B-O approximation) and can only be applied to very small systems. [11] When the Born-Oppenheimer approximation is invoked, application of wave packet dynamics for the nuclei has been demonstrated for systems having up to four atoms. [12]

It is more tractable when the nuclear motion can be determined by using the corresponding classical equations of motion because this reduces computational efforts considerably. Another advantage of classical dynamics is that the results can be inspected visually by means of molecular model animation. The question is then how accurate the classical description is. It would generally be expected that the motion of heavy nuclei are more likely to be well-described in the classical formalism than light nuclei (e.g. hydrogen). To which extent the classical approximation holds can only be determined by direct comparison with a corresponding quantum calculation. It turns out that some properties are poorly reproduced in the classical approximation, while others are close to the quantum result. [13] In most cases it appears that use of classical dynamics can be justified.

The classical motion of the nuclei are determined from Newton's second law:

$$m_i \frac{d^2 q_i}{dt^2} = -\frac{\partial \mathcal{V}(\boldsymbol{q})}{\partial q_i} \tag{17}$$

The definitions of the individual entries in this equation have already been given. It is often more convenient to change to the equivalent Hamilton formalism of the equations of motion. The dynamics of a system is completely specified by the Hamiltonian, $\mathcal{H}(\boldsymbol{p},\boldsymbol{q},t)$. As already indicated, the value of the Hamiltonian is identified as the energy of the system. Assuming that there is no explicit time-dependence, which usually is the case for a molecular system in the absence of time-dependent electromagnetic fields, energy is conserved.

$$E = \mathcal{H}(\boldsymbol{p},\boldsymbol{q}) = \text{constant} \tag{18}$$

We then have (Hamilton's equations of motion) [14]:

$$\frac{\mathrm{d}q_i}{\mathrm{d}t} = \frac{\partial \mathcal{H}}{\partial p_i} \tag{19}$$

$$\frac{\mathrm{d}p_i}{\mathrm{d}t} = -\frac{\partial \mathcal{H}}{\partial q_i} \tag{20}$$

The Hamiltonian is given by $\mathcal{H} = \mathcal{T}(\boldsymbol{p}) + \mathcal{V}(\boldsymbol{q})$ (from now on we drop the subscript n in the kinetic energy of the nuclei.) Any set of momentum and position co-ordinates $\{\boldsymbol{p},\boldsymbol{q}\}$ which obey equations (19) and (20) are called canonical co-ordinates. This gives rise to 6N coupled, first order differential equations. Solution of the equations requires that the initial conditions $(\boldsymbol{p},\boldsymbol{q})$ at time $t = 0$ are specified. Although an analytical solution generally is out of question, one very useful feature of the equations when cast in Hamilton´s form is that they are suited for direct numerical integration by digital computers. Using this approach it is quite straightforward to calculate a reaction trajectory, once a model of the PES is at hand. At the end of the trajectory (when reaction has occurred) the dynamical situation is inspected, and the energy distribution of the products can be determined.

The initial conditions determine the outcome of a trajectory calculation. Great care must be taking in choosing them in accordance with the experiment to be modelled. In order to determine reaction probabilities and statistically averaged energy distributions of the products, a large number of trajectories have to be calculated. Efficient procedures for random sampling of the initial conditions (e.g. from a Maxwell-Boltzmann distribution) exist, and they are called Monte Carlo methods.

Traditionally, analytical potentials (empirical and/or *ab initio*) have been used for classical trajectory calculations. Thanks to progress in computer hardware and more efficient algorithms for the time consuming steps during *ab initio* computation, [15, 16] *ab initio* direct dynamics calculations of gas phase reactions have now become possible. Each reaction trajectory is obtained by step-by-step integration of Newton´s equations for motion of the nuclei. One great advantage of this direct approach is that because $\mathcal{V}(\boldsymbol{q})$, $\boldsymbol{g}(\boldsymbol{q})$ and $H(\boldsymbol{q})$ are calculated *en route*, it becomes unnecessary to construct a complicated multidimensional analytical model potential energy function prior to the dynamical calculation. One can for example apply Equation (14) to obtain a local approximation of the PES.

Neglect of quantum effects is inherent in the classical description. The most important of these are zero point vibrations, potential barrier scattering and tunnelling. To "repair" the classical description various semi-classical approaches have been designed. Zero-point vibrational energy can be included by adding 1/2 $h\nu$ of energy to each mode of vibration initially, but non-trivial mathematical difficulties are encountered when this energy is spread around as the reaction proceeds. This problem and other aspects of semi-classical reaction dynamics are discussed in a relatively recent account. [17]

3.2. PHASE SPACE AND STATISTICAL MECHANICS

We have already introduced the concept of a trajectory; a line $\boldsymbol{q}(t)$ traced out on the PES during time, for a given set of initial conditions. It is clear that this definition does not describe the complete dynamical situation of the system. To achieve this it is also necessary to specify the corresponding $\boldsymbol{p}(t)$ vector. Simultaneous specification of both $\boldsymbol{q}$ and $\boldsymbol{p}$, leads to the 6N dimensional vector, denoted the representative point, or phase-space co-ordinates, $\Gamma(t) = \{\boldsymbol{q}(t),\boldsymbol{p}(t)\}$. The corresponding 6N dimensional space is

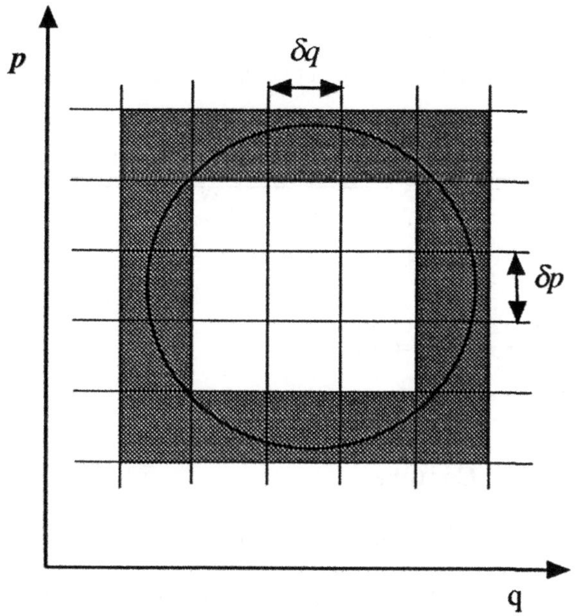

Figure 4. Cells in phase space. The circle (ellipse) represents the trajectory of a harmonic oscillator with energy, E. The Hamiltonian for this system is given by $\mathcal{H} = p^2/2m + 1/2\ kq^2$. The allowed cells for $\mathcal{H} = E$ are indicated by grey shading.

termed the phase space, and is a consequence of Hamilton's formalism. We have already pointed out that it is possible to calculate a set of trajectories $\{\Gamma(t)\}$ from a large number of different statistically sampled initial conditions $\{\Gamma(0)\}$ by solving Hamilton's equations. For large systems this is prohibitive, and statistical theories are needed.

For each pair of canonical co-ordinates q_i and p_i, phase space can be divided into cells of size δp_i times δq_i. [18] The cell size can be chosen arbitrary small, but is usually chosen to be equal to Planck's constant, h, for correspondence to quantum mechanics (Figure 4). This division is equally valid for all components of $\boldsymbol{q}$ and $\boldsymbol{p}$, so a division of phase space into multidimensional cubes of volume h^{3N} can be accomplished. The state of a system is specified by one distinguishable dynamical situation, corresponding to a cell in phase space. It follows from statistical mechanics that the number of states, $G(E)$, of a system is given by the phase space volume divided by h^{3N}. The phase space volume for a system, V, with energy, E, is given by the accessible volume of phase space imposed to the restriction that the energy lies between 0 and E: One therefore has:

$$G(E) = \frac{V}{h^{3N}} = \frac{\int_{\mathcal{H}=0}^{\mathcal{H}=E} \ldots \int dp_1..dp_{3N}dq_1..dq_{3N}}{h^{3N}} \tag{21}$$

The number of states in the energy range between E and $E + dE$ is thereby given by

$$W(E) = G(E + dE) - G(E) = \frac{\int_{\mathcal{H}=E}^{\mathcal{H}=E+dE} \ldots \int dp_1..dp_{3N}dq_1..dq_{3N}}{h^{3N}} \tag{22}$$

and the density of states by

$$N(E) = \frac{G(E + dE) - G(E)}{dE} = \frac{W(E)}{dE} \tag{23}$$

4. Theory of Ion/Molecule Capture

4.1. LANGEVIN THEORY (ION/INDUCED DIPOLE POTENTIAL ONLY)

A reaction between an ion and a neutral molecule is usually considered to proceed in three stages [19]:

$$B_1^+ + C_1 \underset{k_{b.diss}}{\overset{k_{cap}}{\rightleftarrows}} (B_1^+ \cdots C_1) \underset{k_{-1}}{\overset{k_1}{\rightleftarrows}} (B_2^+ \cdots C_2) \underset{k_{b.cap}}{\overset{k_{diss}}{\rightleftarrows}} B_2^+ + C_2 \tag{24}$$

The first step involves formation of a complex by a capture mechanism which is to be described in detail below. The second step is the actual chemical transformation which must take place during the lifetime of the complex. In the third step the product ion molecule complex dissociates to the give the free products. As written in equation (24) all steps are in principle considered to be reversible, although in practice some may be irreversible. The rate of the first step, k_{cap}, formation of the capture complex puts an upper limit to over-all rate, k. As indicated the complex may dissociate back to give the reactants, or react in the following unimolecular step. The efficiency of an ion molecule reaction can be determined by comparing the experimental over-all rate with a theoretical estimate of the capture rate constant:

$$\gamma = \frac{k_{exp}}{k_{cap}} \tag{25}$$

We will now briefly describe the simplest method for estimating k_{cap} which is called the Langevin model after its originator, [20] and the capture rate coefficient is in this case renamed to $k_{\mathcal{L}}$. More detailed treatments given here are found in the literature, and the interested reader should work through the details. [21] In the Langevin model the ion is treated as a point charge and the neutral molecule does not have a permanent dipole moment. To a good approximation (valid at long intermolecular distances, r) the interaction potential is of the ion/induced dipole type

$$\mathcal{V}(r) = -\frac{e^2\alpha'}{8\pi\varepsilon_0 r^4} \tag{26}$$

In this expression e is the elementary charge, α' the polarizability volume of the neutral (in m^3), and ε_0 is the permitivity of free space. This is a central potential, which simplifies the evaluation considerably. The most direct way to obtain the capture rate constant is by analyzing the ion/molecule system in their common centre-of-mass system. Let E be the total energy of the collision system (internal energy not included), given by the reduced mass, μ, and the initial relative velocity, v_0

$$E = \frac{1}{2}\mu v_0^2 \tag{27}$$

and $\mathcal{L}$, the orbital angular momentum, given by

$$\mathcal{L} = \mu v_0 b \tag{28}$$

The impact parameter, b, is defined in Figure 11. At any instance the relative velocity, v, of the two reaction partners has a component along the intermolecular axis, v_r, and a component normal to it, $v_{\mathcal{L}}$. The Hamiltonian is then written in the form

$$\mathcal{H} = \mathcal{T}_r + \mathcal{T}_{\mathcal{L}} + \mathcal{V}(r) \tag{29}$$

with

$$\mathcal{T}_r = \frac{p_r^2}{2\mu} \tag{30}$$

By taking momentum conservation into account we see that the kinetic energy term corresponding to the angular momentum can be reduced to a form which for a given value of E, only depends on b and r :

$$\mathcal{T}_{\mathcal{L}} = \frac{\mathcal{L}^2}{2\mu r^2} = \frac{Eb^2}{r^2} \tag{31}$$

This gives a Hamiltonian in the canonical variables, p_r and r. By having established these relationships it quite straightforward to determine the upper limit of the impact parameter, b_0, which for a given value of E leads to capture (Figure 5). Proper averaging over a Maxwell-Boltzmann distribution of E gives the surprisingly simple expression:

$$k_{\mathcal{L}} = e\left(\frac{\pi\alpha'}{\varepsilon_0\mu}\right)^{\frac{1}{2}} \tag{32}$$

It can be noticed that the capture rate coefficient in this approximation is independent of the temperature.

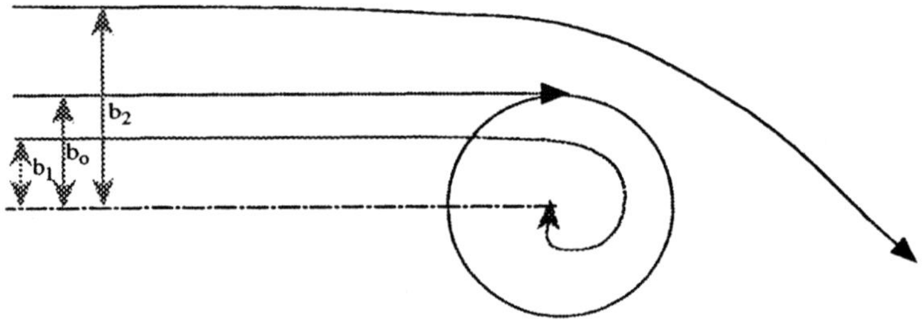

Figure 5. Ion/molecule collisions showing three different trajectories. The trajectories (starting from the left side of the figure) have the same initial energy, E, but differ in the value of the impact parameter, b, and thereby the orbital angular momentum, $\mathcal{L}$. The trajectory with $b = b_0$ represents the upper limit which gives capture. In this case the system is caught in a circular orbit. For any larger value of $b = b_2 > b_0$ capture does not occur, while for $b = b_1 < b_0$ capture occurs and the trajectory collapses towards the centre in a spiral path. This is an unrealistic situation, because real potentials are strongly repulsive at short distances. Capture is a necessary, but not sufficient condition for reaction, because a chemical barrier has to be passed as well.

4.2. CAPTURE THEORIES INCLUDING ION/PERMANENT DIPOLE POTENTIAL

The above result is, however, usually inadequate. It is found that for molecules with permanent dipole moments equation (32) underestimates the capture rate coefficient typically by a factor of two. When the ion/permanent dipole moment term is included the potential appears as follows:

$$\mathcal{V}(r) = -\frac{e^2\alpha'}{8\pi\varepsilon_0 r^4} - \frac{\mu_D e \cos\theta}{4\pi\varepsilon_0 r^2} \tag{33}$$

In this expression μ_D is the permanent dipole moment of the neutral molecule, and θ is the instantaneous angle between the dipole and the line of interaction. Unfortunately, this leads to a non-central potential, for which no simple analytical expression for k_{cap} is at hand. Several approximations have be used to solve the equations in line of thinking with the Langevin approach. Of these the locked dipole approximation (with $\cos\theta = 1$) [22] and the averaged oriented dipole (ADO) approximation [23] should be mentioned. In other approaches $\mathcal{V}(r)$ is averaged by employing statistical mechanics in various ways. [24] These methods turn out to give quite good agreement with experimentally determined rate coefficients. The most widely employed method nowadays originates from a combined microcanonical variational transition state and trajectory calculation study. From this study an empirical parametrisation was obtained which makes it possible to approximate k_{cap} within a limit of ± 10 %. [25] The rate coefficient is then expressed by

$$k_{cap} = k_L G(x) \tag{34}$$

where

$$G(x) = 0.4767x + 0.62, \textit{ when } x \geq 2 \tag{35a}$$

or

$$G(x) = \frac{(x + 0.509)^2}{10.526} + 0.9754, \textit{ when } x < 2 \tag{35b}$$

The auxiliary quantity x is given by

$$x = \frac{\mu_D}{(8\pi\varepsilon_0 \alpha k_B T)^{\frac{1}{2}}} \tag{36}$$

In this equation k_B is Boltzmann constant and T is the temperature.

5. Unimolecular Reaction Theory

5.1. TYPE II REACTIONS: MICROCANONICAL (RRKM) THEORY

In this section we will take a closer look at statistical methods for obtaining rate coefficients for unimolecular reactions. In principle (not always so in practice) it is possible to calculate the rate of a reaction by randomly choosing a large number of initial phase space co-ordinates $\{\Gamma(0)\}$ of the reactant and then perform dynamical calculations to determine the actual trajectory for each of them. By carefully choosing a criterion for when a reaction has occurred, it is possible to monitor the exact lifetime of the reactant, $T(\Gamma(0))$, for each $\Gamma(0)$. The lifetime is the time it takes to reach the product side. The proper statistically averaged lifetime $<T(\Gamma(0))>$ based on possible initial conditions can then be evaluated.

Some assumptions have to be made in order to apply statistical mechanical methods to the problem. The theoretical foundation to be outlined here is what commonly is known as the RRKM theory. [26] The term quasi equilibrium theory is also used for the same theory. The analysis is applied to a microcanonical ensemble of the reacting system. A microcanonical ensemble is a large number of identical systems with a given energy, E, within a narrow range dE. The theory can easily be extended to a canonical ensemble (which is a system with a well-defined temperature) by proper averaging over a Maxwell-Boltzmann energy distribution.

The key quantity to be calculated is the microcanonical rate coefficient, $k(E)$. As indicated the rate coefficient is a function of the energy, and more generally of other constants of motion. The rate coefficient is defined in the traditional phenomenological fashion:

$$\frac{\mathrm{d}[\mathrm{A}^+]}{\mathrm{d}t} = -k(E)\,[\mathrm{A}^+] \tag{37}$$

The number density of reactant molecules, $[A^+]$, is seen to decay exponentially. The rate coefficient therefore represents the fraction of molecules A^+ whose reaction trajectories leave the reactant region per time unit. From this it follows that the average lifetime is given by

$$< T[\Gamma(0)] > \; = \lim_{T \to \infty} \frac{\int_0^T T e^{-k(E)t} dt}{\int_0^T e^{-k(E)t} dt} = \frac{1}{k(E)} \tag{38}$$

To proceed further two basic assumptions must be made. A recent account gives a critical discussion of the validity of these assumptions. [27]

Assumption 1: There exists a dividing surface of dimension 6N-1 in phase space which separates the reactant and product regions (Figure 6). A reaction has occurred as soon as the dividing surface is crossed. In order to be valid this requires that once a trajectory has crossed the dividing surface it never recrosses back to the reactant region. To simplify matters further it is assumed that the dividing surface is a function of the co-ordinates, $\boldsymbol{q}$, only. A species of transient existence called the transition state is now introduced. A transition state is the point where the dividing surface crosses the IRC. A transition state and a transition structure are not necessarily the same. For a type II reaction the dividing surface is usually chosen so it coincides with the transition structure. How the transition state is chosen for a type I reaction will be explained in the next sub-section.

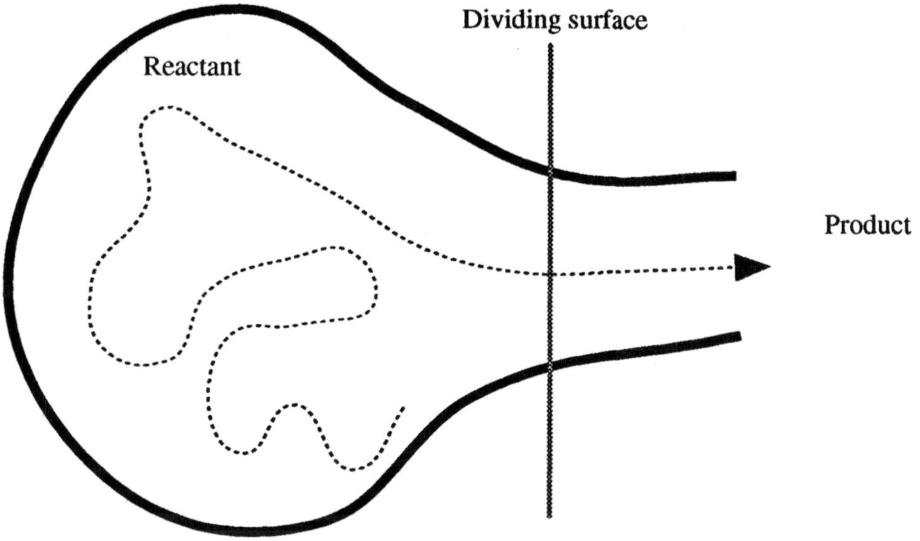

Figure 6. In RRKM theory it is assumed that reaction has occurred as soon as the dividing surface in phase space has been traversed.

Assumption 2 This assumption is often referred to as the ergodicity assumption. It says that for a given energy $\mathcal{H} = E$, all states have equal weight in the microcanonical ensemble. This means that all cells of phase space allowed from the restriction $\mathcal{H} = E$ are uniformly visited in time. Thus the time average of any measurable property, F, can be substituted by the phase space average:

$$\langle F \rangle = \lim_{T \to \infty} \frac{1}{T} \int_0^T F(t)\, dt = \frac{\iint F(q,p) dq\, dp}{\iint dq\, dp}, \quad \text{keeping } \mathcal{H} = E \tag{39}$$

The measurable property which we want to evaluate is the average lifetime - or equivalently the microcanonical rate coefficient. The method by which the reactant molecules are exited (laser excitation, electron impact ionization, thermal collisions) determines the initial state. Irrespective of what the initial states are, the ergodicity assumption requires that a uniform distribution of the representative points, $\Gamma(t)$ is achieved on a time scale short compared to $k(E)^{-1}$. This means that there must exist efficient mechanisms for energy transfer between the system's internal degrees of freedom.

It is also necessary to assume separability of the Hamiltonian in the transition state as discussed previously. We will drop the details of how the expression for $k(E)$is derived. This can be found elsewhere. The result is surprisingly simple, and one needs only to consider the known (e.g. from *ab initio* calculations) properties of the reactant molecule and the transition state (Figure 7):

$$k(E) = \frac{G^{\ddagger}(E - E_0)}{h\,N(E)} \tag{40}$$

$G^{\ddagger}(E - E_0)$ is the number of states (equation 21) of the transition state. The limits of integration are from $\mathcal{H} = E_0$ to $\mathcal{H} = E$. $N(E)$ is the density of states of the reactant molecule, and h is Planck's constant. The dagger is to remind us that we are dealing with the transition state.

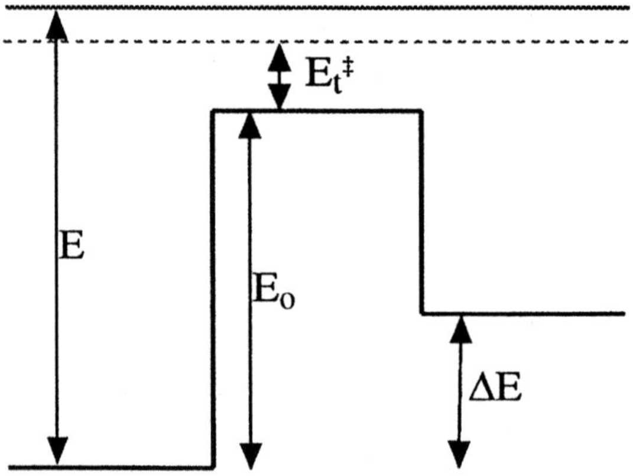

Figure 7. The various energy components which enters the RRKM equations. E_0 is the critical energy, E is the total energy of the system, ΔE is the reaction energy, and $E_t{}^{\dagger}$ is the energy in the reaction co-ordinate in the transition state.

In RRKM theory the internal degrees of freedom are assumed to exchange energy freely. Normally this includes all vibrational degrees of freedom. In addition one of the external rotational modes are sometimes also included. Upon evaluating the numerator and the denominator all internal degrees of freedom which are assumed to exchange energy freely must be included. When external rotational degrees of freedom are included in the transition state they must also be included in the reactant. It is also important to take reaction path degeneracy into consideration when one or both species possess elements of symmetry.

Equation (40) is well suited for practical calculations of $k(E)$. Because the classical formalism used allows for direct correspondence with quantum mechanics, the normal procedure is to substitute the numerator and denominator of equation (40) with the corresponding quantum mechanical analogues. Computer programs used for counting states (to obtain $G^{\ddagger}$ and N) takes advantage of the rapid Beyer-Swinehart algorithm [28] which is based on a well known convolution theorem of statistical mechanics. The reader is advised to obtain one such program, and start to use it guided by the instructions given in the excellent textbooks of Forst, [29] and Gilbert and Smith [30]. After having acquired some experience it is wise to work through the theoretical framework. Rate coefficients evaluated according to the scheme given above are sometimes found to be too large because anharmonic terms in the vibrational Hamiltonian are neglected. It is possible to correct for this.

Some quantum mechanical effects are implicitly taken into account by the procedure explained above. Corrections for differences in zero point vibrational energies of the reactant and the transition state are automatically included in evaluation of the critical energy, E_0. In addition the quantum mechanical $G^{\ddagger}$ and N are used. One important quantum mechanical effect, potential energy barrier tunnelling, has not yet been considered. For light atoms (hydrogen, deuterium) this possibility must always be examined, because the effect often is substantial. Miller has given a description on how equation (40) can be modified to include tunnelling. [31]

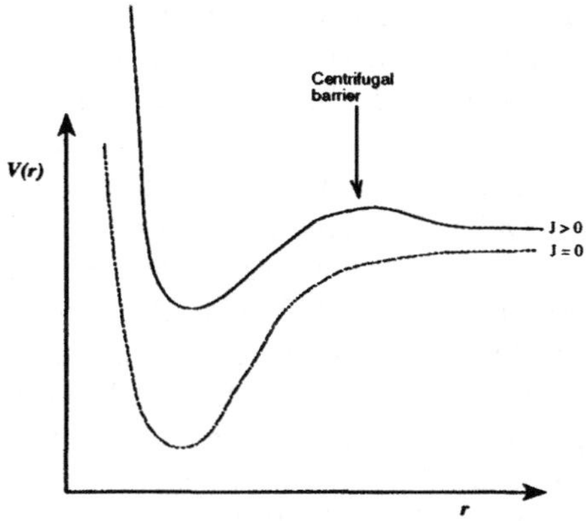

Figure 8. Potential energy profiles for a reaction of type I. As explained in the text the angular momentum translational energy term, $\mathcal{T}_L$ is included in the potential energy, giving an efficient potential energy, $\mathcal{V}_{eff}(r) = \mathcal{V}(r) + \mathcal{T}_L$, which gives different potential energy profiles for different values of L. When the angular momentum is different from zero this gives rise to a maximum along the reaction co-ordinate denoted by centrifugal barrier. It is interesting that in the case of a pure ion/induced dipole potential (according to equation 26) the centrifugal barrier (see section on Langevin theory) coincides with the position of the transition state according to microcanonical variational RRKM theory.

5.2. TYPE I REACTIONS: VARIATIONAL TRANSITION STATE THEORY

In general the rate coefficient for a unimolecular reaction is not only a function of the energy, but also of the total angular momentum, $\mathcal{J}$. For reactions with pronounced barriers (type II), angular momentum is usually of little importance and the dividing surface can successfully be chosen to coincide with the transition structure. Because a type I reaction lacks a transition structure it is not obvious where the dividing surface should be put in order to apply the RRKM equation. We shall see that this can be accomplished by taking the angular momentum explicitly into account.

In the simplest approach the dividing surface is chosen to coincide with the top of the centrifugal barrier (Figure 8). As in ion/molecule capture theory the centrifugal barrier arises from the extra orbital angular momentum term in the Hamiltonian, $\mathcal{T}_{\mathcal{L}}$. It is here assumed that $\mathcal{J} = \mathcal{L}$, which usually is a good approximation. For a given value of $\mathcal{L}$, the distance of separation of the two fragments, $r = r_{\mathcal{L}}^{\ddagger}$, corresponding to the barrier can easily be determined. The rate coefficient $k(E,\mathcal{J})$ is evaluated according to equation (40), with the exception that $G^{\ddagger}(E\text{-}E_0)$ is substituted by the expression $G^{\ddagger}(E\text{-}E_r^{\ddagger}\text{-}E_0)$ which includes the rotational energy term, and $N(E)$ is to be replaced with $N(E\text{-}E_r)$. The vibrational frequencies used to evaluate $G^{\ddagger}(E\text{-}E_r^{\ddagger}\text{-}E_0)$ are often put equal to those of the separated fragments, although a better alternative is to use the frequencies determined from the IRC at $r = r_{\mathcal{L}}^{\ddagger}$.

Although simple and reasonable, this approach has the disadvantage that it is not necessary the case that the optimum dividing surface coincides with the top of the centrifugal barrier. In fact, it is quite likely that many trajectories which pass $r = r_{\mathcal{L}}^{\ddagger}$ in the outward direction may return. To obtain a more realistic estimate for the position of the transition state, $r = r^{\ddagger}$, it has been found that it is better to determine the dividing surface by a variational procedure. Variational transition state theory (VTST) which is available in both microcanonical and canonical versions was developed for this purpose. [32] In microcanonical VTST the rate coefficient, $k(E,\mathcal{J})$ is calculated at given points along the reaction co-ordinate (IRC), and the minimum value is obtained for some particular value of r, which is taken to be the transition state. [33] It is evident that using this procedure a transition state representing a true bottleneck in phase space is found. Computer codes are available which allows the user to calculate a variational transition state once a PES is at hand. [34] In one of the implementations, the PES is supplied in the form of parametric (analytical) form. Alternatively, a direct dynamics procedure can be used where the PES is obtained from a semi-empirical wave function (from the MOPAC package).

There exist alternative statistical theories for calculating $k(E,\mathcal{J})$for type I reactions. The most well-known are the statistically adiabatic channel model (SACM) [35] and phase space theory (PST) [36].

PST assumes an orbiting transition state, which means that the vibrational frequencies of the transition state are those of the separated products. This is generally not the case, since we have stated that the frequencies are found to vary along the IRC. For this reason $k(E,\mathcal{J})$ values obtained by PST are generally overestimated, especially at higher energies.

SACM is a semi-classical approach. This means that the quantum numbers of rotation and vibration are conserved throughout the reaction. Degrees-of-freedom that are of the same type along the complete reaction co-ordinate are called conserved and are treated thereafter. The six degrees of freedom which change from internal vibrations (or hindered rotations) in the reactant into relative rotations and translation in the products are called transitional. They are treated explicitly in SACM. Exact SACM calculations are very tedious, but approximations can be made which makes calculation feasible and still gives reliable results.

During the last years one has observed a trend for hybrid theories. Hybrid theories incorporate elements of the various existing theories mentioned above. An interesting approach which illustrates this combines ideas of RRKM and ion capture rate theories, as well as SACM. [37]

6. Beyond the transition state: Energy partitioning in products

Based on a statistical description of the dynamics of the reactant RRKM theory gives the recipe on how to calculate the rate coefficient of a reaction. However, nothing has so far been stated about the dynamical fate of the system after passage of the dividing surface, and how statistical theories can be extended to incorporate this aspect. A successful theoretical model would make it possible to predict the final state distribution in the products, in particular the relative translational energy of the fragments. This quantity is quite readily obtained experimentally.

A sharp distinction between type I and type II potentials must also be made in this discussion. We will start by looking at reactions of type I which have no potential energy barrier. The fundamental assumption in all statistical theories of unimolecular reaction is of course maintained, namely that the energy is assumed to be statistically distributed among relative translation, rotation and vibration. From this it follows quite straightforward that the probability of finding one particular value of the translational energy release, $E_t^\ddagger$, in the transition state is [38]

$$P(E_t^\ddagger) = \frac{N^\ddagger(E - E_o - E_r^\ddagger - E_t^\ddagger)}{G^\ddagger(E - E_o - E_r^\ddagger - E_t^\ddagger)} \tag{41}$$

This formula is applicable for a given value of the rotational energy. A proper treatment must take the complete rotational energy distribution into account. When this is done phase space theory is used to determine the translational energy release, in some cases in good agreement with experiment. [39] When a Maxwell-Boltzmann distribution of the rotational energy in the reactant is used a decreasing exponential distribution, $P(E_t)$, is often the result. The translational energy in the reaction co-ordinate determined at the transition state is not the same as the relative translational energy measured for the separated fragments. When the surplus energy resulting from the centrifugal barrier is added to the energy already present in the reaction co-ordinate, the distribution function is shifted by this amount.

PST assumes an orbiting transition state (the two fragments are free to tumble relative to each other in the transition state) so there is no coupling between the reaction co-ordinate and the other degrees of freedom, neither in the transition state, nor after passage of it. This description is usually not realistic. For this reason PST is not directly suited for determining the product state distribution. PST may, however, be more useful for type I reactions which have "loose" transition states. A more general approach which is based on variational RRKM theory in combination with the assumption of adiabatic vibrational and rotational motions has been suggested quite recently. [40] The following procedure is followed: The transition state was determined variationally, i.e. it was located for a specific value of the reaction co-ordinate, $r = r^{\ddagger}$ for each E and $\mathcal{J}$ as described above. Conserved degrees of freedom are assumed to be vibrationally adiabatic between $r^{\ddagger} < r < \infty$. The quantum numbers for these modes are therefore the same at infinity as at the transition state. The transitional degrees of freedom are assumed to exchange energy freely between the variatiational transition state, $r = r^{\ddagger}$ and the centrifugal barrier $r = r_{\mathcal{L}}^{\ddagger}$. The centrifugal barrier is determined from the value of the orbital angular momentum, $\mathcal{L}$, which exactly counterbalances the long range radial attractive force between the fragments, as explained in the section on ion/molecule capture theory. The centrifugal barrier corresponds to an orbiting transition state equivalent to that of PST. At $r_{\mathcal{L}}^{\ddagger}$ the distribution of the transitional modes is determined by PST. Comparison with experimental data was successful in the case of photo dissociation of ketene. [41]

This approach can be seen to have many elements in common to SACM. Inherent in SACM is also the ability to predict energy distributions, which generally are in good accordance with experiment.

It must, however, be concluded that despite the success of the above mentioned methods for treating type I reactions they are not likely to give realistic translational energy release distributions for type II reactions which have pronounced barriers. The radical electronic redistribution which takes place from the transition state to the products, will lead to strong coupling between internal modes, and between internal modes and the reaction co-ordinate all the way from the transition state towards the products. For the moment it appears that the only way to treat this is by extensive trajectory calculations. Obviously a reliable potential energy function is needed. Using a direct *ab initio* method it has been demonstrated that translational energy releases can be predicted with good accuracy for reactions with high barriers. [42] Translational energy releases above 80% of the reverse barriers were reported.

7. Examples

In the lectures which form the basis of this written account, some literature examples were incorporated to illustrate how the theories described are applied in practice. The following examples give quite detailed instructions on how to locate transition structures of some type II reactions and how to perform RRKM calculations for some specific reactions:

(i) The two competing reactions of the acetone radical ion. [43]

(ii) Loss of HCl from the radical cation of ethyl chloride. [44]

Good examples which demonstrate reaction dynamic calculations in ion-molecule reactions, including comparison with statistical theories are:

(i) $Cl^{*-} + CH_3Cl \rightarrow Cl^- + CH_3Cl^*$ [45]

(ii) $CH_3O^- + CH_3OH \rightarrow CH_3OH + CH_3O^-$ [46]

A recent and very interesting example on the use of VTST on a large system is:

$Cl^-(H_2O)_n + CH_3Cl^* = CH_3Cl + Cl^{*-}(H_2O)_n$ [47]

8. References

1. See for example: Baym, G. (1969) *Lectures on Quantum Mechanics*, W.A. Benjamin, Inc., Reading, Massachusetts.

2. Heidrich, D., Kliesch, W., and Quapp, W. (1991) *Properties of Chemically Interesting Potential Energy Surfaces*, Vol. 56 in *Lecture Notes in Chemistry*, Springer Verlag, Berlin and Heidelberg.

3. (a) Fukui.K (1970) *J. Phys. Chem.* , **74**, 4161 (b) González, C., and Schlegel, H. B. (1992) *J. Phys. Chem.* **94**, 5523.

4. Wilson, E. B.,Jr., Decius, J. C., and Cross, P. C. (1955) *Molecular Vibrations. The Theory of Infrared and Raman Vibrational Spectra*, MacGraw-Hill, New York.

5. Halgren, T.A., and Lipscomb, W.N. (1977) Chem. Phys. Lett. **49**, 225.

6. Schlegel, H. B. (1994) *J. Chem. Soc. Faraday Trans.*, **90**, 1569.

7. Frisch, M. J., Trucks, G. W., Head-Gordon, M., Gill, P. M. W., Wong, M. W., Foresman, J. B., Johnson, B. G., Schlegel, H. B., Robb, M. A., Replogle, E. S., Gomperts, R., Andres, J. L., Raghavachari, K., Binkley, J. S., Gonzalez, C., Martin, R. L., Fox, D. J., Defrees, D. J., Baker, J., Stewart, J. J. P., and Pople,J.A. (1992) *GAUSSIAN 92* User's Guide, Gaussian Inc., Pittsburgh, PA.

8. Dupuis, M., Watts, J. D., Villar, H. O., and Hurst, G.J.B. *HONDO: Ab Initio HFMO Calculations (Version 7.0)*, IBM Corporation, Scientific Engineering Computations, Kingston, New York 12401.

9. Schmidt, M.W., Baldridge, K.K., Boatz, J.A., Elbert, S.T., Gordon, M.S., Jensen, J.H., Koseki, S., Matsunaga, N., Nguyen, K.A., Su, S., Windus, T.L., Dupuis, M., and Montgomery, J.A. (1993) GAMESS: General Atomic and Molecular Electronic Structure System, *J. Comput. Chem.*, **14**, 1347. See also http://www.msg.ameslab.gov/GAMESS/GAMESS.html

10. Stewart, J.J.P. MOPAC 7: Molecular Orbital Package Version 7 QCMP130, Stewart Computational Chemistry, Colorado Springs, Colorado 80921

11. Diz, A., and Öhrn, Y. (1994) *Int. J. Quantum Chem. Quantum Chem. Symp.* , **28**, 11.

12. (a) Manthe, U., and Köppel, H. (1990a) *J. Chem. Phys.*, **93**, 345. (b) Manthe, U., and Köppel, H. (1990b) *ibid..*, **93**, 1658. (c) Lynch, G. et al. (1991) *ibid.*, **94**, 7150. (d) Pack, R.T., Butcher, E. A., and Parker, G. A. (1993) *ibid.*, **99**, 9310. (e) Manthe, U., Seideman, T., and Miller, W. H. (1993) *ibid.*, **99**, 10078.

13. Reimers, J. R., and Heller, E. J. (1988) *J. Phys. Chem.* , **92**, 3225.

14. (a) Goldstein, H. (1950) *Classical Mechanics*, Addison-Wesley, Reading (Mass.). (b) Porter, K. P., and Raff, L. M. (1976) in: *Dynamics of Molecular Collisions*, Part B., ed. Miller, W.H., Plenum Press, New York.

15. Helgaker, T., Uggerud, E., and Aa. Jensen, H.J. (1990)*Chem. Phys. Lett.*, **173**, 145.

16. (a) Wang, I.S.Y, and Karplus, M. (1973) *J. Am. Chem. Soc.*, **95**, 8160. (b) Leforestier, C. (1978) *J. Chem. Phys.* , **68**, 4406. (c) Chen, W., Hase, W.L., and Schlegel H.B. (1994) *Chem. Phys. Lett.* , **228**, 436.

17. Miller, W.H., Chang, Y.-T., and Makri, N. (1991) in *Computational Advances in Organic Chemistry: Molecular Structure and Reactivity*, Ögretir, C., and Csizmadia, I.G: Eds., Kluwer Academic Publishers, Amsterdam.

18. See for example in: McClelland, B.J. (1973) *Statistical Thermodynamics*, Chapman and Hall & Science Paperbacks, London.

19. (a) Farneth, W. E., and Brauman, J. L. (1976) *J. Am. Chem. Soc.*, **98**, 7891. (b) Olmstead, W. N., and Brauman, J.L. (1977) *ibid.*, **99**, 4219.

20. Langevin, P. (1905) *Ann. Chim. Phys. Ser. 8*, **5**, 245.

21. (a) Giomousis, G., and Stevenson, D.P. (1958) *J. Chem. Phys.* , **29**, 294. (b) Ridge, D.P. (1987) in *Structure/Reactivity and Thermochemistry of Ions*, Ausloos, P., and Lias, S.G., Eds; D. Reidel Publishing Company, Dordrecht, page 1. (c) Steinfeld, J.I., Fransisco, J.S., Hase, W.L. (1989) "Chemical Kinetics and Dynamics", Prentice Hall, Englewood Cliffs (New Jersey).

22. Moran, T.F., and Hamill, W. H. (1963) *J. Chem. Phys.*, **39**, 1413.

23. Su, T., and Bowers,. M. T. (1973) *J. Chem. Phys.*, **58**, 3027 & *Int. J. Mass Spectrom. Ion Phys.*, **12**, 347.

24. (a) Barker, R.A., and Ridge, D.P. (1976) *J. Chem. Phys.*, **64**, 4411. (b) Turluski, J. and Forys, M. (1979) *J. Phys. Chem.*, **83**, 2815. (c) Celli, F., Weddle, G., Ridge, D. P. (1980) *J. Chem. Phys.*, **73**, 801.

25. Su, T., and Chesnavich, W.J. (1982) *J. Chem. Phys.*, **76**, 5183.

26. (a) Hinshelwood, C.N. (1926) *Proc. Roy. Soc. (London)*, **A113**, 230. (b) Rice, O.K., and Ramsperger, H.C. (1927) *J. Am. Chem. Soc.*, **49**, 1617.(c) Marcus, R.A. (1952) *J. Chem. Phys.*, **20**, 359. (d) Rosenstock, H.M., and Krauss, M. (1962) *Adv. Mass Spectrom.*, **2**, 251.

27. Lorquet, J. C. (1994) *Mass Spectrom. Rev.*, **13**, 233.

28. Beyer. T., and Swinehart, D.F. (1973) *Comm. ACM*, **16**, 379.

29. Forst, W. (1973) *Theory of Unimolecular Reactions*, Academic Press, New York.

30. Gilbert, R.G., and Smith, S.C. (1990) *Theory of Unimolecular and Recombination Reactions*, Blackwell Scientific Publications, Melbourne. Includes an instruction how to get an RRKM program and other programs described in this textbook.

31. Miller, W.H. (1979) *J. Am. Chem. Soc.*, **101**, 6810.

32. (a) Wigner, E. (1937) *J. Chem. Phys.* , **5**, 720. (b) Horiuti, J. (1938) *Bull. Soc. Jpn.* , **13**, 210. (c) Eliason, M.A., and Hirschfelder, J.O. (1959) *J. Chem. Phys.*, **30**, 1426. (d) Keck, J. C. (1967) *Adv. Chem. Phys.*, **13**, 85.

33. Reviews: (a) Truhlar, D. G., and Garrett, B.C. (1980) *Acc. Chem. Rev..*, **13**, 440. (b) Hase, W.H. (1983) *ibid.*, **16**, 258.

34. Hu, W.-P, Steckler, R., Lynch, G.C., Liu, Y.-I., Garrett, B.C., Isaacson, A.D., Lu, A.D., Melissas, V., Rossi, I., Stewart, J.J.P., and Truhlar, D.G. (1995) *Quantum Chemistry Program Exchange, QCPE Bulletin* , **15** (2), 32.

35. Quack, M, and Troe, J. (1981) *Int. Rev. Phys. Chem.*, **1**, 97.

36. (a) Nikitin, E.E. (1965) *Teor. Eksp. Khim.*, **1**, 135, 144, 428. (b) Pechukas, P., and Light, J.C. (1965) *J. Chem. Phys.*, **42**, 3281.

37. Smith, S.C., McEwan, M.J., and Gilbert, R.G. (1989) *J. Phys. Chem.*, **93**, 8142.

38. Klots, C. E. (1973) *J. Chem. Phys.*, **58**, 5364.

39. (b) Chesnavich, W.J. and Bowers, M.T. (1976) *J. Am. Chem. Soc.*, **98**, 8301. (c) Klots, C.E. (1988) *Acc. Chem. Res.*, **21**, 16.

40. Klippenstein, S.J., and Marcus, R.A. (1988) *J. Phys. Chem.*, **92**, 3105.

41. Klippenstein, S.J., and Marcus, R.A. (1988) *J. Chem. Phys.*, **91**, 2280.

42. (a) Hvistendahl, G., and Uggerud, E. (1991) *Org. Mass Spectrom.*, **26**, 67. (b) Uggerud, E., Helgaker, T. (1992) *J. Am. Chem. Soc.*, **114**, 4265.

43.Heinrich, N., Louage, F., Lifshitz, C., and Schwarz, H. (1988) *J. Am. Chem. Soc.*, **110**, 8183.

44. Booze, J.A., Weitzel, K.-M., Baer, T. (1991) *J. Chem. Phys.*, **94**, 3649.

45. Hase, W.L. (1994) *Science*, **266**, 998.

46. (a) Lim, K.F., and Brauman, J.I. (1991) *Chem. Phys. Lett.*, **177**, 326. (b) Lim, K.F., and Kier, R.I. (1992) *J. Chem. Phys.*, **97**, 1072.

47. Zhao, X.G., Lu, D.-H., Liu, Y.-P., Lynch, G.C., and Truhlar, D.G. (1992) *J. Chem. Phys.*, **97**, 6369.

ION REACTIONS OF STATE-SELECTED SPECIES: DISSOCIATION OF CLUSTER IONS PREPARED BY MASS SELECTED PULSED FIELD THRESHOLD IONIZATION

Hans Jürgen Neusser, Thomas Ludwig Grebner

Institut für Physikalische und Theoretische Chemie, Technische Universität München,
Lichtenbergstraße 4, D-85748 Garching, Germany

1 Introduction

A basic goal in mass spectrometry has been the production of state-selected ions. Ions with defined internal energy allow the observation of precise energetic onsets for dissociation and isomerization processes. They represent a necessary precondition for the selection of reaction pathways in unimolecular and bimolecular reactions. For the investigation of the decay dynamics it is necessary to observe the decay of ions whose internal energy is well defined. Clear tests of theoretical models are more feasible with energy-selected ions since in this way experiment and theory can be compared with a minimum of averaging. Production of energy-selected ions turns out to be a nontrivial problem: Ionization by electron impact produces ions with an energy distribution. The most promising technique includes the absorption of photons in a one- or multi-photon absorption process. Even if photons of well-defined energy are used for ionization, the ions are generally produced with a broad distribution of internal energies spanning the range from zero to the maximum energy above the adiabatic ionization energy. The emitted electrons carry away a variable amount of energy and leave behind ions with a broad energy distribution. This is shown schematically for a resonance enhanced two-photon ionization in Fig. 1. The shape of the internal energy distribution of the ion is not known apriori since it is a complicated function of Franck Condon factors and autoionization resonances. A clear situation exists only at the first ionization threshold. Here no excess energy is left for the emitted electron and the threshold ions are inherently energy-selected. At higher vibrational ionization thresholds, in addition to the energy-selected threshold ions, a great number of non-energy-selected ions due to the lower ionization potential are produced. This gives rise to a step-like behavior of the total ion current, each step indicating a new ionization threshold. For experiments with energy-selected ions, the separation of threshold

K. R. Jennings (ed.), Fundamentals and Applications of Gas Phase Ion Chemistry, 263–285.

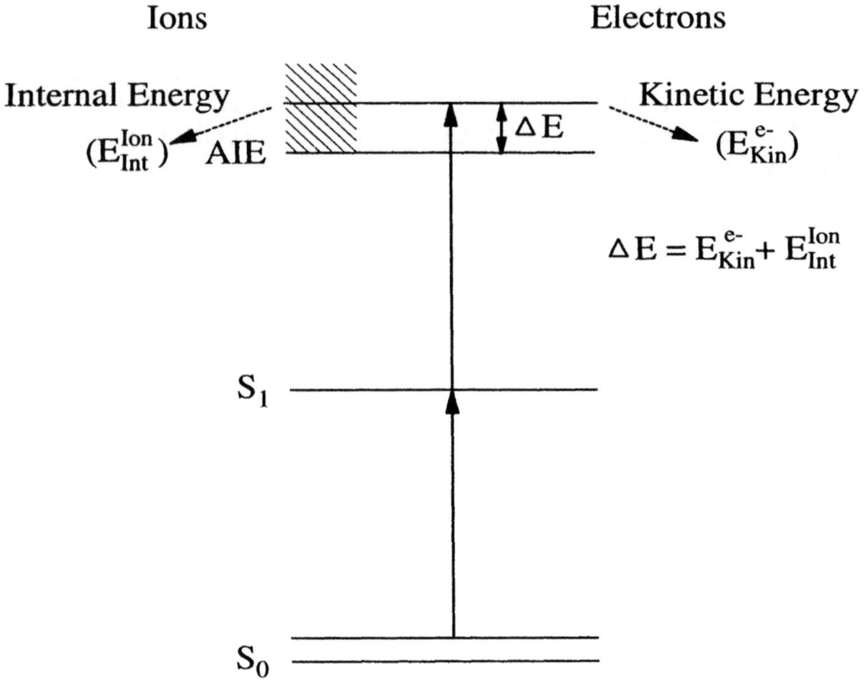

Fig. 1 Resonance enhanced two-photon ionization leading to an excess energy ΔE above the adiabatic ionization energy (AIE). The ions are produced with a varying amount of initial energy since the electrons carry away kinetic energy. In some cases selectivity of ion internal energy is achieved by the selected intermediate state due to suitable Franck-Condon factors.

ions from the strong background of non-energy-selected ions is inevitable. One of the techniques that has been successfully applied is the photoion photoelectron coincidence technique (PIPECO) [1, 2, 3]. Here exclusively ions are monitored that are in coincidence with electrons of low or zero kinetic energy. Naturally, the energy resolution of this technique is limited by the monochromator resolution and the electron energy resolution. The hitherto sharpest energy resolution obtained with this technique was some 20 meV ($\approx$150 cm^{-1}). A better resolution is expected if high repetition rate lasers in the UV become available for ionization.

The selectivity of the ion production in a photoionization experiment is supported by resonance enhanced two-photon ionization. Resonance-enhanced two- and multiphoton ionization was shown to have particular advantages for mass spectrometry [4, 5, 6, 7]. Efficient ionization is achieved by stepwise two- or three-photon absorption via a resonant real intermediate state as is demonstrated in Fig. 1. This resonance enhancement of the multiphoton ionization process leads not only to a drastic increase of the ionization yield but also results in a high selectivity of the ion production [8]. In contrast to a conventional one-photon ionization into the ionization continuum, in resonance-enhanced multiphoton ionization the ionization is controlled by the resonant intermediate state spectrum which often displays sharp structure, even in large polyatomic molecules and small clusters of these molecules. In this case the second absorption step starts from a selected defined excited state and the absorption to the ionization continuum is governed by the Frank Condon principle and the nature of the intermediate state. Often a $\Delta v = 0$ propensity rule leads to some selectivity in ion production: preferentially ions in a vibrational state are produced which equals or resembles the excited intermediate state of the neutral molecule. In addition it is species selective, this being of particular importance for the study of clusters.

In this report we describe a new technique to produce state-selected ions with even higher energy resolution that is based on the pulsed field ionization of high ($n > 100$) long-lived Rydberg states close to an ionization threshold. The optical $\Delta l = \pm 1$ selection rule favors the excitation of low l-states with an asymmetric electron orbit and a large electron- core interaction [9]. However, a series of recent experiments has shown that photoabsorption leads to high Rydberg states with μs lifetime, pointing to a large l quantum number of the produced Rydberg states [10, 11, 12, 13, 14, 15, 16, 17, 18, 19]. As an explanation for this, several processes like l-mixing by collisions and in electric fields have been discussed [20, 21, 22, 23, 24, 25]. The long-lived high Rydberg molecules can be easily separated from the simultaneously produced prompt ions by weak electric fields which affects (accelerates or decelerates) the ions but does not influence the high Rydberg molecules. After their spatial separation from the prompt ions the Rydberg molecules can be converted to threshold ions with a defined amount of internal energy.

We will demonstrate that the production of ions in various vibrational states is possible by this technique. Exclusive detection of state-selected ions is achieved by the kinetic energy analysis of ions in a reflectron mass spectrometer. This leads to vibrationally resolved spectra of molecular ions, like benzene, xylene, aniline, benzonitrile, carbazole, dibenzo-*p*-dioxin etc. and ionic complexes of these molecules with noble gas atoms. The detected threshold cluster ions are state-selected with different internal energies depending on the excited vibrational state. Above a certain vibrational excess energy their decay is observed by the disappearance of vibrational peaks in the threshold ion spectrum at the parent mass and the simultaneous appearance of threshold daughter ions. Dissociation energies of the neutral and the ionic complexes are deduced and interpreted in terms of dispersion interaction and charge induced dipole interaction.

2 Experimental Method

Mass selected pulsed field threshold ionization (MATI) is based on the long lifetime of high Rydberg states. Rydberg series converge to the different vibrational states in the molecular or cluster ion. Therefore high Rydberg states which are within a few cm^{-1} from the respective ionization energy (IE) mark the energetic position of these vibrational states. The lifetime of several μs is long compared to the time necessary to separate high Rydberg molecules from molecules ionized instantaneously by the laser pulses. The general principle of pulsed field threshold ionization combined with resonance enhanced two-photon excitation is illustrated in Fig. 2. The first color $h\nu_1$ is fixed to a vibrational band in the $S_1 \leftarrow S_0$ transition. The second frequency $h\nu_2$ is scanned across the adiabatic ionization energy

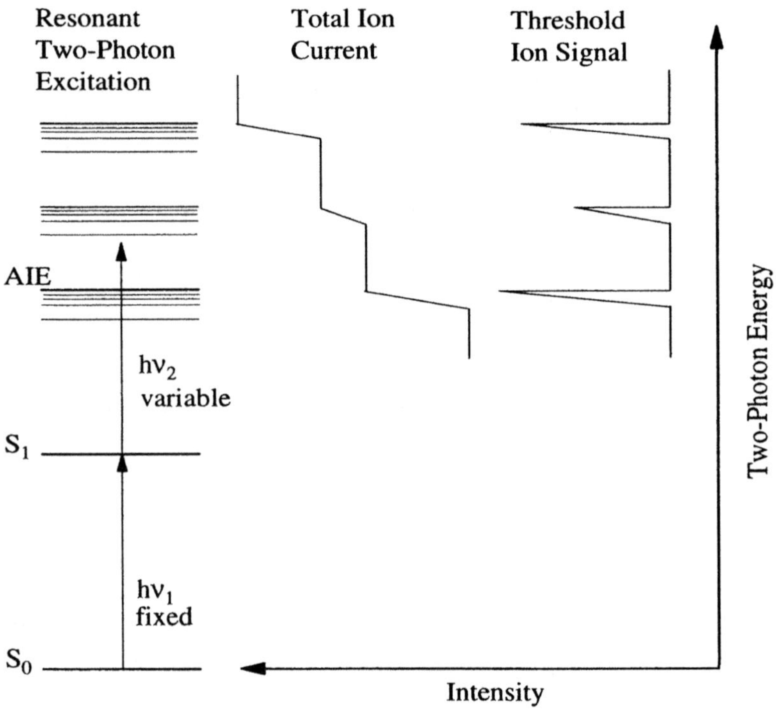

Fig. 2: Excitation scheme of high Rydberg states in a resonance enhanced two-photon two-color process (left). The first color $h\nu_1$ is fixed to a $S_1 \leftarrow S_0$ vibronic transition. The second frequency is scanned around ionization thresholds of the ionic ground state. Collecting all ions leads to the total ion current (middle) which exhibits steps in the intensity. Applying the separation process (described below) leads to the threshold ion spectrum (right).

and to ionic states above the AIE. Recording all ions (field ionized Rydberg molecules and prompt ions) up to several hundreds cm^{-1} leads to the ion efficiency curve which exhibits steps at the individual thresholds (see middle of Fig. 2). When molecules in high Rydberg states surviving the separation process are pulsed field ionized and detected exclusively after the separation process described below sharp peaks are observed at the individual thresholds. This leads to the vibrational spectrum of the ions as shown schematically on the right side of Fig. 2. However, one has to take into account that the ionization threshold is lowered by the weak separation field according to $\Delta IE[cm^{-1}] = \alpha \cdot \sqrt{E[V/cm]}$, with a value for α of ≈ 4 (see below).

First, pulsed field threshold ionization has been used to measure electrons with nearly zero kinetic energy [26]. The technique developed by Reiser et al. displays a considerably increased resolution in photoelectron spectroscopy of the electronic ground state ions [26, 27]. Zhu and Johnson presented a method to combine pulsed field ionization with ion detection [10]. In their technique they separated the threshold ions from ions below this threshold that are responsible for the background in photoionization efficiency spectra by a special technique. This includes three different acceleration regions and a time-of-flight analysis in the mass spectrometer. A similar method was reported by Jouvet et al. [12]. Our technique is based on the ion separation in a reflectron mass spectrometer. This leads: i) to a complete suppression of non-energy selected prompt ions, ii) to a high mass resolution, iii) to the detection of metastable decay processes by a kinetic energy analysis. These points are important for the detection of fragmentation processes of cluster ions leading to a dense mass spectrum.

The principle of our method to separate the ions produced by pulsed field ionization (PFI) from directly ionized molecules is shown in Fig. 3. The molecules enter a reflectron mass spectrometer with the velocity v_{jet} of the molecular beam, which depends on the noble gas used (He or Ar). At this time ($t = 0$) a weak electric field of -0.2 to -0.6 V/cm is present. Its direction is chosen such that the ions instantaneously produced by the laser pulse are decelerated. In the present experiment the pulsed acceleration voltage is delayed by several μs up to 100 μs after the exciting laser pulse. This is the main difference between the conventional mass spectrometric operation of the reflectron and the technique described here.

After the typical delay of 5 μs the highly excited neutral molecules in long-lived Rydberg states and the ions originally produced at the same place are spatially separated by about 1 mm. At this time a pulsed positive voltage of 1000 V is applied to the repeller plate. This leads to an electric field of +333 V/cm, causing the field ionization of the still existing long lived Rydberg states and the rapid acceleration of all ions. Due to the spatial separation the instantaneously produced ions are accelerated by a higher potential difference than the ions produced by delayed PFI. For a typical spatial separation of 1 mm, this energy difference is about 33 eV. It is sufficient to separate both species within the reflector of the

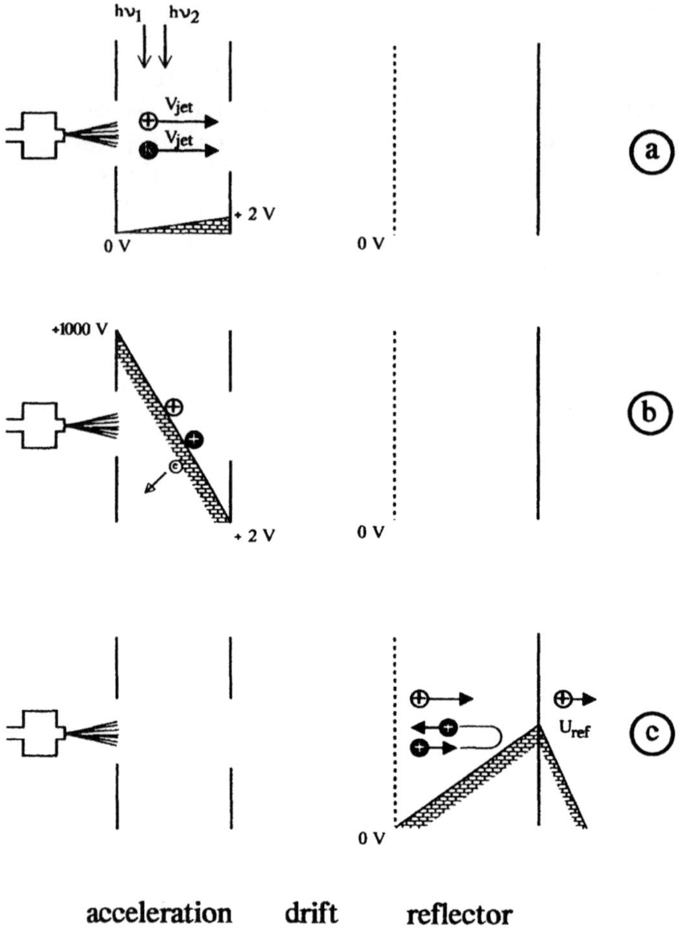

Fig. 3: Principle of mass-selected pulsed field ionization in a reflectron time-of-flight mass spectrometer. a) $t = 0$: molecules are excited by resonance enhanced two-color two-photon excitation. A weak deceleration field leads to the separation of instantaneously ionized molecules ⊕ and the neutral Rydberg molecules Ⓡ. b) $t = 5$ μs: a strong electric field is applied. It ionizes the still existing Rydberg molecules and accelerates all ions. The directly ionized molecules are now separated from the newly produced ions and accelerated by a higher potential difference. c) $t = 25$ μs: Due to their larger kinetic energy they penetrate through the reflector, whereas the ions produced by pulsed field ionization are reflected and monitored separately (taken from [11]).

mass spectrometer. The reflecting voltage U_{ref} is chosen so, that only threshold ions are reflected while the instantaneously produced ions penetrate through the reflector. In this way the threshold ions are detected after reflection and passing the second drift region. See Fig. 4 for the detailed set up of the linear reflectron time-of-flight mass spectrometer [28] which has been used in the experiments described here.

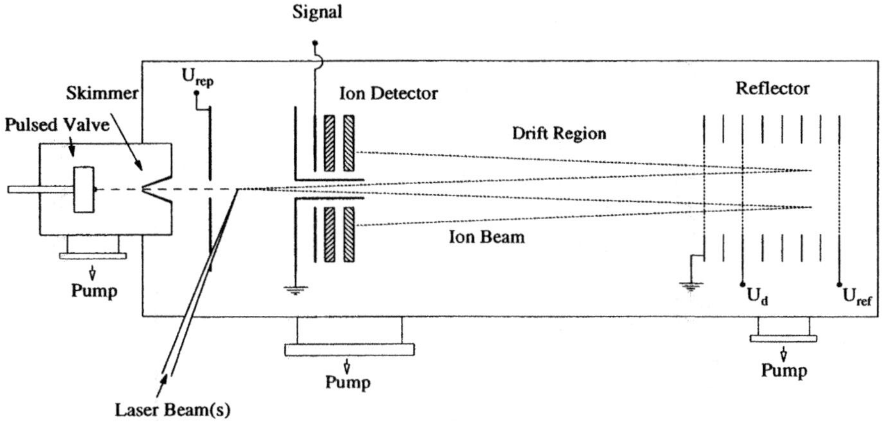

Fig. 4: Schematic drawing of the linear reflectron mass spectrometer that has been used for the pulsed field threshold ionization experiments. The molecular beam produced by a pulsed valve is skimmed and collinear to the ion flight paths. The ions are produced by multi-photon ionization by two laser beams of different color. After acceleration they pass a small hole in the channel plates and are reflected by 180 ^{0} in the reflecting field towards the channel plates (for details, see [28]).

In Fig. 5 we would like to demonstrate the energy analysis in the reflecting field for suppression of the ions produced by direct ionization. These have a somewhat smaller kinetic energy, as demonstrated in Fig. 3. Fig. 5 shows three mass spectra of benzene, C_6H_6, that have been obtained for different reflection potentials U_{ref} and an exciting two-photon energy of 74900 cm^{-1}. For this two-photon energy leading to a higher ionization threshold ($6_1(\pm 3/2)$), prompt ions and threshold ions (shaded peaks) are produced simultaneous. In the mass spectrum (a) at the top of Fig. 5 the reflectron voltage is largest, $U_{ref} = 847$ V, and both types of ions are reflected and detected by the channel plates. In spectrum (b) U_{ref} is decreased to 840 V leading to a smaller signal of the directly produced ions, since only a small part of them is reflected under these conditions. Finally in spectrum (c), at $U_{ref} = 820$ V, the directly produced ions are no longer reflected and do not appear in the mass spectrum, i.e. the observed ions are exclusively produced by pulsed field ionization of long lived Rydberg states. (It should be noted that in the spectra of Fig. 5 the mass resolution is not as high as usual, since the reflectron is operated with incomplete correction so that both ion species are separated in time-of-flight and can be clearly distinguished from each other.) As a result we obtain ion signals originating exclusively from threshold ions, in this case ions in the $6^1(\pm 3/2)$ state,with no background from non-energy-selected directly produced ions.

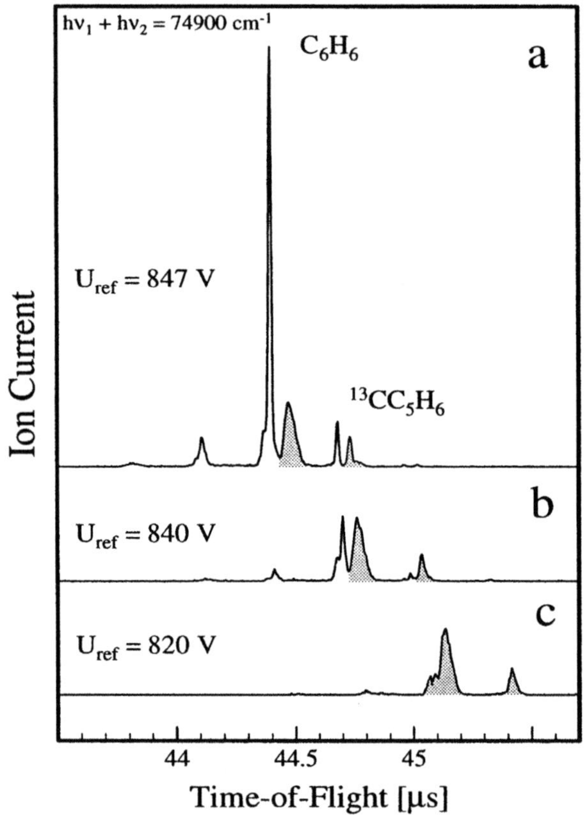

Fig. 5: Time-of-flight mass spectra of benzene obtained for a two-photon energy $h\nu_1 + h\nu_2 = 74900$ cm^{-1} in resonance with the ionization threshold of the $6_1(\pm 3/2)$ ion state. The reflecting voltage of the reflectron mass spectrometer is decreased from a) to c). a) For $U_{ref} = 847$ V threshold ions (shaded) and directly produced ions are reflected and monitored. b) For $U_{ref} = 840$ V directly produced ions are reflected with smaller efficiency due to their higher kinetic energy. c) For $U_{ref} = 820$ V the directly produced ions are no longer reflected due to their higher kinetic energy. Only threshold ions with low kinetic energy are reflected and exclusively monitored. Note that under these conditions the resulting mass spectrum consists of threshold ions (shaded) exclusively without any background from directly produced ions (taken from [29]).

3 Threshold Ion Spectra

After discussion of the procedure leading to an exclusive detection of threshold ions we would like to demonstrate the aspects of this technique for ion spectroscopy: In Fig. 6 three spectra of the benzene cation obtained with three different methods are shown. The trace at the top represents a time-of-flight photoelectron spectrum close to the lowest ionization threshold, with the two-photon energy set to 77210 cm^{-1}, which is 2655 cm^{-1} above the adiabatic ionization energy of benzene. In the middle trace of Fig. 6 the total ion current spectrum of benzene (C_6H_6) is shown. The total ion efficiency curve displays a steplike increase of the ion current

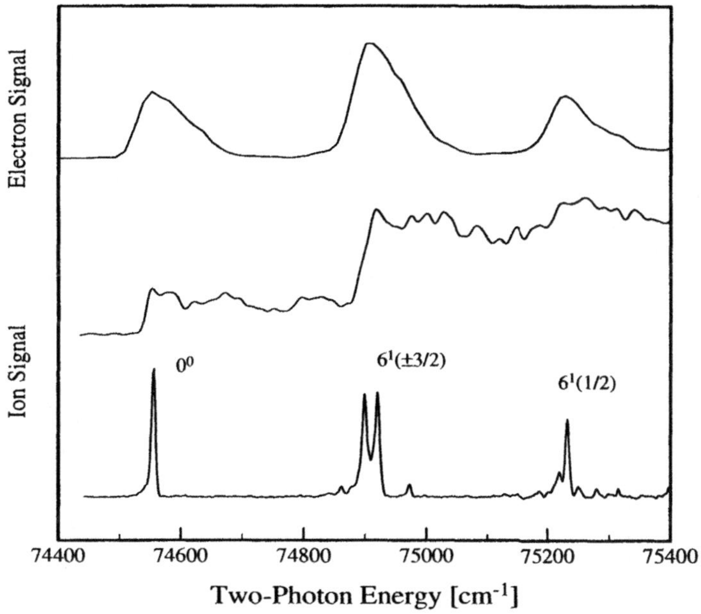

Fig. 6: Three spectra of the benzene cation $C_6H_6^+$ obtained with three different techniques. top: Time-of-flight photoelectron spectrum after ionization of benzene with a two-photon energy of 77210 cm^{-1}. The spectrum reflects the ground state and the lowest vibrational states of the benzene cation. middle: Total ion current as a function of two-photon energy. For each new ionization threshold a step is observed due to the additional ions produced. bottom: Threshold ion signal as a function of two-photon energy. Peaks are observed when a new ionization threshold is reached. For two-photon energies not in resonance with an ionization threshold the threshold ion signal is zero. Note the increased resolution of this spectrum leading to additional features not resolved in the upper two spectra. (for discussion see text, taken from [29]).

whenever a new ionization threshold is reached. The experiment has been performed using a two-photon ionization/excitation process with the first color fixed to the 6^1 intermediate state. The energy of the second photon was scanned from the adiabatic ionization energy up to 1000 cm^{-1} excess energy. In the lower trace the threshold ion current is measured yielding a vibrational spectrum of the ionic ground state of benzene ($\tilde{X}$, $^2E_{2g}$) from zero to 1000 cm^{-1} excess energy. Due to the peaked structure and its flat baseline the threshold ion spectrum contains more spectral information than the total ion efficiency curve. E.g. from the single step in the middle of the ion efficency curve it is not possible to deduce the splitting of the $6^1(\pm3/2)$ peak seen in the threshold ion spectrum and additional peaks with low intensity cannot be observed. The detailed assignment of the vibrational bands has been discussed in our previous work [11]. Briefly, the splitting of the $6^1(\pm3/2)$ band is due to a previously unresolved quadratic Jahn-Teller splitting.

In order to demonstrate the long lifetime of the high Rydberg levels we measured the threshold ion spectra for different delay times between laser excitation and pulsed field ionization. Fig. 7 presents a series of threshold ion spectra of benzene with different time delay from 10 μs to 200 μs. The spectra show the spectral region from the 0^0 to the $6^1(\pm3/2)$ vibrational state also included in the spectrum shown in the lower trace of Fig. 6. Fig. 7 suggests a 'lifetime' of about 100 μs of the high Rydberg states, but it is difficult to interpret the result in terms of a precise microscopic decay, since the decrease of the threshold ion signal is not only

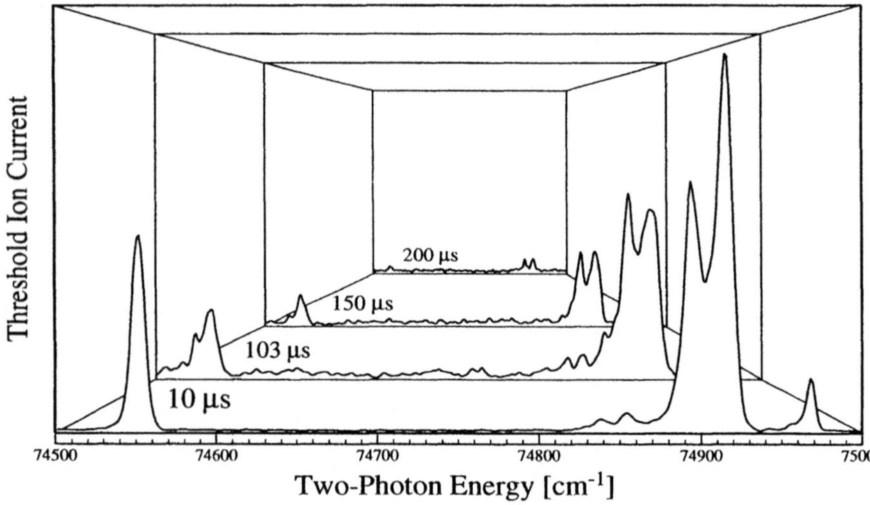

Fig. 7: Threshold ion spectrum of benzene measured for different delays from 10 μs to 200 μs of the pulsed field after laser excitation.

a function of the delay time of the pulsed field but is strongly influenced by the detection efficiency due to geometric effects for different flight times of the Rydberg molecules and ions and by the electric fields in the mass spectrometer which cannot be kept constant during the experiment. The observation of a spectrum for delay times as long as 200 μs demonstrates that even after that time a sufficient number of high Rydberg states is present to guarantee a good signal to noise ratio in the threshold ion current after their field ionization [14].

Recently we have demonstrated that laser-produced ions can be stored for more than 100 ms in a cylindrical ion trap [30]. Here the question arises whether it is possible to combine the MATI technique with a cylindrical ion trap in order to store molecular ions which have been excited to defined vibrational states. This is useful, e.g. for the investigation of ion-neutral molecule reactions on a ms time scale. We incorporated a cylindrical ion trap into a reflectron mass spectrometer as shown in the schematic drawing in Fig. 8. The excitation of neutral benzene molecules to high Rydberg states and the simultaneous excitation of non-energy-selected prompt ions takes place in front of the first end cap (EC_1). A weak collinear electric field is applied to the excitation region affecting only the prompt ions. While the high Rydberg molecules drift into the ion trap through a small hole in the end cap (EC_1) the prompt ions do not enter and are thus separated (see Fig. 8 and top of Fig. 9). In the ion trap the Rydberg molecules are ionized by the storage electrical field, produced by the radio frequency voltage which is applied to the ring electrode (RE). After a storage time of 10 ms the trapped ions

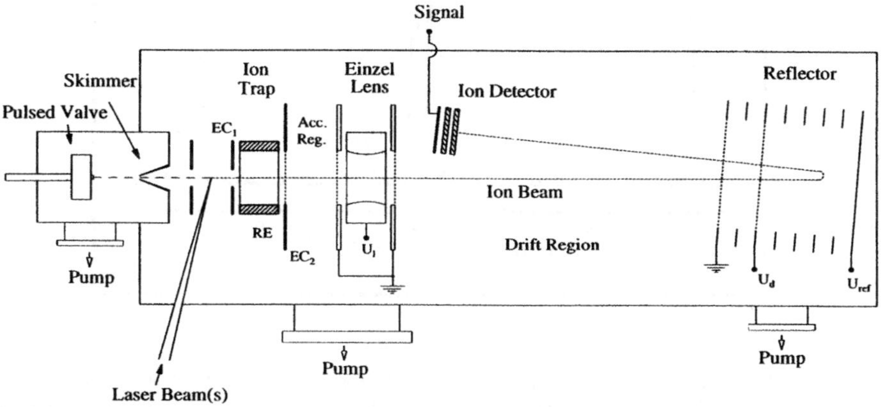

Fig. 8: Schematic drawing of the cylindrical ion trap reflectron time-of-flight (CIT-RETOF) mass spectrometer setup. It consists of two vacuum chambers separated by the skimmer. The first contains the pulsed valve for production of the supersonic molecular beam. In the second chamber the CIT with the two end caps (EC_1 and EC_2) and the ring electrode (RE) is installed. The additional ion optics and the ion detector forms the RETOF mass spectrometer (taken from [30]).

are accelerated into the reflectron mass spectrometer and are thus detected mass selectively. The stored ion current is measured as a function of the two-photon energy. The resulting spectrum is shown on bottom of Fig. 9. It is identical to the threshold ion spectrum in Fig. 6. It exhibits no unstructured background and thus demonstrates the exclusive storage of threshold ions in the trap.

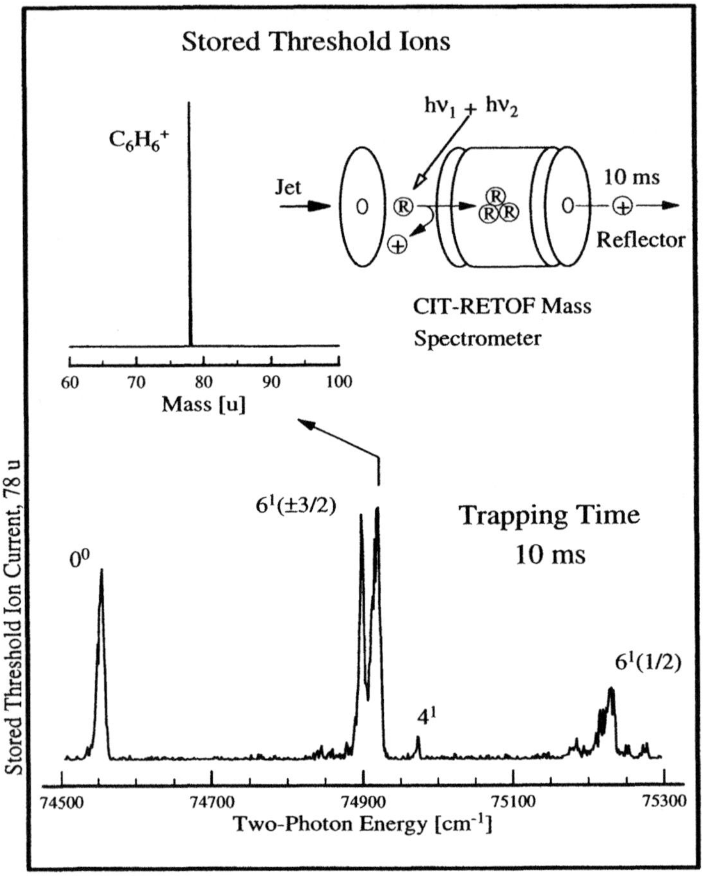

Fig. 9: Threshold ion mass spectrum of benzene (top, left) and part of the benzene cation spectrum (bottom) measured by monitoring threshold ions that have been stored for 10 ms in a cylindrical ion trap. The scheme of the experimental set up is shown on top (right). High Rydberg molecules Ⓡ produced by resonance-enhanced two-photon excitation drift into the cylindrical ion trap where they are ionized by the storage electrical field. Prompt ions ⊕ are spatially separated by a weak static electrical field and do not enter the ion trap. After 10 ms storage time the threshold ions are pulsed out of the ion trap and are detected in a reflectron time-of-flight mass spectrometer.

To demonstrate the broad applicability of mass selective pulsed field threshold ionization and its virtue for ion spectroscopy we present two further examples of molecular ion spectra. In Fig. 10 threshold ion spectra of aniline are shown. Each trace is obtained with the first laser fixed to a certain intermediate state as indicated in the figure. The adiabatic ionization energy is found at 62277 cm^{-1}

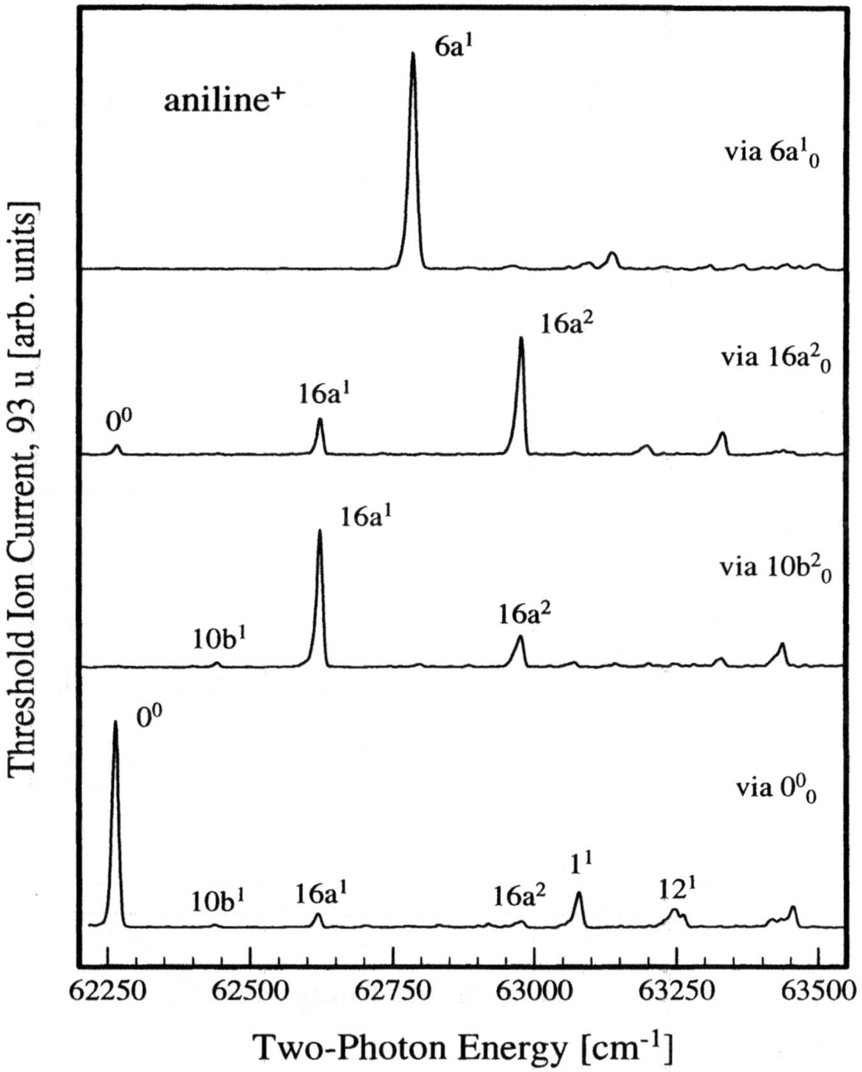

Fig. 10: Threshold ion spectra of aniline (93 u) obtained by resonance enhanced two-photon excitation via different vibrational intermediate states in the S_1 electronic state. The assignment has been adopted from [31].

which is close to the values obtained by threshold photoelectron spectroscopy [31, 32]. In the traces on top and at the bottom a $\Delta v = 0$ propensity rule governs the threshold ion spectra. The strongest peaks are the $6a^1$ and the 0^0 states, when the frequency of the first photon is tuned to the $6a_0^1$ and the 0_0^0 transition, respectively. The situation in the two traces displayed in the middle is more complicated and a preliminary assignment is given.

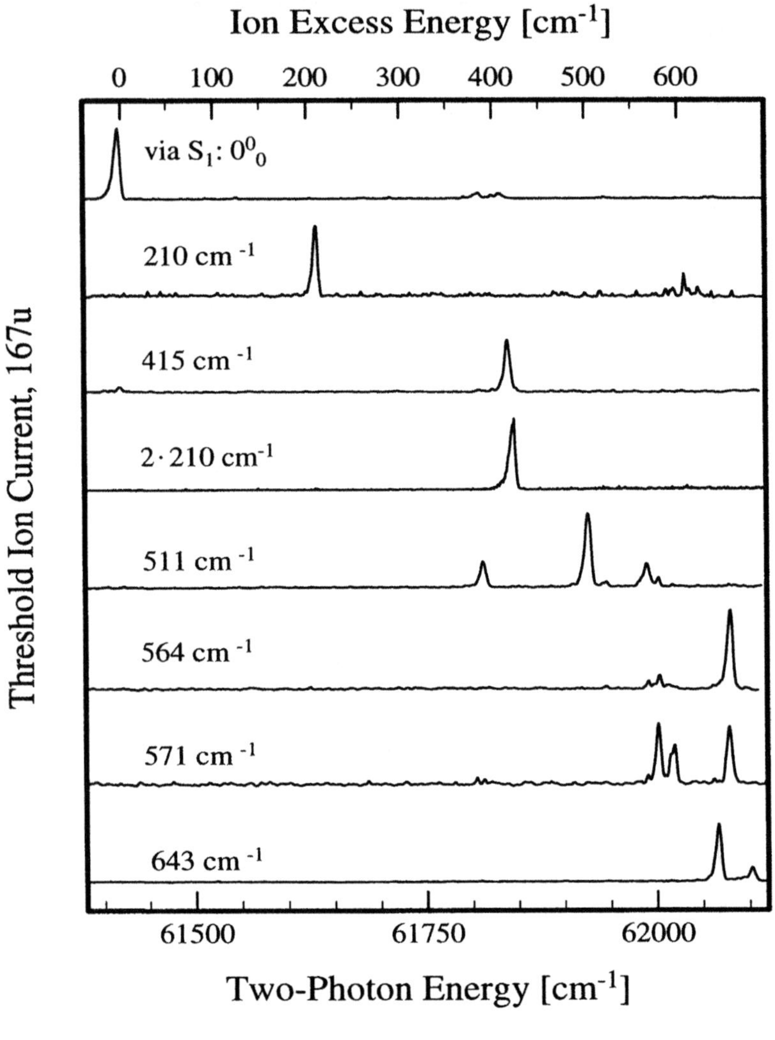

Fig. 11: Threshold ion spectra of carbazole (167 u) obtained by resonance enhanced two-photon excitation via different S_1 intermediate states. The frequencies given for the individual spectra are transition frequencies of the first step relative to the S_1, 0_0^0 *transition at 30815* cm^{-1}.

Fig. 11 displays threshold ion spectra of carbazole, recorded from the adiabatic ionization energy to about 700 cm^{-1} excess energy. A delay time of 40 μs between laser excitation and pulsed field ionization allows for separation of threshold ions from prompt ions in the case of a heavy molecule. The different traces are indicated by the energy (relative to the S_1, 0^0_0 transition) of the resonant intermediate vibrational state in the S_1 state. Carbazole shows a strict $\Delta v = 0$ propensity rule for the transition from the S_1 intermediate state to the Rydberg levels in the case of the first four vibrations. For all other intermediate states this propensity rule is less strict and additional transitions in the vicinity of the $\Delta v = 0$ transition occur. The examples discussed here demonstrate that detailed information about the vibrational states of large polyatomic ions is found which cannot be provided by conventional photoelectron spectroscopy because of its lower resolution.

4 Cluster Ion Dissociation

Benzene·Ar

After demonstration of the virtue of pulsed field threshold ionization for spectroscopic studies of molecular ions, we apply this technique to weakly bound molecular complexes [29, 11]. The frequency of the first laser is now tuned to the intermediate state of the benzene·Ar complex that is produced in the cooled molecular beam. The 6^1_0 transition is red-shifted by 21 cm^{-1} from the corresponding transition in bare benzene. Under these excitation conditions preferentially benzene·Ar complexes are ionized due to the selectivity of the intermediate state. A further selectivity is achieved by the mass selective detection of the threshold ions that is possible in our experiment.

Fig. 12 shows the threshold ion signal spectrum of the benzene·Ar cation (118 u) (middle trace). For comparison, in the lower trace the spectrum of the bare benzene cation is given. It was obtained for a frequency of the first laser tuned to the 6^1_0 transition of bare benzene. The 0^0 transition of the benzene·Ar cation is red-shifted by 172 cm^{-1} due to the stronger bonding of the benzene·Ar cation compared to the neutral complex in the electronic ground state. This represents the shift of the adiabatic ionization energy after complexation. The adiabatic ionization energy of benzene·Ar is found to be AIE = 74383±2 cm^{-1} in good agreement with recent results from high resolution threshold photoelectron spectroscopy [33]. On the blue side of the 0^0 transition small features indicate the additional excitation of van der Waals modes. The striking result is the absence of any vibrational structure above the 4^1 band in the spectrum of benzene·Ar cation. The position of the split $6^1(\pm 3/2)$ band relative to the origin is the same as in the benzene cation. However, in the benzene cation spectrum the $6^1(\pm 3/2)$ mode is the strongest band, while in the benzene·Ar cation the 0^0 band is the strongest. From the missing peaks at higher energy we conclude that for the vibrational bands at excess energies higher than that of the 4^1 band fragmentation of the (benzene·Ar)$^+$ complex occurs. If this conclusion is true, a signal should be observed at the daughter ion mass (i.e., the mass 78 u of benzene). The threshold ion signal monitored at 78 u for the same frequency ν_1 of the first photon is given in Fig. 12 (upper trace) and compared with the spectrum of (benzene·Ar)$^+$

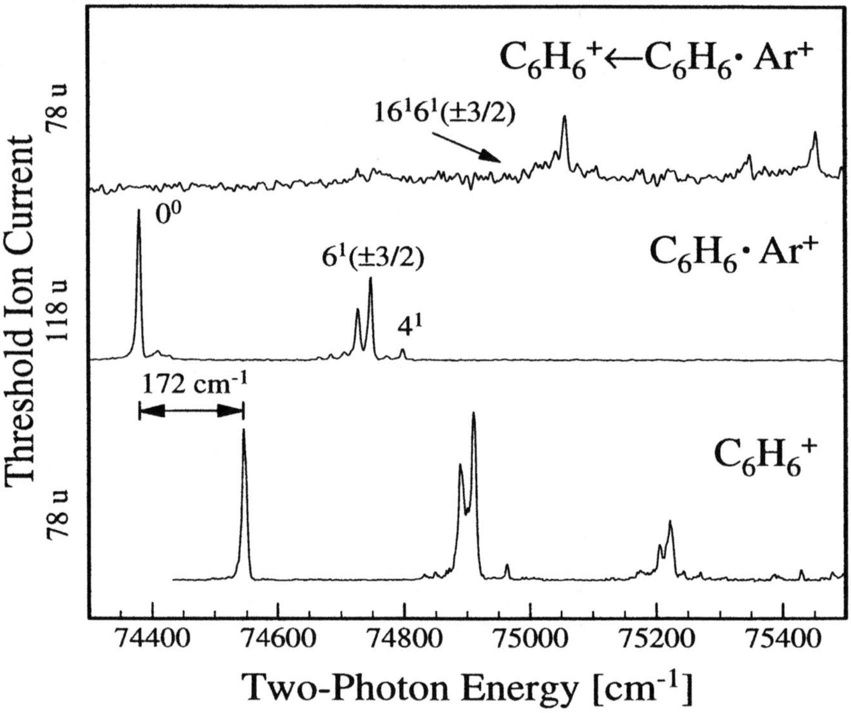

Fig. 12: Three threshold ion spectra obtained for a molecular beam with benzene seeded in Ar under high pressure. Bottom: Threshold ion signal at 78 u for a photon energy $h\nu_1$ in resonance with the 6^1 state of C_6H_6: the vibrational spectrum of the benzene cation is measured. Middle: Threshold ion signal at 118 u for a photon energy $h\nu_1$ in resonance with the 6^1 intermediate state of the neutral benzene·Ar van der Waals dimer. The resulting spectrum is shifted by 172 cm^{-1} to the red of the benzene spectrum (lower trace) due to the increased binding energy in the ion. All peaks with a larger energy than the 4^1 state disappeared. Top: Threshold ion signal at 78 u for a photon energy $h\nu_1$ in resonance with the 6^1 intermediate state of the benzene·Ar dimer. The signal at the benzene mass (78 u) results from a decay of the complex ion and can be observed at the $16^1 6^1(\pm 3/2)$ peak and for peaks with higher energy. For discussion of the dissociation threshold, see text (taken from [14]).

monitored at mass 118 u measured at the same excitation conditions (middle trace). Obviously, the missing bands in the spectrum observed at the parent ion mass (118 u) appear when monitoring the daughter ion signal at 78 u.

By analyzing mass spectra of threshold ions including a metastable analysis at different ion internal energies we conclude that the dissociation of $C_6H_6 \cdot Ar$ complexes in long-lived high Rydberg states occurs within a few μs before they are ionized by the pulsed field [14, 29]. This implies that the dissociation process does not disturb the Rydberg electron (see Fig. 13, for a cartoon-like illustration of this observation).

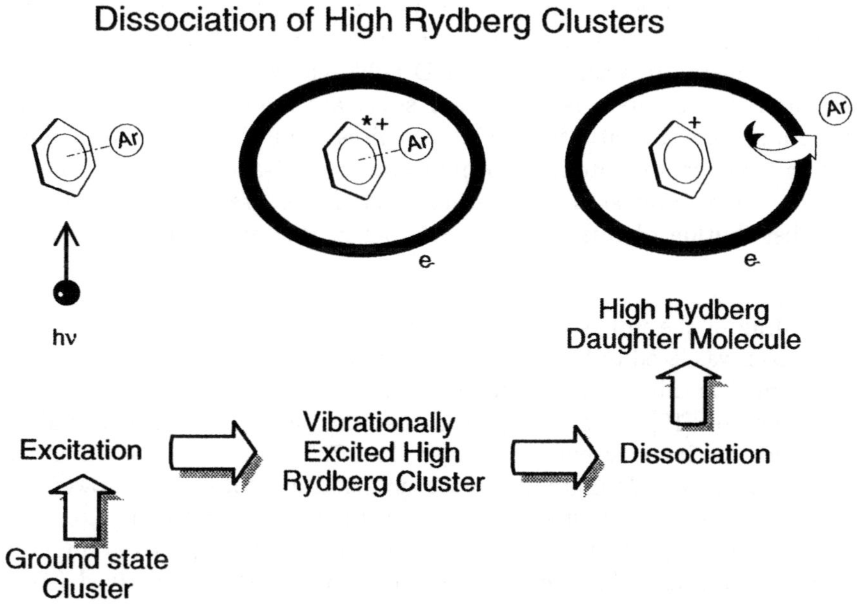

Fig. 13: Cartoon-like illustration of the benzene·Ar complex ion dissociation. The Rydberg electron survives the evaporation of the Ar atom from the benzene substrate.

The benzene·Ar cation dissociates by evaporation of the Ar atom at an internal energy smaller than 629 cm^{-1} (position of the $16^1 6^1(\pm 3/2)$ band). In conclusion, the first observed peak in the daughter ion spectrum at 629 cm^{-1} ($16^1 6^1(\pm 3/2)$ level) yields an upper limit for the dissociation energy of the benzene·Ar cation, i.e., $E_0 < 629 cm^{-1}$ (78 meV). From the 172 cm^{-1} red-shift of the ionization energy of benzene·Ar, an upper limit of the binding energy in the neutral ground state $D_0 \leq 457 cm^{-1}$ (56 meV) results for the dissociation energy of the neutral benzene·Ar complex using the relation $D_0 = E_0 - AIE$. These upper limits can be reduced considerably to $D_0 \leq 340 cm^{-1}$ (42 meV) by comparing the results for the benzene·Ar complex with those of benzene·Kr [14, 34].

Dibenzo-*p*-dioxin·Ar

It is interesting to investigate how the binding energy of the Ar atom varies with increasing size of the aromatic molecule substrate. Furthermore we expect to improve the accuracy of the binding energy for a larger system with more low frequency vibrations than benzene has. Therefore, as a second molecule the large three-ring aromatic substrate dibenzo-*p*-dioxin (DD) has been chosen [19]. It contains a six-atom central ring with two oxygen atoms in para position. Dibenzo-*p*-dioxin has been selected for three reasons: i) it exhibits a dense vibrational structure in the ionic state in an energy range between 500–900 cm^{-1} where the dissociation energy of weakly bound DD-noble gas van der Waals clusters is expected, ii) it represents a larger planar substrate than benzene and is therefore closer to a surface-like situation. This promises to increase our knowledge on the transition from isolated atom-molecule towards atom-surface interaction. iii) it is of environmental interest since intermolecular forces are expected to be relevant for the distribution of this pollutant [35, 36] in the air and its migration in the soil.

For recording of the upper two threshold ion spectra in Fig. 14 the frequency of the first laser was fixed to the first vibronic band in the $S_1 \leftarrow S_0$ intermediate state at 33742 cm^{-1} in dibenzo-*p*-dioxin·Ar which is blue shifted by +17 cm^{-1} from the S_1, 0^0 transition [19]. It is clearly seen that the parent ion signal (middle trace) breaks down for internal energies larger than 731 cm^{-1}. On the other hand no signal appears at the daughter ion mass (upper trace) for energies lower than 717 cm^{-1}. The 'missing' structure in the complex parent ion spectrum expected from the bare molecule spectrum (lower trace) appears at the respective daughter mass for energy larger than 717 cm^{-1}. The dissociation energies derived from these observations is $E_0(DD{\cdot}Ar)^+ = 717\ cm^{-1}$. Considering the -215 cm^{-1} complex AIE red-shift leads to $D_0 = 505\ cm^{-1}$ the neutral ground state.

It is interesting to see that there is a narrow energy range at 724 $\pm 7\ cm^{-1}$ where signal is present in the parent and the daughter threshold ion spectra of (dibenzo-*p*-dioxin·Ar)$^+$ (see the peaks at 724 cm^{-1} located 7 cm^{-1} above threshold appearing in the upper and the middle traces). As the peak at 724 cm^{-1} is observed in the parent and daughter ion spectrum of $(DD{\cdot}Ar)^+$, we conclude that a fraction of the cluster ions has dissociated for that internal energy while others have not and are observed at the parent mass. We exclude dissociation from the three-photon level since experiments with different laser intensities did not show different results.

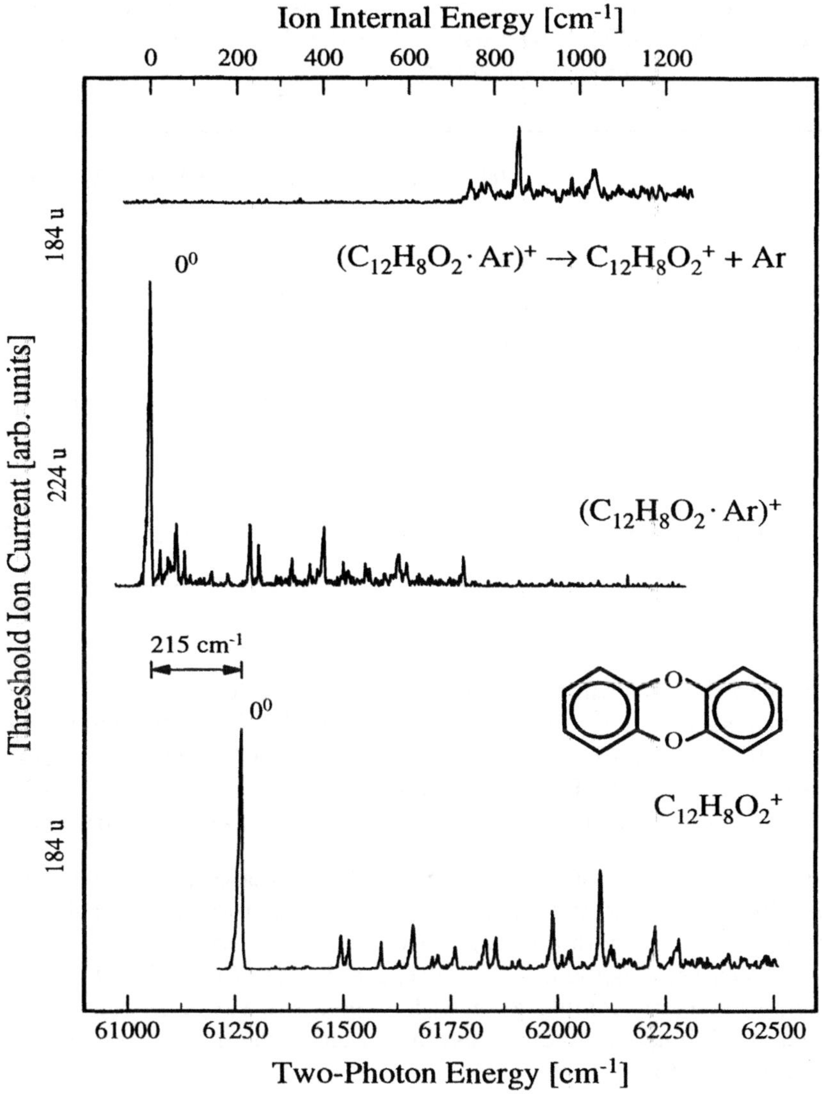

Fig. 14: Bottom: Threshold ion signal at 184 u for a photon energy $h\nu_1$ in resonance with the first vibrational band in the S_1 intermediate state of dibenzo-p-dioxin (DD, $C_{12}H_8O_2$). Middle: Threshold ion signal at 224 u for a photon energy $h\nu_1$ in resonance with the the first prominent vibrational band in the S_1 intermediate state of the neutral DD·Ar van der Waals dimer. The resulting spectrum is shifted by -215 cm^{-1} to the red of the DD spectrum (lower trace). All peaks with a larger energy than 731 cm^{-1} disappeared. Top: Dissociation threshold ion signal at 184 u recorded simultaneously with the complex spectrum (middle trace) The signal at the DD mass (184 u) results from a decay of the complex ion and can be observed for an internal energy larger than 717 cm^{-1}.

Table 1: S_1 intermediate state transition frequencies, S_1 origin redshifts, adiabatic ionization energies (AIE), AIE redshifts (ΔAIE), E_0 ionic ground state, and D_0 neutral ground state dissociation energy [14, 19].

	S_1, 0^0_0	Δ S_1, 0^0_0	AIE [cm^{-1}]	ΔAIE	E_0	D_0
benzene (S_1, 6^1_0)	38606		74555			
benzene·Ar	38585	(-21)	74383	(-172)	512	340
dibenzo-*p*-dioxin	33774		61283			
dibenzo-*p*-dioxin·Ar	33725	(-49)	61071	(-212)	717	505

This leads to the conclusion that for an energy range between 717 cm^{-1} and 731 cm^{-1} dissociation takes place on the μs time scale. Thus we cannot exclude a kinetic shift, but it is expected to be small ($\leq$ 20 cm^{-1}) since the energy range with μs decay is narrow. The relevant values for transition energies, redshifts and dissociation energies in the ionic and neutral complexes are summarized in Tab. 1.

The binding energy D_0 of the neutral complex increases by 165 cm^{-1} when we increase the substrate from a one ring to a three ring aromatic molecule. This is in qualitative agreement with a stronger dispersion interaction due to the increasing number of nearest neighbours. The strong red-shift of the AIE of -212 cm^{-1} of dibenzo-*p*-dioxin after complexation with Ar points to a stronger contribution from charge induced dipole interaction than in benzene, where the charge is delocalized in the aromatic ring.

5 Summary and Conclusion

In this work, we discussed a new technique which is important for internal energy selection in molecular ions. The technique employs a cooled supersonic beam combined with a reflectron time-of-flight mass spectrometer. Resonance-enhanced two-photon processes for ionization are particularly suitable as the resonance enhancement in the first absorption step assists the selection of a single species from a mixture present in the molecular beam and leads to some selectivity in the population of vibrational states in the ion due to suitable Franck Condon factors. The essential part of the new technique is the pulsed field ionization of long-lived Rydberg states close to an ionization threshold. The ions produced in this way only exist when excitation occurs at or close to an ionization threshold. They have a defined internal energy since no energy is converted into kinetic energy of the emitted electrons. First experiments with this powerful technique demonstrate that not only the production of state-selected aromatic molecule-noble gas dimer ions is possible but also their decay can be observed. Future applications will include larger ionic clusters and spectroscopic studies of their ground states.

In conclusion we have shown that recent developments in pulsed field ionization demonstrate the importance of time-of-flight mass spectrometry for the investigation of the spectroscopy and decay kinetics of molecular and particularly cluster ions.

Acknowledgment:

Financial support from the Deutsche Forschungsgemeinschaft and the Fonds der Chemischen Industrie is gratefully acknowledged

References

[1] B. Brehm, E. von Puttkamer, Z. Naturforsch. **22a**, 8 (1967).

[2] T. Baer, *Gas Phase Ion Chemistry* (M.T.Bowers Ed.), Academic Press, New York, 1979).

[3] J.A. Booze, M. Schweinsberger, and Tomas Baer, J. Chem. Phys. **99**, 4441 (1993).

[4] U. Boesl, H.J. Neusser, E.W. Schlag, Z. Naturforsch. **33a**, 1546 (1978).

[5] L. Zandee, R.B. Bernstein, D.A. Lichtin, J. Chem. Phys. **69**, 3247 (1978).

[6] E.W. Schlag, H.J. Neusser, Acc. Chem. Res. **16**, 355 (1983).

[7] H.J. Neusser, Int. J. Mass Spectrom. Ion Proc. **79**, 141 (1987).

[8] For a review see: H.J. Neusser, J. Phys. Chem. **93**, 3897 (1989).

[9] R.S. Freund, *High-Rydberg molecules* (R. F. Stebbings, F. B. Dunning: Rydberg states of atoms and molecules, Cambridge University Press,Cambridge, 1983).

[10] L. Zhu, P.M. Johnson, J. Chem. Phys. **94**, 5769 (1991).

[11] H. Krause, H.J. Neusser, J. Chem. Phys. **97**, 5923 (1992).

[12] C. Jouvet, C. Dedonder-Lardeux, S. Martrenchard-Barra, D. Solgadi, Chem. Phys. Lett. **198**, 419 (1992).

[13] Xu Zhang, J.M. Smith, J.L. Knee, J. Chem. Phys. **99**, 3133 (1993).

[14] H. Krause, H.J. Neusser, J. Chem. Phys. **99**, 6278 (1993).

[15] M.G.H. Boogaarts, P.C. Hinnen, G. Meijer, Chem. Phys. Lett. (1994).

[16] M.C.R. Cockett, K. Kimura, J. Chem. Phys. **100**, 3429 (1994).

[17] W.G. Scherzer, H.L. Selzle, R.D. Levine, E.W. Schlag, Phys. Rev. Lett. **72**, 1435 (1994).

[18] H.J. Dietrich, R. Lindner, K. Müller-Dethlefs, J. Chem. Phys. **101**, 3399 (1994).

[19] Th.L. Grebner, and H.J. Neusser, Chem. Phys. Lett. (1995).

[20] W.A. Chupka, J. Chem. Phys. **98**, 4520 (1993).

[21] S.T. Pratt, J. Chem. Phys. **98**, 9241 (1993).

[22] W.A. Chupka, J. Chem. Phys. **99**, 5800 (1993).

[23] M.J.J. Vrakking, Y.T. Lee, J. Chem. Phys. **102**, 8818 (1995).

[24] F. Merkt, J. Phys. Chem. **100**, 2623 (1994).

[25] M.J.J. Vrakking, Y.T. Lee, Phys. Chem. Rev. **102**, 8833 (1995).

[26] G. Reiser, W. Habenicht, K. Müller-Dethlefs, E.W. Schlag, Chem. Phys. Lett. **152**, 119 (1988).

[27] K. Müller-Dethlefs, E.W. Schlag, Annu. Rev. Phys. Chem. **42**, 109 (1991).

[28] B. Ernstberger, H. Krause, A. Kiermeier, H.J. Neusser, J. Chem. Phys. **92**, 5285 (1990).

[29] H.J. Neusser, H. Krause, Int. J. Mass Spectrom. Ion Proc. **131**, 211 (1994).

[30] Th.L. Grebner, H.J. Neusser, Int. J. Mass Spectrom. Ion Proc. **137**, L1 (1994).

[31] Xu Zhang, J.M. Smith, J.L. Knee, J. Chem. Phys. **97**, 2843 (1992).

[32] M. Takahashi, H. Ozeki, K. Kimura, J. Chem. Phys. **96**, 6399 (1992).

[33] L.A. Chewter, K. Müller-Dethlefs, E.W. Schlag, Chem. Phys. Lett. **135**, 219 (1987).

[34] H.J. Neusser, H. Krause, Chem. Rev. **94**, 1829 (1994).

[35] S. Safe, Crit. Rev. Toxicol. **21**, 51 (1990).

[36] M. Gough, J. Toxicol Environ. Health **32**, 205 (1991).

[illegible] 87 [illegible]

[illegible] Phys. 64, [illegible]

[illegible] (1998).

[illegible]

[illegible]

[illegible]

FUNDAMENTALS OF NEGATIVE ION PHOTOELECTRON SPECTROSCOPY

CAROLINE E. H. DESSENT AND MARK A. JOHNSON
Sterling Chemistry Laboratory
Yale University
New Haven, CT USA

1. Introduction

Since its introduction in 1972 by Levine and Hall,[1] negative ion photoelectron spectroscopy has become one of the most powerful tools in the characterization of atomic and molecular anions as well as negatively charged clusters. The primary experiment simply involves dispersing the kinetic energy distribution of electrons photodetached with a fixed frequency laser:

$$A^- + h\upsilon \rightarrow A + e^- \qquad [1]$$

Several aspects of the experiment combine to account for its current popularity:

i) Species selectivity is readily achieved using mass spectrometry
ii) It is generally applicable to all negative ions
iii) It yields structural information on many electronic states of the neutral formed in photodetachment

In this chapter, we set out to provide a "primer" on the subject, directed at graduate students and those interested in establishing a working understanding of the important aspects of the technique as seen from the perspective of a user.

1.1 GENERAL REMARKS ABOUT THE PHOTODETACHMENT PROCESS

1.1.1 *Excitation with an electric dipole operator*

We shall present the concepts in the most basic terms which nonetheless capture the essence of the physics involved in the process. The general problem involves optical excitation of a valence electron in a molecular anion, described by the approximate electronic wavefunction:

K. R. Jennings (ed.), Fundamentals and Applications of Gas Phase Ion Chemistry, 287–306.

$$\Psi_{A^-} \approx A[\prod_i \phi_i(r_{el})] \qquad [2]$$

where A[] is meant to indicate a Slater-determinant, separated wavefunction into the one-electron orbitals ϕ_i, appropriate for the dominant electron configuration of the ground state anion. In the simplest model for optical excitation, the photodetachment cross section is related to the electric dipole matrix element:

$$\sigma \approx |\langle\phi_i|\boldsymbol{\mu}_{el}|\varphi_c\rangle|^2 \rho_c(E) \qquad [3]$$

where the dipole operator:

$$\boldsymbol{\mu}_{el} = \sum_j e\boldsymbol{r}_j \qquad [4]$$

ensures that detachment is dominated by the sequential removal of one electron from each valence orbital to the continuum described by the wavefunction ϕ_c (i.e. the dipole operator is a one-electron operator). The factor $\rho_c(E)$ is the density of translational states with kinetic energy E, which is proportional (well above threshold) to the asymptotic velocity of the outgoing electron. The simple, one electron picture is universally used in photoelectron spectroscopy as through it, we can sequentially access lower lying orbitals to create the corresponding singly excited states in the electron configuration of the neutral. Note that when detachment occurs from a radical anion, the open shell configuration left after ejection of a core electron from ϕ_n is generally of the form:

$$\psi_A = A[\psi(r_1 - r_l)\,\phi_n(r_j)\,\phi_m(r_k)\,\psi_{spin}] \qquad [5]$$

where the two unpaired electrons can be either single or triplet coupled in Ψ_{spin} (ψ represents the remaining spin-paired electrons). Thus, photoelectron spectroscopy can yield singlet-triplet splittings directly from the spectrum, as demonstrated in methylene.[2]

1.1.2 *Angular distribution of photoelectrons*

Let us begin with the simple case of electron ejection from hydrogenic orbitals, in which case the orbitals involved are:

$$\phi_i = R_{n,l}(r)\,Y_{l,m} \qquad [6]$$

for the bound state and:

$$\varphi_k = \sum_{l',m'} a(l',m')\, Y_{l',m'}(\theta,\phi)\, G_{kl'}(r) \qquad [7]$$

for the continuum wavefunction, where $R_{nl}(r)$ and $G_{kl'}(r)$ denote radial wavefunctions for the bound and continuum states and Y_{lm} are spherical harmonics. Note that the continuum state is not generally in a definite state of angular momentum. The transition matrix element which leads to production of a free electron ejected at an angle θ relative to the electric vector of the laser, I(θ) is proportional to the following terms:

$$I(\Theta) \propto [\langle\phi_i|rY_{10}|\phi_c\rangle]^2$$

$$= [\sum_{l',m'} a(l'm')\,\sigma_l \langle Y_{l'm'}|Y_{10}|Y_{lm}\rangle]^2 \qquad [8]$$

where:

$$\sigma_l = \int_0^\infty rR_{nl}G_{kl'}dr \qquad [9]$$

and the angle between the ejected electron and the electric vector is contained in the a(*l*'m') terms. The angular momentum term in brackets indicates that only l'=l±1 final states are allowed as expected for an electric dipole transition. Bethe[3] treated this problem explicitly, and Cooper and Zare[4] showed that in general the two states l'=l±1are populated and interfere with each other such that the contributions of the highly nodal Y_{lm} functions cancel to leave a remarkably simple expression for the angular distribution for ejected electron:

$$I(\theta) = \frac{1}{4\pi}[1+\beta P_2(\cos(\theta)] \qquad [10]$$

where $P_2\cos(\theta)$ is the second order Legendre polynomial and β is the so-called "asymmetry parameter" which ranges from +2 for the case of perfect (e.g. $\cos^2(\theta)$) alignment along the electric vector and -1 when the electron emerges perpendicular to the electric vector according to a $\sin^2(\theta)$ distribution. The case of *s*-wave ejection clearly corresponds to β=0.

A few cases are worth highlighting. In the case the electron is detached from an s orbital, the only l'=l+1 can be excited which corresponds to a *p* wave, or perfect alignment along the electric vector with a $\cos^2(\theta)$ distribution. This is nearly approached, for example, in the case of photodetachment of the excess electron from anionic water clusters[5] as shown in Fig. 1. Another case of interest is ejection of a p electron, which results in outgoing *s* and *d* waves. These can interfere to create an outgoing electron with a maximum perpendicular to the electric vector, as is found for photodetachment of O^- in the visible region for ejection of the 2p electron.[6]

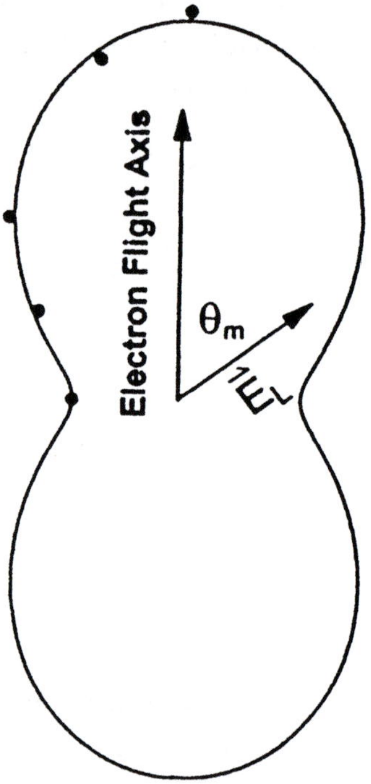

Figure 1. Angular distribution of electrons ejected from $(H_2O)^-_{18}$. Filled circles are experimental points and the solid trace is the calculated fit to $\beta = +0.92$.

In general, however, the photon energy dependence of β is not trivial as the various partial waves, l', interfere and change in intensity and relative phase. The asymmetry parameter depends on the outgoing energy of the electron, and can therefore change even for different vibrational levels of the same electronic state. For example, a calculation of the photon energy-dependent asymmetry parameters for I^- indicates that it indeed starts off at zero, becomes negative and approaches -1 at about 5 eV above threshold, and finally approaches 2 at still higher energy. It is therefore desirable to remove the angular dependence by measuring the total electron yield in the photoelectron spectrum, which is accomplished by exciting the ions with a linearly polarized laser oriented at the "magic angle" of 55° relative to the electron ejection axis, where $P_2[\cos(\theta)]$ is zero.

1.1.3. *Threshold laws for photodetachment*

The overall appearance of the detachment spectrum is governed by both the dipole matrix element and the density of states function, where the density of states controls the shape of the spectrum near the detachment threshold. It is important to point out here that most stable, closed shell molecular negative ions have electron affinities below two electron volts or so,[7] while the lowest unoccupied molecular orbital (LUMO) is usually higher than this value (consider the fact that most closed shell molecules have electronic spectra in the uv). The result is that relatively few negative ions possess electronically excited states, A^{-*}, which are stable with respect to spontaneous ejection of the excess electron (autodetachment):

$$A^{-*} \rightarrow A + e^- \qquad [11]$$

The rates of this spontaneous decay range tend to lie in the range of 1 to 100 fs, and therefore the resulting spectra are usually too lifetime-broadened to be of value for spectroscopic characterization of the anion. Thus, the absorption spectrum of a negative ion typically consists of an onset of the photodetachment process [1]. In the analogous case of photoionization of a neutral, the onset of ionization is sharp at the ionization potential with a continuity of oscillator strength through the converging Rydberg states. Since the negative ion does not, in general possess any electronic states below the electron continuum, there is still a manifestation of continuity such that the cross section for excitation is zero at threshold, and rises according to the density of states function $\rho_c(E)$ in Eq. [3]. In general, the shape of the onset depends on the angular momentum, *l*, of the outgoing electron according to the "Wigner"[8] threshold law:

$$\sigma_D \propto [h\nu - E_{th}]^{(l+1/2)} \qquad [12]$$

A simple case arises for the ejection of an *s*-wave, where there is a discontinuity in the slope of the cross section at zero, as illustrated below for the case of photodetachment of the I^- atomic ion shown in Fig. [2]. Here $\sigma_D \propto (KE_{el})^{1/2}$ according to Eq. [12], which is the same as the translational density of states term in Eq. [3].

For the case of $l \neq 0$, Eq. [12] indicates that σ_D has a much slower onset, as is found experimentally, for example in the case of O_2^- where the excess electron occupies the π^*

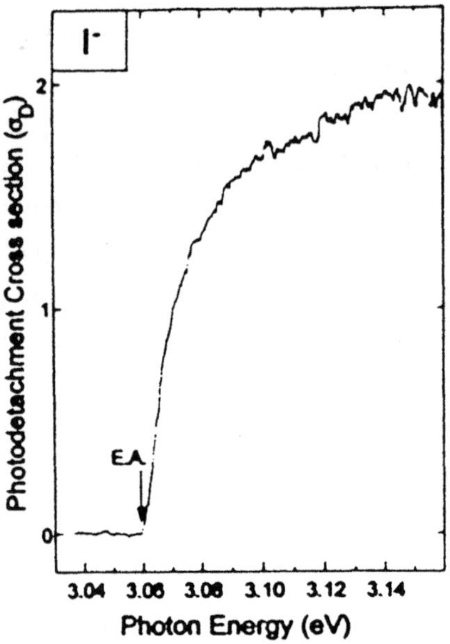

Figure 2. Photodetachment spectrum of I^-. The arrow indicates the location of the electron detachment threshold.

antibonding orbital with a shape closest to the l=2 spherical harmonic.[9] While the faster threshold law for s-wave ejection ensures that excitation near threshold is dominated by this channel, other angular momentum states generally increase in importance with increasing excitation energy.

1.1.4 *Overall shape of the photodetachment spectrum above threshold*

The photodetachment spectrum from I^- shown in Fig. 2 indicates that once above threshold, the photodetachment cross section levels off at an essentially constant value so that the range of the threshold laws is very short (<0.1 eV). Cross sections in this asymptotic regime are typically in the range of 10^{-18} cm^2, making photodetachment a relatively inefficient process. This fact, together with the small target density available from negative ion beam sources, leads to the situation that lasers were required even for the first observation of energy-resolved photoelectrons. It is clear, however, that since the total oscillator strength for excitation of a single electron is finite, the detachment cross section must eventually fall-off with increasing photon energy. The magnitude of the detachment cross section and fall-off are given by the matrix element involving the radial wavefunctions in Eq. [9]. At high energy, the DeBroglie wavelength of the outgoing electron can be smaller than the size of the bound orbital,[10] and we might approximate the continuum function as e^{ikr}. In this case:

$$\sigma_l \approx \int_0^\infty r R_{nl} e^{ikr} dr \qquad [13]$$

and when the e^{ikr} function oscillates much faster than the rR_{nl} factor, the integral vanishes. Another way of viewing this behavior is that it is difficult to excite an electron to a much higher continuum kinetic energy than the electron possesses in the orbital from which it is detached. For most small molecules, these orbitals are a few Å, so that the fall is not expected until the excitation energy reaches the order of 10 volts or so. Thus, for most molecular applications, essentially any visible or uv photon is equally useful for carrying out the experiment; indeed one usually uses the most energetic photon available in order to access as many electronically excited states as possible, as discussed in section 1.1.1.

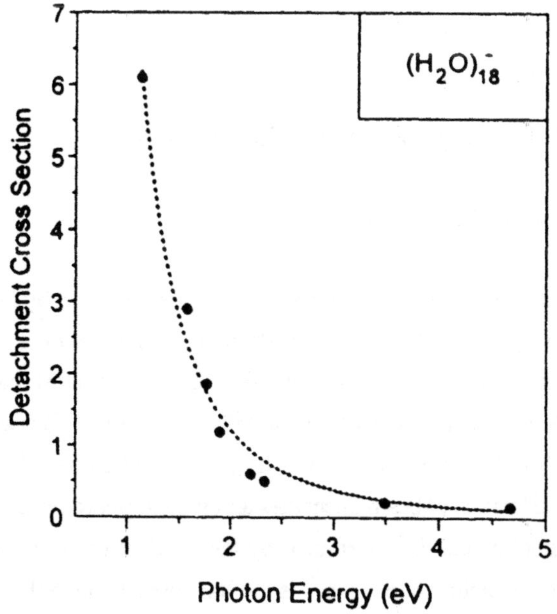

Figure 3. Falling electron detachment cross section with photon energy for $(H_2O)^-_{18}$. The filled circles are data points and the dashed line is a fit to a $1/E^3$ function.

The fall-off in photodetachment cross section can be an issue, however, when the excess electron in the anion is contained in a diffuse orbital. Such diffuse electrons are found when electrons are attached to closed shell molecules in which the LUMO is very high in energy so that the electron is not stable in the valence configuration. Then the electron can only be bound by interaction with the net moments of the neutral system, principally the electric dipole moment. Two examples of this type of binding are the exotic "dipole-bound" anions such as the negative ion of acetonitrile,[11-13] and the famous solvated electrons in water, ammonia, and alcohols.[14,15] In the case of electrons bound to water clusters, Campagnola and co-workers[16] observed that the photodetachment cross sections fell rapidly across the visible region. In this study the first moment of the destruction spectrum, σ(E), was used to determine the mean radius, $<R_{el}^2>$, of the excess electron

$$<R_{el}^2> \propto \int \sigma(E) / E \ dE \qquad [14]$$

The declining cross section for photodetachment with increasing photoexcitation energy for anions with diffuse electrons can seriously hamper the quality of the spectral data, as recently observed in the 1.16, 2.33 and 3.49 eV photoelectron spectra of the diffuse, dipole-bound CH_3CN^- anion. [13,17]

2. Experimental Aspects of Photoelectron Spectroscopy

2.1. HISTORICAL PERSPECTIVE

The first negative ion photoelectron spectrometer was constructed by Hall and Levine[1] and demonstrated on the O_2^- and NO^- ions. This instrument was based on a continuous ion beam which was mass separated using a Wien velocity filter giving a modest mass resolution on the order of 1/50. Photodetachment was carried out with a home built (!) argon ion laser running at 488 nm (2.51 eV). The importance of a high power photodetachment laser is readily understood upon consideration of beam intensities and typical photodetachment cross sections. The number of signal counts is determined by the product of $P \times I \times \phi$, where P is the probability of electron detachment per ion, I is the Ion current and ϕ the collection efficiency (~ 10^{-3}). With photodetachment cross sections on the order of 10^{-18} cm^2 and typical ion beam currents of 1 nA at an interaction energy of 20 eV in the laser interaction region, the laser must provide ~ 10 Watts per count! Following photodetachment, the electrons were energy dispersed using a scanning electrostatic analyzer with a resolution on the order of 30 meV.

The ion source used by Hall and Levine was a plasma discharge and therefore provided ions in a variety of vibrational and electronic states. In 1985, Lineberger and co-workers[2] substantially improved this apparatus for general application to molecular anions by introducing a flowing afterglow ion source that provided vibrational and rotational cooling of the sample ions. One of the first systems attacked with the new apparatus was CH_2^-, resulting in several previously observed spectral features being positively identified as vibrational hot bands, leading to a revised determination of the singlet-triplet splitting in methylene. A few years later, Lineberger[18] implemented photodetachment at 3.532 eV using an external cavity build up system, which extended the range of anions that could be studied.

While neutral PES studies have taken full advantage of electron TOF techniques using broadly tunable pulsed lasers, anion PES has been traditionally carried out with continuous lasers owing to the inherently continuous nature of ion sources. The situation changed dramatically after intense pulsed beams of positive and negative cluster ions were prepared by ionizing a pulsed supersonic jet with either a laser or an electron beam.[19] Ions in these sources are rotationally cooled to the ambient ≈ 20K translational temperature in the jet expansion.[20] The pulsed sources were coupled with techniques to both spatially and temporally focus the ions[21] for optimum overlap with pulsed lasers, using an approach first introduced by Johnson, Alexander and Lineberger[22] in their study of $(CO_2)^+$. This opened the possibility of carrying out mass selected PES experiments with a pulsed laser.[23] The benefit of such an extension results from the capability of pulsed lasers to operate at much shorter wavelengths than those available from cw lasers, creating new possibilities for studying species with high electron affinities. In fact, the first negative ion time-of-flight photoelectron spectra were obtained *by accident* when CH_3I^- and I^- were produced in a pulsed laser multiphoton ionization PES study of CH_3I![24] In the next section, we describe a pulsed photoelectron spectrometer commonly used to study negative ions and clusters.

2.2. A PULSED PHOTOELECTRON SPECTROMETER

A schematic diagram of the typical pulsed PES spectrometer is presented in Fig. 4. Cluster ions, formed by electron impact ionization of a pulsed supersonic jet,[19] are mass selected by injection into a spatially focusing TOF mass spectrometer. A pulsed laser photodetaches the selected ions at the focus of the mass spectrometer, and the resulting photoelectrons are energy analyzed by time-of-flight.

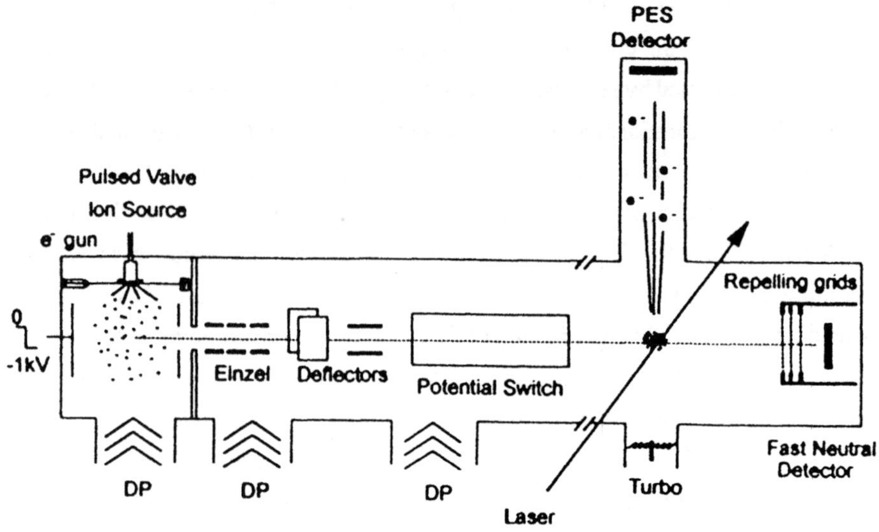

Figure 4. Schematic diagram of a pulsed PES spectrometer.

The cluster anions are generally formed by three body stabilization or dissociative attachment of the slow secondary electrons trapped in the jet. After formation, they drift ~10 cm in a field free region and are then accelerated to an ion beam energy of 2 keV following extraction from the source. One of the difficulties in fast ion beam experiments such as this is that either the ion source or the laser interaction region must be electrically floated to obtain an ion beam with enough velocity to maintain good focusing characteristics. The spectrometer displayed above uses a pulsed feature to overcome this problem so that the ion source and the electron analyzer are *each* kept at ground potential. This is achieved in the ion drift region by electrically isolating the ions in a metallic cylinder (the "potential switch") which is held at 2 kV when the ions enter but rapidly switched to ground while the ions are still inside. Thus, the ions exit the cylinder with 2 keV beam energy but now referenced to ground. The ions are spatially and temporally focused 1.8 m from the source for interaction with a pulsed laser (often Nd:YAG) and mass discrimination is achieved by selective overlap of the laser pulse with the transiently focused ion of interest. The overlap of the laser with ions of a specific mass is optimized by deflecting the ion beam with an "ion reject grid" and selectively detecting the fast neutral photofragments. The neutral peak has the same arrival time as the undeflected parent ion and thus confirms the mass of the photodetached ion. The electron flight tube is orthogonal to the ion beam direction, is ~ 0.6 m long and is shielded by two concentric cylinders of 0.014" thick sheets of μ-metal. The electrons are detected as single counts after amplification, with two successive microchannel plate electron multipliers and the flight times are recorded with a multi-step time-to-amplitude (TAC) converter. A kinematic correction is applied to the electron spectrum to account for the mass-dependent laboratory to center-of-mass transformation.

For a relatively intense parent ion (e.g. 10^4 O_2^- ions/pulse) about 15 photoelectrons are detected per laser shot; 10000 laser shots requiring about 15 min of data acquisition time are typically stored to obtain a reasonable signal-to-noise photoelectron spectrum. The first (intentional) pulsed PES of a molecular anion recorded was again O_2^-.[23] The spectrum, displayed in Fig. 5 (2.33 eV photodetachment energy) reveals vibrational progressions in the $X^3\Sigma_g^-$ and $a^1\Delta_g$ states of O_2 and does not provide any evidence of vibrational hot bands, indicating that the ions are not warmed upon extraction from the jet.

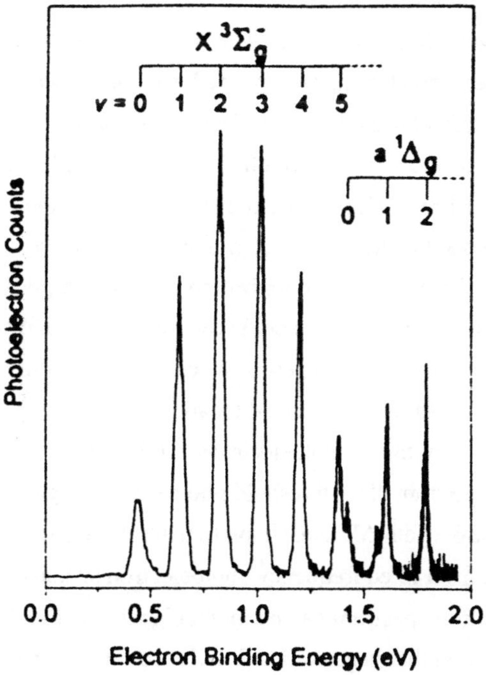

Figure 5. Photoelectron spectrum of O_2^- recorded at 532 nm and "magic angle" (55° with respect to the electron flight axis) laser polarization.

The resolution of the spectrum presented in Fig. 5 is around 30 meV(mainly attributed to electric field imhomogeneity) but has since been improved by approximately a factor of 3. Since the inaugural TOF spectra, pulsed negative ion PES have now been recorded of a host of chemical species including size-selected clusters[25,26] and ion-molecule complexes that represent trapped intermediates along reactive potential energy surfaces.[27,28]

2.3. HIGH RESOLUTION THRESHOLD PHOTODETACHMENT SPECTROSCOPY OF NEGATIVE IONS

A significant drawback of negative ion photoelectron spectroscopy is the relatively poor resolution (compared to optical spectroscopy) of the electron energy analyzer. The energy-dispersive and time-of-flight analyzers used in fixed frequency PES generally have a resolution limit of 5-10 meV. This is sufficient to observe some vibration and electronic structure in small molecules, but for many systems of interest, particularly clusters of moderate size, this limit prevents the observation of vibrational features in the spectrum. In 1989, Neumark[29] constructed a threshold photodetachment spectrometer which combines the advantages of mass-selected negative ion photodetachment with a high resolution *threshold* photoelectron detection scheme (zero electron kinetic energy or ZEKE spectroscopy) developed by Schlag and co-workers[30] for the study of neutral photoionization. A pulsed tunable laser is used as the photoionizing light source and a weak extraction field is applied to the interaction region a short time after the laser fires. This allows the selective detection of the near-zero kinetic energy electrons produced at the laser frequency. Paradoxically, the so called ZEKE scheme applied to neutrals actually involves field ionization of bound, high Rydberg states.[31] Since there are no such states in anions, the threshold scan must necessarily selectively detach free electrons with very low kinetic energy. These slow electrons will only be produced when the photon frequency is at or just above the threshold for an ion to neutral transition. If the electron detection system counts only those photoelectrons produced with a center-of-mass kinetic energy less than ΔE, the spectrum of the threshold photoelectron signal as a function of photon energy should exhibit a peak width ΔE at each ion to neutral transition. In principle, the same information is available from fixed frequency photoelectron spectroscopy, but the resolution in a threshold photoelectron spectrum can be much higher (<0.4 meV). An example of the greatly enhanced resolution that negative ion ZEKE provides over tranditional negative ion PES is presented in Fig. 6 for the $I^{-}\cdot CO_2$ system.[32] Both spectra consist of two distinct bands separated approximately by the $I(^2P)$ spin-orbit splitting (0.943 eV). In the photoelectron spectrum, both bands were found to consist of a short progression of peaks which were assigned to excitation of the CO_2 bend in the neutral complex. The improvement in the spectral resolution of the ZEKE spectra is dramatic. The $I(^2P_{3/2})\cdot CO_2$ envelope is now seen to consist of four bands, with each of the sub-bands displaying low frequency structure which was assigned to the C-I van der Waals stretching modes.

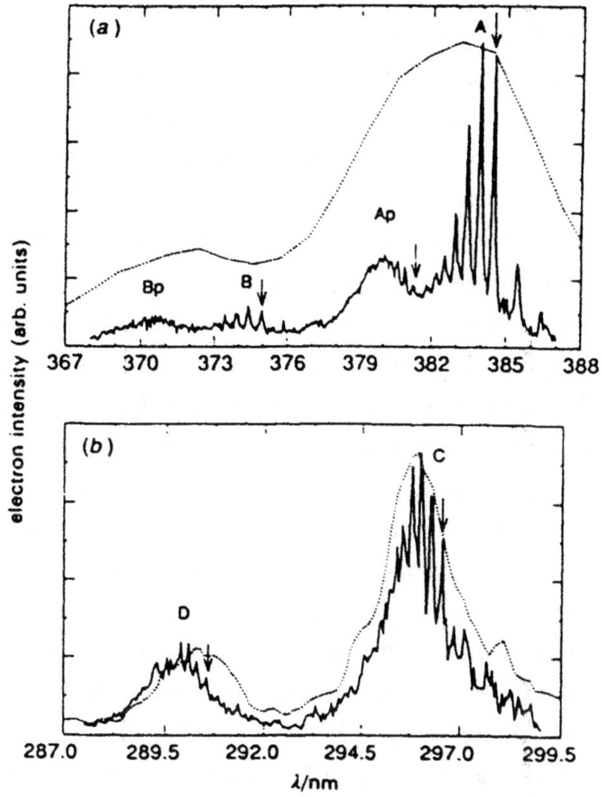

Figure 6. Threshold photodetachment spectrum of $I^-\cdot CO_2$ (solid lines) and photoelectron spectrum of $I^-\cdot CO_2$ (dotted lines). (a) $I(^2P_{3/2})\cdot CO_2$ band and (b) $I(^2P_{1/2})\cdot CO_2$ band. The arrows indicate the band origins.

Note the threshold laws discussed in section 1.1.3 significantly affect the ability to carry out ZEKE spectroscopy in negative ions since photodetachment leading to high angular momentum ejected electrons ($l > 0$) leads to very low cross sections at threshold. It appears, however that the non-spherical symmetry present in molecular and cluster systems may relax this constraint somewhat.

3. Selected applications of negative ion photoelectron spectroscopy

3.1. TRANSITION STATE SPECTROSCOPY

In most negative ion photoelectron spectroscopy experiments, the neutral formed upon photodetachment is stable, and the spectrum consists of resolved bands corresponding to transitions to vibrational levels of the neutral (see for example, Fig. 5) However, this need not always be the case as photoexcitation may access an unbound region on the neutral potential energy surface. The first example was provided by Lineberger and co-workers who used photoelectron spectroscopy of the vinylidene anion to explore the 1,2-hydrogen shift in vinylidene to acetylene.[33] Neumark and co-workers[34] have exploited the phenomenon to study the transition states of neutral bimolecular reactions by photodetachment (either via traditional pulsed PES or higher resolution negative ion ZEKE spectroscopy) of a suitable ground-state negative ion. Although the neutral complex being investigated is unbound, it may be bound along coordinates perpendicular to the reaction coordinate, and will undergo vibrational motion as it dissociates. If this motion occurs on the same time scale as (or faster than) dissociation, a structured photoelectron spectrum in which the peaks correspond to vibrational levels of the "transition-state complex" at or near the transition state, is observed. The peak positions and widths give information on the spectroscopy and dissociation dynamics of the transition-state complex which is extracted from the spectrum quantitatively through a time-dependent wavepacket analysis. The greatest limitation of the technique is that the ground state anion geometry determines the region of the neutral surface that is accessed via photodetachment, so that only certain systems are accessible through this approach.

This technique was applied to the Br + HBr reaction[35] through photodetachment of the stable hydrogen-bonded negative ion, $BrHBr^-$, where the symmetric [BrHBr] complex is formed at or near the top of the potential barrier for the reaction. The photoelectron spectra of $BrHBr^-$ and $BrDBr^-$ ions are presented in Fig. 7 The spectra display progressions in the antisymmetric stretching vibration of the neutral complexes, with the broad peaks being assigned to directly dissociating states of the neutral complex and narrow peaks (15-20 meV) originating from reactive resonance states. Additionally, peaks resulting from an electronically excited state of the [BrHBr] complex were identified (A′ and B′). The spectra were analyzed to yield an effective collinear potential energy surface for the reaction.

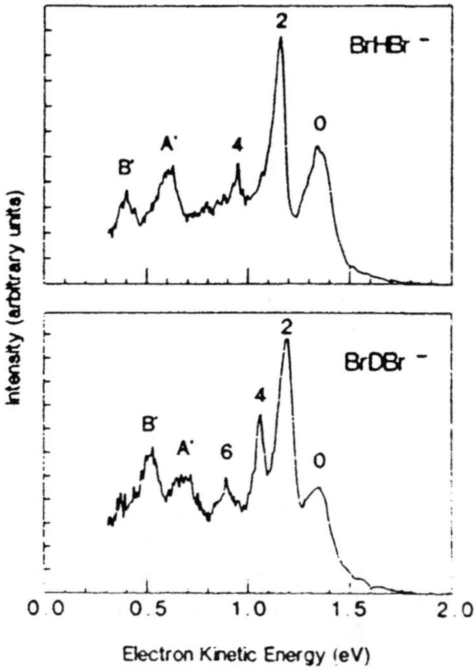

Figure 7. The 213 nm photoelectron spectra of $BrHBr^-$ and $BrDBr^-$

3.2. THE EFFECTS OF NEGATIVE ION RESONANCES ON VIBRATIONAL FEATURES IN ANIONIC PES

In the usual treatment of photoelectron spectra, vibrational excitation results from a distortion of the anion relative to the neutral. It was therefore surprising when vibrational excitation was observed in the photoelectron spectrum of the dipole-bound anion, CH_3CN^-,[17] since such species are thought of as a nominally neutral molecule beside a diffuse electron cloud.[36] An indication that the vibrational excitation was not a result of straightforward Franck-Condon over-laps was implied since the amount of vibrational excitation was found to depend on the photon energy. In particular, the ν_2 mode, which was absent in the 1.165 eV spectrum, was particularly excitation-energy dependent and became pronounced upon 2.331 and 3.496 eV excitation. The photoelectron spectra acquired at these energies are displayed in Fig. 8 .

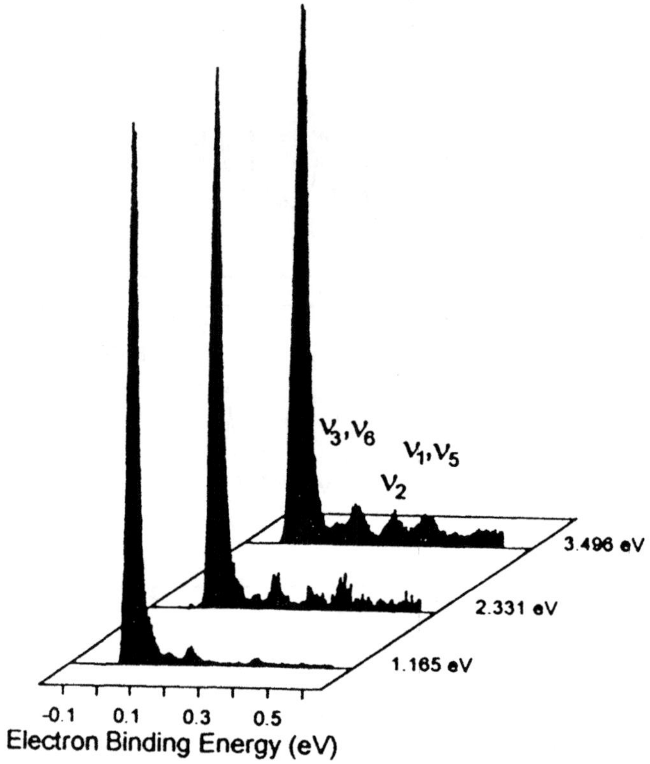

Figure 8. Photoelectron spectra of the CH_3CN^- dipole-bound anion at a) 1.165, b) 2.331 and c) 3.496 eV laser photodetachment energy. The spectra are normalized to the intensity of the $v_{0\text{-}0}$ feature.

The energy dependence of the v_2 vibration was traced to excitation of a shape resonance due to electron capture by the π^*_{CN} orbital[37] (the LUMO of CH_3CN) at 2.9 eV- exactly half way between the two higher PES excitation energies which display the effect. As a result of vibronic coupling between the π^*_{CN} and σ^* orbitals, the CH stretching vibration, v_1, is also enhanced through the resonance. These resonances should be a general feature of dipole-bound anions since, by definition, the anion state corresponding to occupation of the LUMO at the equilibrium geometry of the neutral lies within the electron continuum. However, dipole-bound anions are not the only systems where negative ion resonances may affect the vibrational features and any system that contains a low lying electronic resonance may be susceptible to "resonance distortion".[38] Another aspect of dipole-bound systems is that even off resonance, vibrational excitation can be driven by another non-Franck-Condon mechanism in which the coordinate dependence of the electronic wavefunction leads to vibrational selection rules which yield PES spectra just like the vibrational spectrum of the neutral (in this case CH_3CN).

4. References

1. Celotta, R. J., Bennett, R. A., Hall, J. L., Siegel, M. W. and Levine, J. (1972) Molecular photodetachment spectroscopy. II. The electron affinity of O_2 and the structure of O_2^-, *Phys. Rev. A* **6**, 631-642.
2. Leopold, D. G., Murray, K. K. and Lineberger, W. C. (1984) Laser photoelectron spectroscopy of vibrationally relaxed CH_2^-: A reinvestigation of the singlet-triplet splitting in methylene, *J. Chem. Phys.* **81**, 1048-1050.
3. Bethe, H. A. and Salpeter, E. E. (1977) *Quantum Mechanics of One- and Two-Electron Atoms*, Plenum Publishing Corporation, New York.
4. Cooper, J. and Zare, R. N. (1968) Angular Distribution of Photoelectrons, *J. Chem. Phys.* **48**, 942-943.
5. Campagnola, P. J., Posey, L. A. and Johnson, M. A. (1990) The angular distribution of photoelectrons ejected from the hydrated electron cluster $(H_2O)^-_{18}$, *J. Chem. Phys.* **92**, 3243-3245.
6. Robinson, E. J. and Geltman, S. (1967) Single and double quantum photodetachment of negative ions, *Phys. Rev.* **153**, 4-8.
7. Lias, S. G., Bartmess, J. E., Liebman, J. F., Holmes, J. L., Levin, R. D. and Mallard, W. G., (1988) Gas-Phase Ion and Neutral Thermochemistry, *J. Phys. Chem. Ref. Data* **17**, 647-797.
8. Wigner, E. P. (1948) On the behavior of cross sections near thresholds, *Phys. Rev.* **73**, 1003-1009.
9. Reed, K. J., Zimmerman, A. H., Anderson, H. C. and Brauman, J. I. (1976) Cross sections for photodetachment of electrons from negative ions at threshold, *J. Chem. Phys.* **64**, 1368-1375.
10. Dehmer, J. L., Inokuti, M. and Saxon, R. P. (1975) Systematics of moments of dipole oscillator-strength distributions, *Phys. Rev. A* **12**, 102-121.

11. Compton, R. N., Reinhardt, P. W. and Cooper, C. D. (1978) Collisional Ionization between alkali atoms and some methane derivatives: Electron affinities for CH_3NO_2, CF_3I and CF_3Br, *J. Chem. Phys.* **68**, 4360-4367.

12. Desfrancois, C., Abdoul-Carime, H., Adjouri, C., Khelifa, N. and J. P. Schermann, (1994) Dipole binding to a strongly polar molecule and its homogeneous clusters: Magic distribution of acetonitrile anions, *Europhys. Lett.* **26**, 25-30.

13. Bailey, C. G., Dessent, C. E. H. and Johnson, M. A. (1995) Dipole-bound excited states of the $I^-\cdot CH_3CN$ and $I^-\cdot(CH_3CN)_2$ ion-molecule complexes: Evidence for asymmetric solvation, *J. Chem. Phys.* **103**, 2006-2015.

14. Haberland, H., Ludewigt, C., Schindler, H. G. and Worsnop, D. R. (1985) Clusters of water and ammonia with excess electrons, *Surf. Sci.* **156**, 157-164.

15. Desfrancois, C., Baillon, B., J. P. Schermann, Arnold, S. T., Hendricks, J. H. and Bowen, K. H. (1994) Prediction and Observation of a new, ground state dipole-bound dimer anion: The mixed water/ammonia system, *Phys. Rev. Lett.* **72**, 48-51.

16. Campagnola, P. J., Lavrich, D. J., DeLuca, M. J. and Johnson, M. J. (1991) Photodestruction spectra of the anionic water clusters $(H_2O)_n^-$ n=18 and 30: Absorption to the red of e^-_{aq}, *J. Chem. Phys.* **94**, 5240-5242.

17. Bailey, C. G., Dessent, C. E. H., Bowen, K. H. and Johnson, M. A. (1996) Vibronic effects in the photon-energy dependent photoelectron spectra of the CH_3CN^- dipole-bound anion, *J. Chem. Phys.* in press.

18. Ervin, K. M., Ho, J. and Lineberger, W. C. (1988) Ultraviolet Photoelectron Spectrum of NO_2^-, *J. Phys.Chem.* **92**, 5405-5412.

19. Johnson, M. A. and Lineberger, W. C. (1988) Pulsed methods for cluster ion spectroscopy, *Techniques in Chemistry*, **20**, edited by J. M. Farrar and W. H. Saunders (Wiley, New York) 591-635.

20. DeLuca, M. J. and Johnson, M. A. (1989) Preservation of rotational cooling following pulse compresshion TOF mass selection of N_2O^+ ions created in a supersonic plasma, *Chem. Phys. Lett.* **162**, 255-260.

21. Wiley, W. C. and McLaren, I. H. (1955) *Rev. Sci. Instrum.* **26**, 1150.

22. Johnson, M. A., Alexander, M. L. and Lineberger, W. C. (1984) Photodestruction cross sections for mass-selected ion clusters: $(CO_2)_n^+$, *Chem. Phys. Lett.* **112**, 285-290.

23. Posey, L. A., DeLuca, M. J. and Johnson, M. A. (1986) Demonstration of a pulsed photoelectron spectrometer on mass-selected negative ions: O^-, O_2^- and O_4^-, *Chem. Phys. Lett.* **131**, 170-174.

24. Chupka, W. A., Woodward, A. M., Colson, S. D. and White, M. G., (1985) Electron photodetachment from transient negative ions in the multiphoton ionization of CH_3I *J. Chem. Phys.* **82**, 4880-4885.

25. Coe, J. V., Snodgrass, J. T., Friedhoff, C. D., McHugh, K. M. and Bowen, K. H. (1986) Negative ion photoelectron spectroscopy of N_2O^- and $(N_2O)_2^-$, *Chem. Phys. Lett.* **124**, 274-278.

26. Markovich, G., Pollack, S., Giniger, R. and Cheshnovsky, O., (1994) Photoelectron spectroscopy of Cl^-, Br^- and I^- solvated water clusters, *J. Chem. Phys.* **101**, 9344-9353.

27. Cyr, D. M., Scarton, M. G. and Johnson, M. A. (1993) Photoelectron spectroscopy of the gas-phase S_N2 reaction intermediates $I^-{\cdot}CH_3I$ and $I^-{\cdot}CD_3I$: Distortion of the CH_3I at the "ion-dipole" complex, *J. Chem. Phys.* **99**, 4869-4872.

28. Arnold, D. W. and Neumark, D. M. (1995) Study of N_2O_2 by photoelectron spectroscopy of $N_2O_2^-$, *J. Chem. Phys.* **102**, 7035-7045.

29. Kitsopoulos, T. N., Waller, I. M., Loeser, J. G. and Neumark, D. M. (1989) High resolution threshold photodetachment spectroscopy of negative ions, *Chem. Phys. Lett.* **159**, 300-306.

30. Muller-Dethlefs, K. and Schlag, E. (1991) High resolution zero kinetic energy (ZEKE) photoelectron spectroscopy of molecular systems, *Ann. Rev. Phys. Chem.* **42**, 109-36.

31. Chupka, W. A. (1993) Factors affecting lifetimes and resolution of Rydberg states observed in zero-electron-kinetic-energy spectroscopy, *J. Chem. Phys.* **98**, 4520-30.

32. Zhao, Y., Arnold, C. C. And Neumark, D. M. (1993) Study of the $I{\cdot}CO_2$ van der Waals Complex by Threshold Photodetachment Spectroscopy of I^-CO_2, *J. Chem. Soc. Farad. Trans.* **89**, 1449-1456.

33. Ervin, K. M., Ho, J. and Lineberger, W. C. (1989) A study of the singlet and triplet states of vinylidene by PES of $H_2C{=}C^-$, $D_2C{=}C^-$ and $HDC{=}C^-$: Vinylidene to acetylene isomerization , *J. Chem. Phys.* **91**, 5974-5992.

34. Neumark, D. M. (1993) Transition-State Spectroscopy via Negative Ion Photodetachment, *Acc. Chem. Res.* **26** 33-39.

35. Mertz, R. B., Weaver, A., Bradforth, S. E., Kitsopoulos, and Neumark, D. M. (1990) Probing the Transitoin State with Negative Ion Photodetachment: The Cl + HCl and Br + HBr Reactions, *J. Phys. Chem.* **94**, 1377-1388.

36. Gutsev, G. L., Sobolewski, A. L. and Adamowicz, L. (1995) A theoretical study of acetonitrile (CH_3CN) and its anion CH_3CN^-, *Chem. Phys.* **196**, 1-11.

37. Edard, F., Hitchcock, A. P. and Tronc, M. (1990) π^* and σ^* shape resonances in the vibrational excitation of HCN, CH_3CN and CH_3NC by low-energy electron impact, *J. Phys. Chem.* **94**, 2768-2774.

38. Bailey, C. G., Lavrich, D. J., Serxner, D. and Johnson, M. A., (1996) Autodetachment from vibrational levels of the O_2^- $A^2\pi_u$ resonance across its dissociation limit by photoexcitation from O_2^- $X^2\pi_g$ *J. Chem. Phys.* in press.

METHYL INTERNAL ROTATION IN SUBSTITUTED TOLUENES

Effects of Substituents, Electronic Excitation, and Ionization

KUEIH-TZU LU, ERIK C. RICHARD,
ROBERT A. WALKER, and JAMES C. WEISSHAAR

Department of Chemistry
University of Wisconsin-Madison
Madison, Wisconsin, U.S.A. 53706-1396
Email: weisshaar@chem.wisc.edu

Abstract

Threshold photoionization spectroscopy of substituted toluene cations with 1 cm^{-1} resolution combined with *ab initio* electronic structure calculations has provided new insight into the nature of barriers to methyl group internal rotation. We review recent progress in this area, including an overview of the ZEKE-PFI technique (*zero-kinetic-energy* threshold photoelectron spectroscopy with *pulsed-field ionization* detection). From microwave and optical spectroscopies, we now know methyl rotor barriers for a variety of substituted toluenes in three electronic states: S_0, S_1, and D_0 (the cation ground state). Comparison of barriers for *o*-fluorotoluene, *o*-chlorotoluene, and 2-fluoro-6-chlorotoluene shows an additivity effect. The substantial data base of barrier heights shows several trends which demand theoretical explanation. With the help of *ab initio* electronic structure calculations, we suggest that three different effects contribute to the observed barriers. First, an *ortho*-halogen substituent creates a substantial preference for the pseudo-*trans* conformation, primarily due to steric repulsion. Second, a difference in bond order between the two ring CC bonds adjacent to the methyl rotor favors placement of one CH bond in the plane of the ring *cis* to the ring CC bond of higher order. This is due to attractive, donor-acceptor interactions analogous to those that arise in the acyclic case of ethane. Third, π photoionization often leads to dramatic changes in local ring geometry. The pattern of substitution controls the choice of π orbital from which the electron is removed and also the orbital *orientation*, which in turn governs the conformational preference of the methyl group in the ground state cation.

K. R. Jennings (ed.), Fundamentals and Applications of Gas Phase Ion Chemistry, 307–334.

1. Introduction: The Methyl Rotor Problem

Methyl rotors present relatively simple, fundamental questions about the nature of non-covalent interactions within molecules. The discovery in the late 1930s[1] of the 1025 cm^{-1} potential energy barrier to internal rotation in ethane was surprising, since no covalent chemical bonds are formed or broken as methyl rotates. By now it is clear that the methyl torsional potential depends sensitively on the local chemical environment. The barrier is 690 cm^{-1} in propene,[2] comparable to ethane, but only 5 cm^{-1} in toluene.[3] The pioneering work of Ito and co-workers[4] in the 1980s showed that threefold barriers in substituted toluenes can vary dramatically from S_0 (ground electronic state) to S_1 (first excited singlet electronic state) as well. In addition to presenting subtle structural questions about the relative stability of different molecular conformations, the presence of a methyl rotor can dramatically affect photochemical pathways and quantum yields and also the rate of intramolecular vibrational energy redistribution.[5] Methyl rotors in different chemical environments contribute very differently to the density of states that enters statistical models of chemical reaction rates. Chemists need simple, reliable models based on rigorous quantum mechanics that can unify the seemingly diverse behavior of methyl rotor conformational preference and barrier height.

Microwave spectroscopy can obtain barriers to internal methyl rotation in S_0 from tunneling splittings when the barriers are not too large.[2,3] With the advent of pulsed tunable dye lasers, optical spectroscopy in free jets resolved pure torsional bands and obtained barriers for the S_1 excited electronic state in a variety of substituted toluenes.[4,6,7,8,9] More recently, the new threshold photoionization technique of *pulsed field ionization* (PFI) has extended these data to the corresponding ground state cations.[10,11,12,13] In some cases, Franck-Condon factors reveal *changes* in the preferred methyl group conformation on electronic excitation. The minimum energy conformation in S_0 and D_0 can often best be determined by *ab initio* calculations, as described below. We now have a rich data base encompassing different local chemical structures and different electronic states. Each new example raises the subtle structural question of which methyl conformation lies lowest in energy and why.

In this chapter, we first present a brief overview of the experimental techniques we and others have used to study torsional motion in S_1 and D_0 (Sec. 2). These are resonant two-photon ionization (R2PI) for S_1-S_0 spectroscopy and pulsed-field ionization (commonly known as ZEKE-PFI) for D_0-S_1 spectroscopy. In Sec. 3, we summarize what is known about sixfold methyl rotor barriers in S_0, S_1, and D_0, including a brief description of how the absolute conformational preference can be inferred from spectral intensities. Section 4 describes the threefold example of *o*-chlorotoluene in some detail and summarizes what is known about threefold barriers more generally. The sequence of molecules *o*-fluorotoluene, *o*-chlorotoluene, and 2-fluoro-6-chlorotoluene shows that the effects of *ortho*-fluoro and *ortho*-chloro substituents on the rotor potential are approximately additive in S_0, S_1, and D_0. Finally, in Sec. 5 we present our ideas about the underlying causes of these diverse barrier heights and conformational preferences, based on analysis of the optimized geometries and electronic wavefunctions from *ab initio* calculations.

2. Experimental Techniques

2.1. RESONANT TWO-PHOTON IONIZATION (R2PI)

The use of resonant two-photon ionization (R2PI) and time-of-flight mass spectrometry (TOF-MS) to obtain mass-specific S_1-S_0 absorption spectra is well known. We form an internally cold, skimmed beam of the substituted toluene of choice, typically seeded at its room temperature vapor pressure in a pulsed expansion of Ar gas at ~2 atm stagnation pressure. A frequency-doubled, pulsed tunable dye laser in the near UV intersects the skimmed beam in the extraction zone of the TOF-MS. Two photons of the same frequency ω_1 are absorbed to form molecular cations whenever ω_1 matches an S_1-S_0 resonance; this is known as *one-color* R2PI. The spectrum consists of gated ion detector current at the mass of interest *vs* ionizing frequency ω_1. The resolution is 0.2 cm^{-1}, limited by the dye lasers. S_1-S_0 vibronic bands typical exhibit rotational contours of about 2 cm^{-1} FWHM, indicating that the internal temperature of the molecules is roughly 5 K under our expansion conditions. Details of the technique can be found in the literature,[10-16] and examples of spectra will follow.

Care must be taken not to mistake bands arising from 1:1 or 1:2 van der Waals complexes (e.g., toluene·Ar_x, x = 1 and 2) for monomer bands. Fragmentation of either the neutral excited state or the cation can cause the 1:1 and even the 1:2 bands to appear in R2PI spectra detected at the monomer cation mass. The van der Waals bands usually appear red-shifted by a constant value from corresponding strong monomer bands. The safest way to identify van der Waals bands is to carry out R2PI scans gated on successive toluene·Ar_x^+ masses to see which bands disappear at each step.

2.2. ZERO-KINETIC ENERGY THRESHOLD PHOTOELECTRON SPECTROSCOPY AND PULSED-FIELD IONIZATION (ZEKE-PFI)

ZEKE-PFI is a threshold photoionization spectroscopy capable of providing cation spectra with resolution of 1 cm^{-1} or better, adequate for resolving the cation torsional bands of interest here. The technique was developed in the 1980s by Müller-Dethlefs, Schlag, and co-workers.[14] We present only a brief overview here. The interested reader might well begin with the review article by Müller-Dethlefs and Schlag.[15] An excellent recent monograph[16] is devoted entirely to applications of the new family of high-resolution photoionization and photoelectron techniques.

In standard photoelectron spectroscopy, the ionizing photon energy is *fixed*. Cation energy levels are revealed by kinetic-energy analysis of the resulting photoelectrons. The state of the art in electron energy analysis is about 4 meV = 32 cm^{-1} using hemispherical analyzers and array detectors.[17] Time-of-flight techniques are capable of 1 meV = 8 cm^{-1} resolution at low kinetic energy.[18] Calibration of the electron energy scale can be a time-consuming, repetitive task in both techniques.

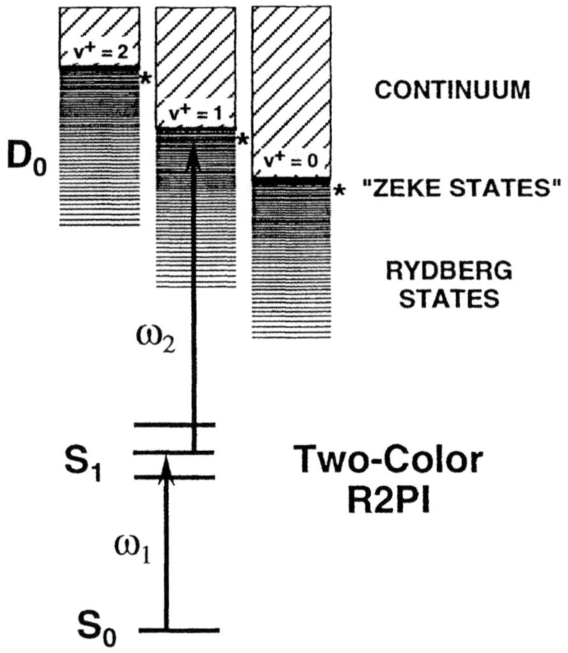

Figure 1. Schematic of two-color resonant two-photon ionization experiment; "ZEKE states" lie below each cation threshold, as indicated by asterisks.

In the ZEKE-PFI technique, the ionizing laser frequency is *scanned* through the manifold of cation states (Fig. 1). A ZEKE-PFI signal is detected only when the scanning laser frequency lies at (or very near) a cation threshold. While ZEKE-PFI spectra have been obtained using one-photon ionization by tunable VUV lasers,[19] the more common experiment uses *two-color* R2PI ($\omega_1 + \omega_2$) to ionize the molecule. In that case, ω_1 is tuned to a particular S_1-S_0 band already observed in the one-color R2PI spectrum. With ω_1 fixed, ω_2 is scanned through the manifold of cation states. The device we use for pulsed field ionization is shown schematically in Fig. 2. The molecular beam and the two dye lasers intersect between a pair of grounded, magnetically shielded stainless steel electron extraction plates. For spectroscopic experiments, both dye lasers are often pumped by harmonics of a single Nd:YAG laser with 5-10 ns pulse width, since the dye laser pulse energies required are modest. At a time delay of 2.0 μs following the laser pulses, a fast voltage step is applied to the lower plate. Extracted electrons travel in a 0.5 V/cm field over 8 mm and then accelerate rapidly to a terminal kinetic energy of 2.6 eV. Gated integration of the delayed electron pulse (60 ns gate width) from a chevron microchannel plate detector 50 cm from the interaction zone produces the PFI spectrum.

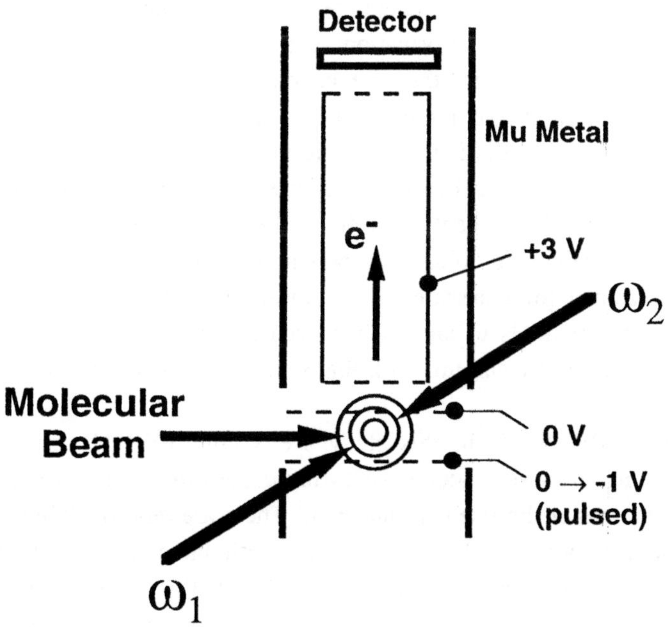

Figure 2. Schematic of apparatus for ZEKE-PFI spectroscopy.

The key to the good resolution of ZEKE-PFI is its discrimination against electrons with ≥ 1 cm^{-1} kinetic energy, which is due to the delayed extraction and time-gated detection. A "bubble" of 1 cm^{-1} electrons (ein "*elektronballon*") expands to a radius of 1.6 cm during the 2.0 μs delay. Such kinetic electrons either miss the detector or arrive outside the signal gate, since they have significantly different terminal velocities than electrons extracted from the center of the interaction region.

The inventors of the technique believed they were detecting true continuum electrons with near-zero (sub-cm^{-1}) kinetic energy,[14] but this has proven not to be the case. Our own experiment probably does not detect any genuine continuum electrons, since stray fields in the apparatus as small as 1 mV/cm will remove them from the interaction region during the 2 μs delay. Rather, we and others detect electrons arising from *pulsed field ionization* (PFI)[19] of high-*n* Rydberg states (Fig. 1) which lie about 3 cm^{-1} below each true cation threshold and survive in any stray electric fields for at least 2.0 μs. The experimental signature of PFI is an asymmetric broadening of bands towards lower energy for larger extraction fields. This explains the hyphenated acronym ZEKE-PFI. The technique was originally called ZEKE, but in fact it is pulsed-field ionization as applied to neutral molecules by most groups.

The ZEKE-PFI practitioner relies on the existence of a narrow band of long-lived, high-*n* Rydberg states ("ZEKE states")[15] lying just below each true cation threshold. If the shift of this band of states relative to the true threshold is essentially constant, then spectroscopic information extracted from *differences* in band frequencies will faithfully reflect the true cation energy levels. This assumption seems to hold to an accuracy of 1 cm^{-1} or better in the many molecules studied, as judged by the success of standard spectroscopic models in fitting ZEKE-PFI spectra. Cation vibrational, rotational, and torsional frequency intervals can thus be measured to a similar accuracy. The absolute shift in the lowest energy ZEKE-PFI band relative to the true adiabatic ionization potential is about 3 cm^{-1} under typical conditions. To account for uncertainties in ionization mechanism, we usually quote adiabatic ionization potentials to an absolute accuracy of ± 5 cm^{-1}, although higher accuracy can be achieved by careful work that systematically varies the ionizing field strength.

The narrowest PFI bands in this study are 3 cm^{-1} FWHM, using 0.5 V/cm extraction field with the lasers attenuated to minimize effects of space charge. We measure band positions at the intensity maxima. These are reproducible to better than ± 1 cm^{-1}. The bandwidth is limited by the rotational contour and also by the ionization process. A major advantage of ZEKE-PFI over more traditional photoelectron techniques is that the energy calibration is that of the tunable dye lasers, which are quite stable from day to day.

The electronic structure of molecules above the first ionization threshold is complex and subtle.[20] Simply put, each cation state supports multiple manifolds of Rydberg series converging to that state. Each Rydberg series might be described in terms of the internal quantum numbers of its cation core (the electronic, vibrational, rotational, and torsional quantum numbers e^+, v^+, J^+, and m^+). Each Rydberg state within a series might be further described by three quantum numbers for the Rydberg electron; these are n (the principal quantum number), ℓ (the orbital angular momentum), and m_ℓ (the projection of the orbital angular momentum onto a space-fixed axis). Rydberg states lying above the adiabatic ionization potential are metastable. They consist of an internally excited ion core to which the Rydberg electron is very weakly bound. The core excitation may be electronic, vibrational, rotational, or some combination of these. Such Rydbergs states have sufficient energy to decay by *autoionization*,[20] which can be viewed as a scattering process in which the ion core transfers energy to the electron. The core loses internal energy and becomes a molecular cation, while the electron becomes free. Rydberg states often decay by dissociation to excited state neutral fragments as well.

The physical nature of the ZEKE states has been the subject of intense experimental and theoretical investigation in the past several years. In the well-studied case of NO,[14,21] we know from the 3 cm^{-1} red shift of the ZEKE-PFI threshold band relative to the true adiabatic ionization potential (extrapolated from highly accurate measurements of Rydberg series) that the ZEKE states have principal quantum number $n \sim 200$ and lifetime of 2 μs or longer. Recent work has found ZEKE states with lifetimes as long as 20 μs.[22]

Such long lifetimes are puzzling in view of the well-known n^3 scaling of Rydberg state lifetimes observed from linewidths of resolvable series and predicted by theory. The n^3 scaling derives from a fundamental property of Rydberg states. The normalization factor in the radial part of the Rydberg orbital scales as $n^{-3/2}$. Thus the probability of finding the Rydberg electron in a small sphere about the cation core scales as n^{-3}. Since either autoionization or dissociation involves interaction of the Rydberg electron with the core, the lifetime lengthens with increasing principal quantum number as n^3. In the scattering picture, the frequency with which the Rydberg electron visits the core scales as the same n^{-3}.

The problem with the observed lifetimes of the ZEKE states is that they are orders of magnitude *too long* compared with estimates obtained by scaling of measured lifetimes of Rydberg states having $n \sim$ 20-30 according to the n^3 law. For $n \sim 200$, it appears the lifetimes scale roughly as n^5. Theoretical ideas due to Chupka[23] and recent experiments[24] appear to be converging on the explanation of the breakdown of n^3 scaling at high n. Briefly, electric-dipole optical excitation from a low-n, low-ℓ, low-m_ℓ valence orbital follows $\Delta\ell = \pm 1$ selection rules. Thus the zeroth-order Rydberg states carrying oscillator strength near a cation threshold that necessarily have $n \sim 200$, low ℓ, and low m_ℓ. However, n, ℓ, and m_ℓ are good quantum numbers only for a truly spherical potential. In the usual laboratory environment of a ZEKE-PFI experiment, the Stark effect from the small *homogeneous* electric field (typically a stray field in the apparatus) mixes states of different ℓ according to the selection rule $\Delta\ell = \pm 1$. Furthermore, the *inhomogeneous* electric field (which apparently arises primarily from the space charge formed by the laser ionization process) mixes states of different m_ℓ. In the limit of complete mixing, each optically active n, low-ℓ, low-m_ℓ state is coupled to *many* states whose quantum numbers ℓ and m_ℓ will be much larger. These background states *do not autoionize or dissociate* on the ZEKE timescale, since the large centrifugal barrier for $\ell \geq 3$ prevents them from visiting the core. This *dilution* of the initially excited low-ℓ character among many long-lived states leads to a corresponding *lengthening* of the lifetime of the ZEKE states beyond the n^3 scaling that applies to low-ℓ, low-m_ℓ states. The lifetime is lengthened in proportion to the number of states with which the optically active state mixes. Since the complete degeneracy of a hydrogenic state is n^2, in the limit of complete ℓ, m_ℓ mixing the lifetimes should scale as n^5, much as observed.

3. Sixfold-Symmetric Methyl and Silyl Rotors

Sixfold barriers to internal rotation occur in molecules such as toluene and *p*-fluorotoluene whose molecular frame has C_{2v} symmetry about the rotor axis. The simplest spectroscopic model of internal methyl rotation assumes a rigid, threefold

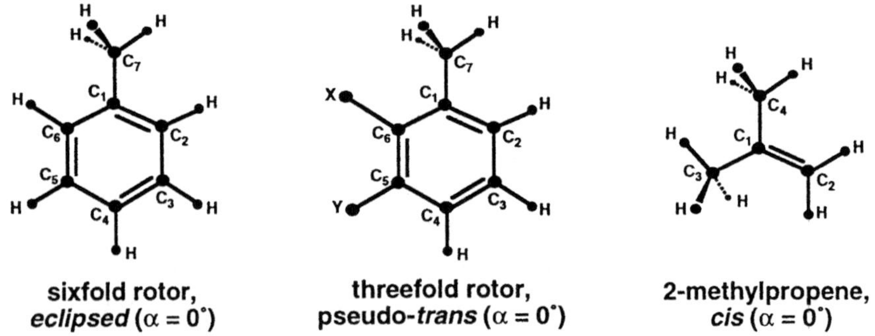

Figure 3a,b,c. Sixfold and threefold rotors and definition of $\alpha = 0$.

symmetric methyl rotor attached to a rigid molecular frame with the C_2 axis coincident with the rotor top axis.[25] The effective one-dimensional sixfold torsional potential takes the traditional form:

$$V(\alpha) = (V_6/2)(1\text{-}\cos 6\alpha), \qquad (1)$$

where α is the dihedral angle of the internal rotor relative to the molecular frame and $|V_6|$ is the barrier height. The sign of V_6 fixes the conformation of minimum potential energy. With $\alpha = 0$ defined as the eclipsed conformation (Fig. 3a), *positive* V_6 gives the *eclipsed* potential minimum while *negative* V_6 gives the *staggered* minimum ($\alpha = 90°$). In addition, the sign of V_6 determines the energetic order of the two $|m| = 3$ torsional states (Fig. 4). A sixfold potential mixes the free rotor wavefunctions $\phi_{+3} = e^{+3i\phi}$ and $\phi_{-3} = e^{-3i\phi}$ to form the symmetric and anti-symmetric linear combinations $|3+\rangle = \cos 3\alpha$ and $|3-\rangle = \sin 3\alpha$. These have, respectively, a_1'' and a_2'' torsional symmetry under G_{12}, the appropriate molecular

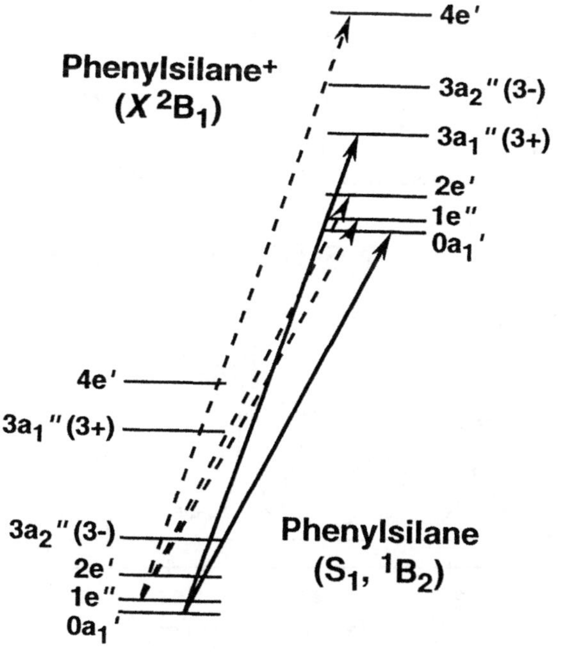

Figure 4. Schematic of torsional energy levels in S_1 and D_0 electronic states of phenylsilane.

symmetry group for this non-rigid molecule.[26] For negative V_6, $|3+\rangle$ lies *above* $|3-\rangle$ and the minimum is staggered, as occurs in S_0 and S_1 for all molecules studied thus far. For positive V_6, $|3+\rangle$ lies *below* $|3-\rangle$ and the minimum is eclipsed, as occurs in D_0 of almost all molecules studied thus far. In either case, the splitting between $|3+\rangle$ and $|3-\rangle$ gives $|V_6|/2$.

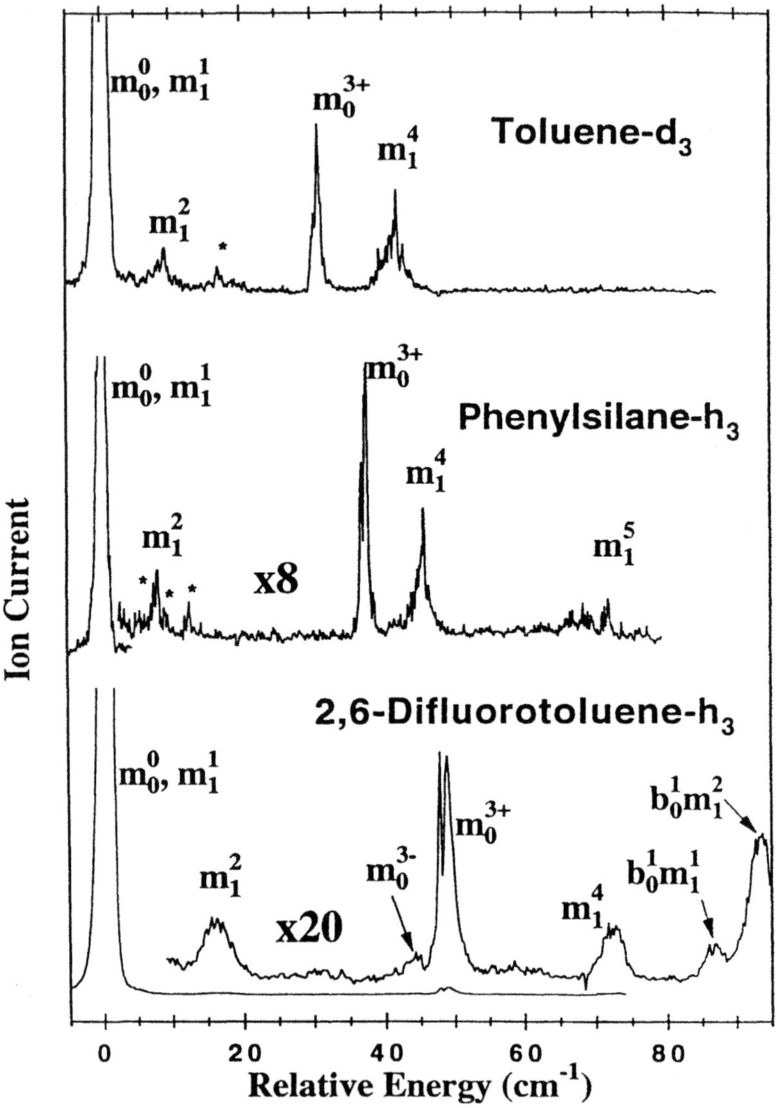

Figure 5. Pure torsional bands near the S_1 origin for three sixfold rotors.

Microwave spectroscopy can determine the magnitude of V_6 in S_0 but not the sign, since the potential well is too small to localize even the $m = 0$ wavefunction. $S_1 \leftarrow S_0$ absorption spectra of cold molecules with 1 cm^{-1} resolution can reveal the magnitude of V_6 in S_1, a technique pioneered by Ito and co-workers.[4] Pratt and co-workers[7] and Miller and co-workers[8] have made major contributions to the *high resolution* optical spectroscopy of rotor-containing molecules.

Figure 5 shows a collection of S_1-S_0 R2PI spectra near the origin. The weak bands at low frequency are pure torsional transitions. The bands labeled $m^2{}_1$, $m^3{}_0$, and $m^4{}_2$ are forbidden in the sense that they do not preserve torsional symmetry; their Franck-Condon factors $|\langle m'|m''\rangle|^2$ are zero by symmetry. These forbidden bands are comparably intense in many spectra, about 1-5% of the intensity of the allowed origin band. The spectra are assigned by fitting observed band positions to the eigenvalues of a one-dimensional model torsional Hamiltonian in S_1, with the S_0 barrier taken from microwave spectra in most cases. Typically only one of the two possible bands $m^{3+}{}_0$ and $m^{3-}{}_0$ is observed with appreciable intensity. This band always lies above the mean of the calculated $m^{3-}{}_0$ and $m^{3+}{}_0$ band positions, but its identity as $m^{3-}{}_0$ or $m^{3+}{}_0$ remains ambiguous. If the observed band can be properly assigned, the sign of V_6 and the most stable conformer in S_1 follow (Fig. 4). All earlier workers have *assumed* the $m^{3+}{}_0$ assignment, yet efforts to calculate the intensity pattern of the forbidden bands based on torsion-overall rotation coupling within S_1 or torsion-electronic coupling between S_1 and S_2 have failed badly. Determination of S_0 conformations from dispersed fluorescence spectra[9] and of cation conformations from PFI spectra[10,11] have relied on the assumed $m^{3+}{}_0$ assignment. If the assignment is correct, it implies that S_1 is always *staggered*.

In collaboration with Prof. Sibert, we have learned to interpret these Franck-Condon forbidden, pure torsional band intensities in S_1-S_0 absorption spectra *quantitatively* and thus place the key $m^{3+}{}_0$ assignment on firm ground.[27] The forbidden bands follow the selection rule $\Delta m = \pm 3$, so we need a perturbation of the form $V_{el}\cos 3\alpha$. Working in an adiabatic representation with the S_0 and S_1 electronic states denoted by $\psi_0(q;\alpha)$ and $\psi_1(q;\alpha)$ and the torsional states by m'' and m', the electric dipole transition moment is:

$$\langle m''|[\langle\psi_1(q;\alpha)|\boldsymbol{\mu}|\psi_0(q;\alpha)\rangle]|m'\rangle, \tag{2}$$

where q represents electronic coordinates, α is the torsional coordinate, μ is the dipole moment operator, the inner integral is over q, and the outer integral is over α. The electronic integral depends on α. The key step involves approximating each of the three components of the electronic integral $\boldsymbol{\mu}_{10}(\alpha)$ as a *Fourier* series in the lowest order terms such as $\cos 3\alpha$, $\cos 6\alpha$, and $\sin 6\alpha$ permitted by the G_{12} symmetry species of each vector component. Details are given elsewhere.[27] The usual Taylor series expansion in powers of α is physically inappropriate because the rotor wavefunctions are delocalized over all α.

Fourier expansion of $\mu_{10}(\alpha)$ allows us to calculate relative intensities of the three forbidden bands $m^2{}_1$, $m^{3+}{}_0$, and $m^4{}_1$ that are in quantitative agreement with experiment. The agreement is excellent over a wide range of ratios of the key model parameters V_6 and F (effective rotational constant), which are taken from experiment. Previous conformational inferences in S_0, S_1, and ground state cation (including our own) were in fact correct, but they now rest on solid ground.

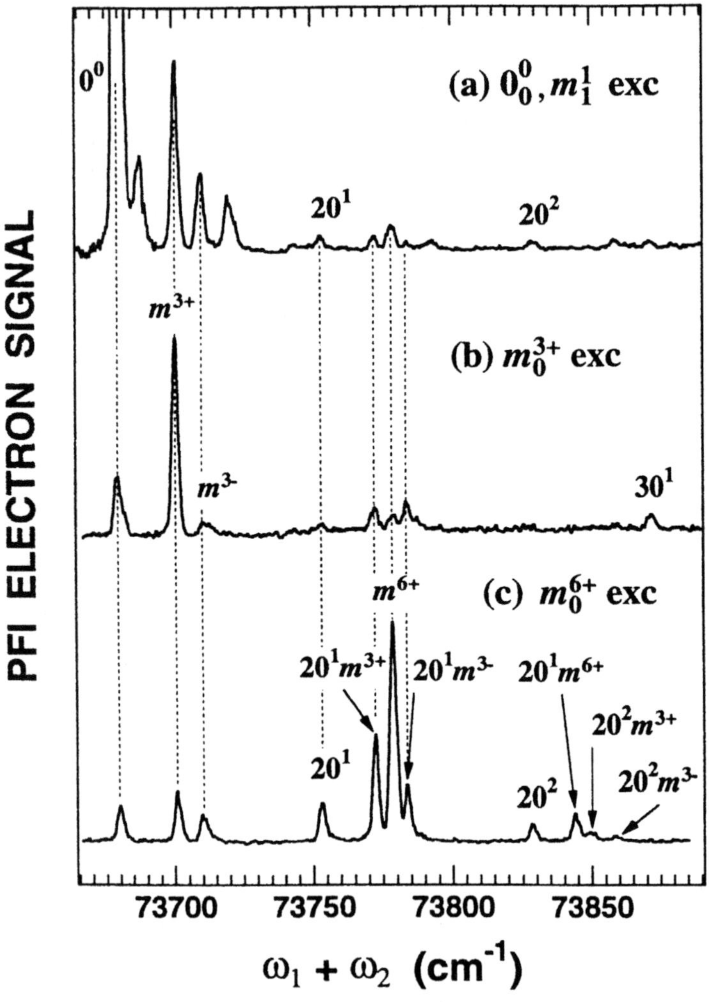

Figure 6. ZEKE-PFI spectra of phenylsilane$^+$ (D_0) through three S_1 intermediate states.

In addition, we have measured new sixfold barrier heights and conformational preferences in S_1 phenylsilane[11] and 2,6-difluorotoluene.[28] For the cation ground state D_0, we have used the 1 cm^{-1} resolution of the ZEKE-PFI technique to determine V_6 for the first time. In Fig. 6, we show PFI spectra of phenylsilane$^+$ obtained through different S_1 intermediate states.[11] Assignment of the spectra is greatly assisted by the rigorous selection rule preserving nuclear spin symmetry. When ω_1 cleanly excites an *a*-symmetry (or *e*-symmetry) state of S_1, we observe transitions only to *a*-symmetry (or *e*-symmetry) states of the cation. Assuming the rigid-frame, rigid-methyl-rotor model Hamiltonian, we obtain the parameters *F* (reduced rotational constant for rotation of methyl relative to the frame) and V_6. Among toluene$^+$-h_3, toluene$^+$-d_3, phenylsilane$^+$, and 2,6-difluorotoluene$^+$, V_6 lies in a narrow range from +12 to +19 cm^{-1}. For *p*-fluorotoluene$^+$, $|V_6|$ is less than 5 cm^{-1}. The most intriguing feature of these data is that they indicate *eclipsed* minima in most ground state cations, since $|3+\rangle$ lies below $|3-\rangle$.

In Ref. 28 (Table 4), we summarize sixfold barrier heights and conformational preferences in S_0, S_1, and D_0 for toluene, *p*-fluorotoluene, 2,6-difluorotoluene, phenylsilane, and *p*-toluidine. In all these sixfold molecules, the barrier height is quite small in all electronic states, less than 5 cm^{-1} in S_0, less than 45 cm^{-1} in S_1, and less than 20 cm^{-1} in D_0. We will explain the uniformly small magnitude of sixfold barriers in Sec. 5. Thus far the preferred conformation is always *staggered* in S_1 and almost always *eclipsed* in D_0. Substitution of silyl for methyl or introduction of two sterically bulky *ortho*-fluoro substituents alters the story only in the quantitative details. We know the conformation in S_0 for only one molecule, *p*-fluorotoluene, from Parmenter's dispersed fluorescence work.[9] It is staggered, like the other neutral (S_1) states.

4. Threefold-Symmetric Methyl Rotors

4.1. OVERVIEW

Chemical substitution in a pattern that breaks the C_{2v} symmetry of the molecular frame introduces a *threefold* component to the one-dimensional model torsional potential:

$$V(\alpha) = (V_3/2)(1 - \cos 3\alpha) + (V_6/2)(1 - \cos 6\alpha). \tag{3}$$

Ethane itself fall into this category with $V_3 = 1025$ cm^{-1}. In *ortho*- or *meta*-substituted toluenes, we define $\alpha = 0$ as the "pseudo-*trans*" conformation (Fig. 3b); $\alpha = 180°$ is then the "pseudo-*cis*" conformation. Typically $|V_3|$ is much larger than

$|V_6|$. The sign of V_3 then determines the minimum energy conformation. For V_3 positive, the minimum is pseudo-*trans*; for V_3 negative, the minimum is pseudo-*cis*. For threefold symmetric cases, the torsional states can be classified under the molecular symmetry group G_6, whose species designations a_1, a_2, and e lack primes or double primes. In the limit of a large threefold barrier, pairs of torsional states (one non-degenerate a-symmetry state plus one doubly degenerate e-symmetry level) converge in energy and the spectrum approaches that of a threefold-degenerate vibrational mode. Again the conservation of nuclear spin symmetry ($a \leftrightarrow a$ and $e \leftrightarrow e$) is strict in both S_1-S_0 R2PI and D_0-S_1 ZEKE-PFI spectra.

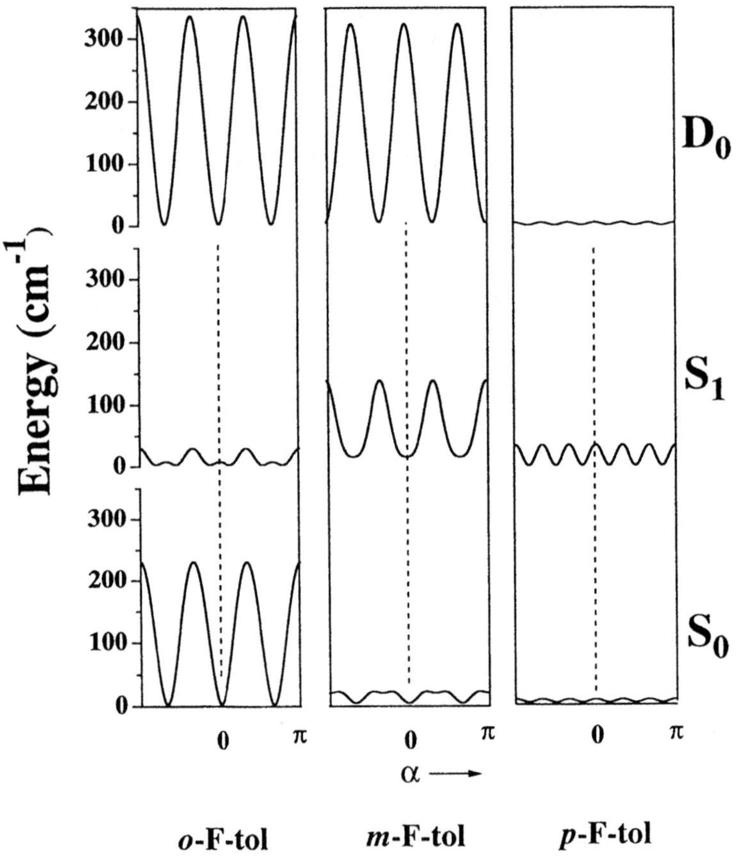

Figure 7. Torsional potentials for *ortho-*, *meta-*, and *para-*fluorotoluene in S_0, S_1, and D_0.

Ito and co-workers[4,12,13] have used both R2PI and more recently PFI spectroscopies to show that in substituted toluenes, the threefold barrier heights and minimum energy conformations depend sensitively not only on the location and nature of substituents but also on the electronic state S_0, S_1, or D_0. Franck-Condon factors in either the S_1-S_0 or D_0-S_1 spectra indicate clearly whether or not the preferred conformation changes on electronic excitation or π-ionization. The absolute conformation in S_0 is available for *o*-fluorotoluene from careful microwave work.[29] More often, *ab initio* calculations[30] yield accurate barrier heights in S_0 and thus make an unambiguous prediction of the S_0 conformation. The spectral intensities then yield absolute conformations in S_1 and D_0. Earlier work in several groups has measured the effects of methyl, amino, and fluorine substituents.[9]

Figure 7 displays the data of Ito[4,13] and others[29] in a 3 x 3 matrix of torsional potentials for *o*-fluorotoluene, *m*-fluorotoluene, and *p*-fluorotoluene in the three electronic states S_0, S_1, and D_0. The matrix reveals patterns that hold for other *ortho*, *meta*, and *para* substituents as well.[9] In S_0, *ortho* substitution creates a large barrier, while *meta* substitution does not. A sensible interpretation invokes steric repulsion between methyl and the *ortho* substituent. The *meta*-substituted cases have very small barriers like toluene itself, apparently for lack of steric effects. However, in *ortho*-substituted toluenes, V_3 *decreases* sharply on $\pi^* \leftarrow \pi$ excitation from S_0 to S_1, while in *meta*-substituted toluenes, V_3 *increases* substantially from S_0 to S_1. This suggests that steric interactions are not the complete story. Most intriguing of all, in *m*-fluorotoluene$^+$ the barrier height is 318 cm^{-1}; the preferred conformer places one CH bond in the plane of the ring on the side nearest F (pseudo-*cis*). Apparently these results cannot be understood in terms of steric repulsion alone. Rather, they clearly signal the importance of methyl rotor interactions with the ring. We are beginning to understand this pattern, as we discuss in Sec. 5.

4.2. ADDITIVITY OF *ORTHO*-SUBSTITUENT EFFECTS

To provide further experimental clues to the nature of substituent effects, we decided to test the additivity of *ortho*-substituent effects. Specifically, we set out to compare methyl torsional potentials in S_0, S_1, and D_0 for the sequence of molecules *o*-fluorotoluene (studied by Ito and co-workers), *o*-chlorotoluene, and 2-fluoro-6-chlorotoluene.

Figure 8 shows the S_1-S_0 R2PI spectrum of *o*-chlorotoluene in the first several hundred cm^{-1} above the origin.[31] ZEKE-PFI spectra obtained with ω_1 tuned to various S_1 intermediate states are shown in Fig. 9. We have obtained comparable spectra for 2-fluoro-6-chlorotoluene as well.[32] *Ortho*-chlorotoluene has 38 vibrational modes which classify as either a_1 (in-plane) or a_2 (out-of-plane) under

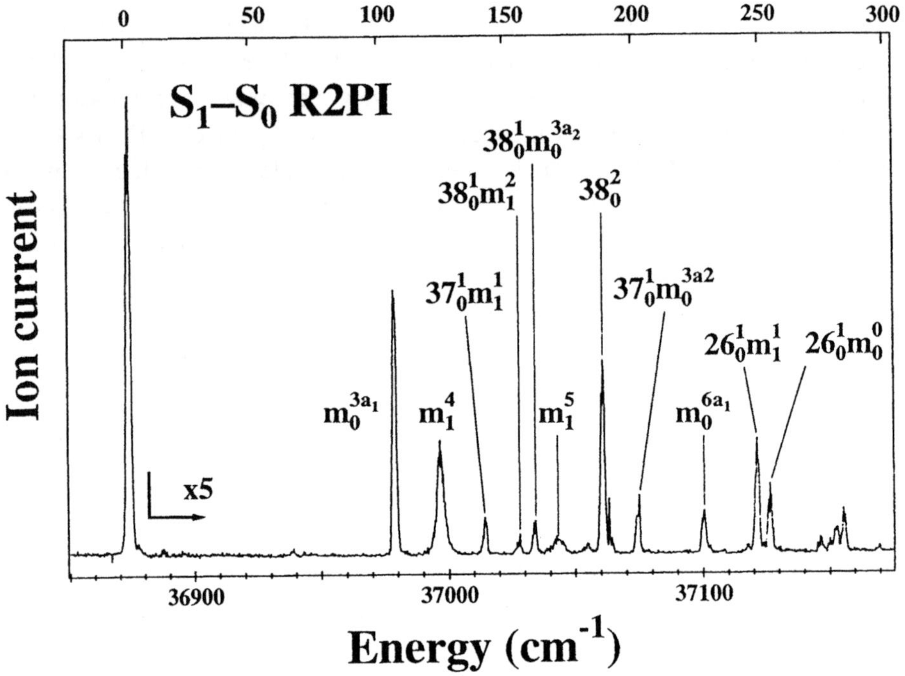

Figure 8. S_1-S_0 absorption spectrum of *o*-chlorotoluene detected by R2PI.

G_6, plus torsion itself. We number the a_1 modes from 1 to 26 in order of decreasing calculated harmonic frequency in S_0; similarly, the a_2 modes are numbered 27 to 38. Torsion is labeled *m*. The band assignments in Figs. 8 and 9 use the standard vibronic notation showing which quantum numbers change in the transition. Thus the band labeled $38^1{}_0m^{3a2}{}_0$ in the S_1-S_0 spectrum is a combination band in which the vibrationless, torsionless S_0 level ($v_{38} = 0$, $m = 0$) is excited to the S_1 level having one quantum of ν_{38} and three quanta of torsion (in the $3a_2$ level, not $3a_1$). For D_0-S_1 PFI spectra, the subscript describes quanta in S_1 (the lower state of the transition) while the superscript describes D_0.

Assignment of these spectra involves solution of a highly redundant, interlocking puzzle. We seek assignments consistent with all available information. We can calculate the torsional energies and wavefunctions for various one-dimensional model torsional Hamiltonians in both S_1 and D_0 to try to fit the observed band frequencies. For the pure torsional transitions, we expect that Franck-Condon factors will predict

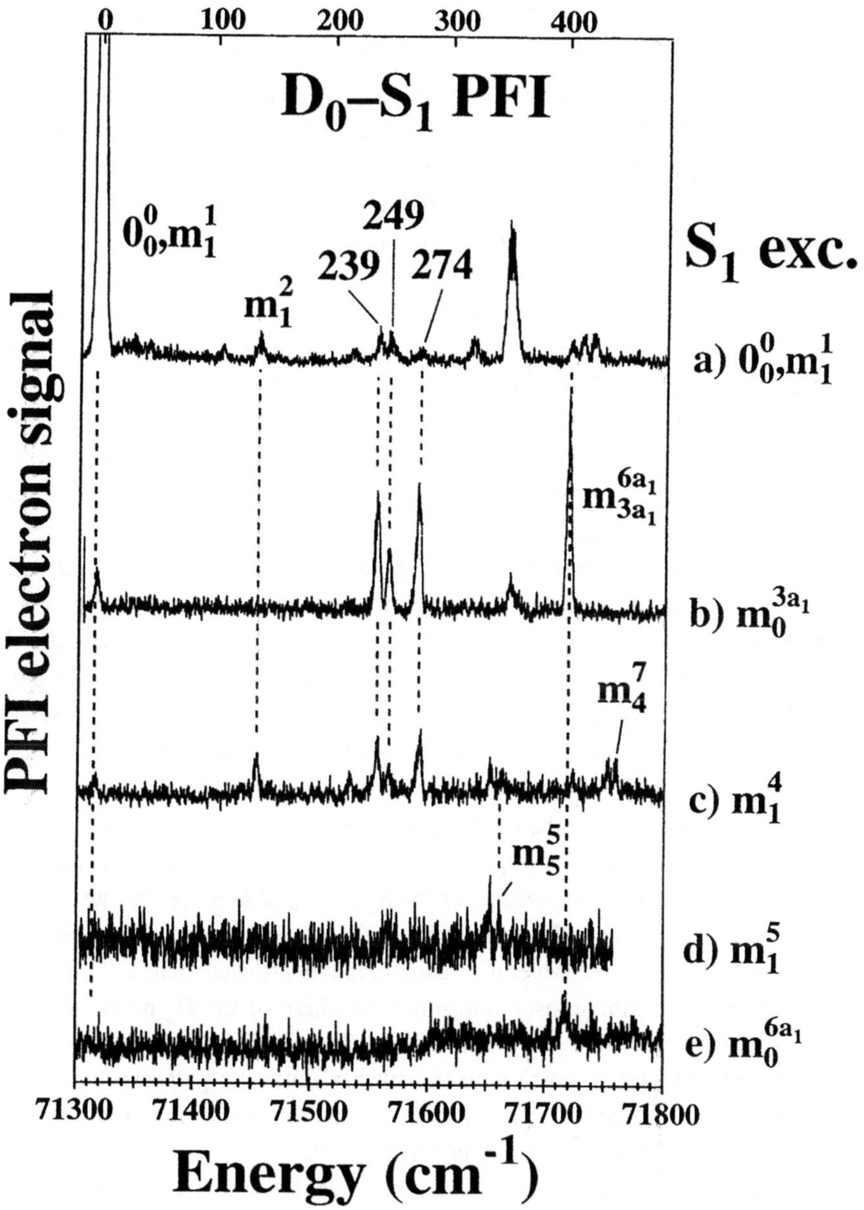

Figure 9. ZEKE-PFI spectra of *o*-chlorotoluene$^+$(D_0) through various S_1 intermediate states.

sensible S_1-S_0 and D_0-S_1 band intensities, which provide a further check on the model torsional potentials. *Ab initio* calculations on S_0 and D_0 provide accurate barrier heights and reliable predictions of the equilibrium conformation that guide construction of the torsional models.[30] The double resonance nature of PFI spectroscopy is a powerful tool that allows us to correlate S_1 levels with cation levels. When the $D_0 \leftarrow S_1$ geometry change is minimal, we expect the most intense transitions to preserve torsional and vibrational quantum numbers. Harmonic frequencies and normal modes from *ab initio* calculations on S_0 and D_0 guide vibrational assignments. Some modes exhibit significant isotope effects from ^{35}Cl to ^{37}Cl in mass-gated R2PI spectra. These shifts can be correlated with the *ab initio* frequencies. Symmetry considerations predict nuclear spin conservation in both S_1-S_0 and D_0-S_1 spectra. In addition, essentially all transitions preserve the overall *torsion-vibration* symmetry, consistent with kinetic energy or potential energy coupling between zeroth-order torsion-vibrational states. We seldom find the need to invoke Coriolis coupling, which could give rise to bands that do not preserve torsion-vibrational symmetry.

Our primary interest here is in the model torsional potentials derived from analysis of the spectra. Figure 10 displays our results for *o*-chlorotoluene[31] and 2-fluoro-6-chlorotoluene[32] as well as the data of Ito and co-workers[4,13] for *o*-fluorotoluene. For S_0 *o*-chlorotoluene and 2-fluoro-6-chlorotoluene, the potentials are calculated from *ab initio* electronic structure theory[30] as described in Sec. 5, since no microwave data was found. Modest levels of theory give quite good accuracy; in fact, for molecules with substantial barriers, *ab initio* calculations now provide the easiest means of determining the conformation of minimum energy. The S_1-S_0 R2PI spectral intensities determine the phase of the S_1 potential relative to S_0. For the substantial barriers of interest, whenever the S_1 and S_0 phases are the same, the origin bands $m^0{}_0$ and $m^1{}_1$ dominate the spectrum. If the phase changed, the intensity would be distributed broadly over several Franck-Condon allowed bands. Similarly, the D_0-S_1 PFI spectra unambiguously determine the phase of the D_0 potential relative to S_1. Combining the absolute S_0 conformation (from either *ab initio* calculations or microwave spectroscopy or both) with the spectral intensity patterns then determines the absolute minimum energy conformation of all three molecules in all three states, as shown in the 3 x 3 matrix of potentials in Fig. 10.

Comparing the potentials across each row, we can test the idea of additivity of *ortho*-substituent effects for 2-fluoro, 6-chloro, and then 2-fluoro-6-chloro substitution. The definition of $\alpha = 0$ changes across each row to permit easy visual addition of the potentials, as shown in the figure. For *o*-fluorotoluene and 2-fluoro-6-chlorotoluene, $\alpha = 0$ denotes the (minimum-energy) pseudo-*trans* conformation; for *o*-chlorotoluene, $\alpha = 0$ denotes the pseudo-*cis* conformation. The notion of additivity has considerable merit in all three electronic states.

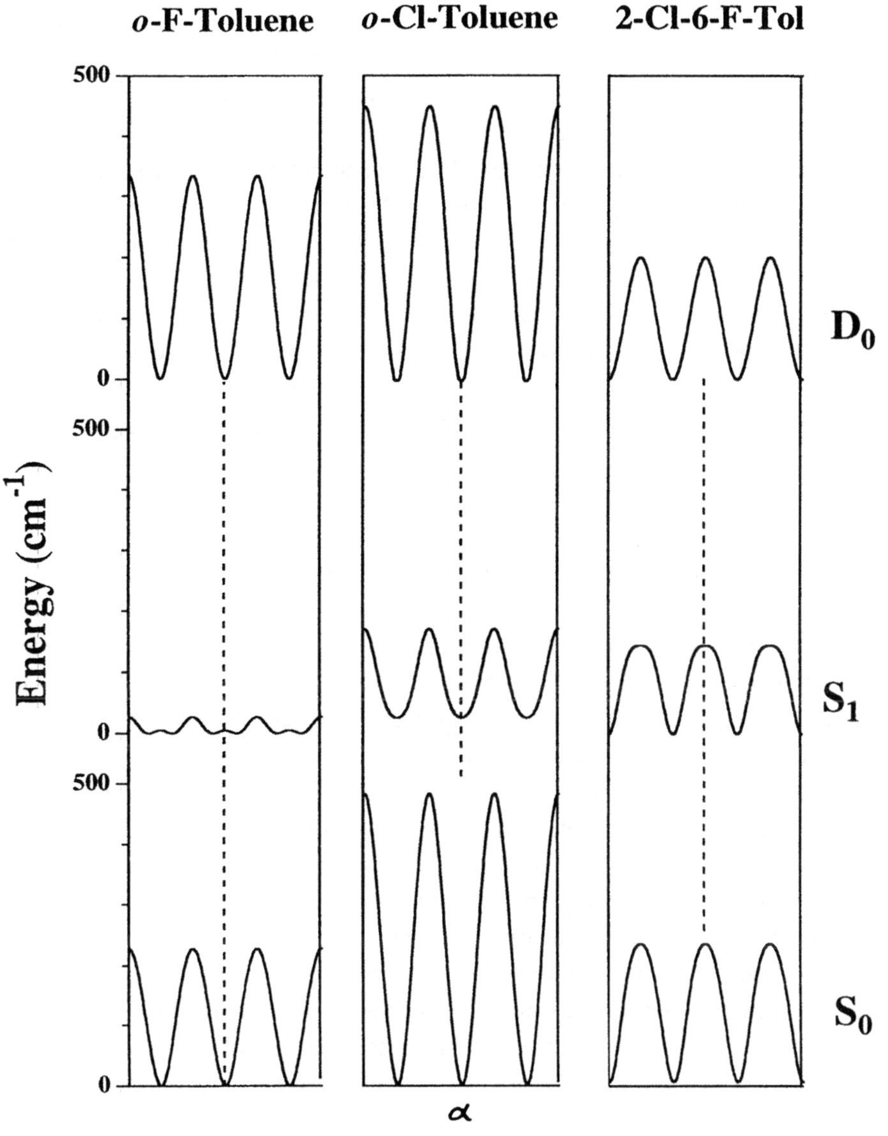

Figure 10. Threefold torsional potentials for *o*-fluorotoluene, *o*-chlorotoluene, and 2-chloro-6-fluorotoluene in S_0, S_1, and D_0. Dashed lines indicate the same reference configuration of methyl in all three molecules.

Comparing the potentials in each column of Fig. 10, we see the effects of S_1 excitation ($\pi^* \leftarrow \pi$) and of π ionization on the methyl rotor potential. In S_0 *o*-fluorotoluene and *o*-chlorotoluene, the minimum energy conformation is pseudo-*trans*, consistent with the idea that halogen-methyl steric repulsion dominates the potential. The larger barrier for *o*-chlorotoluene would then arise from greater repulsion by the larger chlorine substituent. Electronic excitation to S_1 decreases the barrier by a similar amount in both *o*-fluorotoluene and *o*-chlorotoluene, as if the effects of $\pi^* \leftarrow \pi$ excitation in the ring *add* to the steric effects seen clearly in S_0. Intriguingly, π ionization to D_0 essentially restores the barrier to its S_0 value in both molecules. The disubstituted molecule 2-fluoro-6-chlorotoluene shows very different behavior on $\pi^* \leftarrow \pi$ excitation and π ionization, in that effects on the barrier height are modest. We seek qualitative explanations of these electronic effects in the next section.

5. Understanding Barriers to Internal Rotation

In this section, we present a unified picture of the different electronic effects that combine to determine methyl rotor potentials in the S_0, S_1, and D_0 electronic states of different substituted toluenes. Our approach is based on analysis of *ab initio* wavefunctions using the *natural bond orbitals* (NBOs)[33] of Prof. Frank Weinhold. We will attempt to decompose the methyl torsional potential into two dominant contributions. The first is *repulsive steric interactions*, which are important only when an *ortho-* substituent is present. The second is *attractive donor-acceptor interactions* between CH bond pairs and empty anti-bonding orbitals vicinal to the CH bonds. In the NBO basis, these attractive interactions dominate the barrier in ethane (1025 cm^{-1}) and in 2-methylpropene (1010 cm^{-1}). By analogy, donor-acceptor attractions are important in toluenes whenever there is a substantial difference in bond order between the two ring CC bonds adjacent to the C-CH_3 bond. Viewed the other way around, we can use the measured methyl rotor potential as a sensitive probe of *local ring geometry*.

5.1. VIBRATIONALLY ADIABATIC TORSIONAL POTENTIALS

We have used standard *ab initio* electronic structure theory to calculate the methyl rotor potential for a variety of substituted toluenes in S_0 and D_0 at various levels of theory. The first excited state S_1 presents computational difficulties, since it often has the same electronic symmetry as the ground state and may well require a multi-configuration wavefunction. By fixing the dihedral angle α of the rotor relative to the aromatic frame and *optimizing the rest of the geometry*, we obtain a kind of *vibrationally adiabatic*, one-dimensional potential $V(\alpha)$ for comparison with the spectroscopic model potentials derived from experiment. Detailed study of the optimized geometries and how they change with α provides important clues to the electronic interactions that cause the barriers. In favorable cases, we have attempted a quantitative resolution of the calculated barrier into separate contributions from ethane-like, donor-acceptor interactions and from steric repulsion. Details of these calculations are presented elsewhere;[30] we only summarize the most important results here.

For most molecules studied, modest Hartree-Fock calculations yield remarkably accurate barriers that allow confident prediction of the lowest energy conformer in S_0 and D_0. The simplest level of theory that predicts barriers in good agreement with experiment is described as HF/6-31G* for the closed-shell S_0 (Hartree-Fock theory) and UHF/6-31G* for the open-shell D_0 (unrestricted Hartree-Fock theory). The 6-31G* basis set has double-zeta quality, with split valence plus *d*-type polarization on heavy atoms. This is quite modest by current standards. Nevertheless, such calculations reproduce experimental barrier heights within $\pm 10\%$

5.2. STERIC REPULSION AND DONOR-ACCEPTOR ATTRACTION

With accurate calculated barriers in hand, we return to the question of the underlying causes of methyl barriers in substituted toluenes. For simpler acyclic cases such as ethane and methanol, *ab initio* quantum mechanics yields the correct ground state conformer and remarkably accurate barrier heights as well.[34,35,36] Analysis of the wavefunctions in terms of *natural bond orbitals* (NBOs)[33] explains barriers to internal rotation in terms of attractive *donor-acceptor ("hyperconjugative") interactions* between doubly occupied σ_{CH} bond orbitals or lone pairs and unoccupied *vicinal* anti-bonding orbitals.

The NBOs are an optimized set of localized bonding, anti-bonding, and lone pair orbitals. The bond orbitals typically have occupancies of 1.98e in molecules with one simple Lewis structure. In C_2H_6, Weinhold and co-workers[36] have shown that the barrier height $V_3 = 1025$ cm^{-1} can be understood in the NBO basis in terms of $\sigma_{CH} \rightarrow \sigma_{CH}^*$ donor-acceptor interactions. When the corresponding off-diagonal Fock matrix elements are set to zero, the barrier essentially *disappears*. Six donor-acceptor interactions become most favorable in the lowest energy, staggered conformation of C_2H_6. Steric interactions play only a secondary role.

At the next level of complexity, the molecule 2-methylpropene illustrates clearly what happens when methyl is placed between a localized single bond and a localized double bond, without the complication of multiple resonance structures. The calculated barrier is 1010 cm^{-1}, very similar to ethane, with a strong preference for the conformer with one CH bond in the plane of the carbon frame *cis* to the *double* CC bond. To see the analogy between staggered ethane and the *cis* conformer of 2-methylpropene, we can envision the CC double bond as a pair of *banana bonds* lying above and below the plane of the carbon frame. One CH bond of the rotor then lies "staggered" between the two banana bonds, much as it lies staggered between two CH bonds in ethane. Thus the 2-methylpropene barrier is *ethane-like*. We could analyze it in terms of σ, σ^*, π, and π^* donor-acceptor interactions akin to those in acyclic analogues.

Turning now to the substituted toluenes, we know from experiment and the *ab initio* calculations that when the local geometry near methyl has only threefold symmetry, substantial barriers invariably occur. We identify three key effects whose importance differs from case to case. First, a substantial difference in bond order between the local ring CC bonds *a* and *b* (Fig. 13) contributes a threefold potential term favoring the conformation with one CH bond *cis* to the shorter CC bond, analogous to the simple case of 2-methylpropene. For a series of molecules that should suffer essentially no steric effects, we used the natural resonance theory (NRT)[37] to calculate the total bond orders of the ring CC bonds nearest methyl, bonds *a* and *b*. The NRT fits the one-electron density matrix as a weighted sum of one-electron density matrices, each associated with a particular resonance structure (and a corresponding set of NBOs):

$$\Gamma^{(1)} = \Sigma_i \, w_i \, \Gamma_i^{(1)}. \tag{4}$$

The mean bond order computed from the resulting *resonance weights* w_i is called the *natural bond order*. We find a good linear correlation between V_3 and the *difference* in natural bond orders, $O_b - O_a$ (Fig. 11). The best-fit slope is 950 cm^{-1}/bond.

The methyl group responds to the *difference in the three-dimensional electron density distribution* about the two nearest ring CC bonds, and the natural bond orders most simply quantifies the key difference in a unified manner across many molecules. At one extreme, 2-methylpropene has essentially localized single and double bonds ($O_a - O_b \simeq 1$) and a 1010 cm^{-1} barrier. At the other extreme, when the geometry of the ring has good *local* C_{2v} symmetry, as in S_0 toluene, *m*-fluorotoluene, *p*-fluorotoluene, 3,5-difluorotoluene, and 2,6-difluorotoluene, $O_a \simeq O_b$ and the barrier is invariably very small, even for nominal threefold cases. We interpret this equality of bond orders as indicative of essentially equal contributions of the two dominant resonance structures at all α.

Second, *ortho*-halogen substitution places a lone pair (2*p* on F or 3*p* on Cl) in close proximity to the methyl CH bonds. This leads to *steric repulsion* between the lone pair and methyl and also to *attractive donor-acceptor interactions* between the

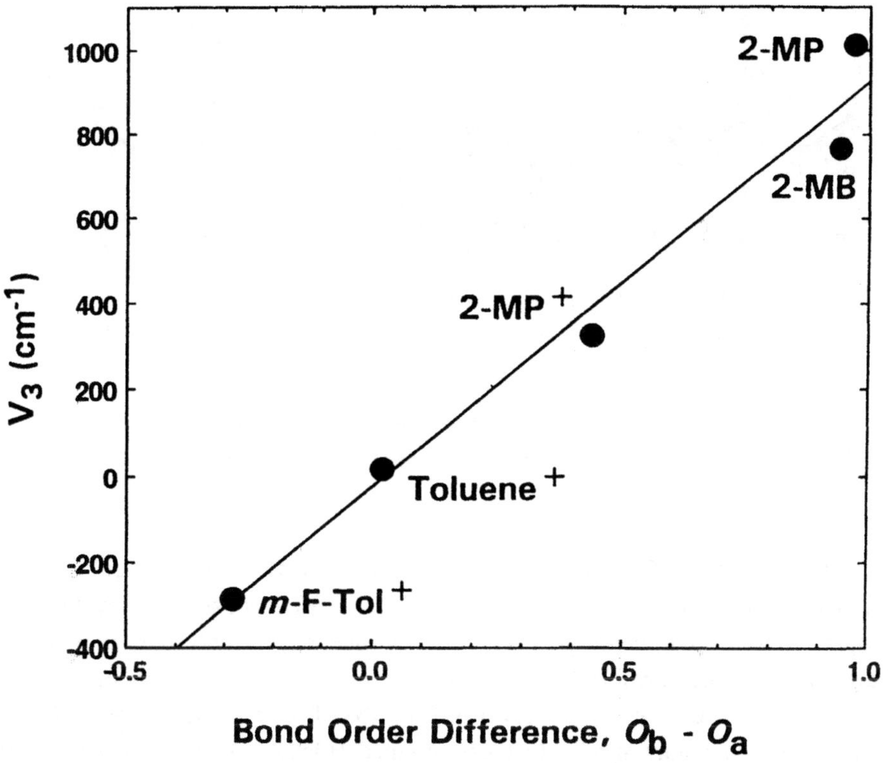

Figure 11. Ab initio potential parameter V_3 *vs* the difference in natural bond order between the two ring CC bonds adjacent to the point of methyl substitution.

halogen lone pair and σ_{CH}^* anti-bonds. These interactions are absent in the simpler case of ethane. We find that the donor-acceptor interactions $n_F \rightarrow \sigma_{CH}^*$ mildly favor the pseudo-*cis* geometry ($\alpha = 180°$), as if the halogen and CH bond engage in a weak hydrogen bond. In *o*-fluorotoluene and *o*-chlorotoluene, we have attempted an approximate quantitative resolution of the barrier into additive contributions from donor-acceptor and steric interactions. We combined the established technique of zeroing Fock matrix elements in the NBO representation[33] with the new technique of *natural steric analysis* (NSA).[30,38] Our results clearly indicate that in S_0, steric repulsion and donor-acceptor attraction *counteract* each other in *ortho*-substituted toluenes, with the steric interactions dominant.

Finally, π-ionization can affect local CC bond order differences and thus methyl rotor barriers in an interesting way. In toluene cations, π-ionization creates a pattern

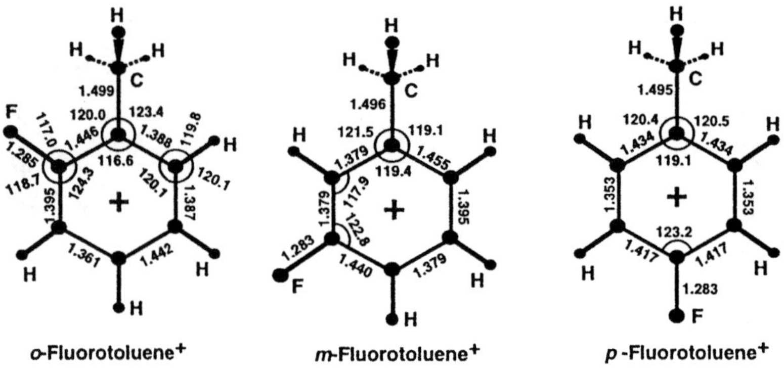

Figure 12. Ab initio optimized geometries with α constrained to 90 degrees. Bond lengths in Angstroms and bond angles in degrees.

of long and short ring CC bonds (Fig. 12). Ionization may occur from either of two nearly degenerate π orbitals, b_1 or a_2 (C_{2v} labels). *Meta*-substitution by the weak π donors fluorine and methyl (as in *m*-fluorotoluene$^+$) *orients* the π molecular orbitals so that they have approximate C_{2v} symmetry about the pseudo-C_2 axis bisecting the two substituents, as shown in Fig. 13. The actual UHF cation LUMO contours look remarkably similar to these oriented Hückel orbitals. In forming the ground state cation, removal of the a_2 electron *maximizes positive charge density at the points of substitution*, where it can best be stabilized by π-donation. This lengthens the two bonds parallel to the pseudo-C_2 axis. Accordingly, the calculated geometry of the ring (Fig. 12) shows two long CC bonds parallel to the methyl axis and four short CC bonds. Almost as a side effect, this pattern of ring CC π-bond orders causes a substantial difference (O_a - O_b) in local CC bond orders, which produces a large threefold barrier in accord with the correlation of Fig. 11! The preferred conformation places one CH bond *cis* to the ring CC bond of higher order, in analogy to the case of 2-methylpropene.

In *o*-fluorotoluene$^+$, the UHF geometries again show the "two-long, four short" pattern of ring CC bonds (Fig. 12), but now the long bonds lie perpendicular to the pseudo-C_2 axis that *bisects* the points of substitution (Fig. 13). Similar reasoning explains the orientation of the pattern. Now the local difference in CC π-bond orders favors the pseudo-*trans* conformer, reinforcing the steric interaction with the fluorine lone pair. The qualitative result is a strong preference for the pseudo-*trans* conformer in *o*-fluorotoluene$^+$.

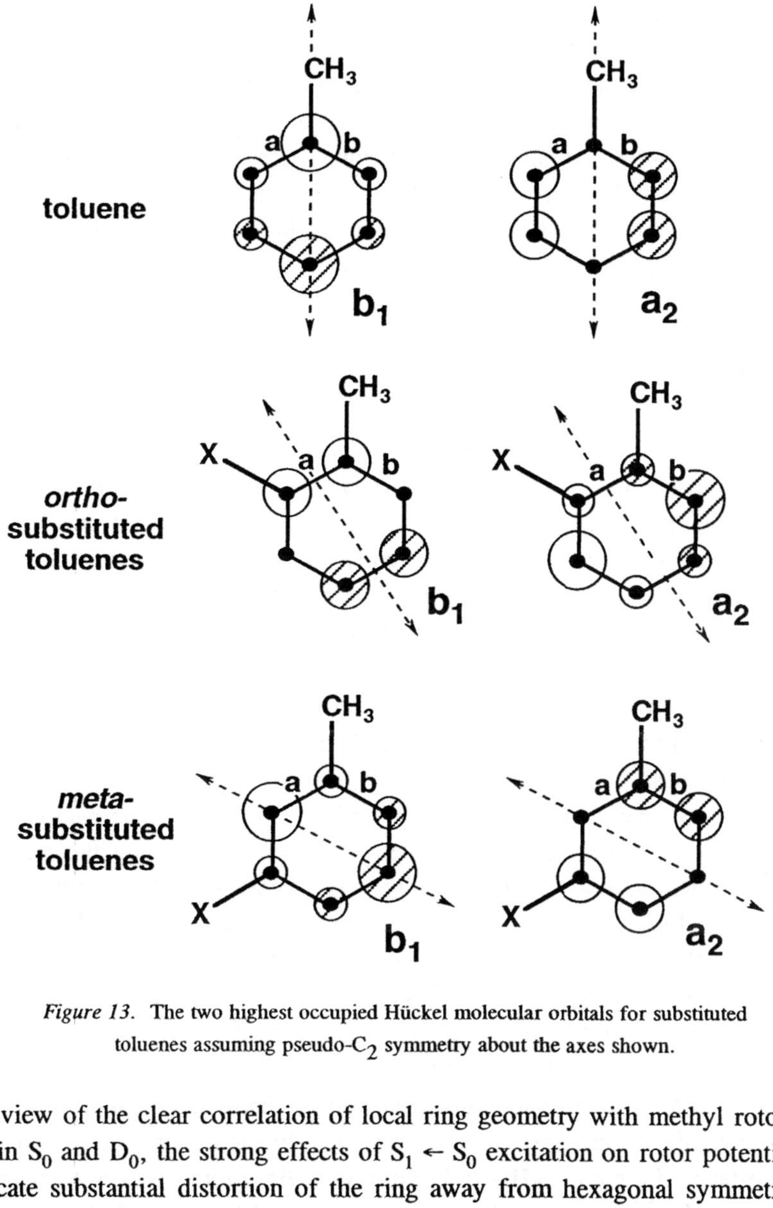

Figure 13. The two highest occupied Hückel molecular orbitals for substituted toluenes assuming pseudo-C_2 symmetry about the axes shown.

In view of the clear correlation of local ring geometry with methyl rotor barrier height in S_0 and D_0, the strong effects of $S_1 \leftarrow S_0$ excitation on rotor potentials seem to indicate substantial distortion of the ring away from hexagonal symmetry in S_1. There is little clear evidence of this from molecular spectroscopy. We have speculated that such a *π-molecular orbital orientation effect* in S_1 (similar to that in the cation) might explain the observed characteristic changes in barrier height on S_1-S_0 excitation of *ortho-* and *meta-*substituted toluenes. However, we have not yet pursued the multi-configuration calculations that might show this effect clearly.

Finally, we return to 2-fluoro-6-chlorotoluene (Fig. 10), for which $\pi^* \leftarrow \pi$ excitation and π ionization cause only minor changes in the rotor potential. In this disubstituted case, the appropriate pseudo-C_2 axis would coincide with the methyl rotor axis itself. We expect the Hückel orbitals to align with the rotor axis, which would make $O_a \simeq O_b$ in all three electronic states, much as in sixfold cases. Accordingly, there is little effect of electronic excitation on the barrier height, which is presumably dominated by steric repulsion in all three electronic states. In terms of π-electron effects, 2-fluoro-6-chlorotoluene behaves like a sixfold case.

6. Conclusion

We have seen how the advent of ZEKE-PFI spectroscopy, with its 1 cm^{-1} resolution for cation spectra, has allowed experimental determination of methyl torsional potentials in large molecular cations for the first time. The strong effect of π ionization in *m*-fluorotoluene$^+$ was sufficiently puzzling to inspire exploration of the electronic structure by *ab initio* methods. The pattern of long and short ring CC bonds in the cation equilibrium geometry then led quite naturally to the notion that weak π-donor substituents direct the orientation of Hückel molecular orbitals along pseudo-C_2 symmetry axes. Our progress provides a good example of how the *combination* of spectroscopic experiments and electronic structure theory can unravel the underlying causes of intramolecular forces in rather complex molecules.

At least two areas require further exploration. The first is the nature of S_1, the first excited singlet state. Excitation to S_1 has a strong effect on the methyl rotor potential, but we do not understand why. We know remarkably little about the geometry of S_1 from either theory or experiment. In the simple Hückel picture, for C_{2v} geometry there are *two* excitations of comparable energy leading to an excited state of the same symmetry as S_1. This suggests that we will need to carry out *multi*-configuration calculations to get a sensible description of the S_1 geometry and thus of the methyl rotor potential. That work is in progress.

The effect of *strong* π-donor substituents on geometry and methyl rotor potentials deserves theoretical and experimental attention. In particular, earlier work shows dramatic effects of amino (-NH_2) and hydroxyl (-OH) groups.[9] The S_0 barriers and the changes in barrier on S_1 excitation are unusually large compared to those of the weak π donors F and Cl. Fujii and co-workers[39] are presently studying the corresponding cations by ZEKE-PFI. Hydroxyl and especially amino π orbitals probably mix substantially with the ring π orbitals, so we must extend our thinking beyond the subtle orientation effects that work for fluorine and chlorine. *Ab initio* calculations on these systems should again be revealing.

Acknowledgments

The theoretical aspects of this work were carried out in collaboration with Prof. Frank Weinhold, Prof. Ned Sibert, and Dr. Jay Badenhoop. We have benefitted enormously from the skill and enthusiasm of these colleagues. JCW thanks the U.S. Department of Energy, Division of Chemical Sciences, Office of Basic Energy Research (No. DE-FG02-92ER14306) for generous support of this research. The National Science Foundation provided equipment used in the experimental work. The IBM Corporation generously donated two RS6000 workstations under the IBM-SUR program.

References

1. Kemp, J.D. and Pitzer, K.S. (1936) *J. Chem. Phys.* **4**, 749. Smith, L.G. (1949) *J. Chem. Phys.* **17**, 139. Wilson, E.B., Jr. (1959) *Adv. Chem. Phys.* **2**, 367.
2. Durig, J.R., Guirgis, G.A., and Bell, S. (1989) *J. Phys. Chem.* **93**, 3487.
3. Rudolph, H., Dreizler, A., Jaeschke, W., and Wendling, P. (1967) *Z. Naturforsch.* **22A**, 940.
4. Murakami, J., Ito, M., and Kaya, K. (1981) *Chem. Phys. Lett.* **80**, 203. Okuyama, K., Mikami, N., and Ito, M. (1985) *J. Phys. Chem.* **89**, 5617. Mizuno, H., Okuyama, K., Takayuki, E., and Ito, M. (1987) *J. Phys. Chem.* **91**, 5589.
5. Trinkers, P.J., Parmenter, C.S., and Moss, D.B. (1994) *J. Chem. Phys.* **100**, 1028; and references therein.
6. Breen, P.J., Warren, J.A., Bernstein, E.R., and Seeman, J.I. (1987) *J. Chem. Phys.* **87**, 1917.
7. Spangler, L.H. and Pratt, D.W. (1994) in J.M. Hollas and D. Phillips (eds.), *Jet Spectroscopy and Molecular Dynamics*, Chapman and Hall, London.
8. Lin, T.-Y., Tan, X.-Q., Cerny, T.M., Williamson, J.M., Cullin, D.W., and Miller, T.A. (1992) *Chem. Phys.* **167**, 203; and references therein.
9. Zhao, A.-Q., Parmenter, C.S., Moss, D.B., Bradley, A.J., Knight, A.E.W., and Owens, K.G. (1992) *J. Chem. Phys.* **96**, 6362. See Tables VI and VII for summaries of known threefold and sixfold barriers.
10. Lu, K.-T., Eiden, G.C., and Weisshaar, J.C. (1992) *J. Phys. Chem.* **96**, 9742.
11. Lu, K.-T. and Weisshaar, J.C. (1993) *J. Chem. Phys.* **99**, 4247.
12. Takazawa, K., Fujii, M., Ebata, T., and Ito, M. (1992) *Chem. Phys. Lett.* **189**, 592.
13. Takazawa, K., Fujii, M., and Ito, M. (1993) *J. Chem. Phys.* **99**, 3205.
14. Müller-Dethlefs, K., Sander, M., and Schlag, E.W. (1984) *Chem. Phys. Lett.* **112**, 291.
15. Müller-Dethlefs, K. and Schlag, E.W. (1991) *Ann. Rev. Phys. Chem.* **42**, 109.
16. I. Powis, T. Baer, and C.-Y. Ng (eds.), *High Resolution Laser Photoionization and Photoelectron Studies*, John Wiley, Chicester (1995).

17. See, for example: Ho, J., Polak, M.L., and Lineberger, W.C. (1992) *J. Chem. Phys.* **96**, 144.

18. Sanders, L., Hanton, S.D., and Weisshaar, J.C. (1990) *J. Chem. Phys.* **92**, 3485.

19. Tonkyn, R.G., Winniczek, J.W., and White, M.G. (1989) *Chem. Phys. Lett.* **164**, 137.

20. R.F. Stebbings and F.B. Dunning (eds.) *Rydberg States of Atoms and Molecules*, Cambridge University, New York (1983).

21. Biernacki, D. Th., Colson, S.D., and Eyler, E.E. (1988) *J. Chem. Phys.* **88**, 2099.

22. Pratt, S.T. (1993) *J. Chem. Phys.* **98**, 9241.

23. Chupka, W.A. (1993) *J. Chem. Phys.* **98**, 4520.

24. Vrakking, M.J.J. and Lee, Y.T. (1995) *J. Chem. Phys.* **102**, 8818. Vrakking, M.J.J., Fischer, I., Villeneuve, D.M., and Stolow, A. (1995) *J. Chem. Phys.* **103**, 1.

25. Wilson, E.B., Lin, C.C., and Lide, D.R. (1955) *J. Chem. Phys.* **23**, 136. Gordy, W. and Cook R.L. (1984) *Microwave Molecular Spectra*, 3rd Edition, Wiley-Interscience, New York.

26. Longuet-Higgins, H.C. (1963) Mol. Phys. **6**, 445. Bunker, P.R. (1979) *Molecular Symmetry and Spectroscopy*, Academic, New York.

27. Walker, R.A., Richard, E., Lu, K.-T., Sibert III, E.L., and Weisshaar, J.C. (1995) *J. Chem. Phys.* **102**, 8718.

28. Walker, R.A., Richard, E.C., Lu, K.-T., and Weisshaar, J.C. (1995) *J. Phys. Chem.* **99**, 12422.

29. Schwoch, D. and Rudolph, H.D. (1975) *J. Mol. Spectrosc.* **57**, 47.

30. Lu, K.T., Weinhold, F., and Weisshaar, J.C. (1995) *J. Chem. Phys.* **102**, 6787.

31. Walker, R.A., Richard, E.C., and Weisshaar, J.C. (1996) *J. Chem. Phys.*, in press.

32. Richard, E.C., Walker, R.A., and Weisshaar, J.C. (1996) *J. Phys. Chem.*, in press.

33. Reed, A.E., Curtiss, L.A., and Weinhold, F. (1988) *Chem. Rev.* **88**, 899.

34. Pitzer, R.M. (1983) *Accts. Chem. Res.* **16**, 207.

35. Dorigo, A.E., Pratt, D.W. and Houk, K.N. (1987) *J. Am. Chem. Soc.* **109**, 6591. Goodman, L., Leszczynski, J. and Kundu, T. (1993) *J. Am. Chem. Soc.* **115**, 11991; and references therein.

36. Corcoran, C.T. and Weinhold, F. (1980) *J. Chem. Phys.* **72**, 2866.

37. Glendening, E.D. and Weinhold, F. (1995) *J. Am. Chem. Soc.*, in press.

38. Badenhoop, J.K. and Weinhold, F., work in progress. Badenhoop, J. K. (1994) Ph.D. Thesis, Dept. of Chemistry, Univ. of Wisconsin-Madison.

39. Takazawa, K. and Fujii, M., work in progress.

ION MOLECULE REACTIONS AND RADIATION CHEMISTRY

Maurizio Speranza

Dipartimento di Studi di Chimica e Tecnologia
delle Sostanze Biologicamente Attive,
Università degli Studi di Roma "La Sapienza",
P.le A. Moro 5 - 00185 Rome, Italy.

Introduction.

The first modern application of radiolysis as a tool for generating specific ions in the gas phase and for investigating their behavior towards neutral substrates dates back to the late fifties (Meisels, Hamill, and Williams, 1957), namely several decades after Lind's (Lind, 1912a-c; 1928; 1935; 1939; Glocker and Lind, 1939; Livingstone and Lind, 1936) and Mund's (Mund, 1934; 1956) first seminal studies on radiation chemistry. Few years later, Ausloos and Lias first provided firm evidence of the occurrence of an ion-molecule reaction in the gas-phase radiolysis of propane (Ausloos and Lias, 1962). Since then, Ausloos' group (Ausloos, Lias, and Gorden, 1963; Ausloos, Lias, and Sandoval, 1963; 1970, Ausloos, Lias, and Scala, 1966; Ausloos, 1966; 1969; Ausloos and Lias, 1967; 1972) made itself conspicuous by investigating the behavior of radiolytic ions towards isotopically labeled substrates, whose highlights have been surveyed by Lias in her lecture held at the 1974 NATO Advanced Study Institute on Interactions between Ions and Molecules, Biarritz, France (Lias, 1975). Along the same stream, Cacace and his group (Cacace and Speranza, 1972, 1976a, 1976b; Cacace and Giacomello, 1973; Cacace and Possagno, 1973; Attinà, Cacace, Ciranni, and Giacomello, 1976) applied extensively stationary radiolysis to a variety of gas-phase ion chemistry problems related to the most classical organic reactions in the condensed phase, whose continuous progress has been outlined is successive stages (Cacace, 1982; Speranza, 1983; Speranza; 1987; Cacace, 1988; Speranza, 1992), beginning from Cacace's lecture at the 1978 NATO Advanced Study Institute on Kinetics of Ion-Molecule Reactions, La Baule, France (Cacace, 1979).

The growing impact of the stationary radiolytic technique to the study of ionic reactions in the gas phase is justified by its relative simplicity and its unique features that set it apart from most common mass spectrometric approaches. As in the source of most common mass spectrometers, radiolytic ions are generated from the interaction of energetic electrons with suitable neutral gases, contained in glassy bulbs at pressures from several tens of torrs to several atmospheres. Under such conditions, the radiolytic ions may undergo efficient thermal equilibration by collisions with the bath gas prior to their conversion to products. The reactivity of the thermalized ions is measured by standard kinetic procedures, including pressure- and temperature-dependence studies and competition experiments. The actual isolation and structural characterization of their neutral products provide valuable information upon the regio- and stereochemistry of the

K. R. Jennings (ed.), Fundamentals and Applications of Gas Phase Ion Chemistry, 335–380.

ionic process investigated. These basic features confer upon radiolysis a role in gas-phase ion chemistry complementary to that covered by conventional mass spectrometry, in which thermalization of ionic intermediates is prevented by the low-pressure operating regime and investigation of their reactivity is severely hampered by the inherently limited structural resolution.

As to the organization of the chapter, first the basic principles of radiation chemistry are concisely surveyed and the theoretical and experimental aspects of stationary radiolysis of gases outlined. Then, selected recent applications of the radiolytic technique to the investigation of the chemo-, regio-, diastereo-, and enantioselectivity of gas-phase positive ion-molecule reactions are presented.

Principles of Radiation Chemistry.

The study of gas-phase ion-molecule reactions by the radiolytic method and its comparison with the classical methodologies of investigation require a detailed knowledge of the physical and chemical events occurring when a gaseous mixture sealed in a glassy bulb is exposed to ionizing radiations, such as x- ($\lambda=10^{-2}$-10^{2} nm; energy=10-1x10^{5} eV) or γ-rays (λ=3x10^{-4}-3x10^{-2} nm; energy=4x10^{4}-4x10^{6} eV) (Spinks and Woods, 1990).

Cobalt-60 sources are commonly used to produce γ-rays of discrete energy (1.332 and 1.173 MeV). X-rays are produced whenever high-speed electrons are rapidly decelerated through metal targets. The energy of the emitted x-radiation ranges from near zero to the maximum energy of the incident electrons. A continuous energy spectrum is obtained with superimposed energy spikes, characteristic of the elements present in the metal target, produced when electron vacancies created in the element are filled by electrons from an orbital with a lower energy.

Under narrow-beam conditions, transmission of monoenergetic ionizing radiation through thin absorbers obeys the following equation:

$$N=N_i e^{-\mu l} \quad (1)$$

where N is the number of transmitted photons, N_i is the number of incident photons, l is the thickness of the absorber, and μ is the total attenuation coefficient of the material for photons of the appropriate energy. On passing through thin (dl) absorbers, the reduction in radiation intensity (dI), defined as the product of the number of absorbed photons and their average energy, is given by:

$$dI=-I_i \mu dl \quad (2)$$

where I_i is the intensity of the incident radiation. By integration, the transmitted monoenergetic radiation intensity is given by the following equation:

$$I=I_i e^{-\mu l} \quad (3)$$

Dividing the total attenuation coefficient μ by the density ρ of the absorbing material gives the mass attenuation coefficient μ/ρ, which is independent of the density and physical state of the material. Also useful are the atomic attenuation coefficient $_a\mu$ (eq. 4a) and the electronic attenuation coefficient $_e\mu$ (eq. 4b). In these equations, A is the atomic weight and Z the atomic number of the stopping material, and N_A is the Avogadro constant.

$$_a\mu = \mu A/\rho N_A \quad (4a)$$

$$_e\mu = \mu A/\rho N_A Z \quad (4b)$$

The atomic and electronic coefficients have the units of an area and are often referred to as cross sections for radiation absorption. Attenuation coefficients for several compounds for 1-MeV γ-radiation are reported in Table I.

Table I. Total Attenuation Coefficients for 1-MeV γ-Radiation.

Coefficient	Units[a)]	Hydrogen	Oxygen	Carbon	Silicon	Water	Methane[b)]	Silica
Electronic, $_e\mu$	b electron^{-1}	0.211	0.211	0.211	0.211	0.211	0.211	0.211
Atomic, $_a\mu$	b atom^{-1}	0.211	1.69	1.27	2.95	2.11	2.11	6.33
Mass, μ/ρ	cm^2· g^{-1}	0.127	0.064	0.064	0.064	0.071	0.079	0.063
Linear, μ	cm^{-1}	—	—	—	—	0.071	5.2×10^{-5}	0.140

a) 1b (barn) = 10^{-28} m^2; b) T = 298 K; P = 1 Atm.

The energy released from the radiation to the absorber can be derived by considering the absorber made up of a large number of very thin layers perpendicular to the direction of the incident radiation and each dl thick.The energy absorbed by any one layer is a function of the reduction of intensity as the radiation passes through it as given in eq. 2:

$$dI_a = -I'\mu_a dl \quad (5)$$

with μ_a as the energy absorption coefficient and I' as the radiation intensity at the position occupied by the layer. This is given by eq. 3. Thus, at any depth l:

$$dI_a = -I_i\mu_a e^{-\mu l} dl \quad (6)$$

The total reduction in intensity due to energy absorption as the radiation passes through the entire thickness of absorber is found by integrating eq. 6:

$$I_a = \int -I_i\mu_a e^{-\mu l} dl = I_i(1-e^{-\mu l})\mu_a/\mu \quad (7)$$

The energy absorbed by the stopping material (E_a) is found by integrating eq. 7 with respect to time:

$$E_a = E_i(1-e^{-\mu l})\mu_a/\mu \quad (8)$$

where E_i is the total radiation energy incident on the absorber.

By assuming $\mu_a=\mu$, in passing through the 2-mm-thick walls (l=0.2) of a 1-l round silica bulb containing CH_4 at 1 atm and 298 K, the average 1.25-MeV ^{60}Co-γ-radiation releases ca. 3% of its energy to the walls and only ca. 0.07% to the gaseous content of the bulb (eq.8).

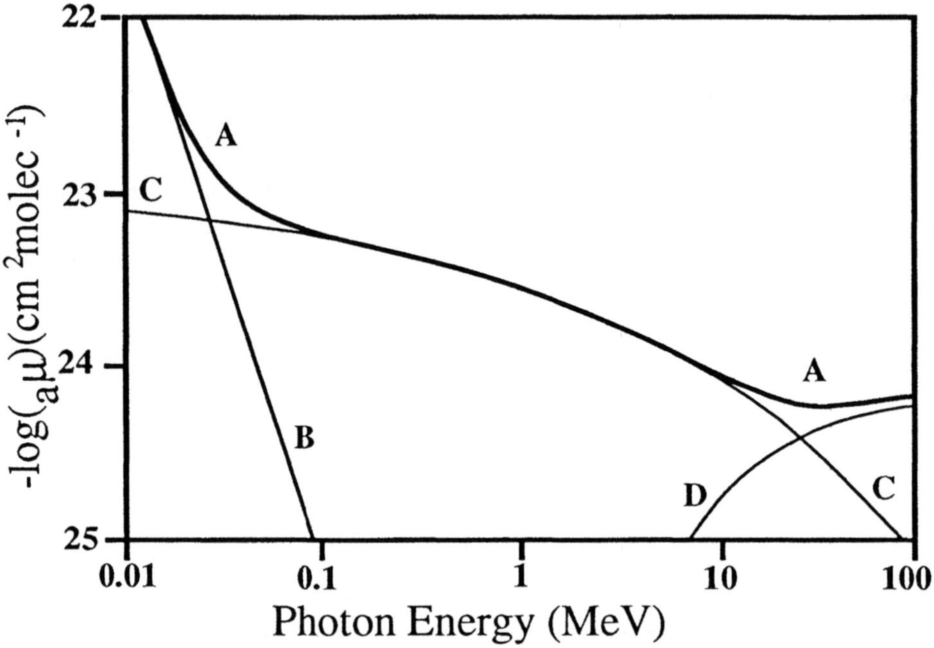

***Figure1.** Atomic attenuation coefficient for water. A) Total attenuation coefficient; B) photoelectric attenuation coefficient; C) total Compton coefficient; D) pair production coefficient.*

Energetic photons lose energy to the matter through the photoelectric effect, the Compton effect, and pair production. In the photoelectric interaction, the entire energy of the photon (E°) is transferred to a single atomic electron (often from the K or the L shell), which is ejected from the atom with an energy (E_e) equal to the difference between the photon energy and the binding energy (E_s) of the electron in the atom:

$$E^\circ = E_e + E_s \qquad (9)$$

The vacancy created by the loss of the electron from an inner orbital will be filled by an electron from an outer orbital with emission of characteristic x-radiation, which in turn may be absorbed yielding low-energy Auger electrons. Photoelectric interaction predominates with low-energy photons in high-Z materials (Fig.1).

In Compton scattering, the photon (E°) interacts with a loosely bound electron of the absorber, so that the electron recoils from the atom with an energy (E_e) and the photon is deflected with a reduced energy (E_γ). The energy of the recoil electron can be expressed by:

$$E_e=E°-E_\gamma \tag{10}$$

and may have any value from zero up to a maximum value, when the deflection angle between the incident photon and the scattered photon is 180°. Compton interaction predominates with 1-5-MeV photons in high-Z materials. With low-Z materials, e.g. water, Compton scattering predominates with 0.03-20 MeV photons.

Pair production involves the complete absorption of a photon (E°) in the vicinity of an atomic nucleus with the formation of two particles, an electron and a positron. The combined kinetic energy of the electron (E_e) and the positron (E_p) amounts to the difference between the energy of the incident photon and the rest energy (m_ec^2) of the two particles:

$$E_e+E_p=E°- 2m_ec^2 \tag{11}$$

The positron eventually combines with an electron, the two particles being replaced by two 0.51-MeV γ-rays (annihilation radiation) emitted in opposite directions. Pair production requires incident photon energies above 1.02 MeV.

When a beam of electromagnetic radiation penetrates the external walls of a glassy bulb, the secondary photoelectric, Compton, and pair-production radiation builds up to a maximum distance below the wall surface that is roughly equal to the greatest range of the secondary electrons, and then falls as the primary beam is attenuated. Since for thin absorbers (i.e. wall thickness=0.2 cm) the build-up of the secondary electrons follows the total reduction in intensity I_a (eq.7), the inside face of the bulb wall releases a number of secondary electrons inversely dependent upon the thickness of the wall. This secondary radiation (δ-rays), which has a wide spectrum of energy (up to hundreds of eV), are almost entirely absorbed by the content of the bulb (e.g. gaseous CH_4 at 1 atm) and are responsible of most of the ionization and excitation events occurring in it. Thus, while the primary electromagnetic radiation (e.g. the cobalt-60 γ-rays) escaping the bulb walls does not significantly interact with the gaseous content of the bulb, the secondary electrons emitted from the bulb walls by virtue of the small fraction of absorbed radiation acts toward the gaseous content of the bulb in the same way as the beam of high-energy electrons emitted from the hot filament in the ion source of a mass spectrometer. The correspondence is even stricter, if it is considered that experimental results and theoretical calculations indicate that the energy transferred to a gaseous molecule is fairly insensitive to variations in the energy of the ionizing electron as long as the energy is above twice the ionization potential of the molecule (Eyring and Wahrhafting, 1961; Vestal, Wahrhafting, and Johnston, 1962; Kebarle and Godbole, 1962; Monahan and Stanton, 1962). Since this normally ranges below 20 eV, the distribution and the energy content of the primary activated species formed in the radiolysis of gaseous systems is fully comparable to those obtained in the source of a mass spectrometer in which the ionization is effected by 70-eV electrons.

Radiation Dosimetry.

Likewise in the source of a mass spectrometer the number of ions produced per time unit from a given precursor depends on the electron beam flux and energy, in radiation chemistry quantitative measurements require a knowledge of the energy transferred from the radiation field to the absorber per time unit and the distribution of the absorbed energy within the material.

Radiation fields are described in terms of exposure (X_A), which defines its ability to produce ionization in air. It is the total charge of ions of one sign produced in air when all the electrons and positrons are completely stopped in air. Its unit is coulomb per kilogram (C Kg^{-1}). The absorbed dose (D_A) in air exposed to the radiation field can be calculated from the exposure, X_A, if the mean energy required to form an ion pair in air, W, is known:

$$D_A = WX_A/e \qquad (12)$$

where e is the charge on the electron. If e is expressed in coulomb and W in J, the absorbed dose is given in J Kg^{-1} (or gray, Gy). The absorbed dose rate (Gy s^{-1} or Gy h^{-1}) is the absorbed dose per unit time. Ionization measurements can be used to determine the absorbed dose in materials other than air by measuring the exposure in air at a given point, using a ionization chamber, and then irradiating the material at the same point.

Radiation-chemical yields are described in terms of G values, where G(X) is the number of molecules of product X formed (n(X)), or of starting material Y changed (shown as G(-Y)), per J of energy absorbed:

$$G(X) = n(X)/D_A \qquad (13)$$

$$G(-Y) = n(-Y)/D_A \qquad (14)$$

The G(X)/G(-Y) ratio provides the yield factor of the product X.

Production of Ionization and Excitation.

Electrons ejected from the inside bulb walls as a consequence of the ionization produced by radiation absorption are generally sufficiently energetic (up to several hundreds of eV) to produce further ionization and excitation in the gaseous content of the bulb.

Ionization is a distinctive consequence of the absorption of ionizing radiations by matter and a key reaction in all radiolysis mechanisms:

$$A \rightarrow A^+ + e^- \qquad (15)$$

$$A \rightarrow A^{*+} + e^- \qquad (16)$$

While at dose rates lower than ca. 25 Gy h^{-1}, the probability is high that ions will drift freely to the bulb walls and be neutralized, at dose rates typical of stationary γ-irradiations (10^4-10^6 Gy h^{-1}), the normal fate of ions is to be neutralized by an ion of opposite sign, that is:

$$A^+ + e^- \rightarrow A^{**} \qquad (17)$$

$$A^+ + B^- \rightarrow A^* + B^* \qquad (18)$$

The electron ejected during ionization may have considerable energy, which can be lost through interactions with other molecules. The electron is reduced to thermal or near thermal energy and possibly captured by some gaseous component with the high electron affinity (B) before recombination with a positive ion or discharge on the bulb walls. The immediate product of the recombination are the neutral, highly excited molecules A** and A*, which, unless collisionally or radiatively quenched, eventually dissociate to more or less excited molecular products (eq. 19) or, more often, free radicals (eq. 20).

$$A^* \text{ (or } A^{**}) \rightarrow M^* + N \qquad (19)$$

$$A^* \text{ (or } A^{**}) \rightarrow \cdot R^* + \cdot S \qquad (20)$$

In the gas phase, the mean lifetime of a charge is expressed by:

$$\tau=(\alpha n^{\pm})^{-1/2} \qquad (21)$$

with $n^{\pm}$ as the number of ions formed per time unit and volume unit, and α is the rate coefficient of the recombination reactions 17 and 18. Most values of α reported in the literature fall in the range of 10^{-6}-10^{-5} cm^3 $molecule^{-1}$ s^{-1}. At 10^4-10^6 Gy h^{-1} dose rates, the $n^{\pm}$ term ranges from 10^8 to 10^{14} ions cm^{-3} s^{-1}, so that the lifetime τ of an ion under those conditions will span from 10^{-4} to 10^{-1} s. Since the collision interval between ions and their neutral precursors ranges from 10^{-8} to 10^{-11} s over the 1-1000 torr pressure interval, it is sufficient to add to the gaseous precursor 0.1 mol% of a neutral substrate (S) to make the ion react with it well before its recombination with its negatively charged counterpart.

$$A^+ + S \rightarrow [A{\cdot}S]^+ \rightarrow C^+ + D \qquad (22)$$

On the whole, stationary irradiation of a gaseous mixture by high-energy radiation results in a random distribution of positive ions (electrically paired by far-removed electrons or anions) as well as electronically excited molecules, both possibly fragmenting in smaller radicals. The fate of these reactive species is determined by their concentration, which is in turn determined by the secondary electron flux (and, thus, by the absorption dose rate) as well as by the nature and the composition of the irradiated gaseous medium.

Energy Degradation of Excited Species.

Electronically excited species, produced either directly from the ionizing radiation (e.g. A^{*+}) or by ion neutralization (e.g. A**, M*, ·R*, etc), may convert unimolecularly to lower-energy states either radiatively (fluorescence, ca.10^{-8} s; phosphorescence, ca.10^{-3}-10^{-4} s) or by internal conversion (ca.10^{-13} s). Bimolecular radiative and nonradiative energy transfers to another molecule are also rather efficient processes, provided that the transferred energy exceeds the excitation potential of the acceptor. After reaching the electronic ground state, radiolytic ions, molecules, and radicals still present a large excess of vibrational energy which can be rapidly dissipated by collisions with other molecules and converted in thermal energy. Considering ca. 10^{-9} s the collision time between a neutral gas at 100 torr and the excited species from radiolysis, their complete energy degradation can be readily achieved prior of interacting with the neutral substrate (S), provided that this latter is present in the gaseous mixture at concentrations of 0.1 mol% or less.

Ion and Radical Scavenging.

A technique that has proved helpful in studying the mechanism of radiolytic reactions is the deliberate addition of small quantities of materials that have a high affinity for active transient species, either ions or radicals. Addition of the materials, described as ion or radical scavengers, sometimes makes possible to identify the origin of the neutral radiolytic products as well as role of a particular ion or radical within the overall reaction mechanism. In irradiated gaseous systems, ions can be also removed by applying a sufficiently high potential across the system. The technique was employed by Essex and his colleagues in a series of early experiments that showed that chemical changes in irradiated gases are not due to the action of ions alone (Essex, 1954).

Cation Scavengers.

Substances that have been used as positive-ion scavengers can be divided in two broad classes, namely: i- Lewis bases, such as ammonia, amines, water, alcohols, esters, etc., able to accept a proton or other cationic fragments from the donor (RH^+) to form stable closed-shell ionic derivatives (e.g. eq.23), and ii- electron donors, i.e. compounds of low ionization potential, such as hydrogen sulfide, nitric oxide, etc., able to release an electron in a charge exchange reaction with open-shell cations ($\cdot R^+$) (e.g. eq.24).

$$RH^+ + NH_3 \rightarrow R + NH_4^+ \qquad (23)$$

$$\cdot R^+ + H_2S \rightarrow R + \cdot H_2S^+ \qquad (24)$$

Their effect upon the yield and distribution of the neutral radiolytic products may provide precious indications about their actual origin, whether involving a cationic process or not. Indeed, competition of the cation scavenger with the substrate S causes a drastic decrease of the yields of products from cation-molecule reactions, leaving those from noncationic processes unaffected.

An example of the application of a cation scavenger of type i (i.e. B=NEt_3, $Me_2N(CH_2)_2OH$, and c-$C_6H_{10}O$) is given in Figure 2, concerning the gas-phase attack of $sC_3H_7^+$ ions, obtained by ^{60}Co-γ-radiolysis of C_3H_8 at 37.5°C and 1 atm, on trimethylsilylbenzene (TSB) in the presence of variable concentrations of B, as the cation trap (Attinà, Cacace, and Ricci, 1991). The figure reports the dependence of the G(P)/G°(P) ratio, expressed by the ratio between the total absolute yield of products P, i.e. isomeric $sC_3H_7C_6H_4SiMe_3$, $sC_3H_7C_6H_5$, and C_6H_6, at any given concentration of the base B and that measured in the absence of added bases, upon the r factor, given by:

$$r = k_S[TSB]/(k_S[TSB]+k_B[B]) \qquad (25)$$

Taking $k_S \approx k_B$, as expected for a highly reactive and indiscriminate $sC_3H_7^+$ electrophile, the r factor reduces to:

$$r \approx [TSB]/([TSB]+[B]) \qquad (26)$$

namely simply the mole fraction of TSB, relative to all nucleophiles present in the gaseous mixture.

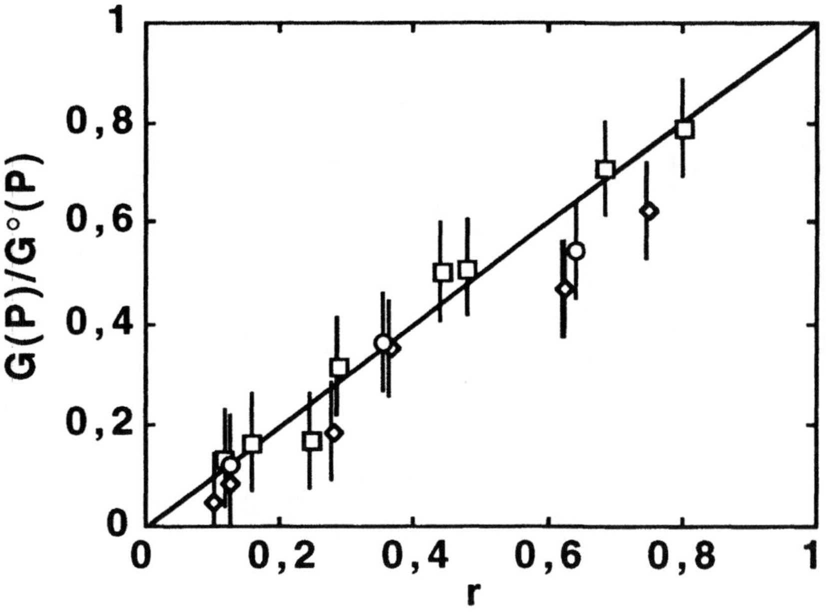

Figure 2. Dependence of the absolute yields of products from the reaction of $sC_3H_7^+$ ions with TSB on the r factor (NEt_3 (squares), $Me_2N(CH_2)_2OH$ (circles), and $c\text{-}C_6H_{10}O$ (diamonds)) (see text).

The excellent linear correlation of Fig.2 points to the exclusive ionic origin of the radiolytic products P, whose formation obeys the simple kinetic model of network 27, involving the competition of TSB with B for the charged electrophile.

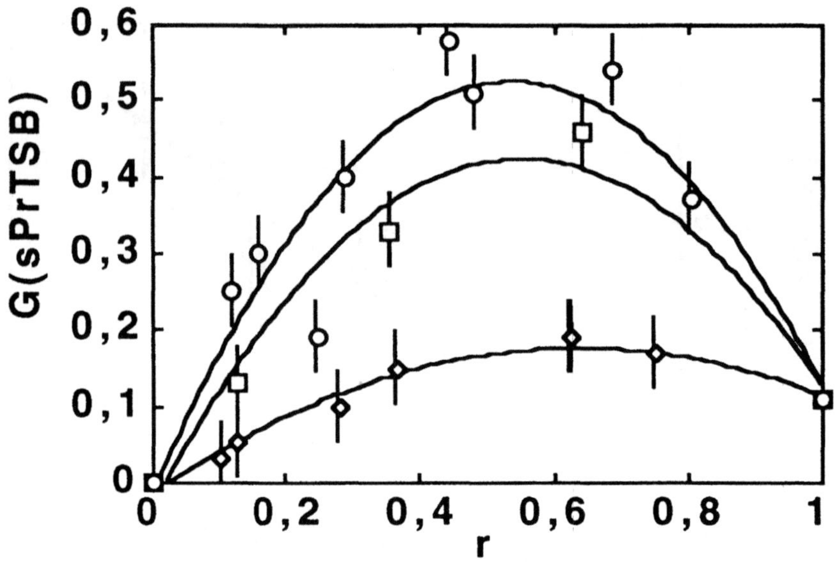

Figure 3. *Dependence of the absolute yield of isomeric s-propylated TSB (sPrTSB) as a function of the r factor (NEt$_3$ (squares), Me$_2$N(CH$_2$)$_2$OH (circles), and c-C$_6$H$_{10}$O (diamonds)) (see text).*

However, inspection of the dependence of the absolute yield of isomeric s-propylated TSB (G(sPrTSB)) upon the nature and the concentration of the bases B reveals the peculiar bell-shaped trends, reported in Figure 3.

This behavior arises from the distinct tendency of arenium intermediates **1**, unless rapidly neutralized, to rearrange via consecutive proton shifts into the stable *ipso*-substituted isomer **2** (Cacace, Crestoni, De Petris, Fornarini and Grandinetti, 1988) (eq.28).

All bases B can react exothermically according to eq. 28b. However, once **2** is formed, its fate depends on the nature of B, in that oxygenated bases, e.g. c-$C_6H_{10}O$, act as efficient desilylating reagents (eq.29), whereas nitrogen bases, e.g. NEt_3, react as deprotonating agents (eq.30).

In the absence of added B, the inefficiency of deprotonation 28b allows extensive **1**$\rightarrow$**2** isomerization and subsequent complete desilylation of **2** by the oxygenated bases present in the gaseous mixture, e.g. H_2O. Small concentrations of B enhance G(sPrTSB) by deprotonating **1** before its isomerization to **2**. This effect is counterbalanced by the interception of $sC_3H_7^+$ by B, which tends to depress G(sPrTSB) especially at high B concentrations. This accounts for the peculiar maximum that characterizes the curves of Fig.3. The different efficiency of the bases B, clearly illustrated in Fig.3, is traced to their different affinity for the *ipso*-hydrogen and the silyl centers of ions **2**.

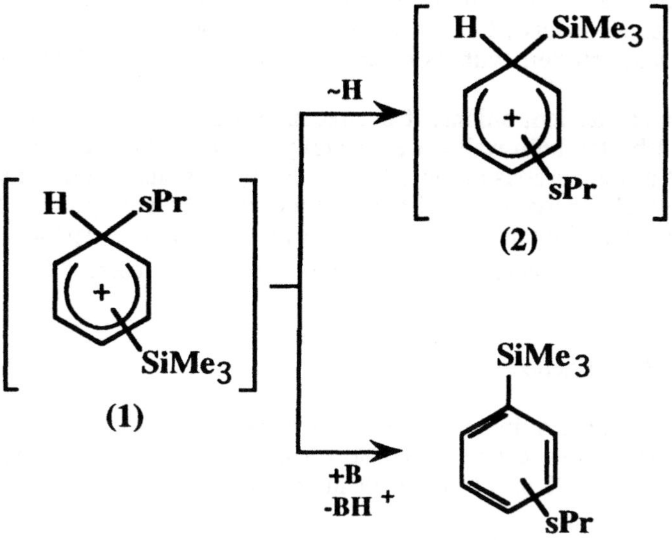

(28a)

(28b)

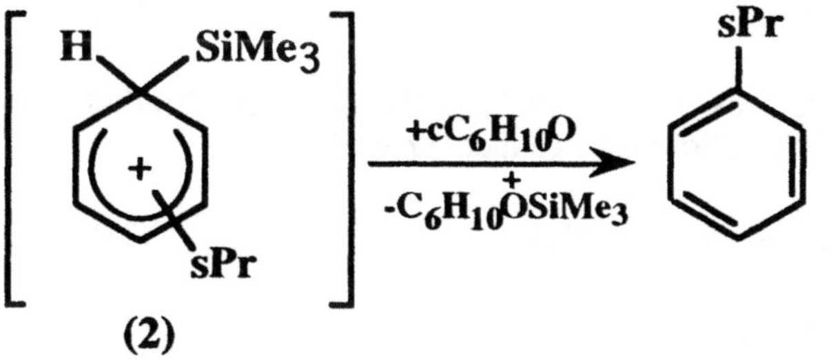

(29)

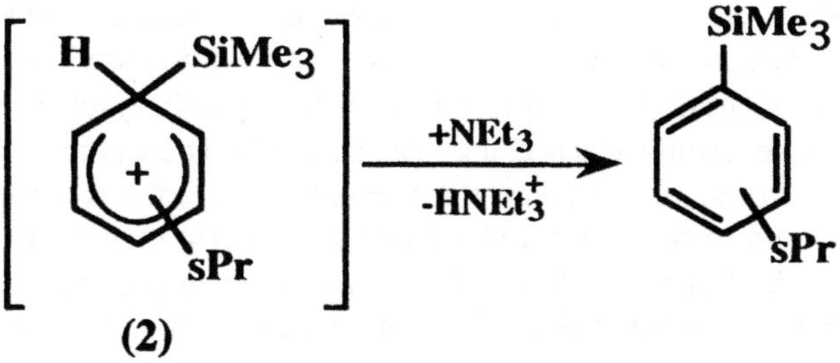

(30)

Electron and Anion Scavengers.

Since the absorption and scattering of secondary electrons from high-energy radiation is only slightly affected by changes in atomic number of the absorber, the probability of their interaction with each individual component of a gaseous mixture will depend essentially upon the concentration of the component. Thus, use of limited concentrations of S ($\leq$0.1 mol%) minimizes its undesired direct radiolysis, possibly leading to neutral radiolytic products which may interfere with those from the ionic processes of interest (eq.22).

Epithermal electrons and transient anions, if not recombining with positive ions (eqs.17 and 18) or trapped by the bulb walls, are generally captured via a resonance process by the component of the gaseous mixture having the highest electron affinity, eventually producing negatively charged ions. If this component is a neutral product P from the ion-molecule reaction of interest (eq.22), it gives rise to the corresponding long-lived anions P^-, which may be eventually neutralized by the cationic reactant A^+ by a process similar to eq.18:

$$A^+ + P^- \rightarrow \text{Excited fragments} \tag{31}$$

Products showing such tendency include halogenated and nitrated compounds, aromatic hydrocarbons, ketones, sulfur compounds (Chowdhury and Kebarle, 1986), etc., whose absolute yield from reaction 22 may be dramatically reduced by occurrence of the parasitic sequence: $P \rightarrow P^- \rightarrow$fragments.

The action of epithermal electrons and transient anions in irradiated mixtures can be efficiently minimized by using electron scavengers, namely gaseous additives with exceedingly high electron affinity. One of them is sulfur hexafluoride, SF_6, which scavenges electrons to form a stable negative ion:

$$SF_6 + e^- \rightarrow SF_6^- \tag{32}$$

eventually neutralized by positive ions. For instance, in the gas-phase irradiation of ammonia, electron capture by SF_6 is followed by (Sauer, 1976):

$$SF_6^- + NH_4^+ \rightarrow \cdot SF_5 + NH_4F \tag{33}$$

$$2\ \cdot SF_5 \rightarrow S_2F_{10} \tag{34}$$

An example of the effect of a powerful electron scavenger, such as SF_6, upon the radiolytic products yield and distribution concerns the gas-phase electrophilic nitration of arenes by the $MeNO_3H^+$ ions, generated at 740 torr from ^{60}Co-γ-radiolysis of a $CH_4/MeNO_3$= 45/1 gaseous mixture (Attinà, Cacace, and Yanez, 1987). The k_T/k_B rate constant ratio measured in toluene/benzene=1/1 competition experiments, carried out in the absence of added SF_6, amounts to 4.8, with a nitrotoluene isomeric distribution *ortho:meta:para*=19.7:1.0:12.7. Addition of SF_6 (5 torr) was found to cause some changes in both the k_T/k_B rate constant ratio, which increases to 5.1, and the nitrotoluene isomeric distribution, which becomes *ortho:meta:para*=8.4:1.0:4.9. These variations demonstrate that SF_6 efficiently prevents attachment of epithermal electrons, as well as electron transfer from anions, to the nitrated radiolytic product with the highest electron affinity, namely the *meta* isomer of nitrotoluene.

Radicals Scavengers.

In radiation chemistry, free radicals are produced by the dissociation of excited molecules (eq.20) and by ionic processes, including ion dissociation (e.g. eq.35), ion-molecule reactions (e.g. eq.36), and ion neutralization (eq.18).

$$[\cdot C(CH_3)_4^+]^* \rightarrow C(CH_3)_3^+ + \cdot CH_3 \quad (35)$$

$$\cdot CH_4^+ + CH_4 \rightarrow CH_5^+ + \cdot CH_3 \quad (36)$$

It has been estimated that for many systems the radiation-chemical yield of radicals, G(radicals), is about 0.6 μmol J^{-1} or 6 radicals per 100 eV of absorbed energy. Thus, in general, irradiation of gases leads to roughly equal numbers of ions and radicals.

A characteristic property of free radicals is the instability associated with the presence of an unpaired electron. Radicals are often extremely reactive, reacting in such a manner that the unpaired electron becomes paired with a similar electron in another radical (e.g. eq.34) or is eliminated by an electron-transfer reaction. Alternatively, the radical may react so as to form a second, more stable, radical (e.g. eq.24).

Highly reactive radicals are generally very unselective (e.g. the ·F atom). Less reactive radicals may be rather selective, showing either nucleophilic (e.g. $\cdot CH_3$) or electrophilic properties (e.g. the ·Cl atom). Radicals are among the most versatile reactive species, undergoing a variety of unimolecular and bimolecular reactions, such as unimolecular rearrangements and fragmentation, or addition to unsaturated molecules, hydrogen-atom abstraction, recombination, and disproportionation. Therefore, they generally play an important role in the formation and the distribution of neutral products in irradiated mixtures and, hence, their control is essential in the study of ionic processes by radiolytic methods.

Scavengers for free radicals include oxygen, nitric oxide, iodine, hydrogen iodide, organic iodides, and unsaturated organic compounds. In general, few mol% of these compounds in the gaseous systems are sufficient to scavenge all thermal radical formed by radiation.

Oxygen, O_2, readily adds to free radicals and, if added to the gaseous mixture, will almost invariably play part in radiation-induced reactions. In its ground state ($^3\Sigma_g$), O_2 has two unpaired electrons in the π^*2p antibonding orbital and behaves as a molecule with two free-radical centers. It efficiently reacts with free radicals yielding relatively stable peroxy radicals:

$$\cdot R + \cdot O{=}O\cdot \rightarrow R\text{-}O\text{-}O\cdot \quad (37)$$

Peroxy radicals are generally less reactive than the radicals from which they are formed, and are thus longer-lived and more selective in the groups or molecules they attack. Reactions of peroxy radicals include hydrogen abstraction from suitable donors:

$$R\text{-}O\text{-}O\cdot + DH \rightarrow RO_2H + \cdot D \quad (38)$$

which initiates a chain reaction if the donor DH coincides with the precursor RH of the radical itself. The hydroperoxide RO_2H may break down to give both the peroxy ($RO_2\cdot$) and alkoxy (RO·) radicals which may propagate the chain, unless abstracting a H atom from a donor, yielding, in the RO· case, the corresponding alcohol ROH.

Molecular iodine has been widely used as a radical scavenger and reacts as follows:

$$\cdot R + I_2 \rightarrow R\text{-}I + \cdot I \quad (39)$$

$$2 \cdot I \rightarrow I_2 \qquad (40)$$

The reaction is favorable energetically because the R-I bond energy is generally greater than the I-I one and it is an advantage that the iodine atoms produced dimerize without initiating any additional radical reaction. Radicals that are scavenged by iodine can be identified by identifying the iodides RI formed. However, there are also disadvantages to the use of iodine; for example, it enters into complex reactions with aromatic substrates, and the hydrogen iodide produced when hydrogen atoms are scavenged tends to react with unsaturated molecules or bases present in the mixture, giving rise to undesired effects. Besides, iodine and organic iodides can also efficiently trap free electrons yielding stable I^-.

Table II. Radiolysis of Mixtures Containing Methane.

Product	$G_{(P)}$ (μmol J^{-1})		Relative Yields		
	Pure CH_4	CH_4 + NO	CH_4 + Ar	CH_4 + Ar + I_2	CH_4 + I_2
H_2	0.66	0.37	100	100	
C_2H_2			0	1.6	
C_2H_4	0.013	0.066	0.5	14	
C_2H_6	0.22	0.033	45	0.6	
C_3H_6	0	0.003	0.2	0	
C_3H_8	0.027	0.001	7	0	
n-C_4H_{10}	0.013	0	3.5	0	
i-C_4H_{10}	0.006	0			
i-C_5H_{10}	0.005	0			
$(CH_2)_n$	0.22				
HI					8.5
CH_3I				34	70
C_2H_5I				18	4.5
C_3H_7I				0.5	
C_4H_9I				0.1	
CH_2I_2				9	8.5

Other important radical scavengers are nitric oxide, ·NO, which adds to radicals to form stable nitroso compounds:

$$\cdot R + \cdot NO \rightarrow R\text{-}NO \qquad (41)$$

or alkenes, which react by radical addition:

$$\cdot R + CH_2{=}CH_2 \rightarrow R\text{-}CH_2\text{-}CH_2\cdot \qquad (42)$$

Unsaturated radiolysis products behave in the same way as alkene scavengers and can cause marked changes in radiation yields as they build up during irradiation.

The effects of radical scavengers in radiolysis of gases are eminently illustrated in Table II, concerning the neutral product yield and distribution from radiolysis of methane (Yang and Manno, 1959; Meisels, Hamill, and Williams, 1957; Gevantman and Williams, 1952).

Radiolytic products are accounted for by the following sequence of ionic and radicalic reactions:

$$CH_4 \rightarrow CH_4^+ + e^- \qquad (43)$$

$$CH_4 \rightarrow CH_3^+ + \cdot H + e^- \qquad (44)$$

$$CH_4 \rightarrow \cdot CH_2^+ + H_2 + e^- \qquad (45)$$

$$CH_4 \rightarrow CH_4^* \qquad (46)$$

$$CH_4^* \rightarrow \cdot CH_3 + \cdot H \qquad (47)$$

$$CH_4^* \rightarrow :CH_2 + H_2 \qquad (48)$$

$$CH_4^+ + CH_4 \rightarrow CH_5^+ + \cdot CH_3 \qquad (49)$$

$$CH_3^+ + CH_4 \rightarrow C_2H_5^+ + H_2 \qquad (50)$$

$$\cdot CH_2^+ + CH_4 \rightarrow C_2H_5^+ + \cdot H \qquad (51)$$

$$CH_5^+ + e^- \rightarrow \cdot CH_3 + H_2 \qquad (52)$$

$$C_2H_5^+ + e^- \rightarrow \cdot C_2H_5 \text{ (or } C_2H_4 + \cdot H) \qquad (53)$$

$$2\cdot H \rightarrow H_2 \qquad (54)$$

$$\cdot H + CH_4 \rightarrow H_2 + \cdot CH_3 \qquad (55)$$

$$2\cdot CH_3 \rightarrow C_2H_6 \qquad (56)$$

$$\cdot CH_3 + \cdot C_2H_5 \rightarrow C_3H_8 \text{ (or } C_2H_4 + CH_4) \qquad (57)$$

$$2\cdot C_2H_5 \rightarrow C_4H_{10} \text{ (or } C_2H_6 + C_2H_4) \qquad (58)$$

$$\cdot R + C_2H_4 \rightarrow RC_2H_4\cdot \text{ (+ } C_2H_4 \rightarrow \text{polymer)} \qquad (59)$$

Principal radical products detected by the appearance of the corresponding iodides were $\cdot H$, $\cdot CH_3$, $\cdot C_2H_5$, and $:CH_2$. The increase in the yield of ethene in the presence of scavenger is probably due to removal of radicals which would otherwise add to the alkene to form polymeric products. The experiments with mixtures of methane and an inert gas in which methane is the minor component illustrate the effects of charge transfer processes between the primary noble gas ions and methane on the radiolytic product pattern.

When methane is irradiated in the presence of oxygen, as the radical scavenger, oxidation products, such as carbon oxides, formaldehyde, formic acid, alcohols, and peroxides, are formed via radicalic processes.

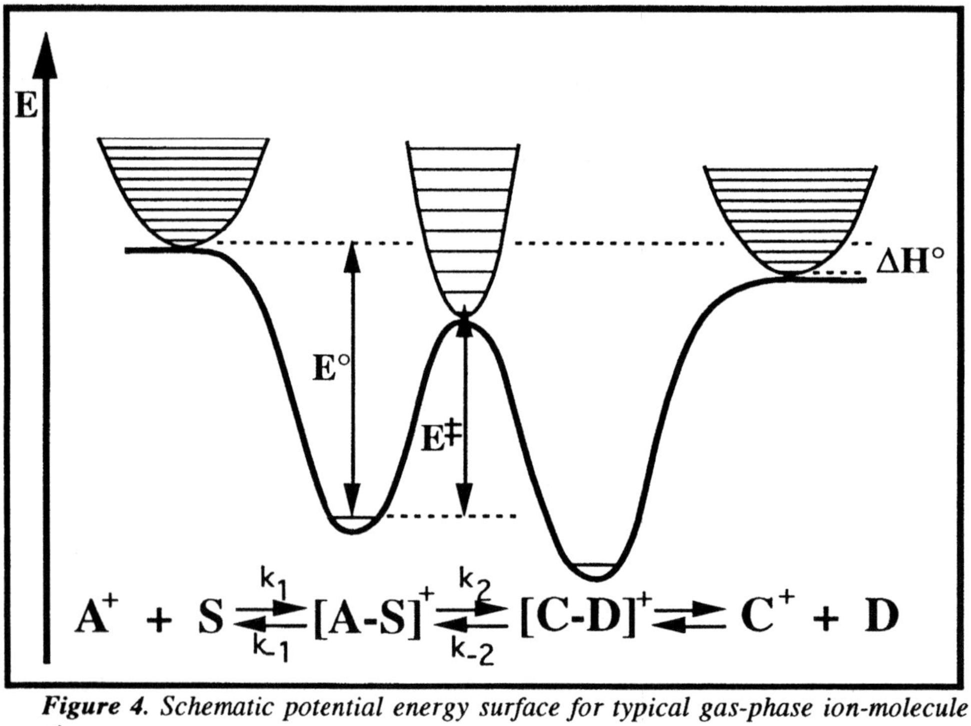

Figure 4. *Schematic potential energy surface for typical gas-phase ion-molecule reactions.*

Gas-Phase Ion-Molecule Reactions in Radiolytic Experiments.

Radiation-chemistry investigation on gas-phase ion-molecule reactions integrates and completes the kinetic and mechanistic information obtained by mass spectrometric methodologies. The greatest contribution of radiation chemistry to gas-phase ion-molecule reactions research concerns: i- the study of their kinetics and mechanisms under conditions not limited to the rather narrow density range covered by conventional mass spectrometry ($<10^{16}$ molecule cm^{-3}); ii- the fast interception and neutralization of the relevant ionic intermediates (after ca.10^{-9}-10^{-7} s), which strongly reduce their isomerization and fragmentation generally observed in mass spectrometry within the long time interval between the ion generation and detection (ca.10^{-5} s); iii- the characterization of the ion-molecule intermediates through the structural and stereochemical analysis of their neutral products, which normally represents an inherent limitation of purely mass spectrometric methods.

In both gas-phase radiolysis and mass spectrometry, ionization of a neutral molecule A (eqs.15 and 16) is a vertical process frequently leading to ions A^+ in several electronic states (i.e. A^{*+}), with an excess of internal energy often characterized by a non-Boltzmann distribution. In media containing ≤0.1 mol% of the substrate of interest (S), ions A^+ can readily dissipate their excess energy either by radiative emission or by unreactive collisions with the gaseous medium before interacting with S. Interaction of thermalized ions A^+ with molecule S eventually leads to products C^+ and D (eq.22), possibly via the intermediacy of one or more persistent complexes (e.g. $[A{\cdot}S]^+$) (Farneth and Brauman, 1976; Olmstead and Brauman, 1977; Dodd and Brauman, 1986). A schematic potential energy profile for eq. 22 is reported in Figure 4.

Under collision-free conditions, approached in the low-pressure mass spectrometric sources, the energy E° gained in the interaction between thermalized A^+ and S remains stored in the internal degrees of freedom of the ion-molecule complex $[A{\cdot}S]^+$ and is available to overcome the energy barrier $E^{\ddagger}$ to the products. Thus, in case where $E°>E^{\ddagger}$, unless the excess internal energy is removed *before* the reactive event by thermalizing collisions with the gaseous medium, most exothermic ion-molecule reactions occur at collision rate, with *no* ($E°>>E^{\ddagger}$) or *negative* temperature coefficients ($E°\geq E^{\ddagger}$). A negative temperature dependence of the ion-molecule reaction efficiency arises from the large difference in the density of states at the transition structures for the back-dissociation of $[A{\cdot}S]^+$ to reactants (high density of states reflecting the "loose" transition structure involving free rotations of the reactants) and the conversion of $[A{\cdot}S]^+$ to products (low density of states reflecting the "tight" transition structure involving in the unimolecular isomerization of the primary complex). Even in those cases where $E°<E^{\ddagger}$, the kinetically effective barrier corresponds to the $E^{\ddagger}$-E° difference, rather than to the intrinsic energy barrier $E^{\ddagger}$. This situation has been defined by Olmstead and Braumann (1979) as "chemically featureless", since their rates reveal little about their mechanistic details.

Resort to high-density media ($>10^{19}$ molecule cm^{-3}; pressure≥1 atm) is the most direct mean to carefully estimate the intrinsic energy barrier $E^{\ddagger}$ of a gas-phase ion-molecule reaction and, thus, to confer a mechanistic significance to the relevant *thermal* kinetics. Evaluation of $E^{\ddagger}$ requires thermal equilibration of the primary complex $[A{\cdot}S]^+$ prior to its conversion to products. This depends upon many factors, including the relative activation free-energies for its back-dissociation to reactants and its conversion to products, as well as the nature and the temperature of the gaseous medium and its collision frequency with the ionic complex. A precise evaluation of the pressure/temperature conditions ensuring thermal equilibration of a given encounter complex could be obtained only for few representative ion-molecule reactions, such as (Speranza, 1992):

$$NH_4^+ + NH_3 \underset{k_{-1}}{\overset{k_1}{\rightleftarrows}} [NH_4\text{-}NH_3]^+_{exc} \xrightarrow{k[M]} [NH_4\text{-}NH_3]^+_{ground} \qquad (60)$$

$$sPr^+ + iBuH \underset{k_{-1}}{\overset{k_1}{\rightleftarrows}} [sPr\text{-}iBuH]^+_{exc} \xrightarrow{k[M]} [sPr\text{-}iBuH]^+_{ground} \qquad (61)$$

$$[sPr\text{-}iBuH]^+_{exc} \xrightarrow{k_2} PrH + tBu^+$$

whose individual kinetic parameters have been determined in independent studies (Meot-Ner and Field, 1975, 1976; Olmstead, Lev-On, Golden, and Brauman, 1977; Sunner, Hirao, and Kebarle, 1989).

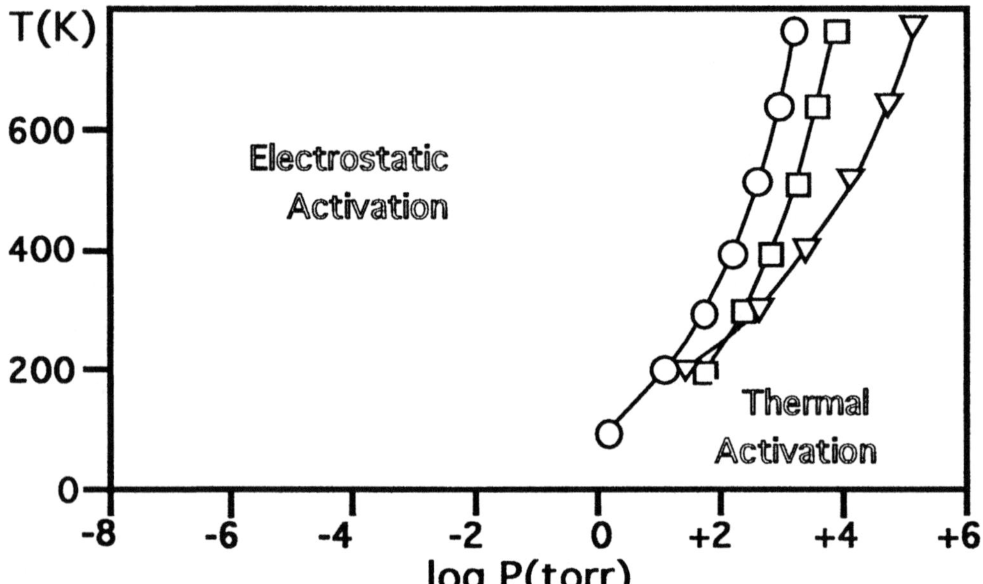

Figure 5. *Isokinetic curves representing equal values for the collision rate constant of the excited adduct* $[A\text{-}S]^+$ *and that of its conversion to products: (*$[A\text{-}S]^+$*, products): (circles)* $[NH_4\text{-}NH_3]^+$, $NH_4^+ + NH_3$*; (triangles)* $[sC_3H_7\text{-}iC_4H_{10}]^+$, $sC_3H_7^+ + iC_4H_{10}$*; (squares)* $[sC_3H_7\text{-}iC_4H_{10}]^+$, $tC_4H_9^+ + C_3H_8$.

The isokinetic curves of Fig.5 represent the limiting pressure and temperature of the gaseous medium beyond which the relevant ion-molecule complexes $[A.S]^+$ undergo thermal equilibration prior to their conversion to products. Thus, if occurring under pressure/temperature conditions falling on the left of the curves, the ion-molecule reactions involve excited complexes and, therefore, are said to be *electrostatically activated*. If instead occurring under pressure/temperature conditions falling on the right of the curves, the same ion-molecule processes involve thermally equilibrated complexes and, hence, are said to be *thermally activated.* At sufficiently high pressures and low temperatures, ions may establish clustering equilibria with the present neutral molecules (A), regulated by the corresponding equilibrium constants ($K_{n,n+1}$):

$$A^+(A)_n + A \rightarrow A^+(A)_{n+1} \qquad (62)$$

Thus, for instance, ca. 85% of CH_5^+ generated in pure methane at 1 atm and 298K is present in the monosolvated $CH_5^+{\cdot}CH_4$ form, the remainder being unsolvated (Hiraoka and Kebarle, 1975). Under the same conditions, the accompanying $C_2H_5^+$ ion is present predominantly (ca.87%) in the unsolvated state, with the monosolvated $C_2H_5^+{\cdot}CH_4$ form amounting to only 13% (Hiraoka, Mori, and Yamabe, 1993). No clustering occurs under the same conditions between $sC_3H_7^+$ and CH_4 (Hiraoka, Mori, and Yamabe, 1993). In 1 atm of C_3H_8 and at 298K, ca. 100% clustering of $sC_3H_7^+$ with one molecule of C_3H_8 takes place (Sunner, Hirao, and Kebarle, 1989). In 1 atm of iC_4H_{10} and at 298K, nearly one half of the $tC_4H_9^+$ ions remains unsolvated, while the other half adds to an iC_4H_{10} molecule forming a loosely-bonded adduct (Sunner, Hirao, and Kebarle, 1989).

In the interaction between the ionic cluster $A^+(A)_n$ and the neutral substrate S, departure of all or some of the coordinated molecules A may take place (eq.63), providing an additional means of thermal equilibration of the ion-molecule complex $[A{\cdot}S]^+$.

$$A^+(A)_n + S \rightarrow [A.S]^+ + nA \qquad (63)$$

In fact, the $[A{\cdot}S]^+$ complex is generated with a comparatively low excess energy, amounting to the difference between E° and the combined bond energy of the A molecules evaporating from the reacting ion cluster. In this case, it follows that the thermal activation limit of Fig.5 can be somewhat shifted to the left, thus increasing the pressure/temperature domain ensuring thermal activation.

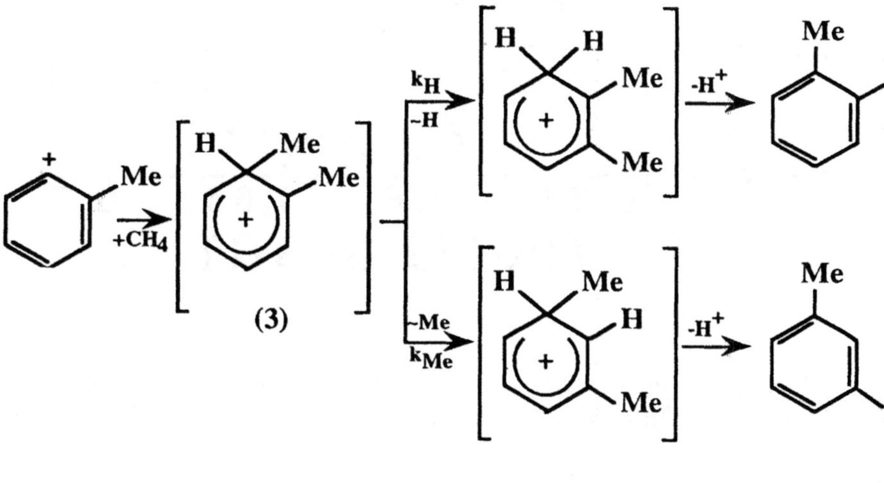

(64a)

(64b)

For these reasons, direct evidence of attainment of thermal activation in an ion-molecule reactions must be obtained in an operational sense, by analyzing the temperature dependence of the reaction rate over a relatively wide pressure range. As mentioned above, the efficiency of electrostatically activated processes is determined by entropic, rather than by energetic factors. Their reaction rates are essentially temperature-independent or follow a negative temperature dependence. The same processes, carried out under thermal-activation conditions, is expected to follow a positive temperature dependence, whose trend must be unaffected above the corresponding system-pressure limit. Beyond this limit, the reaction is thermally activated, and its activation parameters can be safely inferred from the relevant Arrhenius plot.

These above aspects have been throughly examined in a recent attempt to evaluate the relative activation parameters for the gas-phase intramolecular isomerization network 64, concerning the *ipso*-protonated *ortho*-xylenium ion **3** (Speranza and Ursini, 1996). Insertion into a C-H bond of methane by the *ortho*-tolylium ion, generated by fluoride-ion transfer from *ortho*-fluorotoluene to radiolytic SiF_3^+ ions, provides an convenient route to desired ion **3**.

The k_{Me}/k_H ratio was measured from the [*meta*-xylene]/[*ortho*-xylene] product ratio after a given reaction time, determined by the addition of an appropriate concentration of base (c-$C_6H_{10}O$; 5 torr). Figure 6 reports the experimental results.

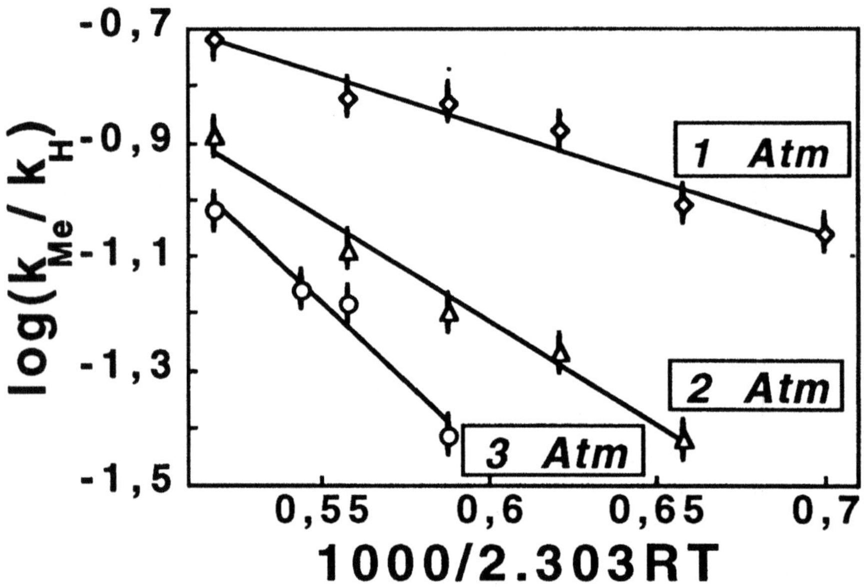

Figure 6. Pressure dependence of the phenomenological Arrhenius plot relative to the competing methyl vs. hydrogen shifts in ipso-protonated ortho-xylenium ion (eq. 64).

The steady increase of the slope of the relevant Arrhenius plot with the pressure demonstrates that complete thermal equilibration of the *ipso*-protonated *ortho*-xylenium intermediate is not attained before its isomerization even at pressures above 3 atm. This is due to the substantial energy excess acquired by the *ipso*-protonated *ortho*-xylenium ion, owing to the large exothermicity of its formation process ($-\Delta H°$=ca.55 kcal mol^{-1}), which makes the time required for its isomerization by intramolecular 1,2-H shifts fully comparable to that for its first quenching collision with a SiF_4 molecule (ca.$2x10^{-11}$ s at 3 atm).

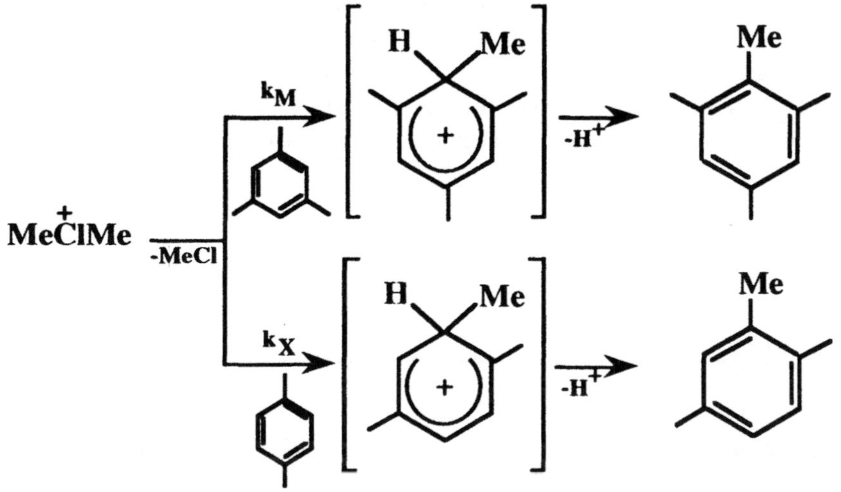

(65a)

(65b)

The smooth transition from electrostatic to thermal activation is illustrated by the pressure effect on the intermolecular selectivity of radiolytic $(CH_3)_2Cl^+$ ions toward mesitylene (M) and *para*-xylene (X) (network 65) (Speranza and Laguzzi, 1988).

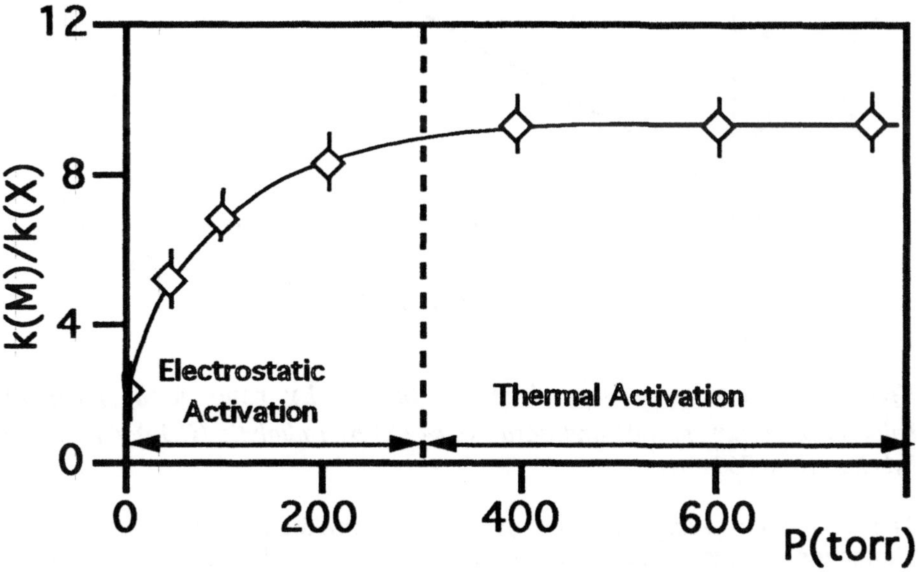

Figure 7. *Pressure dependence of the k_M/k_X ratio in the gas-phase methylation of mesitylene (M) and para-xylene (X) by radiolytic $(CH_3)_2Cl^+$ ions in methyl chloride at 100°C.*

The significant increase of the k_M/k_X ratio measured at 100°C as a function of the CH_3Cl pressure (1-760 torr) reflects the transition between the two activation regimes, with thermal activation at pressures above 300 torr (Fig.7).

Beyond this limit, the slope of the Arrhenius plot of Fig.8 provides an estimate of the activation energy difference of the competing methylation reactions ($E^{\ddagger}_X$-$E^{\ddagger}_M$), which amounts to a value of 2.2 kcal mol^{-1}, and a preexponential factor ratio (A_X/A_M) of ca.2. These values largely exceed those measured at 1 torr ($E^{\ddagger}_X$-$E^{\ddagger}_M$=0.6 kcal mol^{-1}; A_X/A_M=ca.0.8) (Stone, Lin, and Varah, 1981; Stone, Splinter, and Splinter, 1984), namely under conditions where thermal activation is not safely achieved. It is therefore concluded that any correlation between the kinetic parameters of the $(CH_3)_2Cl^+$-methylation of the selected arenes and the phenomenological Arrhenius plot obtained at 1 torr is illegitimate.

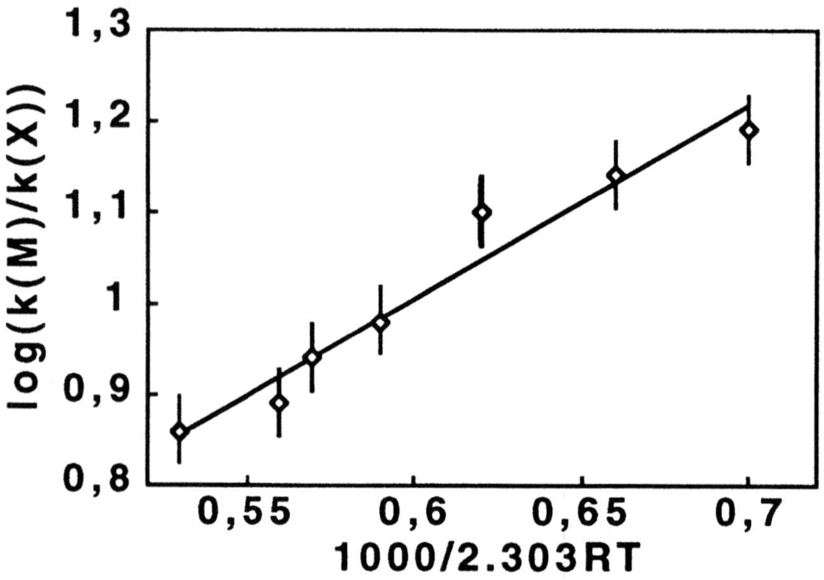

Figure 8. Temperature dependence of the k_M/k_X ratio in the gas-phase methylation of mesitylene (M) and para-xylene (X) by radiolytic $(CH_3)_2Cl^+$ ions in methyl chloride at 760 torr.

The very short sampling time of ionic intermediates and their structural discrimination emerge as most valuable features of the radiolytic technique in the detection and characterization of very elusive ionic species, such as free cyclohexyl cation (Attinà, Cacace, and Di Marzio, 1989). In fact, all attempts to generate this ion both in nonnucleophilic solution and in the source of a mass spectrometer invariably failed, the 1-methylcyclopentyl cation being the only species observed.

The formation of the neutral cyclohexyl derivatives from a gaseous cyclohexyl cation, generated at 1-2 atm pressure in the presence of a sufficient concentration of a nucleophile (MeOH; 2-21 torr) (network 66), demonstrates the existence of the free ion with a lifetime exceeding 10^{-7} s, namely the interval between its radiolytic formation and its trapping by the nucleophile. The high tendency of gaseous cyclohexyl cation to isomerize to 1-methylcyclopentyl cation is testified by the formation of significant yields of methyl 1-methylcyclopentyl ether.

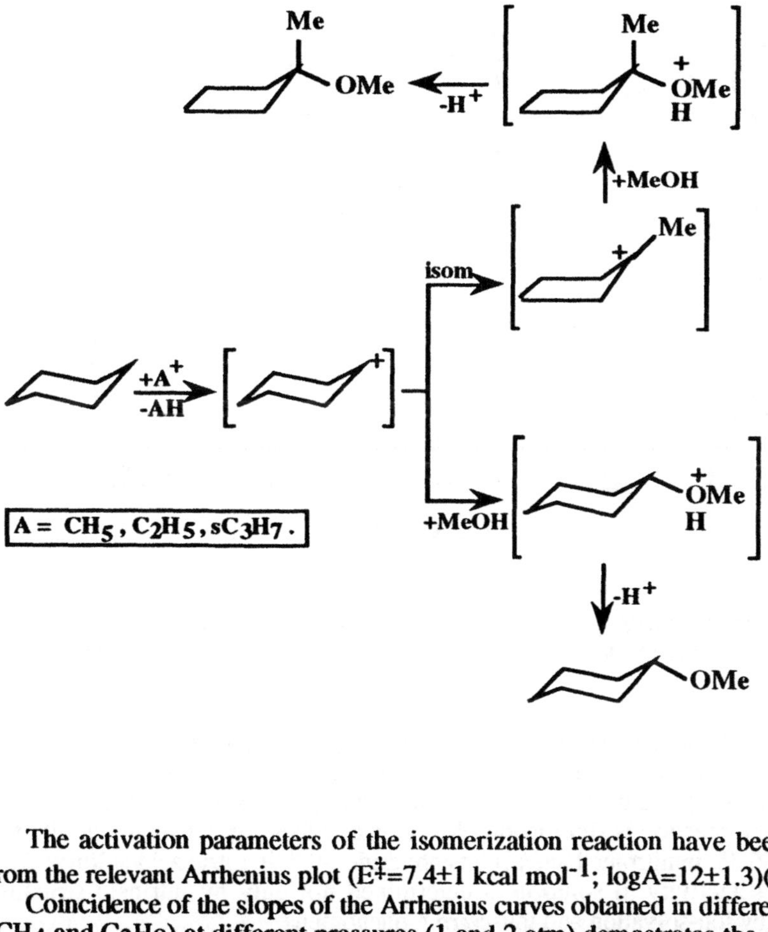

(66a)

(66b)

The activation parameters of the isomerization reaction have been inferred from the relevant Arrhenius plot ($E^{\ddagger}$=7.4±1 kcal mol^{-1}; logA=12±1.3)(Fig.9).

Coincidence of the slopes of the Arrhenius curves obtained in different gaseous media (CH_4 and C_3H_8) at different pressures (1 and 2 atm) demostrates the complete thermal equilibration of the cyclohexyl cation before its unimolecular isomerization to 1-methylcyclopentyl cation.

Evaluation of kinetics and mechanisms of gas-phase ionic processes under thermal activation conditions may allow direct comparison with the same processes carried out in the condensed phase, providing information on solvation and ion-pairing effects upon reactivity. In some instances, gas-phase investigations can shed light on reactions which do not occur in solution at all, such as, for instance, the ionic silylation of arenes.

An enlightening example is provided by the gas-phase aromatic nitration data. $CH_2ONO_2^+$- and $EtO(NO_2)_2^+$-nitration of arenes under low-pressure mass spectrometric conditions, beside accompanied by extensive charge exchange and oxygen-atom transfer processes, display a paradoxical selectivity, reacting at higher rates with deactivated substrates (Dunbar, Shen, and Olah, 1972; Ausloos and Lias, 1978).

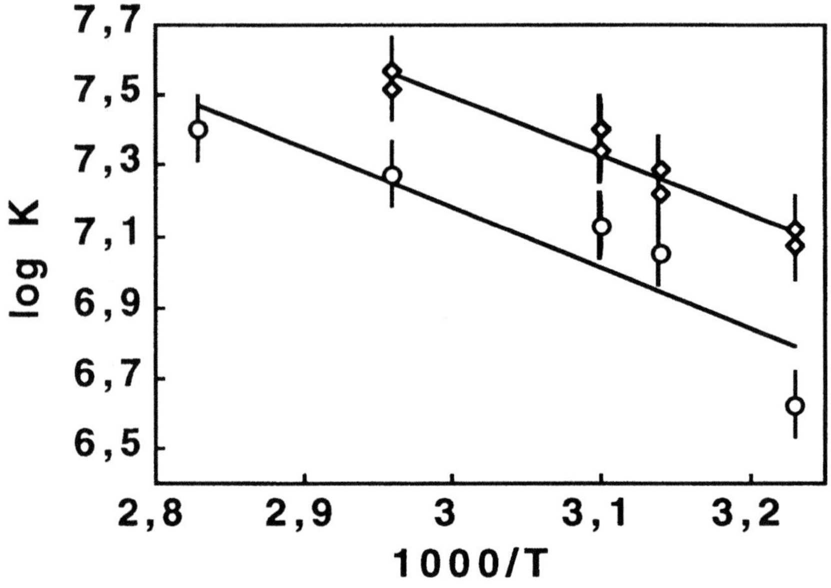

Figure 9. *Arrhenius plots for the cyclohexyl→1-methylcyclopentyl cation rearrangement in propane at 1480 torr (circles) and in methane at 750 torr (diamonds).*

Atmospheric-pressure arene nitration by radiogenic protonated methyl nitrate follows instead the classical reactivity model of solution studies (Attinà, Cacace, and Yanez, 1987) (Fig. 10). The ρ value of the Hammett curve (ρ = -3.87) for the gas-phase thermally activated reaction is significantly less negative than for conventional nitration in solution (ρ = -6.35, in nitromethane/acetic anhydride; -9.7, in nitric acid/sulfuric acid), suggesting that selectivity in solution is determined not only by intrinsic structural factors, but also by ion-solvation and ion-pairing phenomena.

The wealth of quantitative information accessible to the radiolytic methodology, especially if used in strict conjunction with the complementary mass spectrometric background, allows definition of basic aspects of ion-molecule reactions in the gas phase. In particular, a comparison can be made between kinetics and mechanisms of the ionic process under different activation regimes (electrostatic activation at low pressures; thermal activation at high pressures). A unique feature of the radiolytic experiments, i.e. the possibility of investigating the kinetics and the mechanisms of thermally activated ionic processes within extended pressure and temperature ranges, allows meaningful comparison with the kinetics and the mechanisms of the corresponding reactions in solution and their sensitivity to solvation and ion-pairing factors. Another distinctive feature is identified in the high structural resolution of the radiolytic method, which has been repeatedly pointed out in the past years (Cacace, 1979; Cacace, 1982; Speranza, 1983; Speranza; 1987; Cacace, 1988). The following sections report a survey of the most recent radiolytic studies on the chemo-, regio-, diastereo-, and enantioselectivity of ionic processes in the gas phase.

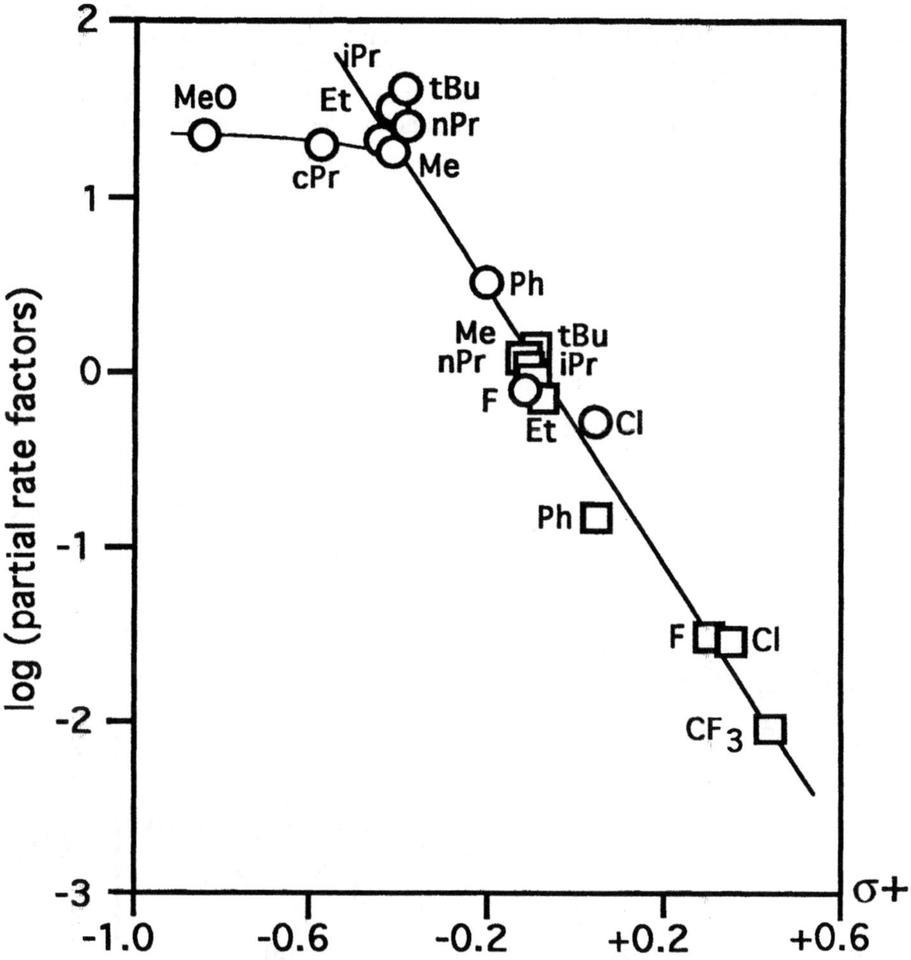

Figure 10. Logarithmic plot of the high-pressure gas-phase nitration partial rate factors vs. the σ+ constants of the arene substituents (para (circles), meta (squares)).

Ion Chemoselectivity.

Complex organic molecules normally have more than one reactive site or functional group. The ability of an ion to discriminate among the reactive sites is referred as *chemoselectivity*. The most recent case of ion chemoselectivity investigated by radiolysis concerns discrimination between functional groups belonging to two different classes, i.e. a OH and a C=C, towards the exceedingly strong and generally believed unselective $C_nH_5^+$(n=1,2) acids. Two different substrates, i.e. but-1-en-3-ol (racemic mixture) and *trans*-but-2-en-1-ol, were used to probe the chemoselectivity properties of these ions (network 67)(Dezi, Lombardozzi, Pizzabiocca, Renzi, and Speranza, 1995).

The criterion followed to discriminate between channels a) and b) of network 67, and thus to evaluate the chemoselectivity of $C_nH_5^+$, is based on the fact that the oxonium ion **4** readily undergoes nucleophilic displacement by an appropriate nucleophile (e.g. MeOH) to yield eventually the corresponding ethers, whereas the carbocation **5** would isomerize to the most stable O-protonated carbonyl structure **6** via fast 1,2-H shift (Bowen and Williams, 1978). The exclusive formation of isomeric 1-methoxybut-2-enes and 3-methoxybut-1-ene, and the complete absence of carbonylic compounds among the radiolytic products, demonstrate the pronounced chemoselectivity of the strong $C_nH_5^+$(n=1,2) acids for the n-type center of the selected unsaturated alcohols.

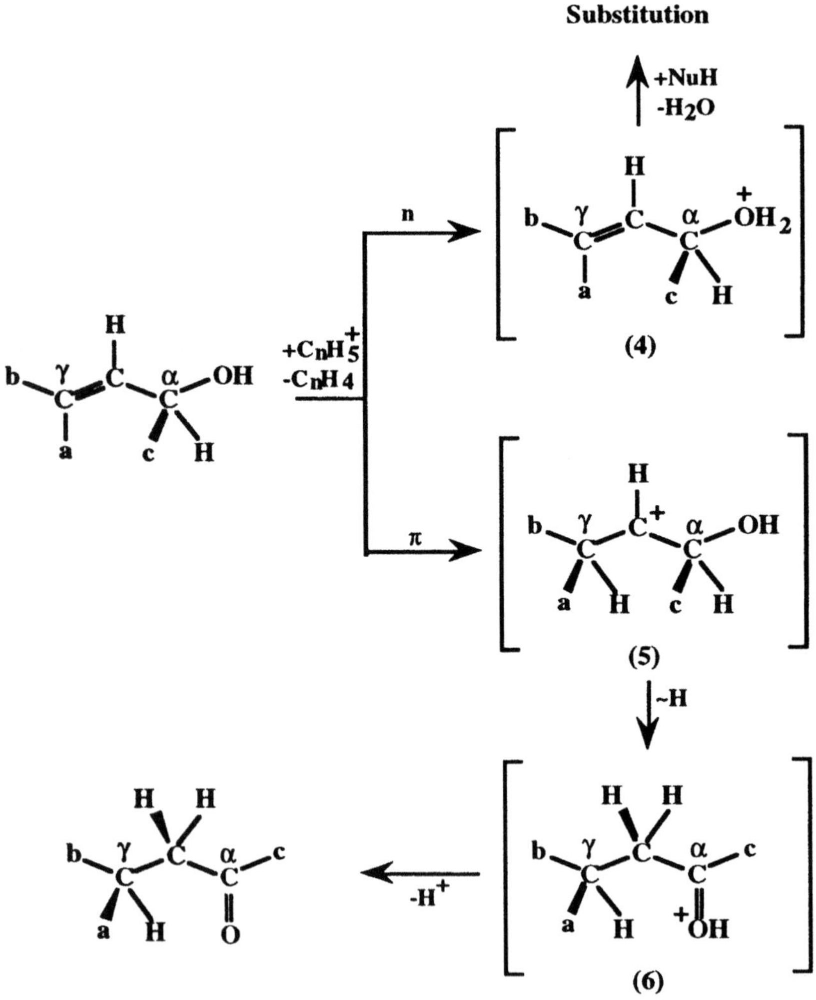

(67a)

(67b)

Ion Regioselectivity.

At a given reactive site of an organic molecule, there may be several orientations by which a reagent may approach the reactive site; this defines the *regioselectivity* of the reactant. Nucleophilic attack of MeOH on O-protonated but-1-en-3-ol and but-2-en-1-ol

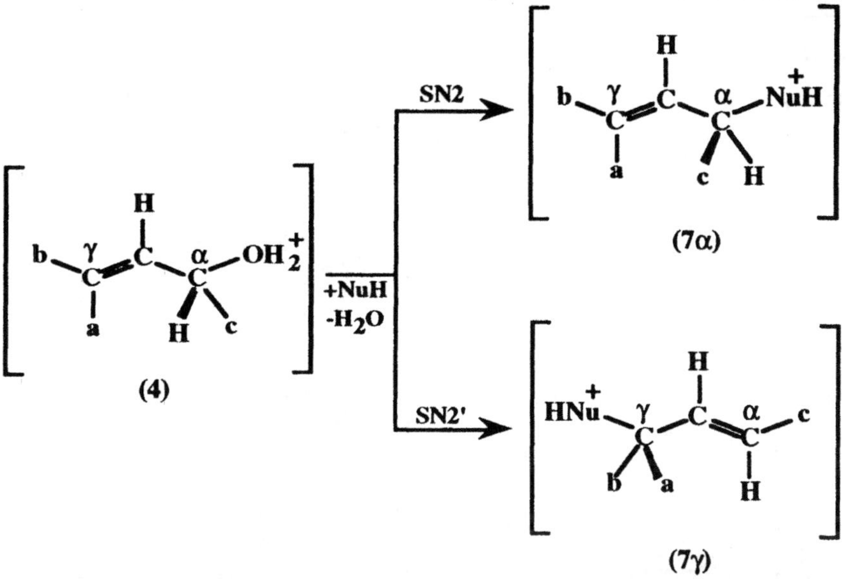

(68a)

(68b)

presents itself as one of the most recent example of regioselectivity in a gas-phase ion-molecule reaction (Dezi, Lombardozzi, Pizzabiocca, Renzi, and Speranza, 1995; Renzi, Lombardozzi, Dezi, Pizzabiocca, and Speranza, 1995). Bimolecular nucleophilic attack on allylic compounds may in fact occur at the α center (a S_N2 process)(eq.68a) or at the γ one, with concomitant migration of the double bond (a S_N2' process)(eq.68b).

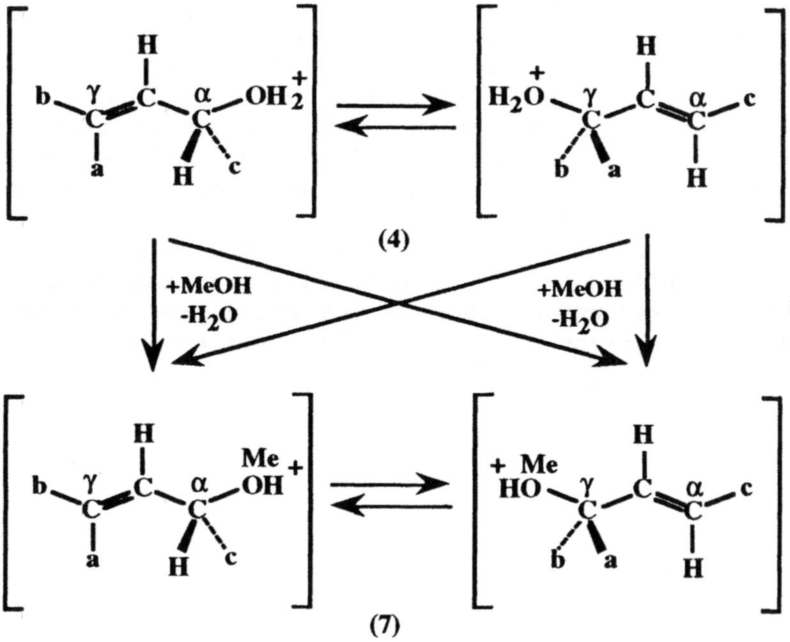

(69)

In solution, the search of authentic examples of the S_N2' mechanism is intense but lacking of unequivocal issues. The main reason is the occurrence of a variety of competing substitution mechanisms (see Chart I), which make the unambiguous assignment of the S_N2' one extremely rare.

Chart I

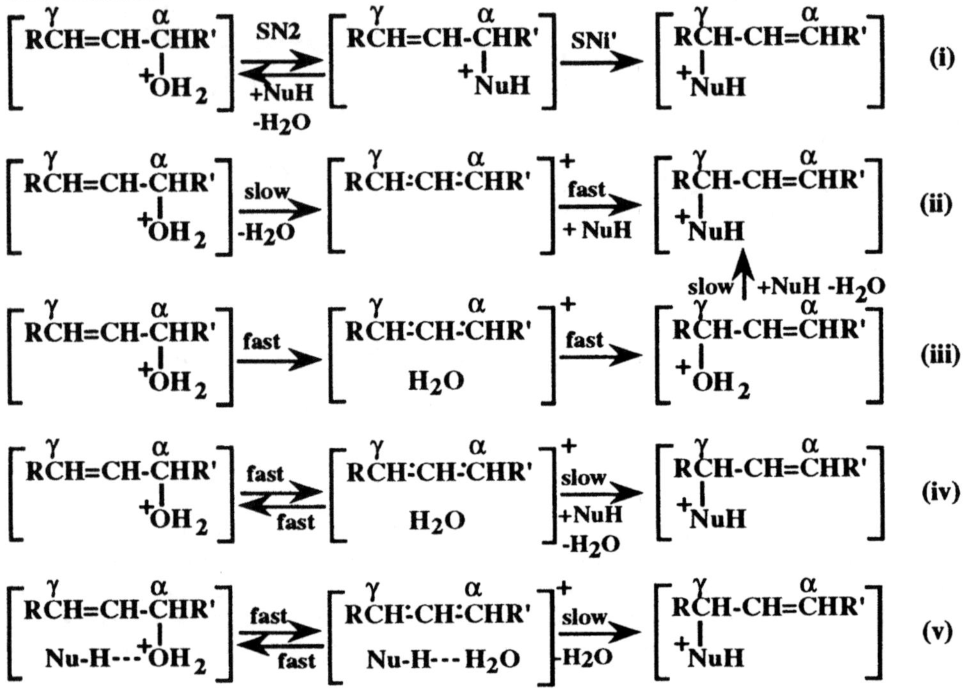

Accordingly, the isolation of the substitution product of eq.68a is *per se* insufficient for assigning to its formation mechanism the S_N2' label, since this demands that: i-the reaction follows second-order kinetics; ii- the formation of the substitution product of eq.68a does not proceed through rearrangement of both the starting substrate before substitution (path iii of Chart I) and the initially formed S_N2 product (path i of Chart I); iii- the process does not involve the intermediacy of a free allylic cation (path ii, iv, and v of Chart I).

In the search of a genuine example of S_N2' reactions without the complicating interference of solvation and ion-pairing, the acid-induced nucleophilic attack of MeOH (ca.0.6 torr) on but-1-en-3-ol (racemic mixture) and isomeric but-2-en-1-ols has been investigated at 100-760 torr pressure. The experimental results are consistent with the reaction pattern 69.

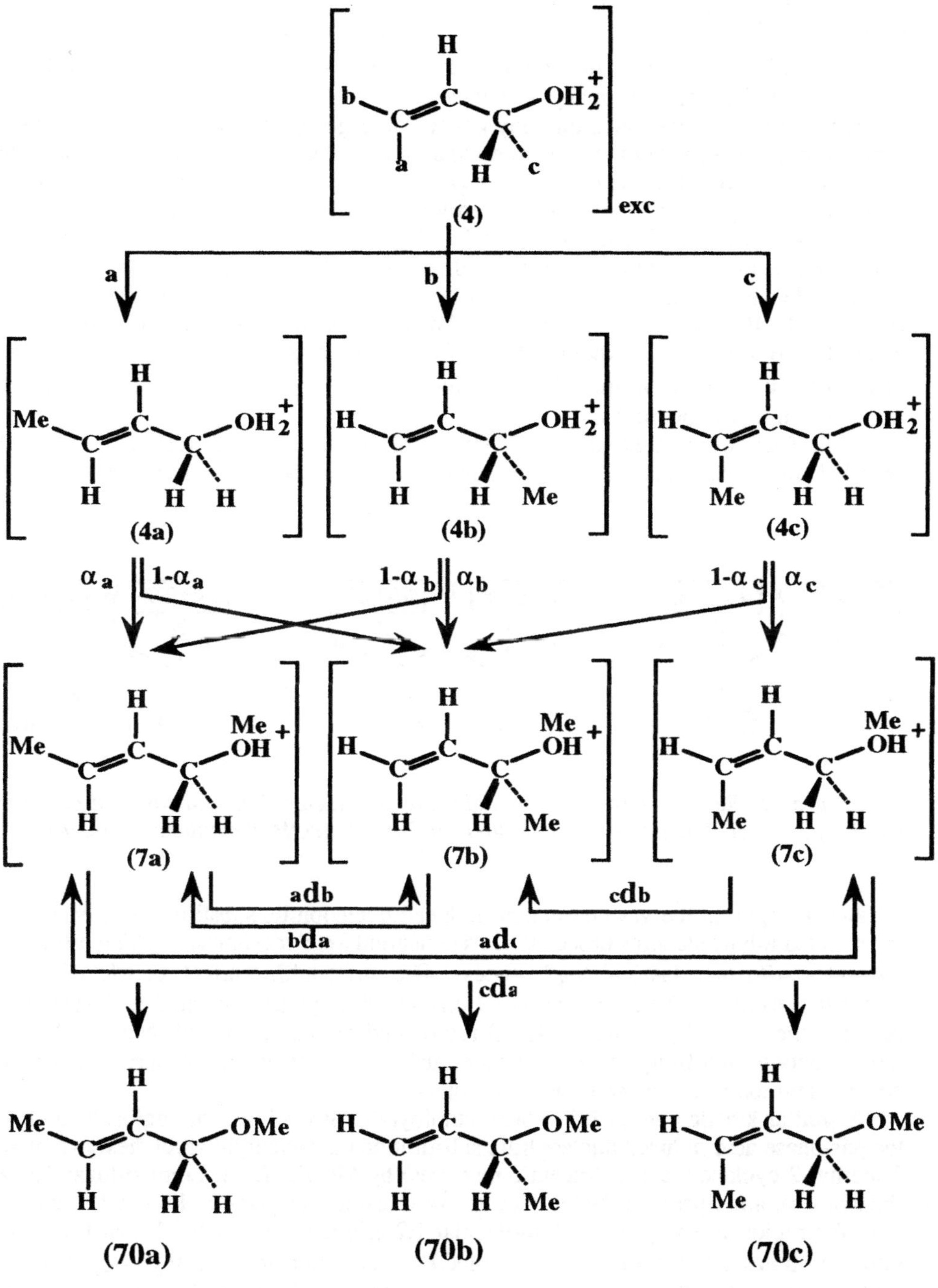

The conceivable intermediacy of free 1-methylallyl cation in the substitution reaction is safely ruled out by the observation that the ion, obtained by $C_nH_5^+$(n=1,2)-protonation of 1,3-butadiene, adds to $H_2{}^{18}O$ (^{18}O-content>97%)(ca.2 torr) yielding exclusively large quantities (yield>83%) of ^{18}O-but-1-en-3-ol (^{18}O-content>70%), whereas no incorporation of the ^{18}O-label was observed by $C_nH_5^+$(n=1,2)-protonation of the selected allylic alcohols in the presence of the same amount of $H_2{}^{18}O$. This observation simultaneously satisfies points i and iii of the above S_N2' requirements, while excluding occurrence of bimolecular $H_2{}^{18}O$-to-$H_2{}^{16}O$ displacement in the O-protonated allylic substrates to any detectable extent.

This latter point indicates that any conceivable gas-phase isomerization of the O-protonated allylic substrates is to be referred as a unimolecular process, whose extent (the a-c terms in the kinetic scheme 70) under given experimental conditions was determined by the absolute yield and the relative distribution of the isomeric allylic alcohols (point ii). As an example, in CH_4 at 760 torr and 37.5°C, O-protonated *cis*-but-2-en-1-ol gives rise to 8.3% of *trans*-but-2-en-1-ol and 3.7% of but-1-en-3-ol. Concerning the conceivable isomerization of O-protonated 1-methoxybut-2-enes and 3-methoxybut-1-ene from the MeOH attack on the O-protonated allylic substrates (point ii), a measure of its extent (the d terms in the kinetic scheme 70) directly arises from the methoxybutenes distribution obtained from the $(CH_3)_2F^+$-methylation of each allylic alcohol under the same experimental conditions.

In the framework of the kinetic pattern 70, all the experimental results are mutually satisfied only if the values of the α unknowns are those reported in Figure 11.

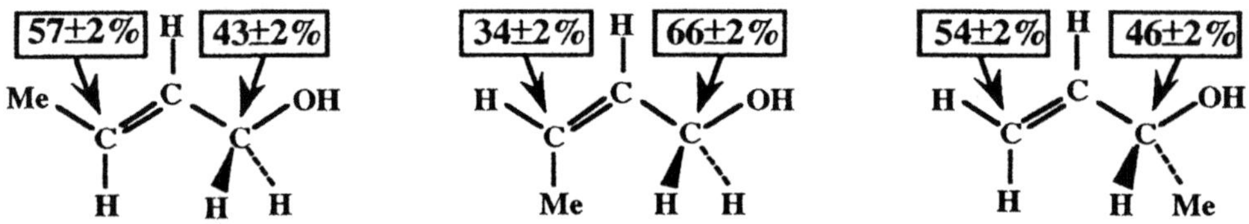

Figure 11. *Regioselectivity of the acid-induced nucleophilic substitution on trans-but-2-en-1-ol, cis-but-2-en-1-ol, and but-1-en-3-ol by MeOH in methane at 760 torr.*

Accordingly, in the gas phase, acid-induced nucleophilic substitution by MeOH on the selected allylic alcohols proceeds via a concerted S_N2' mechanism, which efficiently competes with the classical S_N2 process. The nucleophilic attack of MeOH on O-protonated *trans*-but-2-en-1-ol and but-1-en-3-ol displays almost equal regioselectivity between their α and γ center. With O-protonated *cis*-but-2-en-1-ol, MeOH shows a more pronounced affinity for the α center, attributed to the enhanced steric strain in the *cis* oxonium ion due to its *endo*-methyl moiety.

A similar kinetic approach has been employed to investigate the regioselectivity of the gas-phase acid-induced nucleophilic substitution on 3-methyl-2-cyclohexen-1-ol and 1-methyl-2-cyclohexen-1-ol (racemic mixtures) by MeOH (Dezi, Lombardozzi, Renzi, Pizzabiocca, and Speranza, 1995). The relevant results are given in Figure 12, together with the regioselectivity of the bimolecular E2 elimination induced by NMe_3 on the same substrates. Nucleophilic attack of MeOH on O-protonated 3-methyl-2-cyclohexen-1-ol proceeds almost exclusively at the α center (a S_N2 process), while with on O-protonated 1-methyl-2-cyclohexen-1-ol the reaction takes place predominantly at the γ center (a S_N2' process). Instead, the regiochemistry of the NMe_3-induced elimination process conforms to a more favored attack on the H_δ atom of both oxonium ions (a E2' process) and appears only slightly affected by the nature of both the nucleophile and the leaving group.

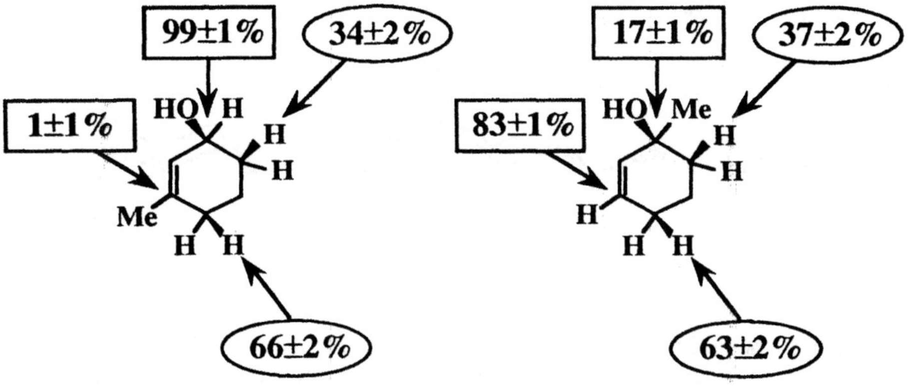

Figure 12. Regioselectivity of acid-induced nucleophilic substitution (squares) and elimination (ellipses) reactions on 3-methyl-2-cyclohexen-1-ol and 1-methyl-2-cyclohexen-1-ol by MeOH and NMe_3, respectively, in methane at 760 torr.

Ion Stereoselectivity.

Stereochemical orientation of a reagent approaching the reactive site of an organic substrate comprises two types of selectivity. First, the control of relative stereochemistry is exerted between two or more chiral or prochiral centers and leads to different diasteroisomers. In this case, the stereochemical control is referred to as *diastereoselectivity*. In each of these diastereoisomers there are further stereoisomers that differ from each other as mirror images. *Enantioselectivity* refers to discrimination between such mirror-image isomers or enantiomers.

Ion Diastereoselectivity.

The radiolytic technique appears particularly suitable for the study of the stereochemical course of gas-phase ion-molecule reactions, where positive discrimination of structurally very close species, such as diastereoisomeric and enantiomeric intermediates and products, is essential. Indeed, the first direct evidence of ion diastereoselectivity in a gas-phase ion-molecule reactions refers to the predominant formation of the tritiated *meso*-1,2-dichloro-1,2-difluoroethane (77%), over the *d,l*-form (23%), when *meso*-1,2-dichloro-1,2-difluoroethane was submitted to the electrophilic attack by radiolytically generated $C_nH_4{}^3H^+$(n=1,2) (Cacace and Speranza, 1972). The study was then extended to the stereochemical course of the gas-phase nucleophilic displacement reactions shown in network 71.

$$NuH + RXH^{+} \longrightarrow HNuR^{+} + HX \qquad (71)$$

RX =

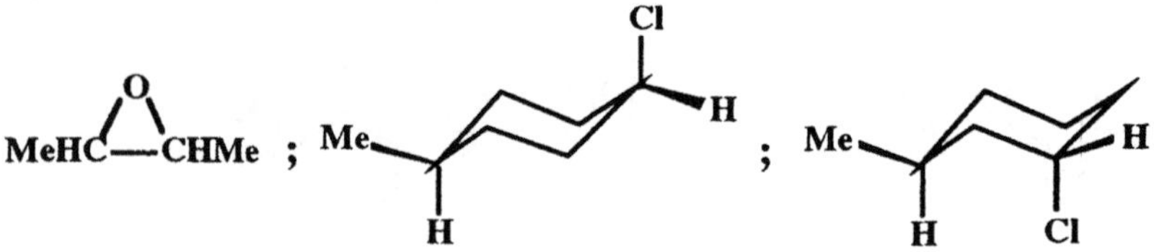

Analysis of the neutral end products demonstrates occurrence of predominant *backside* attack of the nucleophile with inversion of the reaction center (64-98%)(Angelini and Speranza, 1978, 1981a, 1981b; Speranza and Angelini, 1980). On bifunctional substrates (Angelini and Speranza, 1981a, 1981b; Fornarini, Sparapani, and Speranza, 1988a, 1988b), e.g. halohydrins, the acid-induced nucleophilic displacement leads to mostly inverted epoxybutanes (80-93%), accompanied by predominantly *retained* 2,3-butanediols (73-98%). A fast HO-3 process accounts for the formation of *inverted* O-protonated epoxybutanes (and eventually their neutral derivatives), which may further undergo *backside* displacement by the nucleophile (e.g. H_2O) yielding eventually the *retained* glycols (network 72).

In the area of ring-opening of O-protonated epoxides by nucleophiles, a puzzling problem concerns the regio- and diastereoselectivity of the reaction on 1,2-epoxysilane, in which substitution occurs exclusively at the epoxy carbon bearing the silyl group. In order to establish the origin of this behavior, whether reflecting intrinsic electronic properties of the silylated substrate or some influence of the reaction medium as well, the reaction has been investigated in the several gaseous media at 100-760 torr. The oxonium ion **8** (eq.73) was generated from the attack of radiolytically-formed ionic electrophiles ($E^{+}=D_3^{+}$, $C_nH_5^{+}$(n=1,2), and $(CH_3)_2F^{+}$) on 1,2-epoxy-1-trimethylsilylcyclohexane (racemic mixture) in the presence of the nucleophile NuH (e.g. MeOH)(Crotti, Macchia, Pizzabiocca, Renzi, and Speranza, 1986).

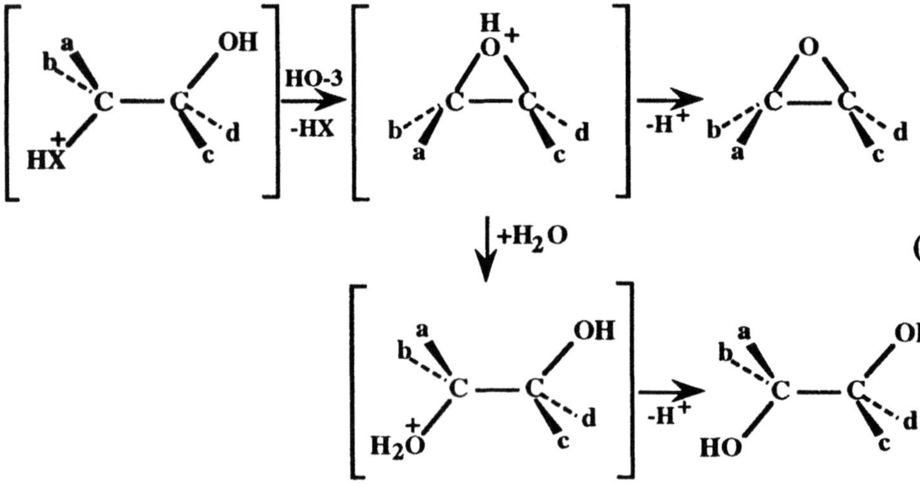

(72)

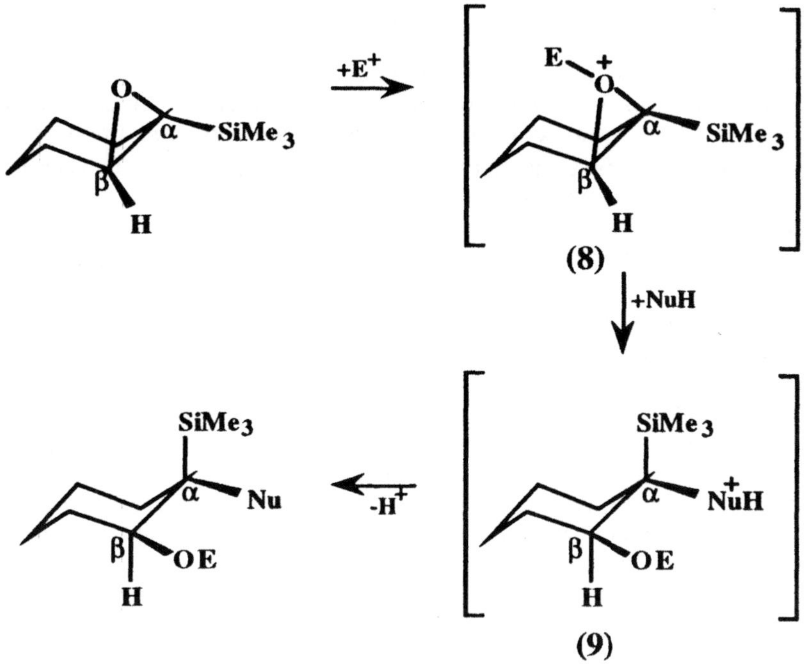

(73)

The oxonium ion **8** offers at least three different reaction sites to nucleophilic attack, namely the Si atom in $SiMe_3$, the C_α and the C_β centers, yielding the corresponding substitution products. Furthermore, the diastereoselectivity of the nucleophilic attack by MeOH at the bridgehead C atoms can be directly determined by analysis of the reaction products. The exclusive formation of the *trans* product **9** is consistent with a nucleophilic attack occurring *exclusively* at the C_α atom and involving its *complete* inversion of configuration. The regio- and diastereoselectivity of the gas-phase reaction fully conform to those observed in solution, whose stereochemical course is therefore determined exclusively by the intrinsic electronic properties of the $SiMe_3$ substituent. An reason for this peculiar effect of the $SiMe_3$ substituent can be found in terms of the frontier orbital theory, if not only the LUMO coefficients, but also the spatial expanse of the LUMO wave function is considered (Tomoda and Ohno, 1991). Owing to the large exterior electron density near the silicon center, the first interaction between MeOH and **8** involves overlapping between donor's HOMO and acceptor's LUMO wave functions outside molecular surface. This favors coordination between MeOH and the Si atom before nucleophilic displacement, which therefore takes place exclusively at the adjacent C_α center with inversion of configuration.

Replacement of the $SiMe_3$ substituent with the CMe_3 group changes dramatically the gas-phase nucleophilic substitution pattern in the corresponding O-protonated epoxide (Crotti, Macchia, Pizzabiocca, Renzi, and Speranza, 1987a). Gas-phase protonation of 1,2-epoxy-1-tertbutylcyclohexane (racemic mixture) in the presence of variable concentrations of MeOH produces comparable amounts of 2-tertbutylcyclohexanone, 1-tertbutylcyclopentancarboxaldehyde, as well as a *trans*:*cis*>10:1 mixture of 2-methyl-2-isopropenyl-cyclohexanol. No appreciable formation of the expected substitution products was observed even at the highest MeOH concentrations (3 torr). This points to an O-protonated epoxide unimolecularly isomerizing well before its first interaction with the MeOH (ca.10^{-7} s). Concerning the isomerization mechanism, two possible pathways can be envisaged, the first involving the C_α-O bond fission with formation of the corresponding cyclohexyl ion (eq.74), the second involving neighboring Me group participation in the C_α-O bond cleavage (eq.75).

The latter mechanistic hypothesis is substantiated by the high diastereoselectivity characterizing the formation of 2-methyl-2-isopropenyl-cyclohexanols. The large predominance of the *trans* isomer over the *cis* one provides strong evidence in favor of *backside* participation of a methyl of the CMe_3 group to the unimolecular C_α-O bond fission of the oxonium ion, with prevailing *inversion* of configuration at the C_α reaction center. The predominance of *trans*-2-methyl-2-isopropenyl-cyclohexanol is at variance with a mechanism involving a planar cyclohexyl ion intermediate, which would eventually produce comparable amounts of *trans*- and *cis*-2-methyl-2-isopropenyl-cyclohexanols.

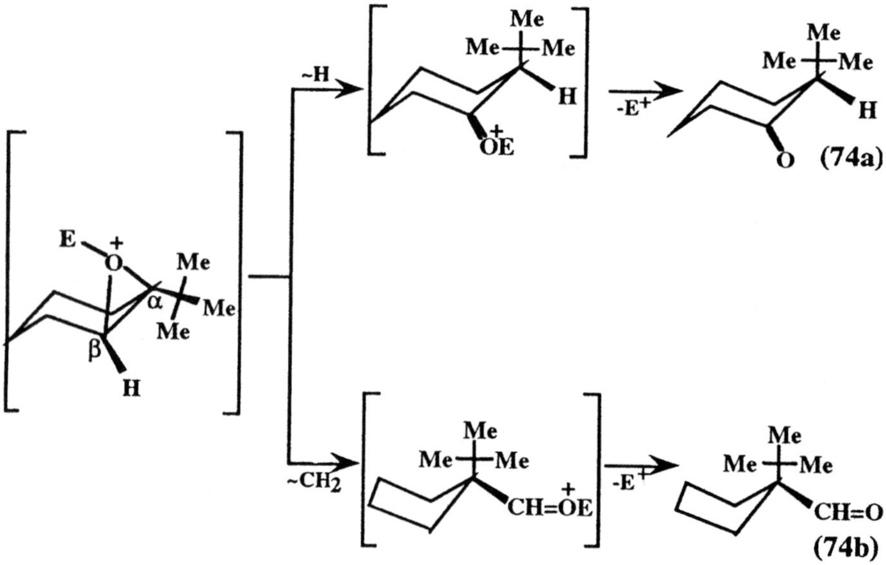

(74a)

(74b)

Unimolecular isomerization 75 is thermochemically prevented in the O-protonated 1,2-epoxy-1-methylcyclohexane (racemic mixture), which therefore undergoes exclusive nucleophilic attack at its C_α center with complete *inversion* of configuration.

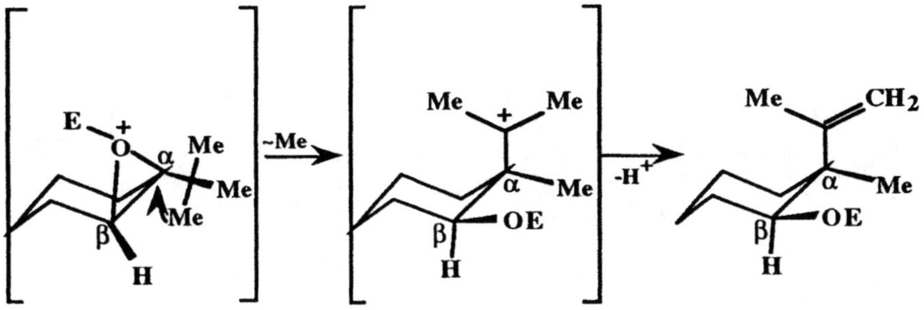

(75)

When extended to 1,2-epoxy-1-phenylcyclohexane (racemic mixture), the acid-induced nucleophilic displacement by MeOH leads to a substitution pattern characterized by the predominant formation of the *cis* product **10** (>91%)(eq.76), which is consistent with a nucleophilic attack occurring *exclusively* at the C_α atom, *but in this case involving predominant retention of configuration* (Crotti, Macchia, Pizzabiocca, Renzi, and Speranza, 1987b).

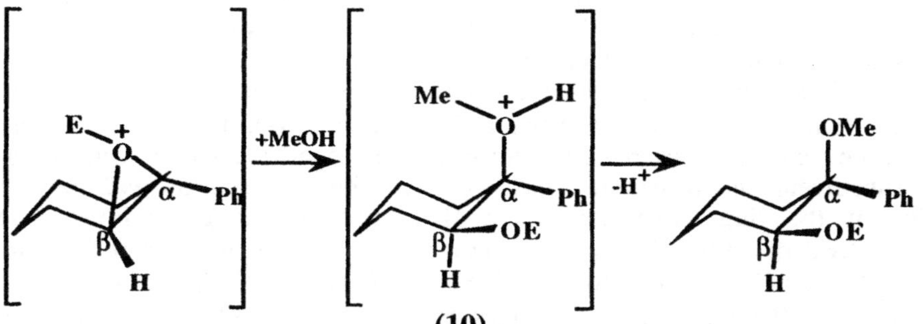

(76)

(10)

In analogy with the conclusions reached in the above systems, the pronounced C_α regioselectivity observed in the MeOH attack on O-protonated 1,2-epoxy-1-phenylcyclohexane reflects a certain extent of positive charge development at the C_α center in the substitution transition state, favored by the stabilizing effects of the phenyl group. The energy released in the proton-bound encounter complex **11** (eq.77) between the oxonium ion and MeOH may enhance charge localization at the C_α center so as to allow its isomerization to the less strained structure **12**, yielding eventually the *retained* substitution intermediate.

Isomerization 77 must takes place well before intervention of a second MeOH molecule (ca.10^{-7}s), which would *backside* attack the proton-bound encounter complex **11** producing the *inverted* substitution intermediate. This occurs with O-protonated 1,2-epoxy-1-methylcyclohexane, in which an isomerization process alike 77 is slowed down owing to the minor positive-charge stabilizing effects of the α-CH_3 group.

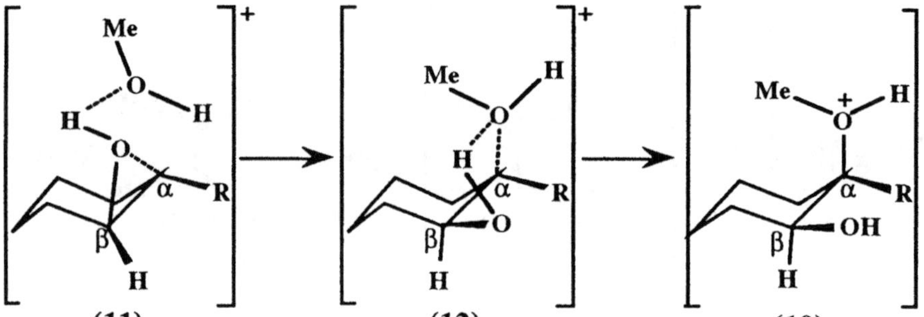

(11) (12) (10) (77)

As a consequence, O-protonated 1,2-epoxy-1-methylcyclohexane is long-lived enough ($>>10^{-7}$s) to undergo attack by another MeOH molecule at its unshielded backside.

The gas-phase results fully conform with solution data concerning the same nucleophilic substitution carried out in apolar aprotic solvents (e.g.CH_2Cl_2), where 90% retention of configuration is observed . In polar protic solvents (e.g.MeOH), backside attack efficiently competes with mechanism 77, leading to lower *syn*-diastereoselectivities (*trans* **10**/*cis* **10**=ca. 1) (Battistini, Crotti, Damiani, and Macchia, 1979).

Investigation on the gas-phase acid-induced nucleophilic substitution by MeOH on 1-aryl cyclohexene oxides (racemic mixtures), containing electron-withdrawing groups (*meta*-Cl or *para*-NO_2) in the aromatic ring, allowed to substantiate experimentally the hypothesis that the regio- and diastereoselectivity of the nucleophilic attack is regulated by the charge localization at the benzylic C_α carbon in the transition state (Cecchi, Pizzabiocca, Renzi, Chini, Crotti, Macchia, and Speranza, 1989). Indeed, a significant variation of the regio- and diastereoselectivity of the nucleophilic attack as a function of the ring substituent was observed. Thus, by increasing the electron withdrawing properties of the ring substituent from H to *meta*-Cl to *para*-NO_2, the diastereoselectivity of the MeOH attack changes dramatically from predominant *frontside syn* (>91%(H); 70-87%(*meta*-Cl)) to exclusive *backside anti* (100%(*para*-NO_2)). The regioselectivity of the MeOH attack changes as well, the displacement taking place exclusively at the C_α carbon when no or the *meta*-Cl substituent is present in the aromatic ring, while, with *para*-NO_2, 20-32% of the attack occurs at the C_β center as well.

In the framework of mechanism 77, the presence of an electron-withdrawing group on the aromatic ring inhibits in the transition state **11** the development of sufficient positive charge at the C_α center so as to allow C_α-O bond rupture and *frontside syn* attack by the MeOH moiety. Hence, the isomerization 77 rate is depressed to the point that it becomes comparable (*meta*-Cl) or much smaller (*para*-NO_2) that the collision frequency of the proton-bound adduct **11** with another MeOH molecule at its unshielded backside (ca.10^7 s^{-1}). With the strong electron-withdrawing *para*-NO_2 substituent, destabilization of the positive charge at the C_α center is so pronounced that the C_β center effectively competes for substitution.

The mechanistic model 77 finds interesting analogies with analogous solution processes, except for the diastereoisomeric 9,10-oxides of *trans*-1,2,3,4,4a,10a-hexahydrophenanthrene **13a** and **13b** (eq.78). In fact, in solution, the totally different diastereoselectivity of the hydrolysis of epoxides **13a** (100% *trans* diol) and **13b** (mixture of *cis* and *trans* diol) was tentatively rationalized in terms of a completely developed benzylic carbenium ions **14a** and **14a**, which are preferentially attacked by the nucleophile in a pseudoaxial fashion (eq.78)(Sayer, Yagi, Silverton, Friedman, Whalen, and Jerina, 1982).

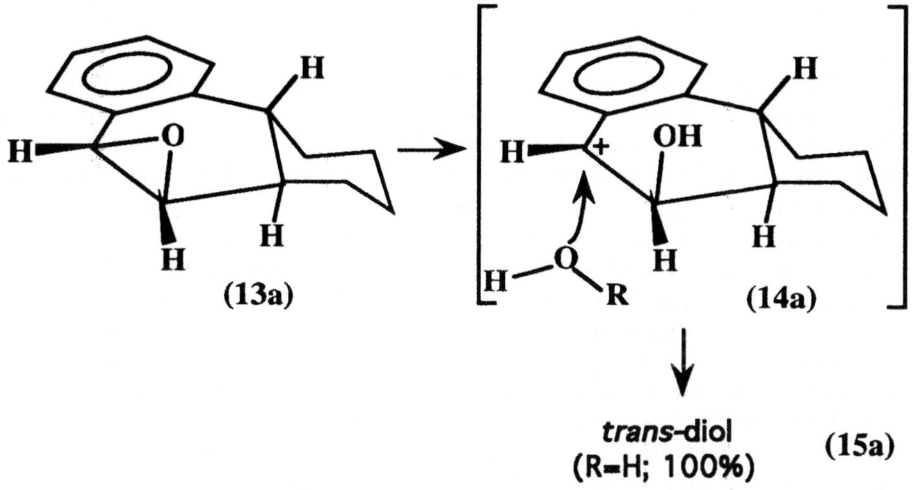

(78a)

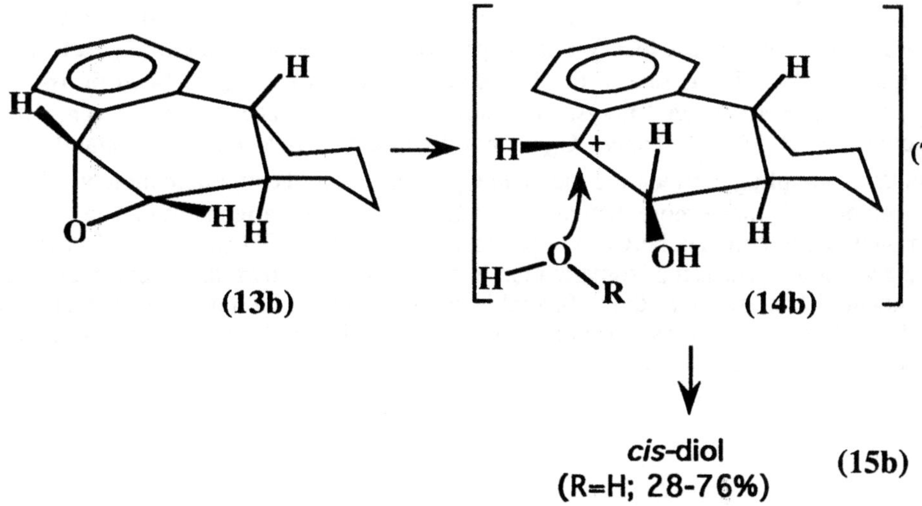

(78b)

In an attempt to unify the above data in a single reactivity model, acid-induced nucleophilic attack on **13a** and **13b** (racemic mixtures) by MeOH was investigated in the gas phase (100-760 torr), under conditions where the complicating effects of solvation and ion-pairing are excluded (Cecchi, Chini, Crotti, Pizzabiocca, Renzi, and Speranza, 1991). The *cis*-hydroxyether **15b** (R=Me) is predominantly formed from **13b** (*syn* attack), in yields increasing at high pressures and in the presence of a base (NMe_3)(73-81%). The opposite trend is observed with **13a**, where the *trans*-hydroxyether **15a** (R=Me) is prevailing (78-82%) (*anti* attack). At low pressures (100 torr) and in the absence of added bases, almost equal amounts of the *syn* and *anti* products are formed. Complete regiospecificity is observed for both substrates, pointing to the development of significant positive charge at the benzylic C_α carbon. Competition experiments demonstrated that isomer **13b** is almost twice as reactive as **13a** in 760 torr of CH_4 and in the presence of NMe_3 (5 torr).

The results obtained at 100 torr and in the absence of bases, namely under long-lived *excited* ion conditions, are consistent with a reaction pattern involving the unimolecular C_α-O bond fission with formation of the carbenium ions **14a** and **14b** of eq.78. In the gas phase, ions **14a** and **14b** may undergo equally efficient attack by MeOH on both sides of the quasi-planar C_α^+ center, as demonstrated by the almost equimolar amounts of the corresponding *syn* and *anti* adducts. At high pressure (760 torr) and in the presence of NMe_3, unimolecular C_α-O bond fission is efficiently prevented by collisional stabilization of the excited oxonium ions **16a** and **16b**, the first undergoing preferentially a *backside* displacement (57-82%) and the latter a *frontside* MeOH attack (58-81%). Comparison of the gas-phase diastereoselectivity of the displacement processes with that measured for the same reactions in various condensed media (100% *anti* adduct from **15a**; 28-76% *syn* adduct from **15b**) calls into question the soundness of the mechanistic model 78. In fact, if a preferential pseudoaxial attack of MeOH on ions **14a** and **14b** was operating, it is difficult to explain why this pathway becomes more important for **14b** (increased amount of the *syn* adduct) and contemporary less important for **14a** (decreased amount of the *anti* adduct) on passing from condensed- to gas-phase operating conditions. This apparent anomaly is cleared up by using a reaction model analogous to eq.77 and shown in network 79.

The more pronounced *syn* diastereoselectivity of **16b** relative to that of **16a** is attributed to a more intense location of the positive charge at the benzylic C_α carbon of **16b** than of **16a**. This is due to the fact that the pseudoaxial C_α-O bond of **16b** is nearly parallel to the phenyl π-system, whereas this alignment is absent in **16a**. As a consequence, in **16b**, extensive C_α-O bond rupture is favored by the stabilizing conjugative coupling between the incipient C_α^+ center and the aromatic moiety, whilst the same event is made more difficult by unfavorable orbital alignment in **16a**. Thus, the proton-bound complex between O-protonated **13b** and MeOH can readily isomerize to the *syn* adduct. The same isomerization is greatly slowed down in the complex with O-protonated **13a**, owing to the unfavorable conformation of the phenyl π-system, and therefore it is preceded by the attack of an external MeOH molecule proceeding predominantly at the unshielded backside, yielding the *trans* adduct.

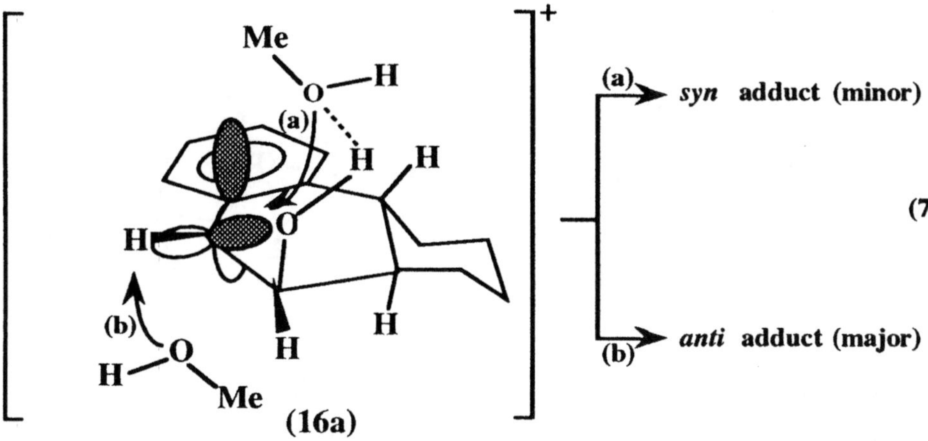

(79a)

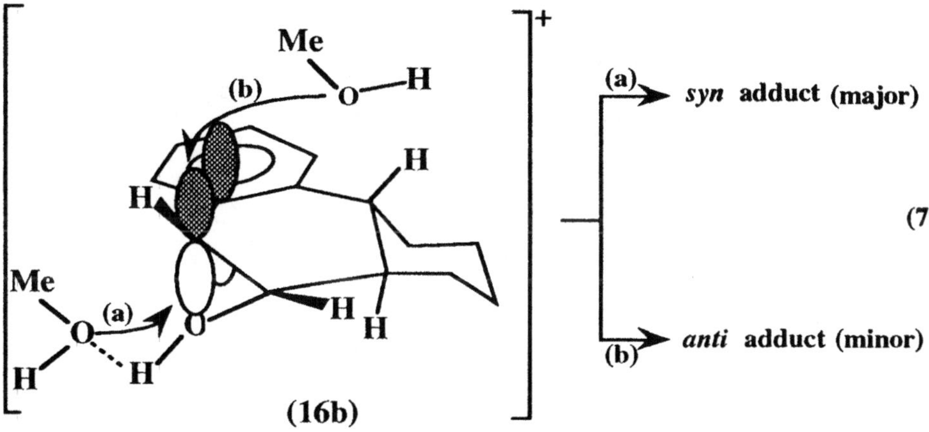

(79b)

Analysis of the ring substituent effects on the stereochemical outcome of the oxirane ring opening in diastereoisomeric intermediates **17a-18a** and **17b-18b** (Chart II) fully conforms to the mechanistic hypothesis depicted in eqs.79a,b (Chini, Crotti, Minutolo, Dezi, Lombardozzi, Pizzabiocca, and Renzi, 1993). The *syn/anti* diastereoselectivity observed in the gas phase is strictly dependent on the substituent present on the aromatic ring, the larger amounts of *syn* adduct being obtained with the electron-releasing MeO group and the lower ones with the electron-withdrawing Br substituent. The satisfactory Hammett-type linear correlations, shown in Figure 13 together with those concerning hydrolysis and methanolysis of **13b-17b-18b** in the condensed phase, was found. The epoxy ring-opening reaction in epoxides **13b-17b-18b** in the gas phase is more *syn*-stereoselective than the corresponding reaction of the same epoxides in the condensed phase. This indicates that opening processes in the gas phase possess a higher carbocationic character than when the same reactions are carried out in the condensed phase. In the gas phase, diastereoisomers **13b-17b-18b** display a sensitivity to the electronic effects of the ring substituent which is significantly higher than that found for diastereoisomers **13a-17a-18a** under the same conditions. This responds to a higher extent of charge dispersal in the transition states involving **13b-17b-18b** due to the favorable orientation between their π system and the benzylic oxirane C_{α}-O bond, which is instead absent in their diastereoisomers **13a-17a-18a**.

Chart II

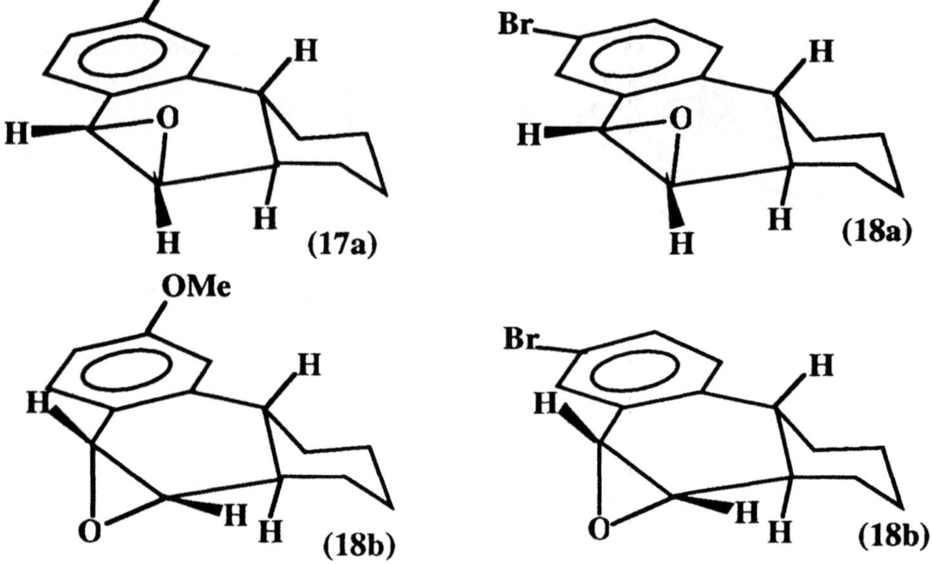

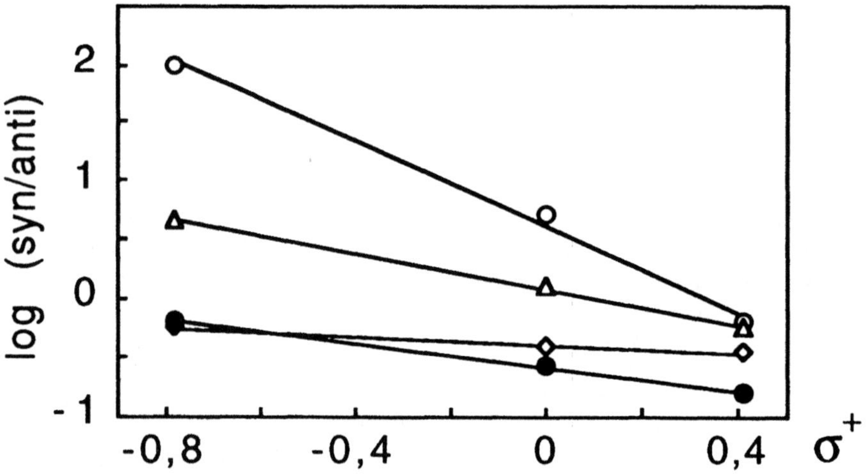

Figure 13. *Hammett-Brown $\rho\sigma^+$ plot for the acid-catalyzed ring opening of epoxides* ***13a-17a-18a*** *and* ***13b-17b-18b*** *with the indicated nucleophile:* ***13a-17a-18a*** *with MeOH, gas phase (full circles);* ***13b-17b-18b*** *with: MeOH, gas phase (open circles); MeOH, condensed phase (diamonds); H_2O, condensed phase (triangles).*

Ion Enantioselectivity.

The study of the enantioselectivity of the gas-phase acid-catalyzed nucleophilic displacement on chiral alcoholic compounds requires a detailed knowledge of the extent of conceivable isomerizazion and racemization of the chiral substrate by interaction with the gaseous acid catalyst prior to substitution. Direct information on these points has be achieved by using the S-enantiomer of *trans*-4-hexen-3-ol (**19S**) as the chiral starting compound (Troiani, Gasparrini, Grandinetti, and Speranza, 1996). When the chiral alcohol is submitted to $C_nH_5^+$(n=1,2)-protonation (CH_4; 760 torr), both the R-enantiomer of *trans*-4-hexen-3-ol (**19R**) as well as the racemic mixture of its isomer, i.e. the *trans*-3-hexen-2-ol (**20R** and **20S**), are formed in appreciable yields (eq. 80).

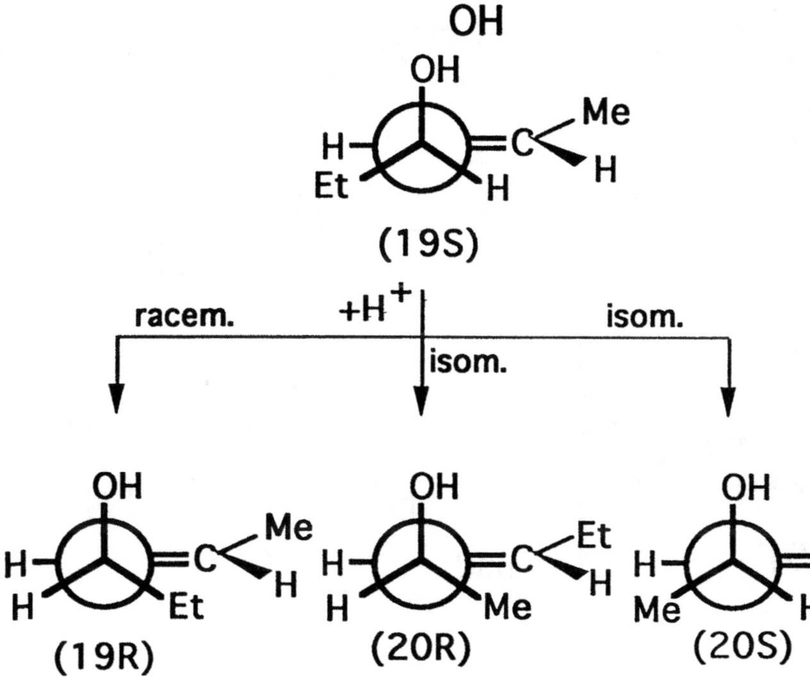

(80)

Formation of these products can involve either unimolecular isomerization and racemization of the O-protonated *trans*-4-hexen-3-ol (**21S**) or the intermediacy of a free hexenyl cation (**22**), arising from partial unimolecular fragmentation of the excited oxonium ion **21S** (eq. 81). The extent of the unimolecular process 81b can be estimated by considering that the free hexenyl cation, generated in the gas phase by $C_nH_5^+$(n=1,2)-protonation of *cis, trans*-2,4-hexadiene (CH_4; 760 torr) in the presence of 2 torr of $H_2{}^{18}O$ (^{18}O-content>70%), yields almost equal proportions of the racemic mixtures of *trans*-4-hexen-3-ol (**19R** and **19S**) and *trans*-3-hexen-2-ol (**20R** and **20S**), containing over 50% of the ^{18}O-label. Under the same experimental conditions, $C_nH_5^+$(n=1,2)-protonation (CH_4; 760 torr) of **19S** in the presence of 2 torr of $H_2{}^{18}O$ (^{18}O-content>70%) yields a different pattern of isomerized and racemized products, with a ^{18}O-content not exceeding 18%. When several torrs of a powerful base, such as

$(CH_3)_3PO_4$, is added to the mixture, the ^{18}O-label incorporation into the reaction products falls to vanishingly small levels (<2%). Under these latter conditions, the contribution of the free hexenyl cation **22** to the formation of the isomerization and racemization products is similarly negligible and, therefore, the formation of the R-enantiomer of 4-hexen-3-ol (**19R**) as well as of the R- and S-forms of the 3-hexen-2-ol isomer (**20R** and **20S**) must predominantly involve unimolecular rearrangement of the collisionally stabilized O-protonated starting alcohol (**21S**) (eq. 81a).

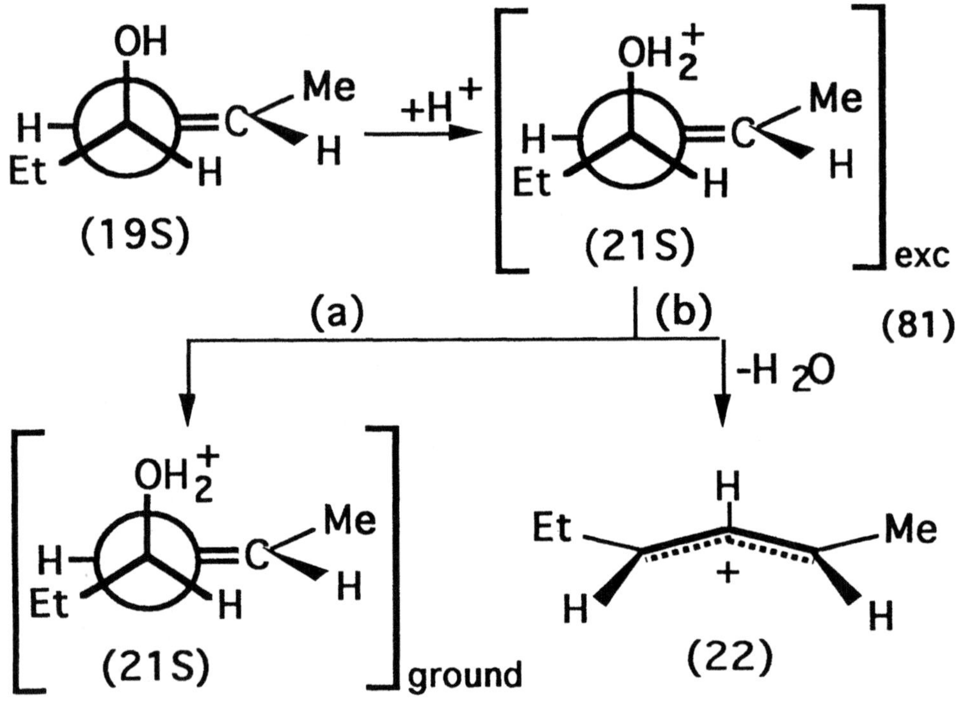

(81)

Ab initio calculations indicate that rearrangement of **21S** involves the rate-limiting formation of a common structure (**23**), with a H_2O molecule lying in the plane of the hexenyl cation skeleton and coordinated to an allylic hydrogen, which may evolve via independent routes either to the inverted structure **21R** or to the racemic mixture of the O-protonated **20R** and **20S** isomers. The rate constant of rearrangement of **21S** (k_{over}) can be inferred from the overall absolute yield of the products and the time interval between the formation of the primary oxonium ion and its capture by the base. Its value is reported as a function of the reaction temperature in Figure 14. Regression analysis of the data leads to the equation : $\log k_{over}=(12.3\pm0.8)-[(8.6\pm1.2)\times10^3)/2.303RT]$. Figure 15 shows the temperature dependence of the branching ratio of racemization vs. isomerization open to intermediate **23**. Regression analysis of the data leads to the preexponential factor ratio $A_{racem}/A_{isom}\approx200$ and to an activation barrier difference $E^*_{racem}-E^*_{isom}\approx3.6$ kcal mol^{-1}.

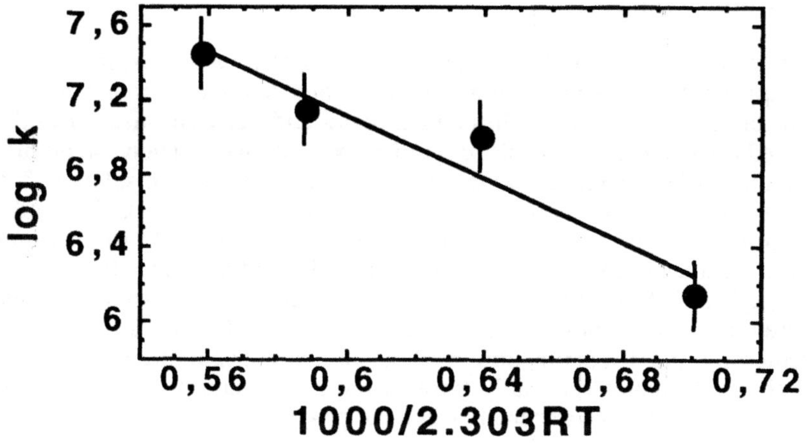

Figure 14. Arrhenius plot for the acid-induced intramolecular rearrangement of 19S.

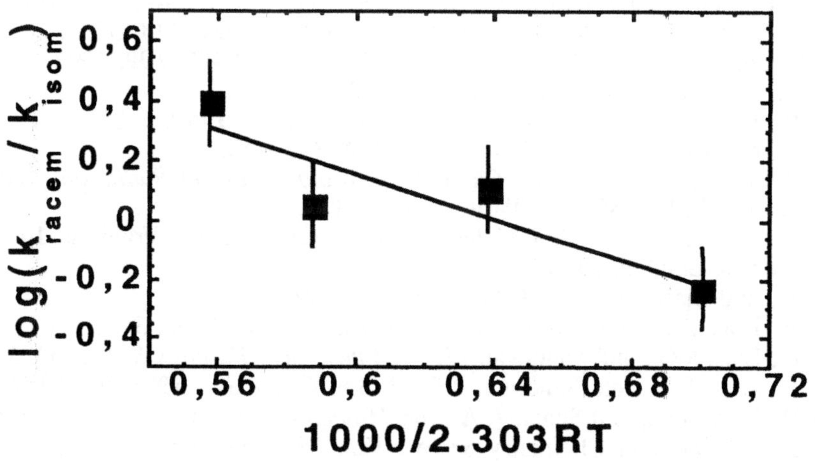

Figure 15. Temperature dependence of the rate constant ratio of gas-phase intramolecular racemization vs. isomerization of 19S.

Concluding Remarks.

An attempt has been made here to provide the basic foundations of the stationary radiolytic approach to gas-phase ion chemistry and a flavour of the most recent applications of the technique to studies on ion regio- and stereochemistry. Results have been reviewed for a number of gas-phase positive ion-molecule reaction and discussed by contrasting their behavior relative to that of analogous processes in solution. Many of the underlying measurements were instigated by a general interest in understanding ion kinetics and mechanisms in the isolated state and their sensitivity to environmental factors, which characterized in the last decades the activity of a group at the University of Rome "La Sapienza", headed by F. Cacace, and a group at the University of Camerino with G. Renzi, A. Pizzabiocca, and their colleagues. The radiolytic technique, complemented by mass spectrometric methods, are now beginning to be extended to problems of ion dynamics especially through the use od chiral compounds and will provide precious insights into extremely fast ionic processes.

References.

Angelini, G., and Speranza, M., *J. Chem. Soc. Chem. Commun.* **1978**, 213.
Angelini, G., and Speranza, M., *J. Am. Chem. Soc.* **1981a**, *103*, 3792.
Angelini, G., and Speranza, M., *J. Am. Chem. Soc.* **1981b**, *103*, 3800.
Attinà, M., Cacace, F., Ciranni, G., and Giacomello, P., *J. Chem. Soc. Chem. Commun.* **1976**, 466.
Attinà, M., Cacace, F., and Di Marzio, A., *J. Am. Chem. Soc.* **1989**, *111*, 6004.
Attinà, M., Cacace, F., and Ricci, A., *J. Am. Chem. Soc.* **1991**, *113*, 5937.
Attinà, M., Cacace, F., and Yanez, M., *J. Am. Chem. Soc.* **1987**, *109*, 5092.
Ausloos, P., *Ann. Rev. Phys. Chem.* **1966**, *17*, 205.
Ausloos, P., *Progr. React. Kinetics*, Vol. 5, Porter, G., Ed., **1969**, p. 113.
Ausloos, P., and Lias, S.G., *J. Chem. Phys.* **1962**, *36*, 3163.
Ausloos, P., and Lias, S.G., *Action Chimique et Biologiques des Radiations*, Vol. 11, Haissinsky, M., Ed., Masson, Paris, **1967**, p. 1.
Ausloos, P., and Lias, S.G., *Ion-Molecule Reactions*, Vol. II, Franklin, J.L., Ed., Butterworths, London, **1972**.
Ausloos, P., and Lias, S.G., *Int. J. Chem. Kinet.* **1978**, *10*, 657.
Ausloos, P., Lias, S.G., and Gorden, R., Jr., *J. Chem. Phys.* **1963**, *39*, 3341.
Ausloos, P., Lias, S.G., and Sandoval, I.B., *J. Chem. Soc. Faraday Soc.* **1963**, *36*, 66.
Ausloos, P., Lias, S.G., and Sandoval, I.B., *J. Am. Chem. Soc.* **1970**, *92*, 5037.
Ausloos, P., Lias, S.G., and Scala, A.A., *Ion-Molecule Reactions in Gases*, Ausloos, P., Ed., Advances in Chemistry Series, Vol. 58, American Chemical Society, Washington, DC, **1966**, p. 264.
Battistini, C., Crotti, P., Damiani, D., and Macchia, F., *J. Org. Chem.* **1979**, *44*, 1643.
Bowen, R.D., and Williams, D.H., *J. Am. Chem. Soc.* **1978**, *100*, 7454.
Cacace,F., *Kinetics of Ion-Molecule Reactions*, Ausloos, P., Ed., **1979**, p. 199.
Cacace,F., *Radiat. Phys. Chem.*, **1982**, *20*, 99.
Cacace,F., *Acc. Chem. Res.*, **1988**, *21*, 215.
Cacace, F., Crestoni, M.E., De Petris, G., Fornarini, S., and Grandinetti, F., *Can. J. Chem.* **1988**, *66*, 3099.
Cacace,F., and Giacomello, P., *J. Am. Chem. Soc.* **1973**, *95*, 5851.
Cacace,F., and Possagno, E., *J. Am. Chem. Soc.* **1973**, *95*, 3397.
Cacace,F., and Speranza, M., *J. Am. Chem. Soc.* **1972**, *94*, 4447.

Cacace,F., and Speranza, M., *J. Am. Chem. Soc.* **1976a**, *98*, 7229.

Cacace,F., and Speranza, M., *J. Am. Chem. Soc.* **1976b**, *98*, 7305.

Chowdhury, S., and Kebarle, P., *Structure/Reactivity and Thermochemistry of Ions*, Ausloos, P., and Lias, S.G., eds., Reidler Publish. Co., Dordrecht, **1986**, p. 321.

Cecchi, P., Chini, M., Crotti, P., Pizzabiocca, A., Renzi, G., and Speranza, M., *Tetrahedron* **1991**, *47*, 4683.

Cecchi, P., Pizzabiocca, A., Renzi, G., Chini, M., Crotti, P., Macchia, F., and Speranza, M., *Tetrahedron* **1989**, *45*, 4227.

Chini, M., Crotti, P., Minutolo, F., Dezi, E., Lombardozzi, A., Pizzabiocca, A., Renzi, G., *Tetrahedron* **1993**, *49*, 5845.

Crotti, P., Macchia, F., Pizzabiocca, A., Renzi, G., and Speranza, M., *J. Chem. Soc. Chem. Commun.* **1986**, 485.

Crotti, P., Macchia, F., Pizzabiocca, A., Renzi, G., and Speranza, M., *Gazz. Chem. Ital.* **1987a**, *117*, 739.

Crotti, P., Macchia, F., Pizzabiocca, A., Renzi, G., and Speranza, M., *Tetr. Letters* **1987b**, *28*, 3393.

Dezi, E., Lombardozzi, A., Pizzabiocca, A., Renzi, G., and Speranza, M., *J. Chem. Soc. Chem. Commun.* **1995**, 547.

Dezi, E., Lombardozzi, A., Renzi, G., Pizzabiocca, A.,and Speranza, M., *Chemistry Eur. J.* **1995**, in press.

Dodd, J.A., and Brauman, J.L., *J. Phys. Chem.* **1986**, *90*, 3559.

Dunbar, R.C., Shen, J., and Olah, G.A., *J. Am. Chem. Soc.* **1972**, *94*, 6862.

Essex, H., *J. Phys. Chem.* **1954**, *58*, 42.

Eyring, E.M., and Wahrhafting, A.L., *J. Chem. Phys.* **1961**, *34*, 23.

Farneth, W.E., and Brauman, J.L., *J. Am. Chem. Soc.* **1976**, *98*, 7891.

Fornarini, S., Sparapani, C., and Speranza, M., *J. Am. Chem. Soc.* **1988a**, *110*, 34.

Fornarini, S., Sparapani, C., and Speranza, M., *J. Am. Chem. Soc.* **1988b**, *110*, 42.

Gevantman, L.H., and Williams, R.R., Jr., *J. Phys. Chem.* **1952**, *56*, 569.

Glocher, G., and Lind, S.C., *The Electrochemistry of Gases and Other Dielectrics*, Wiley, New York, **1939**.

Hiraoka, K., and Kebarle, P., *J. Am. Chem. Soc.* **1975**, *97*, 4179.

Hiraoka, K., Mori, T., and Kebarle, P., *Chem. Phys. Lett.* **1993**, *207*, 178.

Kebarle, P., and Godbole, E.W., *J. Chem. Phys.* **1962**, *36*, 302.

Lias, S.G., *Interactions between Ions and Molecules*, Ausloos, P., Ed., **1975**, p. 541.

Lind, S.C., *Monatsh. Chem.* **1912a**, *33*, 295.

Lind, S.C., *Am. Chem. J.* **1912b**, *47*, 397.

Lind, S.C., *J. Phys. Chem.* **1912c**, *16*, 564.

Lind, S.C., *L'Action Chimique des Rayons Alpha en Phase Gazeuse*, Hermann, Paris, **1935**.

Lind, S.C., *The Chemical Effects of Alpha Particles and Electrons*, Chemical Catalog Co., New York, **1939**.

Lind, S.C., *J. Chem. Phys.* **1939**, *7*, 790.

Livingstone, R.S., and Lind, S.C., *J. Am. Chem. Soc.* **1936**, *58*, 612.

Meisels,G.G., Hamill, W.H., and Williams, R.R., Jr., *J. Phys. Chem.* **1957**, *61*, 1456.

Meot-Ner, M., and Field, F.H., *J. Am. Chem. Soc.* **1975**, *97*, 5339.

Meot-Ner, M., and Field, F.H., *J. Chem. Phys.* **1976**, *64*, 277.

Monahan, J.R., and Stanton, H.E., *J. Chem. Phys.* **1962**, *37*, 2654.

Mund, W., *Bull. Soc. Chem. Belg.* **1934**, *43*, 49.

Mund, W., *Action Chimique et Biologiques des Radiations,* Vol. 2, Chap. 1, Haissinsky, M., Ed., Masson, Paris, **1956**.

Olmstead, W.N., and Brauman, J.L., *J. Am. Chem. Soc.* **1977**, *99*, 4219.

Olmstead, W.N., Lev-On, M., Golden, D.M., and Brauman, J.L., *J. Am. Chem. Soc.* **1977**, *99*, 992.

Renzi, G., Lombardozzi, A., Dezi, E., Pizzabiocca, A., and Speranza, M., *Chemistry Eur. J.* **1995**, in press.

Sauer, M.C., Jr., Advances in Radiation Chemistry, Burton, M., and Magee, J.L., eds., Wiley, New Yoprk, **1976**, Chap. 2.

Sayer, J.M., Yagi, H., Silverton, J.V., Friedman, S.L., Whalen, D.L., and Jerina, D.M., *J. Am. Chem. Soc.* **1982**, *104*, 1972.

Speranza, M., *Gazz. Chim. Ital.* **1983**, *113*, 37.

Speranza, M., *Spectrosc. Int. J.* **1987**, *5*, 1.

Speranza, M., *Mass Spectrom. Review* **1992**, *11*, 73.

Speranza, M., and Angelini, G., *J. Am. Chem. Soc.* **1980**, *102*, 3115.

Speranza, M., and Laguzzi, G., *J. Am. Chem. Soc.* **1988**, *110*, 30.

Speranza, M., and Ursini, O., **1996**, in press.

Spinks, J.W.T., and Woods, R.J., *An Introduction to Radiation Chemistry*, Wiley, New York, **1990**.

Stone, J.A., Lin, M.S., and Varah, J., *Can . J. Chem.* **1981**, *59*, 2412.

Stone, J.A., Splinter, W.M., and Splinter, D.E., *Can . J. Chem.* **1984**, *62*, 1373.

Sunner, J.A., Hirao, K., and Kebarle, P., *J. Phys. Chem.* **1989**, *93*, 4010.

Tomoda, S., and Ohno, K., *Proceedings of the Kyushu International Symposium on Physical Organic Chemistry*, Fukuoka and Ube, **1991**, p. 390.

Troiani, A., Gasparrini, F., Grandinetti, F., and Speranza, M., **1996**, in press.

Vestal, M., Wahrhafting, A.L., and Johnston, W.H., *ARL Technical Documentary Report* **1962**, 62-426.

Yang, K., and Manno, P.J., *J. Am. Chem. Soc.* **1959**, *81*, 3507.

TECHNIQUES FOR THE PREPARATION AND STUDY OF CLUSTER IONS

A.J. STACE
School of Molecular Sciences,
University of Sussex,
Falmer, Brighton, BN1 9QJ, U.K.

1. Introduction

In recent years there has been a considerable research effort directed towards studying the chemical and physical properties of clusters [1,2]. In part, this activity has reflected a need to understand the basic building-blocks of bulk materials in terms of their constituent atoms or molecules; however, there is also the desire to search for properties which may be uniquely characteristic of finite-sized systems, for example, structures based on stable icosahedra. Many experiments on clusters have concentrated on the study of ions rather than neutrals [3]; the reason being that all known techniques for generating clusters produce a distribution of sizes which makes it difficult to probe size-specific properties. Although significant advances have been made using resonance ionisation methods to limit the effects of a neutral distribution [4], there are still uncertainties created by the possibility of fragmentation. In contrast, the study of ions, means that tandem mass spectrometric methods can be used to analyse the effects of spectroscopic or chemical interrogation on clusters of a single size.

This article will review some of the methods available for generating clusters ions and briefly illustrate the types of experiment that can be undertaken to investigate their chemical and physical properties. The emphasis will be on mass spectrometric based experiments and, in particular, the use of tandem mass spectrometry in the study of unimolecular, collision-induced, and laser-induced fragmentation patterns.

K. R. Jennings (ed.), Fundamentals and Applications of Gas Phase Ion Chemistry, 381–400.

2. Techniques for preparing cluster ions.

Despite numerous variations, the techniques available for generating clusters and their ions fall into three basic categories [5]:

1) Supersonic expansion.
2) Gas aggregation.
3) Sputtering (laser-induced or particle impact).

Although each method has distinct advantages for the production of particular types of clusters, there is some overlap regarding the materials that can be studied using one or more of the above techniques.

2.1 SUPERSONIC EXPANSION

Since this is the most common technique used to form clusters, the basic principles will be covered in some detail. Neutral clusters can be generated by the adiabatic expansion of a gas at high pressure (reservoir) through a small hole (nozzle) into a vacuum. For an ideal gas, two limiting conditions can be identified, each of which is classified in terms of a "Knudsen" number, K_n, where:

$$K_n = \frac{\text{Mean free path in the reservoir}}{\text{Diameter of the nozzle}} \qquad (1)$$

When $K_n > 1$, the atoms or molecules pass through the nozzle without undergoing any collisions, and there is little or no redistribution of energy between translational and internal degrees of freedom. Such behaviour is often referred to as "effusive" flow and is the type of behaviour normally associated with a Knudsen cell. The latter are frequently used as sources of metal atoms and small metal clusters, i.e. dimers and trimers, however, Knudsen cells are not efficient at generating intense beams of larger clusters.

Supersonic flow is characterised by the condition $K_n < 1$ and corresponds to a situation where atoms or molecules passing through a nozzle experience very large numbers of collisions. There is extensive energy redistribution, with the forward translational energy of the beam increasing at the expense of any local or internal energy. This pattern of behaviour has two effects: (i) the most probable forward velocity of the beam increases beyond that appropriate for a Maxwellian velocity distribution at the temperature of the reservoir; (ii) the velocity spread associated with atoms or molecules in the beam is reduced considerably. Since temperature is defined in terms of the width of a velocity distribution, it is possible to equate the reduction in spread with a drop in local temperature.

For the isentropic expansion ($\Delta S = 0$) of an ideal gas, the temperature reduction from an initial value T_i to a final value T_f during an expansion can be calculated from the expression [5]:

$$\frac{T_f}{T_i} = \left[1 + \frac{(\gamma - 1)}{2} M^2\right]^{-1} \quad (2)$$

and the corresponding drop in pressure is given by:

$$\left(\frac{P_f}{P_i}\right)^{\frac{\gamma-1}{\gamma}} = \left[1 + \frac{(\gamma - 1)}{2} M^2\right]^{-1} \quad (3)$$

where γ is the heat capacity ratio, C_p/C_v, and M is the Mach number, which is the ratio of the final flow velocity to the local velocity of sound. Since the velocity of sound varies as $T^{1/2}$, Mach numbers > 1 are typical and, as a result, the technique is frequently referred to as "supersonic" expansion. For a continuous gas flow at a distance greater than a few nozzle diameters from the nozzle, the Mach number can be approximated as:

$$M = A\left(\frac{x}{D}\right)^{\gamma-1} \quad (4)$$

where x is the distance from a nozzle of diameter D and A is a constant which depends on γ and is equal to 3.26 ($\gamma = 5/3$, monatomic gas) or 3.64

(γ = 7/5, diatomic gas). A useful approximation to equation (2) is the equation [6]:

$$\frac{T_f}{T_i} = \left[1 + 5896(pD)^{0.8}\right]^{-1} \qquad (5)$$

where p is the reservoir pressure in atmospheres and D in cm; this equation is based on an expansion of pure argon. Table 1 lists some the characteristics associated with the adiabatic expansion of an ideal gas. The temperature drops steadily as a function of distance x from the nozzle, but reaches a terminal value (not shown by the equations) at x/D ~ 60 because the collision frequency becomes too low for any further redistribution of energy. For the most part, the predictions provided by the model outlined above are supported by experiment. Thus, the supersonic expansion of a molecule heavily diluted in an excess of helium will display spectroscopic features characteristic of a very low internal temperature.

Table 1. **Predicted conditions for an ideal gas at 300K expanded through a 200μm nozzle at a reservoir pressure of 10^6Pa, as given by equations (2) - (4).**

x/D	M	P_f/P_i	T_f/T_i
1	3.3	0.022	0.22
5	9.5	1.8×10^{-4}	0.032
10	15	1.9×10^{-5}	0.013
20	24	1.7×10^{-6}	5.0×10^{-3}
40	38	1.8×10^{-7}	2.0×10^{-3}
60	50	3.0×10^{-8}	1.0×10^{-3}

It is not possible to extend this simple model to cluster formation; in fact there are no analytical models capable of predicting either the conditions necessary for the onset of cluster formation or the expected size distribution. However, some clues as to the mechanism are provided by Table 1. It can be seen that at extremely small distances from the nozzle the temperature is already very low, but the pressure is still quite high. Thus, the conditions are ideal for the three-body processes that are required to initiate cluster formation. Since the relative velocities of the atoms or molecules are extremely low, long-range interactions will be particularly significant in their contribution to the initial nucleation process. Typically, the first step will be dimer formation, i.e.

$$Ar + Ar + Ar \rightarrow Ar_2 + Ar \qquad (6)$$

This step is likely to be rate-determining and once there is significant dimer and trimer formation, further clustering can proceed via two-body collisions, i.e.

$$Ar_n + Ar \rightarrow Ar_{n+1}^* \qquad (7)$$

$$Ar_{n+1}^* + Ar \rightarrow Ar_{n+1} + Ar \qquad (8)$$

where step (7) promotes growth to form Ar_{n+1}^* with a comparatively high internal energy and a second collision (8) will then stabilise the cluster. It is, therefore, evident that cluster formation is an exothermic process and that large cluster may have significantly higher internal temperatures than their atomic or molecular counterparts. In turn this raises the interesting question as to whether or not clusters exist in a liquid-like or solid-like form. In the case of rare gases the clusters are probably always solid-like because the liquid existence range is very small; however, for more complex molecular systems the physical state of the cluster could vary according to the severity of the expansion conditions.

An interesting (but often inconvenient) aspect to cluster formation is the fact that as they grow in size they retain a significant fraction of the high forward velocity predicted by the simple theory given above. Hence, a large cluster can have a very high laboratory frame kinetic energy. The limiting velocity of monomers in a supersonic beam can be approximated by

$$v_{max} = \left(\frac{\gamma}{[\gamma - 1]}\right)^{\frac{1}{2}} \left(\frac{2kT_i}{m}\right)^{\frac{1}{2}} \tag{9}$$

Hence for a cluster composed of N atoms or molecules, the maximum kinetic energy will be

$$E_{kin} = \frac{1}{2} N m v^2_{max} \tag{10}$$

which, for a monatomic gas reduces to

$$E_{kin} = 0.0645\, N \left(\frac{T_i}{300 / K}\right) / eV \tag{11}$$

Thus, a cluster of 1000 argon atoms generated from a reservoir at 300K will have a laboratory-frame kinetic energy of 64eV. If a cluster source is situated at right-angles to the ion source of a mass spectrometer, then the ion extraction optics have to be capable of accommodating a wide variation in kinetic energies if the instrument is to operate over a significant mass range. This type of problem is not usually encountered in the mass spectral analysis of other high molecular weight materials, such as peptides.

Supersonic expansion can be used to form clusters from any gaseous atomic or molecular substance. In the case of the rare gases, only helium and neon require some form of additional cooling in order to promote cluster formation. Most volatile molecular liquids require the presence of a carrier gas, i.e. helium or argon, and materials with low vapour pressures often need the nozzle to be heated in order to achieve high molecular cluster number densities. If the latter are expanded at room temperature with argon then a frequent result is the formation of mixed clusters consisting of one or two molecules and large numbers of rare gas atoms. Although there are qualitative ground rules, the efficient formation of a high intensity cluster beam is still very much a "black art", particularly when it comes to generating mixed clusters.

2.2 GAS AGGREGATION

By creating conditions whereby a vapour comes into contact with a background gas at a much lower temperature, it is possible to grow clusters through condensation [5]. This is the underlying principle behind the gas aggregation or smoke source as it is sometimes called. In contrast to some of the other cluster sources discussed, the smoke source has the advantage that it is capable of generating high number densities of very large clusters, frequently containing up to 10^6 atoms. The design of a typical source is very simple, consisting of a mechanism for creating a vapour, i.e. a small oven or a discharge, and a flow of cold gas, i.e. helium cooled with liquid nitrogen. The clusters do not normally experience any form of supersonic expansion as they emerge from the source which, in turn, means that their kinetic energies are comparatively low. A variant of this type of source has been designed to prepare alkali metal cluster ions with variable temperatures [5]. In this way it is possible to avoid the indeterminate heating effects associated with photon or electron impact ionisation.

2.3 SPUTTERING SOURCES

There are many variations on this central theme, ranging from direct particle bombardment, i.e. similar to SIMS or FAB, through to pulsed arc sources and finally laser ablation techniques [5]. All these methods rely on using some form of energy source to generate plasmas from what are, under normal conditions, involatile materials; typically, the latter are metals, semi-conductors or ceramics.

The most widely used version of this type of source, is that developed by Smalley and co-workers [7] and combines laser ablation with supersonic expansion. Material is ablated from a rotating disc or rod into the path of an intense pulse of helium. The mixture undergoes turbulent motion in a small "waiting room" before expanding through a 10° conical nozzle. This approach is capable of generating beams of cold clusters (neutral, anionic, and cationic) from most refractory materials.

2.4 PICK-UP METHODS FOR MIXED CLUSTER IONS

One of the more difficult aspects of cluster preparation is that of being able to generate mixed species in a controlled manner. Probably the most obvious approach is to mix the constituents prior to expansion and then adjust the gas ratio in order to achieve clusters of the desired composition. However, this method is only effective over a very narrow size range and works best when the vapour pressures of the two components are comparable.

Under circumstance where the requirement is for clusters consisting of single atoms or molecules solvated in an excess of a second component, i.e. $Mg.Ar_n$ or $H_2O.(CH_3OH)_n$, then the most successful approach appears to be the pick-up method. A beam of clusters is allowed to interact with a background gas downstream from the nozzle; if the gas has a partial pressure of ~10^{-5}mbar then the clusters will, on average, pick up one atom or molecule. Scoles and co-workers [8] were the first to use the pick-up technique for creating mixed clusters, and the method has been used with considerable success for a broad range of materials. Two aspects of the technique are of interest: first, the method appears to operate more effectively when a pulsed nozzle [9] is used, the rational being that the background gas has an opportunity to replenish during the closed cycle of the supersonic expansion. Secondly, it would appear that to produce mixed clusters of molecules, i.e. $SF_6.(C_6H_6)_n$, which are then stable to ionisation, it is necessary to first generate mixed clusters of benzene with argon [10]. It would appear that the following evaporative mechanism operates to stabilise the mixed clusters both during the pick-up collision and subsequent ionisation

$$Ar_m.(C_6H_6)_n + SF_6 \rightarrow SF_6.(C_6H_6)_n.Ar_j + (m\text{-}j)Ar \qquad (12)$$

$$SF_6.(C_6H_6)_n.Ar_j + e^- \rightarrow SF_6.(C_6H_6)_n^+ + jAr + 2e^- \qquad (13)$$

Without an initial step involving the formation of mixed molecular clusters with argon it is frequently not possible to observe mixed cluster ions from pick-up collisions.

3. Mass spectrometric analysis

Figure 1 shows a schematic diagram of an apparatus that has been designed specifically to study the fragmentation patterns of large cluster ions [11]. A supersonic nozzle contained within a differentially pumped vacuum chamber is coupled to a high resolution, double focusing mass spectrometer (VG ZAB-E). Neutral clusters are prepared via the adiabatic expansion of a gas or vapour through a 200μm pulsed conical nozzle. The central core of the expansion is collimated using a 1mm diameter skimmer and the resultant beam passes into the ion source of the mass spectrometer where the clusters are ionised by electron impact. A channel with a cross-section 0.5x1.0cm has been machined through the ion source block in order to allow the unimpeded entry and exit of the cluster beam; the objective being to create as little disturbance as possible and avoid any scattering.

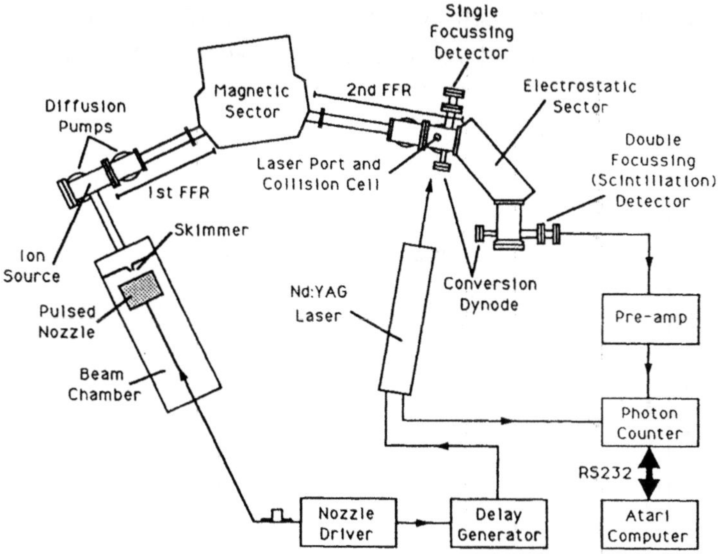

Figure 1. Schematic diagram of an apparatus for generating cluster ions and analysing their fragmentation patterns.

Following ionisation, cluster ions are extracted from the source with a potential of 3-8kV. The double focusing detection system consists of a scintillation (Daly) detector fitted with a photon counting tube. When operated in analogue mode for the purposes of recording mass spectra, signals of up to 10^{-6}A can be detected via phase sensitive detection. In photon counting mode, which is used to study fragmentation, the lower detection limit is < 1 ion s^{-1}. Figure 2 shows a typical mass spectrum that has been recorded following the adiabatic expansion of a benzene/argon mixture at a very low reservoir pressure (~10psi) [12]. In addition to singly charged ions of the form $(C_6H_6)_n^+$, there are also present odd-sized doubly charge ions, $(C_6H_6)_n^{2+}$; the corresponding even-sized ions are hidden beneath the singly charged clusters. What is of particular interest is the fact that the doubly charged ions have a lower limit to the size at which they are stable. Thus, $(C_6H_6)_{23}^{2+}$ can clearly be seen, but there is no $(C_6H_6)_{21}^{2+}$, the reason being that doubly charged benzene ions with $n \leq 21$ break apart because Coulomb repulsion between the two positive charges becomes greater than the binding energy between the individual constituents.

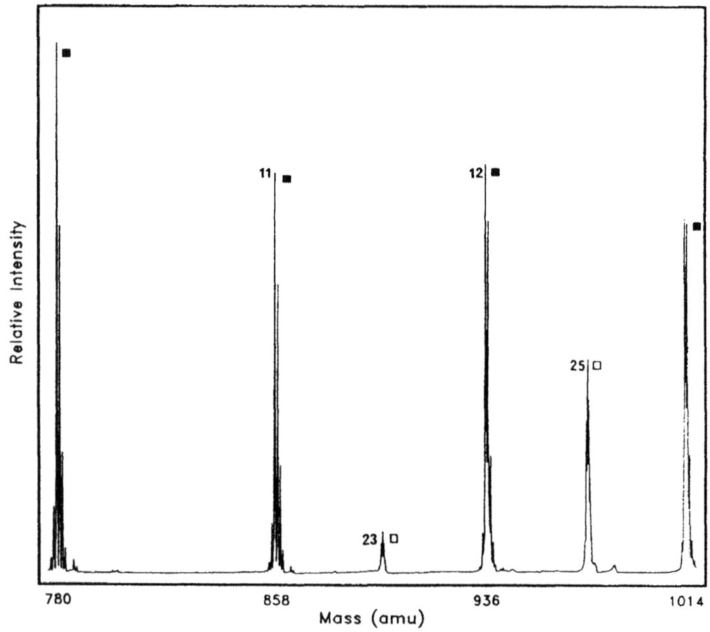

Figure 2. Mass spectrum of benzene cluster ions. Singly charged ions are shown thus (■) and doubly charged ions thus (□).

3.1 UNIMOLECULAR AND COLLISION-INDUCED FRAGMENTATION

Electron impact ionisation is quite a violent process and, compared with typical binding energies found in clusters, the excess energy is often sufficient to cause quite extensive fragmentation. It is for this reason that normal (electron impact) mass spectra cannot be used to characterise neutral cluster beams. However, the excess energy does provide for an opportunity to study fragmentation using the MIKES (Mass Analysed Ion Kinetic Energy) technique. Although cluster ions lose a significant fraction of their initial internal energy through the evaporation of monomer units in the ion source, large numbers of ions still reach the 2ffr with sufficient energy to undergo further fragmentation.

Figure 3 shows an example of fragmentation in argon cluster ions. Ar_{22}^+ has been mass-isolated using the magnetic sector and then, as the ion passes through the 2ffr, the loss of argon atoms has been monitored using the electrostatic analyser [13]. In this particular example, the energy resolution has been increased to a point where kinetic energy release associated with each fragmentation has an influence on peak width. The peaks labelled c, d and e are examples of collision-induced processes where, in each case, fragmentation steps have been recorded with the background pressure in the 2ffr collision cell increased to $\sim10^{-6}$mbar. The unimolecular and collision-induced (CID) loss of a single argon atom have been separated by floating the cell at a voltage of -400V. Two peaks are recorded for the CID loss of two argon atoms: (d) corresponds to the sequential loss of two atoms, one in the cell followed by one outside, and (e) corresponds to the loss of both atoms in the cell.

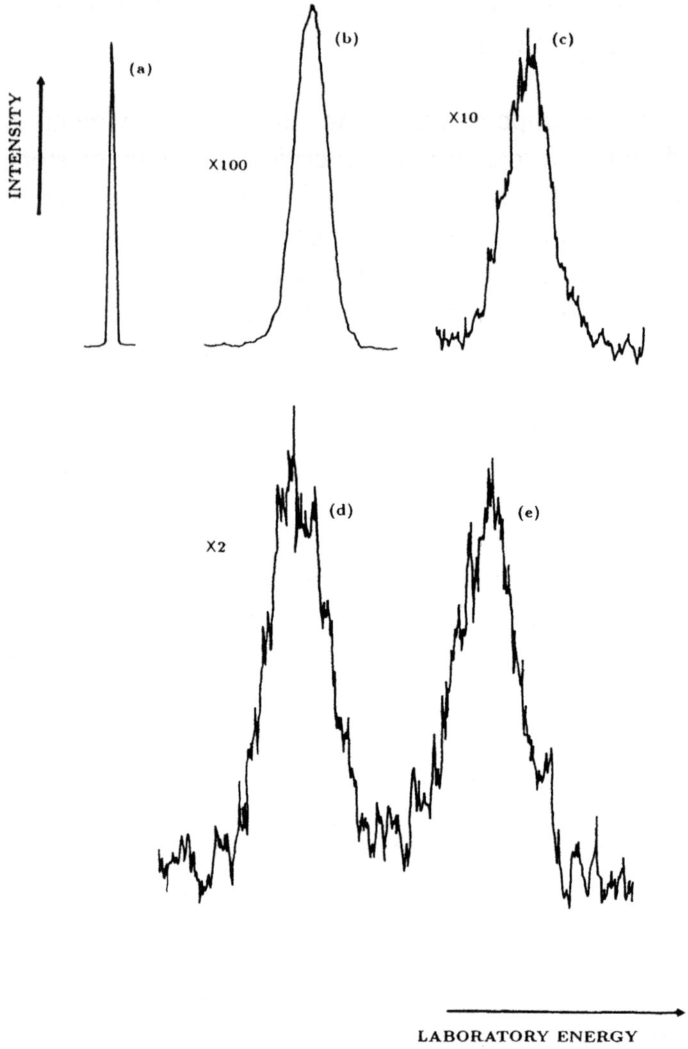

Figure 3. Unimolecular and collision-induced fragmentation pattern of Ar_{22}^+ recorded following a MIKE scan. (a) is the parent ion, (b) the unimolecular loss of one argon atom, (c) the collision-induced loss of one argon atom, (d) and (e) collision-induced loss of two argon atoms, (d) sequential loss of one atom in the cell followed by one outside, (e) loss of both atoms in the cell.

The MIKE technique can also be used to follow chemical reactions in cluster ions. A particularly nice example comes from the reactions of $NO^+.(H_2O)_n$ cluster ions and has some relevance to the chemistry of the upper atmosphere [14]. In a sequence of exchange reactions proposed to account for the formation of proton hydrates in the upper atmosphere, the final two steps are thought to be [15]:

$$NO^+.(H_2O)_2 + H_2O \rightarrow NO^+.(H_2O)_3 \tag{14}$$

$$NO^+.(H_2O)_3 + H_2O \rightarrow (H_2O)_3H^+ + HONO \tag{15}$$

However, no direct observation of reaction (15) appears to have been made. The cluster analogue of (15) is:

$$NO^+.(H_2O)_4 \rightarrow (H_2O)_3H^+ + HONO \tag{16}$$

Where the unimolecular step could be viewed as a half-collision, representing the separation of reaction products from a transition state appropriate for (15), and the cluster ion itself could be considered as having a structure similar to that adopted by the reactants just prior to entering a transition state.

Mixed clusters of NO and water were prepared using the pick-up technique with NO gas being bled into a stream of water clusters generated via the supersonic expansion of an argon/water mixture at 60psi through a 200μm diameter conical nozzle. In practice, D_2O rather than H_2O was used because of a mass coincidence between $NO^+(H_2O)_5$ and Ar_3^+.

Figure 4 shows the fragmentation pattern established for the $NO^+(H_2O)_n$ cluster ions as a function of size using the MIKES technique. For small cluster ions the dominant fragmentation route is loss of a single water molecule:

$$NO^+.(H_2O)_n \rightarrow NO^+.(H_2O)_{n-1} + H_2O \tag{17}$$

However, at n = 4 there is a marked change in behaviour and the cluster ions begin to lose HONO which is, as already stated, the analogue of the bimolecular reaction proposed for the formation of proton hydrates. Beyond n= 4 loss of HONO becomes the dominant fragmentation route.

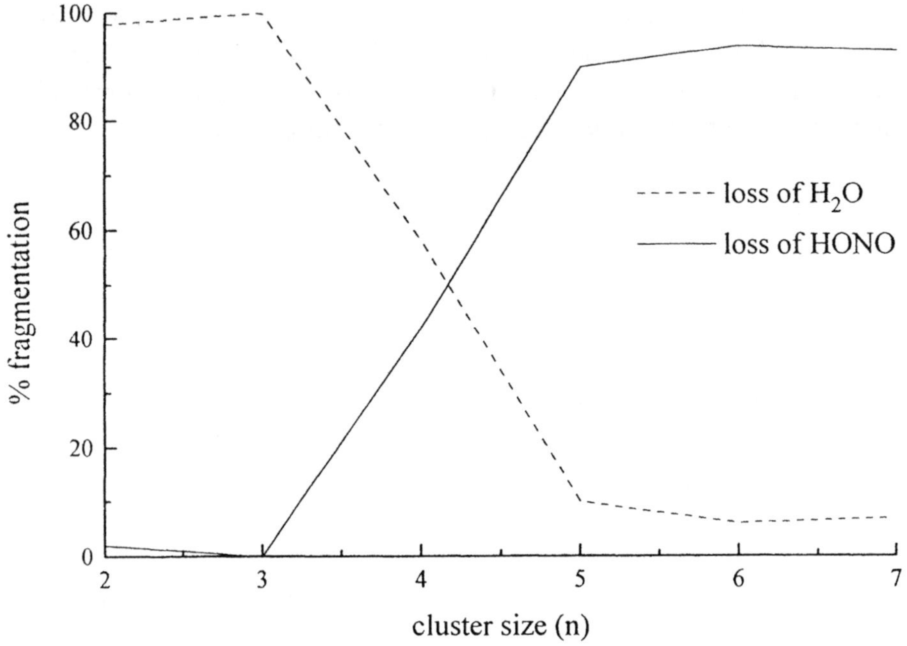

Figure 4. **Fragmentation pattern recorded for $NO^+.(H_2O)_n$ and plotted as a function of cluster size (n)**

Following from these experiments, it was possible to identify a series of reactions that were initiated in mixed clusters by the presence of NO^+ and, for some of the more complex systems, it was suggested that the processes observed might be unique to clusters [16]. For example, mixed clusters of the composition $CH_3OH.(H_2O)_n.NO^+$ were found to lose either H_2O or CH_3ONO except for one specific composition, where the following reaction was observed:

$$CH_3OH.(H_2O)_3.NO^+ \rightarrow [CH_3OH.(H_2O)_2]H^+ + HONO \qquad (18)$$

Since this single process involves the simultaneous presence of five species, it is unlikely to proceed in the gas phase between isolated molecules and ions. Similarly, the critical composition required to drive reaction (18) is unlikely to occur in the liquid phase. NO^+ would have to created, for example, by radiolysis in the close proximity of one methanol and three water molecules; one more or one less molecule and the reaction pathway would be quite different.

3.2 LASER-INDUCED FRAGMENTATION

The very low number densities associated with cluster ions, means that traditional spectroscopic techniques, such as LIF, cannot be used. For the most part, the only route available is to record photofragment spectra as a function of laser wavelength. Since cluster ions are very fragile, it requires a particularly gentle form of photoexcitation if, in the study of these laser-induced processes, complete fragmentation is to be avoided. A good example of the latter is to be seen in the excitation of argon cluster ions, Ar_n^+, at visible wavelengths [17]. For $n > 25$, up to 20 argon atoms are lost following the absorption of a single photon; whilst such behaviour can help in the measurement of absorption cross-sections, it does not serve to reveal some of the more subtle features of cluster structure.

At the other extreme, infrared excitation can provide a very gentle probe of structure because the energy being deposited into cluster ions is comparable to the binding energies of single monomer units. However, there are still problems: first, there are very few chromophores which fall within the wavelength range of the lasers available and, secondly, lasers that have a broad wavelength range, i.e. CO, CO_2, and N_2O, are line-tuneable which restricts their resolution.

The problem of finding a suitable chromophore has recently been resolved in a series of experiments involving SF_6 [10,18-21]. This molecule not only has a very large infrared transition moment which falls within the range of the CO_2 laser, it also has a comparatively high ionisation energy. Hence, when mixed clusters containing SF_6 are ionised, the charge remains located within the cluster leaving the molecule neutral and thus capable of continued infrared absorption from a CO_2 laser. The same philosophy applies to negative ions: SF_6 has a positive electron affinity, but this is resonant attachment with a maximum cross-section at ~ 0eV. If the molecule is associated with a cluster with a high electron affinity, such as $(NO)_n$ [19], then negatively charged clusters can be generated in the knowledge that the excess electron resides on the cluster.

The application of this method is illustrated using the SF_6/NO system [19]; other examples can be found in [10,18,20,21]. Mixed clusters were generated using the pick-up technique with SF_6 being bled into the path of an $(NO)_n.Ar_m$ beam. Positively charged ions were formed using 70eV electron impact and negatively charged ions were generated with the electron beam energy fixed at 20eV; it is probable that the latter creates low energy secondary electrons which then became attached to the clusters. Referring to figure 1, ions were extracted from the ion source with a potential of 6kV and mass-selected using the magnetic sector. When the ions entered the 2ffr they were irradiated with photons from a line-tuneable carbon dioxide laser aligned coaxially with the ion flight path. This configuration provided an overlap between the ion and laser beams of approximately 150cms. The masses of any ionic photofragments were then identified by scanning the electrostatic analyser. The pick-up procedure proved very successful in generating stable ion beams with almost uniform intensity for clusters consisting of a single chromophore and up to 100 host molecules.

An interesting aspect to the SF_6/NO system, is the very pronounced difference in fragmentation pattern between the odd- and even-sized cluster ions (anions and cations). For even-sized ions the preferred fragmentation path is as follows:

$$SF_6.(NO)_n^{+(-)} + h\nu \rightarrow SF_6.(NO)_{n-1}^{+(-)} + NO \qquad (19)$$

and for odd-sized cluster ions the preferred path is as follows:

$$SF_6.(NO)_n^{+(-)} + h\nu \rightarrow SF_6.(NO)_{n-2}^{+(-)} + 2NO \qquad (20)$$

This marked difference is as result of electron pairing in the clusters. NO has an unpaired electron; hence, even-sized cluster ions (+ or -) have an odd electron and can become even-electron by losing a single NO molecule. In contrast, odd-size cluster ions are even-electron systems and to maintain that state they prefer to lose 2NO molecules. This electron pairing effect is also reflected in the mass spectra, with odd-sized ions being a factor of 10 more intense than their even-sized counterparts [19].

By summing fragment ion intensities as a function of laser wavelength, infrared absorption profiles can be recorded for size-selected cluster ions. Figure 5 shows two examples of infrared profiles recorded for $SF_6.(NO)_{15}^+$ and $SF_6.(NO)_{15}^-$. The difference in shape between the profiles is believed to be due to a repulsive interaction between the excess electron and the closed electron orbitals of the SF_6 molecule. Although the latter can form a stable negatively charged ion, the resonant condition necessary to create that state does not appear to exist in the mixed clusters. The net result is that the cluster is destabilised with the SF_6 seeking a site as far away as possible from the excess electron [19].

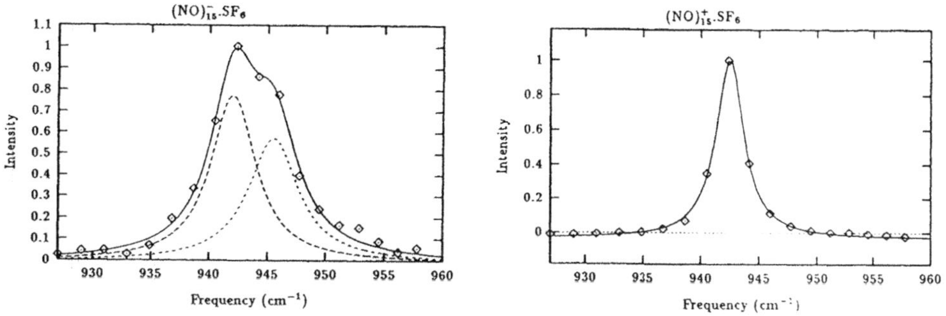

Figure 5 Infrared absorption profiles recorded by summing over all fragment ion intensities as a function of laser wavelength.

4. Conclusion

It has been shown that tandem mass spectrometric methods in association with supersonic expansion techniques for the generation of clusters, can provide a very effective combination for studying the fragmentation patterns of cluster ions. Particularly successful has been use of the pick-up technique to generate mixed species consisting of single molecules attached to clusters of atoms or molecules. In conjunction with laser-induced fragmentation, the pick-up techniques has provided the means for studying the infrared spectra of molecules solvated in a broad range of clusters [18-21]. This general approach has recently been extended to solvated metal ions [22].

5. Acknowledgement

The author would like to thank the EPSRC for financial support and also to acknowledge the contribution Drs A.B. Jones, J.F. Winkel and C.A. Woodward have made to material discussed in this article.

6. References

1. Haberland, H. (ed) (1994) *Clusters of Atoms and Molecules I*, Springer-Verlag, Berlin.
2. Haberland, H. (ed) (1995) *Clusters of Atoms and Molecules II*, Springer-Verlag, Berlin.
3. Märk, T.D. and Echt, O., in ref. 2, Internal Reactions and Metastable Dissociations After Ionisation of van der Waals Clusters, 154-182.
4. Venturo, V.A., Maxton, P.M. and Felker, P.M. (1992) Raman/vibronic double resonance spectroscopy of benzene doped argon clusters, *Chem. Phys. Letts.* **198**, 628-636.
5. Harberland, H. in ref. 2, Experimental Methods, 207-252.
6. Flygare, W.H. (1982) Pulsed Fourier transform microwave spectroscopy, in Marshall, A.G. (ed.) *Fourier, Hadamard and Hilbert Transforms in Chemistry*, Plenum Press, New York, 207-270.
7. Maruyama, S., Anderson, L.R. and Smalley, R.E. (1990) Direct injection supersonic cluster source for FT-ICR studies of clusters, *Rev. Sci. Instrum.* **61**, 3686-3693.
8. Levandier, D.J., Goyal, S., McCombie, J., Pate, B. and Scoles, G. (1990) Infrared spectroscopy and dimer formation at the surface of medium - large argon clusters, *J. Chem. Soc. Faraday Trans.* **86**, 2361- 2368.
9. Schulz, C.P. and Hertel, I.V., in ref 2, Solvated Atoms in Polar Solvents, 7-18.
10. Jones, A.B., Woodward, C.A., Winkel, J.F. and Stace, A.J.(1994) Infrared laser-induced photofragmentation of size-selected $SF_6.(CO_2)_n^+$ clusters, for n in the range 2-11, *Int. J. Mass Spectrum. Ion Processes* **133**, 83-95.
11. Lethbridge, P.G. and Stace, A.J. (1988) Structure-reactivity correlations in ion clusters. An investigation into the fragmentation patterns of Ar_n^+ clusters, for n in the range 10-200, *J. Chem. Phys.* **89**, 4062-4073.
12. Gotts, N.G., Lethbridge, P.G. and Stace, A.J. (1992) Observations of Coulomb explosion in doubly charged atomic and molecular clusters, *J. Chem. Phys.* **96**, 408-421.
13. Woodward, C.A. and Stace, A.J. (1991) Measurements of kinetic energy release following the unimolecular and collision-induced decay of argon cluster ions, Ar_n^+, for n in the range 2-60, *J. Chem. Phys.* **94**, 4234-4242.
14. Stace, A.J., Winkel, J.F., Lopez Martens, R.B. and Upham, J.E. (1994) Observations on the behaviour of $NO^+(H_2O)_n$ clusters and their relevance to the ion chemistry of the upper atmosphere, *J. Phys. Chem.* **98**, 2012-2014.

15. Ferguson, E.E., Fehsenfeld, F.C. and Albritton, D.L. (1979) Ion Chemistry of the Earth's Atmosphere, in Bowers, M.T. (ed.), *Gas Phase Ion Chemistry*, Academic Press, New York, Vol. 1, 45-82.

16. Winkel, J.F. and Stace, A.J. (1994) Switching reactions on and off in NO-containing cluster ions: evidence of chemical processes that are probably unique to systems of a finite size, *Chem. Phys. Lett.* **221**, 431-435.

17. Levinger, N.E., Ray, D., Alexander, M.L. and Linberger, W.C. (1988) Photoabsorption and photofragmentation studies of Ar_n^+ cluster ions, *J. Chem. Phys.* **89**, 5654-5662.

18. Woodward, C.A., Winkel, J.F., Jones, A.B. and Stace, A.J. (1993) Infrared predisociation in cluster ions containing a guest chromophore: a gentle route to the dynamics and spectroscopy of heterogeneous systems, *Chem. Phys. Lett.* **206**, 49-56.

19. Winkel, J.F., Jones, A.B., Woodward, C.A., Kirkwood, D.A. and Stace, A.J. (1994) Infrared laser-induced photofragmentation of the positive and negative ions of size-selected $SF_6.(NO)_n$ clusters, *J. Chem. Phys.* **101**, 9436-9446.

20. Jones, A.B., Lopez-Martens, R. and Stace, A.J. (1995) Laser-induced infrared photofragmentation of $SF_6.(C_6H_6)_n^+$ cluster ions, *J. Phys. Chem.* **99**, 6333-6339.

21. Winkel, J.F., Woodward, C.A., Jones, A.B. and Stace, A.J. (in press) Infrared photofragmentation spectra of size-selected $SF_6.Ar_n^+$ cluster ions, *J. Chem. Phys.*.

22. Woodward, C.A., Dobson, M. and Stace, A.J. (to be published).

REACTION MECHANISMS : DISCUSSION SUMMARY

HANS-FRIEDRICH GRÜTZMACHER
Lehrstuhl I für Organische Chemie
Fakultät für Chemie
Universität Bielefeld
D-33501 Bielefeld, Germany

The knowledge of reaction mechanism is a prerequisite for an understanding and description of macroscopic chemical changes of matter on a molecular basis in all areas of chemistry and biology. In particular, the mechanisms of organic reactions have been studied for more than 40 years in much detail both by experiment and by theory, leading to the development of advanced models for structure/reactivity correlations. Mass spectrometry and gas phase ion chemistry have made important contributions to this field of research not only by providing very versatile analytical tools but also by allowing the study of the reactions of isolated ionic species in the gas phase without any interference of solvent effects. Recently, the mechanisms of reactions in inorganic, organometallic and elementorganic chemistry came into focus, and again the advanced techniques of modern mass spectrometry are expected to play an important role in the investigation of these reactions.

Accordingly, there has been a broad interest for the presentation and discussion of recent results from studies of specific ionic reactions in the gas phase during the NATO ASI. The corresponding workshops were held on two afternoon meetings during the conference. These workshops covered a variety of aspects of ionic gas phase reactions, including both negative and positive ions and studies of the fragmentation mechanisms of ionized compounds valuable for the analytical application of mass spectrometry as well as more fundamental studies of the mechanisms of unimolecular reactions and of bimolecular ion/molecule reactions. A summary of the contributions follows.

1. Generation of Distonic Radical Anions with Molecular Fluorine.

Robert R. Squires, P. G. Wenthold, and Ju Hu
Department of Chemistry, Purdue University, West Lafayette, Indiana 47907, U.S.A.

A regioselective method has been developed for generating intense beams of structurally-defined biradical negative ions ("distonic radical anions") based on gas-phase ion/molecule reactions between organic negative ions and fluorine (F_2). Biradical negative ions are essential subjects for negative photoelectron spectroscopic (NIPES) measurements of electron affinities and singlet-triplet energy gaps in the corresponding neutral biradicals, and are also of interest for mass spectrometric studies of free radical chemistry and for investigations of the acid-base properties of free radicals. The new gas-phase synthetic procedure has been used to generate trimethylenemethane anion [1], *meta*- and *para*-benzyne anions [1,2], acetate radical anion [3], *meta*-dehydrobenzyl anion, and a variety of other novel distonic radical anions. Triradical negative ions ("distonic biradical anions") can also be formed by ion/molecule reactions between F_2 and appropriate precursor ions.

K. R. Jennings (ed.), Fundamentals and Applications of Gas Phase Ion Chemistry, 401–417.

The experiments described were carried out at room temperature in a flowing afterglow - triple quadrupole apparatus [4]. The new synthetic method is a variant of the fluorodesilylation procedure developed by DePuy and coworkers for regioselective formation of gas-phase carbanions from trimethylsilyl compounds, $RSi(CH_3)_3$ [5]. First, a trimethylsilyl-substituted carbanion is formed from the corresponding *bis*-trimethylsilyl derivative. This ion is then allowed to react with *molecular* fluorine, F_2, in order to effect the displacement of the remaining trimethylsilyl group (equ. 1).

(1)
$$Me_3Si\text{-}C{-}C\text{-}SiMe_3 \xrightarrow{F^-} Me_3Si\text{-}C{-}C^- \xrightarrow{F_2} \cdot C{-}C^-$$

Thus, two trimethylsilyl groups determine the positions of the charged and odd-spin sites in the radical anion products. The mechanism is believed to involve electron transfer from the trimethylsilyl-substituted carbanion to F_2 (EA = 3.01 ± 0.07 eV [6]) within the initial ion/molecule collision complex, followed by desilylation of the resulting radical, with either the $F_2^{\cdot -}$ or F^- (formed by dissociation of $F_2^{\cdot -}$) serving as the nucleophile.

This synthetic method is general in scope, inasmuch as distonic radical anions derived from many different types of hydrocarbons and heteroatom-containing biradicals have been made. *Bis*-trimethylsilyl precursors are not essential: any method of forming a $(CH_3)_3Si$-containing carbanion is applicable. For example, oxyallyl radical anion, $\cdot CH_2C(O)CH_2^-$, has been formed by deprotonation of the trimethylsilylenol ether of acetone followed by treatment with F_2. The nascent F^- formed by dissociative electron transfer to F_2 within the initial collision complex can participate in reactions other than fluorodesilylation, such as proton abstraction, nucleophilic substitution at carbon, and elimination. For instance, the acetate radical anion, $\cdot CH_2CO_2^-$, can be formed in high yield by reacting F_2 with the enolates of either acetic acid or ethyl acetate(equ. 2; [3])

(2)
$$CH_3CO_2CH_2CH_3 \xrightarrow{^-OH} {}^-CH_2CO_2CH_2CH_3 \xrightarrow{F_2} CH_2CO_2^- + HF + CH_2{=}CH_2$$

The α-dehydrophenoxide, $C_6H_4O^{\cdot -}$, is formed when F_2 reacts with deprotonated anisole, presumably by way of S_N2 displacement by F^- at the methyl group of the intermediate *ortho*-methoxyphenyl radical. Fluorine appears to affect multiple dehydrogenations of certain carbanions. For example, C_3^-, C_3H^-, and $C_3H_2^-$ are all produced from the reaction between F_2 and the propargyl anion, $HCCCH_2^-$. Triple desilylations with *tris*-trimethylsilyl compounds have also been observed: treatment of both 1,3,5-*tris*(trimethylsilyl)benzene and 1,3,5-*tris*(trimethylsilylmethyl)benzene with F^- followed by F_2 yields the corresponding distonic biradical anions $C_6H_3^-$ and $C_6H_3(CH_2)_3^-$.

2. Competing Electron Transfer and Nucleophilic Substitution in Gas-Phase Reactions of (Radical) Anions with Halomethanes

P. O. Staneke, G. Groothuis, Steen Ingemann, and Nico M. M. Nibbering
Institute of Mass Spectrometry, University of Amsterdam, Nieuwe Achtergracht 129, NL-1018 WS Amsterdam, The Netherlands

The product ion distributions and rates of the gas-phase reactions of two series of (radical) anions with chloro- and bromomethanes (CH_3Cl, CH_2Cl_2, $CHCl_3$, CCl_4, CH_3Br, CH_2Br_2, $CHBr_3$, and CBr_4) have been determined by FT-ICR mass spectrometry. The first series consists of anions (HO^-, CH_3O^-, $C_2H_5O^-$, $C_3H_7O^-$, and CH_3S^-), for which the corresponding neutral radical has a relatively high electron affinity (EA > 150 kJ·mol^{-1}). The second series consists of (radical) anions ($CH_2S^{\cdot-}$, $CH_2{=}CH{-}CH_2^-$, $CH_2{=}C(C_3){-}CH_2^-$, $C_6H_4^{\cdot-}$, and $C_6H_5^-$), for which the corresponding neutral species has a relatively low electron affinity (EA < 100 kJ·mol^{-1}). These (radical) anions react mainly with the halomethanes ($CH_{4-a}X_a$; a = 1 - 4) to afford i) halide ions (equ. 1), ii) halomethyl anions if proton transfer is exothermic (equ. 2), and iii) halomethyl anions with one halogen atom less than the parent substrate (equ. 3). The latter process involves nucleophilic attack on a halogen atom and is particularly important for the reactions with substrates containing three or four halogen atoms.

$$A^- + CH_{4-a}X_a \longrightarrow X^- + \text{neutral products} \qquad (1)$$

$$A^- + CH_{4-a}X_a \longrightarrow CH_{3-a}X_a^- + AH \qquad (2)$$

$$A^- + CH_{4-a}X_a \longrightarrow CH_{4-a}X_{a-1}^- + AX \qquad (3)$$

The halide ions may arise by a number of different pathways such as S_N2 (equ. 4) , α-elimination (equ. 5), halogen attack followed by dissociation of the halomethyl anions (equ. 6), and overall dissociative electron transfer (equ. 7). The S_N2 process is held responsible for the formation of the halide ions in the reactions with the monohalomethanes and may play a significant role in the formation of halide ions in the reactions with dihalogen substituted substrates. The α-elimination pathway is likely to be of importance only for the reactions with the trichloro- and tribromomethanes. In the reactions of the second series of ions with these two halomethanes as well as with CCl_4 and CBr_4, attack on a halogen atom followed by dissociation of the initially generated product ions may also play a role in the formation of the halide ions.

$$A^- + CH_{4-a}X_a \longrightarrow X^- + ACH_{4-a}X_{a-1} \qquad (4)$$

$$A^- + CH_{4-a}X_a \longrightarrow X^- + CH_{3-a}X_{a-1} + AH \qquad (5)$$

$$A^- + CH_{4-a}X_a \longrightarrow X^- + CH_{4-a}X_{a-2} + AX \qquad (6)$$

$$A^- + CH_{4-a}X_a \longrightarrow X^- + A\cdot + \cdot CH_{4-a}X_{a-1} \qquad (7)$$

The plurality of pathways which may lead to the halide ions hampers a direct assessment of electron transfer in the reactions of the (radical) anions with the halomethanes. Nevertheless, the occurrence of overall dissociative electron transfer cannot be excluded since this process is energetically feasible for the second series of ions (with the exception of $C_6H_5^-$) and most of the halomethanes. However, direct evidence for the occurrence of electron transfer is obtained for the reactions of the $CH_2S^{\cdot-}$ ion with $CHCl_3$, CCl_4, and CBr_4. In these systems, minor amounts of stable molecular radical anions of the halomethanes are generated in addition to CH_2SCl^- or CH_2SBr^- ions and abundant halide ions. The initial step in these reactions is likely an electron transfer and formation of an ion/molecule complex of thioformaldehyde and the halomethane radical anion. This complex may dissociate to the observed radical anion of the halomethane, or a halide ion transfer may occur followed by the dissociation into CH_2SCl^- or CH_2SBr^- ions.

The radical anions of the halomethanes react readily with the parent compound by halide ion transfer as exemplified by the reaction of $CCl_4^{\cdot-}$ with CCl_4 which yields a CCl_5^- ion. Based on the occurrence of this process, the halomethane radical anions are proposed to exist as a halide ion weakly bonded to a halomethyl radical[7]. In addition, it is suggested that the halide ion formed in the reactions with the $CH_2S^{\cdot-}$ ion arises by subsequent dissociation of the halomethane radical anions. In conclusion, the results indicate that electron transfer is a likely pathway if this process is exoergic even though other processes such as α-elimination or halogen atom attack followed by dissociation also contribute significantly to the generation of the halide ions, in particular for the methanes containing three or four halogen atoms.

3. Gas-Phase Nucleophilic Reactions of Mono-halogen Substituted Carbene Radical Anions ($CHX^{\cdot-}$; X = F, Cl, and Br) and the Related Carbanions, CH_2X^-.

Monique Born, Steen Ingemann, and Nico M. M. Nibbering
Institute of Mass Spectrometry, University of Amsterdam, Nieuwe Achtergracht 129, NL-1018 WS Amsterdam, The Netherlands

The gas-phase reactions of mono-halogen substituted carbene radical anions, $CHX^{\cdot-}$, (X = F, Cl, Br) and the corresponding carbanions, CH_2X^-, (X = Cl, Br) with selected substrates have been examined by FT-ICR mass spectrometry. The chlorine and bromine containing ions react only by S_N2 substitution with chloro- and bromomethanes whereas competing $B_{AC}2$ and S_N2 reactions occur with the ester substrates, dimethyl carbonate and methyl trifluoroacetate.

The experimental efficiencies reveal that the anions react more readily than their corresponding radical anions in the S_N2 substitutions with the halomethanes. In addition, the overall efficiencies of the reactions of the anions with dimethyl carbonate are also higher than those of the reactions of the radical anions. With methyl trifluoroacetate both types of ions react with collision rate.

Table 1 : Normalized and corrected distributions of the yields of the product ions formed in the reactions of the radical anions $CHX^{\cdot-}$ (X = F, Cl, Br) and the anions CH_2X^- (X = Cl, Br) with $CH_3OCO_2CH_3$ and $CF_3CO_2CH_3$.

substrate	product ions	reaction	$CHF^{\cdot-}$	$CHCl^{\cdot-}$	$CHBr^{\cdot-}$
$CH_3OCO_2CH_3$	X^-	$B_{AC}2$	-	40	75
	$COCHX^{\cdot-}$	$B_{AC}2$	35	60	25
	$CH_3OCOCX^{\cdot-}$	$B_{AC}2$	5	-	-
	$CH_3OCO_2^-$	S_N2	60	-	-
$CF_3CO_2CH_3$	X^-	$B_{AC}2$	-	15	30
	CF_3COCHX^-	$B_{AC}2$	15	30	40
	$CF_3CO_2^-$	S_N2	85	55	30
				CH_2Cl^-	CH_2Br^-
$CH_3OCO_2CH_3$	X^-	$B_{AC}2$		25	80
	$CH_3OCOCHX^-$	$B_{AC}2$		50	10
	$CH_3OCO_2^-$	S_N2		25	10
$CF_3CO_2CH_3$	X^-	$B_{AC}2$		20	45
	$CF_3CO_2^-$	S_N2		80	55

The product ion distributions are given in Table 1. The S_N2 substitution gives rise to only one product ion, i.e. $CH_3OCO_2^-$ and $CF_3CO_2^-$, respectively, whereas the $B_{AC}2$ reaction leads to a number of product ions as indicated in Table 1 and exemplified in Scheme 1 for the reactions of the radical anions with dimethyl carbonate. Summation over the abundances of the $B_{AC}2$ product ions shows the relative importance of this pathway compared to substitution. This reveals that i) the S_N2 process is less important for radical anions than for carbanions with the esters, and ii) for both the radical anions and the carbanions the $B_{AC}2$ pathway is more important for the heavier halogens.

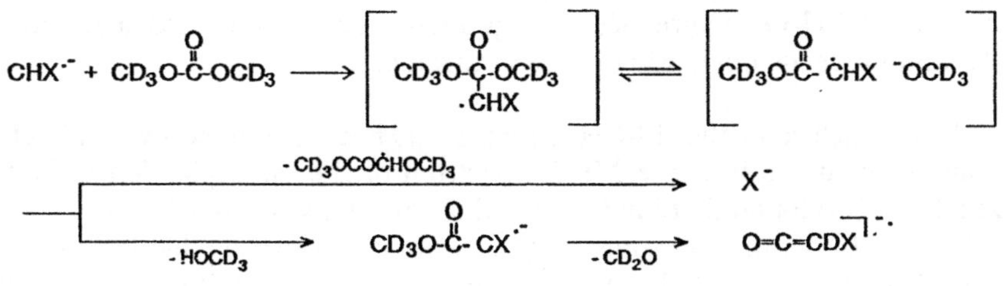

4. Fragmentation of ionized N-(Alkoxymethyl)sulfonamides: An Alkyl Group Migration

Witold Danikiewicz, and Krzysztof Wojciechowski
Institute of Organic Chemistry, Polish Academy of Sciences, ul. Kasprzaka 44/52, 01-224 Warsaw, Poland.

During our project directed towards the synthetic applications of aza-*ortho*-xylylenes[8] generated *via* thermal SO_2 expulsion from 2,1-benzisothiazoline-2,2-dioxide derivatives (benzosultams) we observed a rather surprising fragmentation reaction in the 70 eV mass spectrum of 1-methoxymethyl-2,1-benzisothiazoline-2,2-dioxide **1**. Contrary to other molecular ions of benzosultams studied previously[9] $\mathbf{1}^{+\cdot}$ do not eliminate SO_2 but gives rise to an intense signal at $(M\text{-}30)^{+\cdot}$, ion *b*. Deuterium labelling experiments using **1** with a OCD_3 group confirmed unequivocally the elimination of a molecule of formaldehyde, CH_2O, generated exclusively from the CH_2O- segment attached to the N atom. This observation prompted us to study the fragmentations of **1** and the related acyclic N-methoxymethylsulfonamide **2** more in detail. The main mass spectrometric fragmentation routes of **1** and **2** are presented elesewhere.

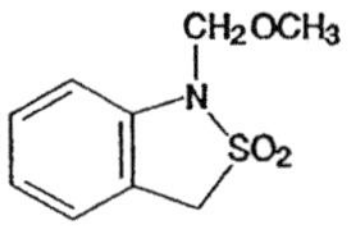

1

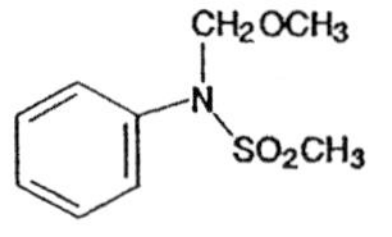

2

The second important fragmentation route of $\mathbf{1}^{+\cdot}$ besides elimination of CH_2O starts by the loss of the CH_3O group yielding ion m/z 182 and proceeds by the sequential losses of SO_2, HCN, and C_2H_2. The abundance of the ion m/z 182 is about 60% of the abundance of ion *b*, m/z 183. The general fragmentation pattern of **2** is very similar to that of **1**.
The most intense peak in the EI mass spectra of both compounds corresponds to the ion *a*, $[CH_3\text{-}O{=}CH_2]^+$, formed by homolytic N-C bond cleavage.

The MIKE spectrum of the molecular ions of **1** exhibits only two abundant peaks. The one peak at m/z 45 corresponds to ion *a* while the other one at m/z 183 results from the elimination of CH_2O yielding ion *b*. In the MIKE spectrum the relative abundance of the signal of ion m/z 182, formed by loss of a CH_3O radical from $\mathbf{1}^{+\cdot}$ is less than 10% of the ion *b*, m/z 183. Hence, the ion m/z 182 is significantly less abundant than in the EI mass spectrum. These results indicate that the elimination of CH_2O requires less energy than a direct bond cleavage and loss of $CH_3O\cdot$ and proceeds obviously by a rearrangement.

The structure of the $[M\text{-}CH_2O]$ ion *b* has been examined by its MIKE spectrum and comparison with the MIKE spectrum of the molecular ion of N-methylbenzosultam. It is found that ion *b* originating from **1** gives two product ions m/z 151 and 119 in a 10 : 1 ratio. Accurate mass determination reveals that the ion m/z 152 is formed by the loss of a CH_3O radical from *b*. The peak at m/z 119 is a doublet consisting of 85% $C_8H_9N^{+\cdot}$ (loss of SO_2 from *b*) and 15% of a species $C_7H_5NO^{+\cdot}$. In contrast, the MIKE spectrum of the molecular ion of N-methylbenzosultam exhibits only one peak at m/z 119. These results demonstrate that the ions *b* formed by loss of CH_2O from $1^{+\cdot}$ are at least two isomers. To rationalize these observations we propose a mechanism which involves an ion/neutral complex[10] composed of the methoxymethyl cation *a* and a benzosultamyl radical and which gives rise to the isomeric ions b_1 and b_2 by intramolecular methyl cation transfer either to the N- or O-atom of the sultamyl radical.

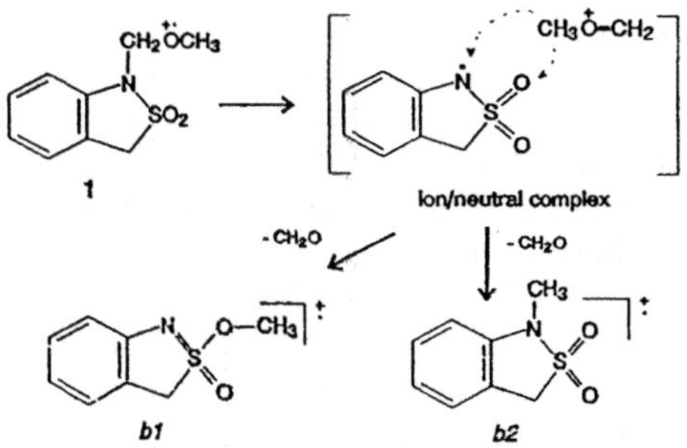

The MIKE spectrum of the molecular ions of **2** show the peaks for loss of CH_2O and of CH_3O in a ratio of approximately 1 : 2. Thus, contrary to **1** the loss of CH_3O by direct bond cleavage is preferred by metastable $2^{+\cdot}$ and the elimination of CH_2O by the rearrangement is less favourable from the acyclic sulfonamide **2**. Further, the structure of the product ion $[M\text{-}CH_2O]^{\cdot +}$ is not so clear as in the case of *b* generated from **1**. Metastable ions spectra show that the ion $[M\text{-}CH_2O]^{\cdot +}$ fragments along several pathways, the most important ones being the losses of the radicals $CH_3O\cdot$, $C_2H_5O\cdot$, $CH_3SO_2\cdot$, and $C_2H_5SO_2\cdot$. The first process is consistent with a structure analogous to b_1 formed by the proposed mechanisms *via* an ion/neutral complex. Elimination of the fragment CH_3SO_2 can be rationalized by both structures analogous to ions *b* from the fragmentations of metastable $1^{+\cdot}$. However, the two other processes require some additional rearrangements and will not be discussed here.

5. Anchimeric Assistance in the Gas-Phase Hydride Transfer Reactions

Maria. E. Crestoni, Simonetta Fornarini, Massimo Lentini, and Maurizio Speranza
Dipartimento di Studi di Chimica e Tecnologia delle Sostanze Biologicamente Attive, Università di Roma "La Sapienza", P. le A. Moro, 5, I-00185, Roma, Italy.

The interaction between an ion A^+ and a molecule S in the gas phase leads to the formation of a long-lived complex AS^+ which may either evolve to products crossing the activation energy barrier E^* or relocate all of the interaction energy E^o into a single mode and dissociate back to the reactants. The reaction efficiency is determined by the branching between these competing channels and depends on the free energy profile and the experimental conditions. The complex at the top of the internal barrier E^* is a tight complex (low density of states, negative activation entropy) whereas the complex involved in the back dissociation is a loose one (high density of states, positive activation entropy)).

At the collision-free limit the complex AS^+ is characterized by a constant total energy and angular momentum. Its evolution to reactants or products is governed essentially by entropic rather than by enthalpic factors. Hence, the crossing of the internal barrier is rather unlikely and the reaction efficiency is generally low. Increasing the pressure of a buffer gas favors the disposal of excess energy E^o of

AS^+ by unreactive collisions with the gas. The lifetime of AS^+ increases, and is evolution is now progressively governed by enthalpic rather than entropic factors and follows the lowest activation energy pathway. For exothermic processes with $E^* > E^o$ (Fig. 1) this favors back-dissociation over conversion to products. Therefore, a negative pressure dependence of the reaction efficiency is observed. Instead, for exothermic processes with $E^* < E^o$ (Fig. 2) increasing the pressure of the system favors the conversion of AS^+ to the products, and hence, a positive pressure dependence of the reaction efficiency is observed. Thus, the pressure dependence of the reaction efficiency can be used as a criterion for characterizing the general features of the potential energy profile of a ion/molecule reaction and for investigating its detailed reaction mechanism. This technique has been applied here to demonstrate the intervention of a so called anchimeric assistance in some gas-phase hydride transfer reactions

Figure 1:

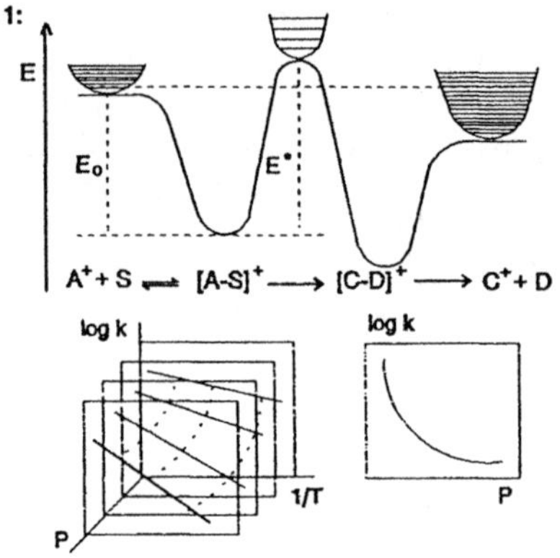

Figure 2:

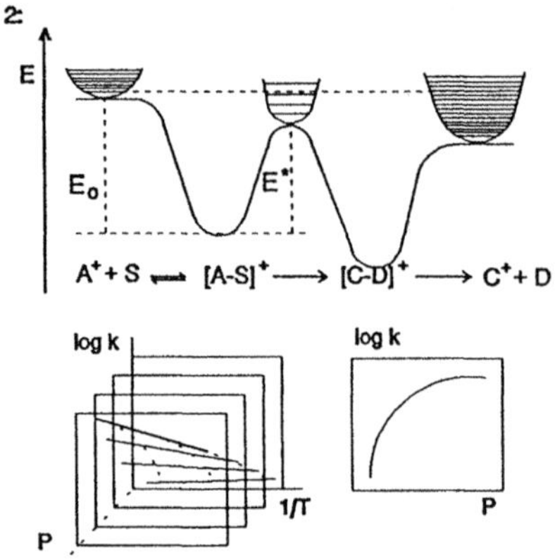

The reaction investigated concerns the hydride transfer from 1,1-dimethylyclopentane to *tert.*butyl and *sec.*propyl cations under the low pressure (*c.* 10^{-8} mbar) typical of FT-ICR experiments and the high pressure of radiolytic generation of the carbenium ions. Using the *tert.*butyl cation as the hydride acceptor no reaction was observed by FT-ICR but a quite intense hydride transfer of 7 - 10% in the radiolytic experiments. If radiolysis was carried out in the presence of methanol as a trap for intermediate carbenium ions the reaction yields predominantly equal portions of *cis* and *trans* 1,2-dimethylcyclopentyl methyl ether and minor amounts of 1-methylcyclohexyl methyl ether. If the 1,2-dimethylcyclopentyl cation is generated independently by protonation of 1,2-dimethylcyclopentene under comparable experimental conditions, trapping of the ion so formed with methanol leads only to a 1:1 mixture of *cis* and *trans* 1,2-dimethylcyclopentyl methyl ether, excluding an isomerization of the 1,2-dimethylcyclopentyl cation as the source of the 1-methylcyclohexyl cation which is more stable by 1 kcal/mol.

The reaction pattern observed in the radiolytic experiments is not consistent with the schematic potential profiles shown in Fig. 3 for a stepwise mechanism of the hydride transfer from the 1,1-dimethylcyclopentane to the *tert.*butyl cation. In fact, the stepwise mechanism imposes the overcoming of a transition state lying above the total potential energy given by the reactants, which should result in a very low reaction efficiency as observed in the FT-ICR experiments. However, the reaction efficiency should exhibit *a negative pressure dependence*, contrasting the distinct *positive* pressure dependence observed by comparing the outcome of the FT-ICR and radiolytic experiments. This indicates that the reaction proceeds through a *concerted mechanism involving anchimerically assisted transition states* the potential energy of which must be lowered by interaction with neighboring methyl or methylene groups.

Figure 3:

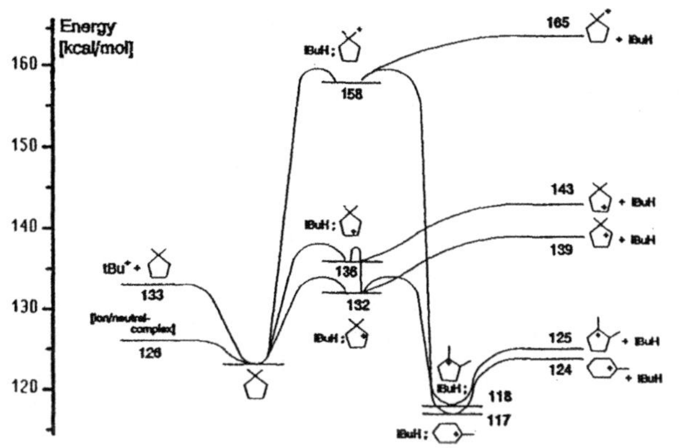

Accordingly, the hydride ion may depart from the α-methylene group, anchimerically assisted by the vicinal methyl group. Alternatively, the hydride ion may depart from the methyl group. Here, the anchimeric assistance is provided by the neighboring methylene moiety rather than by the methyl groups. This is shown by the absence of the corresponding derivative, namely 1-ethylcyclopentyl methyl ether, among the reaction products. A hydride departure from the ß-methylene group is exluded since this leads to the enthalpically unfavored 2,2-dimethylcyclopentyl ion as the intermediate.

In the case of a hydride abstraction from 1,1-dimethylcyclopentane by the *sec.*propyl cation the analysis of the potential energy profiles does not allow any straightforward conclusions regarding the processes involving ring hydrogen atoms, since non of the internal energy barriers protrudes above the total energy given by the reactants. However, in view of the exceedingly low energy barriers involved in exothermic hydride transfer in the gas phase, a stepwise mechanism of the hydride transfer would proceed through a fast preequilibrium between the collision complex *sec.*propyl ion/1,1-dimethylcyclopentane and a product complex consisting of 2,2-dimethylcyclopentyl cation and propane, followed by the rate-limiting isomerization of the 2,2-dimethylcyclopentyl ion into the more stable 1,2-dimethylcyclopentyl cation. This should give rise to H/D-exchange between the reactants in the case of perdeuterated *sec.*propyl cations which is not observed. Thus, the experimental results agree with a hydride transfer from the α-methylene group to the *sec.*propyl cation *via* a concerted mechanism as well.

The operation of anchimeric assistance in these gas-phase hydride transfers is supported by an evaluation of the kinetic isotope effects using 2-deutero-1,1-dimethylcyclopentane.

6. Protonated *para*-Benzoquinone

Karl K. Irikura[a], Michael Meot-Ner (Mautner)[a], L. Wayne Sieck[a], Andrew D. Fant[b] and Joel F. Liebman[b]
[a] *Chemical Kinetics and Thermodynamics Division, National Institute of Standards and Technology, Gaithersburg, MD 20899, U.S.A.*
[b] *Department of Chemistry and Biochemistry, University of Maryland Baltimore County, Baltimore, MD 21228, U.S.A.*

The chemistry of quinones is dominated by their electron-accepting character [11,12]. In this contribution the rather different chemistry was discussed, in which the quinone serves as an electron donor. In particular, the structure and stability of protonated *para*-benzoquinone (pBQ) was determined using high-pressure mass spectrometry and ab initio calculations.

The principal experimental difficulty is the low volatility of pBQ (mp = 114 °C), which was overcome by injecting a dilute aqueous solution of pBQ into an inlet reservoir heated to 180 °C. Gas-phase titration[13] of the acidic protons of $pBQH^+$ with D_2O indicates only a single acidic site, which implies that pBQ is protonated on oxygen (structure **1**) and not on the ring (**2**). The proton affinity of pBQ could not be measured in equilibrium because of the ill-defined pressure of pBQ, but bracketing experiments indicate PA(EtCN) < PA(pBQ) < PA(i-C_4H_8). Using the most recent proton affinity scale[14] then yields PA(pBQ) = 798.5 ± 6.7 kJ/mol at 300 K (standard uncertainty, 1σ).

$$(pBQH^+)\bullet(H_2O) \rightarrow pBQH^+ + H_2O \qquad \textbf{(1)}$$

The equilibrium constant for the hydration reaction 1 was measured from 148 to 270 °C, yielding ΔH°_D = 90.0 ± 2.3 kJ/mol and ΔS°_D = 123.4 ± 4.9 J/mol.K (95% confidence).

Exploiting the linear correlation between proton affinities and OH^+...O bond enthalpies[15] leads to PA(pBQ) = 804.3 ± 7.7 kJ/mol (standard uncertainty) after adjusting for the revised proton affinity scale.[14] Weighting the bracketing and hydration methods equally leads to our final experimental value PA(pBQ) = 801.4 ± 8.9 kJ/mol (standard uncertainty).

Scheme 1

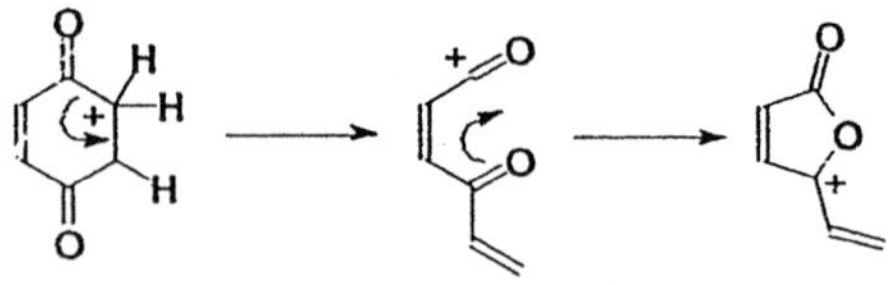

2   Z-3   4

Ab initio calculations were done both to verify the proton affinity measurement and to determine the role of irreversible rearrangements such as **2** → **3** and **3** → **4**, which would affect the interpretation of the experiments. Proton affinities in (kJ/mol) were calculated at the MP2/6-311-G**//HF/6-31G* level (including vibrational zero-point energy) for pBQ (805.3), acetone (815.7), and benzene (736.0).

Comparison with current[14] values for acetone (801.4 ± 8.9) and benzene (744.8 ± 5.0) is good. Our best theoretical estimate for PA(pBQ), from an empirical adjustment based upon acetone, is 807.2 ± 11.2 kJ/mol, in agreement with the present experimental value. The computed geometries of the keto moieties pBQ and acetone, both neutral and protonated, are very similar. Combined with the very similar proton affinities of pBQ and acetone, this indicates that pBQ behaves as an aliphatic ketone in its protonation chemistry.

Relative energies, calculated at the same level as the proton affinities, are 0, 201, 61, and 8 kJ/mol for **1**, **2**, **3**, and **4**, respectively, indicating that protonation on the ring is about 200 kJ/mol less favorable than protonation on oxygen. Difficulties in locating an energy minimum corresponding to **2**, and failure to locate a minimum corresponding to the *Z*-isomer of **3**, lead us to conclude that protonation of pBQ on the ring will lead to fairly facile ring-opening to acylium ion *Z*-**3** followed by closure to form the ionic, conjugated lactone **4** (Scheme 1).

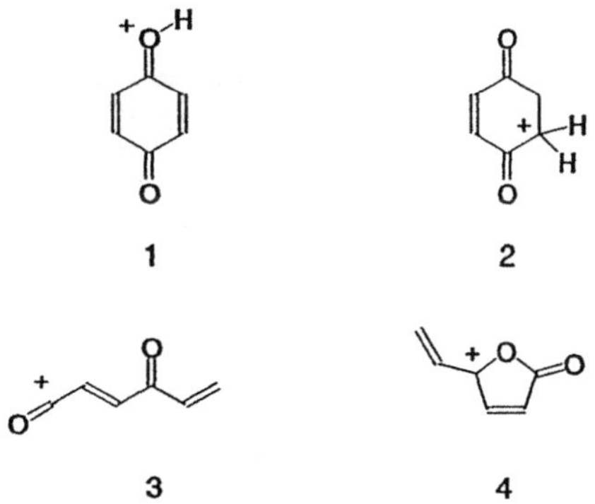

The present experimental PA(pBQ) implies an enthalpy of formation $\Delta_f H^\circ_{298}(\mathrm{pBQH^+})$ = 605.6 ± 9.8 kJ/mol (standard uncertainty). This is much lower than values of about 678 kJ/mol derived from appearance energy (AE) measurements from *p*-hydroxyanisole (pHA).[16] However, since there is no least-motion pathway leading from $\mathrm{pHA^+}$ to singlet $\mathrm{pBQH^+}$, the AE experiments probably produce $\mathrm{pBQH^+}$ in its triplet excited state. This interpretation is consistent with our measurements only if the singlet-triplet gap in $\mathrm{pBQH^+}$ is about 72 kJ/mol. For comparison, the singlet-triplet gap in (un-hydroxylated) phenoxyl cation is 83 kJ/mol.[17]

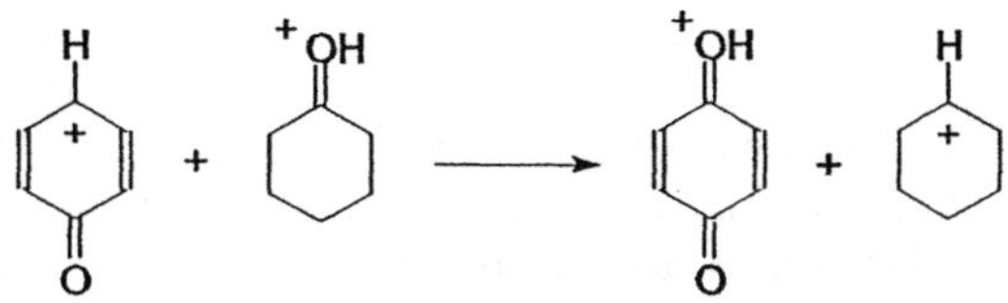

Analogies with related molecules support our measured value for PA(pBQ). For example, the correlation between ionization energies and proton affinities of ketones, combined with IP(pBQ) = 10.04 ± 0.18 eV,[18] suggests that PA(pBQ) = 795 ± 19 kJ/mol (adjusted for revised PA scale), in agreement with our value of 801.4 ± 8.9. The enthalpy of formation of $\mathrm{pBQH^+}$ can also be estimated by approximating the isodesmic reaction shown at right as thermoneutral. This leads to $\Delta_f H_{298}(\mathrm{pBQH^+}) \approx 602.5$ kJ/mol, in good agreement with our value of 605.6 ± 9.8 kJ/mol.

7. Heavy Atom Isotope Effects of Bromine

Allan Petersen
Department of Chemistry, H. C. Ørsted Institute, University of Copenhagen, Denmark.

A series of unusually large heavy-atom kinetic isotope effects is reported for a variety of bromine containing compounds (Table 1). The magnitude seems to depend on the number of carbon atoms and fluorine substitution increases the effect remarkably. The isotope effects were determined from the relative peak heights (MIKE spectra, corrected for the decomposition in the ESA) for the competing loss of ^{79}Br and ^{81}Br:

$$C_2H_4{}^{79}Br^+ \ + \ {}^{81}Br\cdot \qquad {}^{79}BrC_2H_4{}^{81}Br^{+\cdot} \qquad C_2H_4{}^{81}Br \ + \ {}^{79}Br\cdot$$

Table 1: Kinetic isotope effects $k_{79Br}/k_{81}Br$ observed for Br loss from metastable precursor ions.

compound/precursor ion		isotope effect
C_1 - C_2 derivatives	CH_2Br_2	
	CD_2Br_2	1.03 - 1.06
	$BrCH_2CH_2Br$	
	$BrCD_2CD_2Br$	
	CF_2Br_2	1.15
	$BrCF_2CF_2Br$	1.09
C_3 - C_5 derivatives	$Br(CH_2)_3Br$	
	$Br(CH_2)_4Br$	
	$Br(CH_2)_5Br$	< 1.025
	$(CH_3)_2CBr_2$	
	$CH_3CHBrCHBrCH_3$	

The peak shapes for the loss of Br in the MIKE spectra are extremely dependent on the relative position of the bromine atoms in the precursor ion (Fig.1). If the bromine atoms are located at different carbon atoms, "normal" peak shapes are observed, whereas the loss of bromine atoms placed at the same carbon atoms gives rise to peak shapes "with skirts" (except for the decomposition of $CF_2Br_2{}^{+\cdot}$). In the case of $CH_3CH^{79}Br^{81}Br^{+\cdot}$, "strange" peak shapes are observed: the skirts are extremely pronounced and the peak due to loss of ^{79}Br seems to have an extra component exhibiting a narrow sharp singnal, which is completely absent for the loss of ^{81}Br from the same precursor ions. This difference in peak shapes disappears if the loss of ^{79}Br and of ^{81}Br from $CH_3CH^{79}Br_2{}^{+\cdot}$ and $CH_3CH^{81}Br_2{}^{+\cdot}$, respectively, is measured in separate experiments.

Ionized 1,4-dibromobutane reacts by loss of a bromine atom to form a five membered cyclic bromonium ion. This reaction is accompanied by a large kinetic energy release (T_{50} = 150 meV) causing overlapping of the two broad peaks appearing in the MIKE spectrum of $C_4H_8{}^{79}Br^{81}Br^{+\cdot}$ for the loss of ^{79}Br and ^{81}Br. A large kinetic energy release is commonly explained by the energy gain due to the formation of the stable cyclic bromonium ion. Surprisingly however, an identical kinetic energy release and a superimposable MIKE spectrum is observed for the loss of bromine atoms from ionized 2,3-dibromobutane. Since the ion source EI mass spectra are also identical we hypothesized a rearrangement of the molecular ions of 2,3-dibromobutane into the 1,4-isomer in the ion source. However, CID of the $C_4H_8{}^{79}Br^+$ isomers formed by the reaction of the metastable dibromobutane isomers exhibit a significant difference for the signal at m/z 106/107 in the respective MIKE spectra.We also found that the metastable product ions yield CID spectra identical with that of the corresponding ions generated in the ion source. This excludes rearrangement of ionized 2,3-dibromobutane into ionized 1,4-dibromobutane, and no explanation was found for the considerable kinetic energy release during the decomposition of metastable molecular ions of 2,3-dibromobutane.

8. Base Catalyzed Tautomerization of Gaseous Vinylammonium Ions: A Shuttle Mechanism in a Collision Complex

Andreas Nixdorf, Hans-Friedrich Grützmacher
Lehrstuhl I für Organische Chemie, Fakultät für Chemie der Universität Bielefeld, POB 10 01 31, D-33501 Bielefeld, Germany

The gas-phase reactions of radical cations of unsaturated organic halides with ammonia and methylamines continue to be of interest as model systems for the interactions of electron deficient radical cations with electron rich nucleophiles[19]. In the case discussed here, reactions of the radical cations of vinyl chloride and vinyl bromide with ammonia in the cell of a Bruker CMS 47X FT-ICR spectrometer yield $C_2H_6N^+$ product ions with high reaction efficiencies besides some deprotonation of the radical cations by NH_3[19e]. The formation of a $C_2H_6N^+$ product ion corresponds to the substitution of the halogen atom of the radical cation by the NH_3 and occurs *via* an addition/elimination mechanism[19]. The structure of the product ion may correspond to the vinylammonium ion **v** or to the more stable protonated acetaldimine **a**. The formation of **v** is the result of a simple replacement of the halogen atom of the vinyl halide radical cation while the formation of **a** requires additional H migrations. Ab-initio calculations of the reaction pathways favor the formation of **v**[19e], but a subsequent isomerization of **v** to **a** would necessitate either a 1,3-proton shift forbidden by orbital symmetry requirements or two succeeding energy demanding 1,2-hydrogen shifts. To get experimental confirmation of the structure of the $C_2H_6N^+$ ions, these ions were isolated in the FT-ICR cell at the end of the substitution reaction and studied by CID and structural specific ion/molecule reactions.

The CA mass spectrum of the $C_2H_6N^+$ product ions is distinguished clearly from the well known CA mass spectra of the $C_2H_6N^+$ isomers N-methyl formiminium ion, protonated acetaldimine, and protonated aziridine by a significant peak at m/z 17 ($NH_3{}^{+\cdot}$) and a peak of doubly charged $C_2H_6N^{2+}$ generated by charge stripping. Thus, the $C_2H_6N^+$ product ions contain obviously a new $C_2H_6N^+$ species, but it is not evident from CID whether this species is indeed the vinylammonium ion **v** and whether the product ions present only **v** as the single species.

This information can be obtained by using deprotonation of the $C_2H_6N^+$ product ions as a structure specific ion/molecule reaction in gas-phase titration experiments[20]. Here, one makes use of the fact that an efficient proton transfer is only observed for exoergic processes, i.e. to bases of a higher proton affinity (PA) than the corresponding base (vinylamine or acetaldimine) of the $C_2H_6N^+$ product ion. The PA(acetaldimine) = 895 kJ/mol is known[21] and the distinctly smaller PA(vinylamine) 866 kJ/mol obtained by ab-initio calculations[19e]. Table 1 presents the results of the gas-phase titration of the $C_2H_6N^+$ product ions with reference bases of a PA[21] ranging from 861 to 985 kJ/mol, and Figure 1 presents a typical example of an titration experiment.

Table 1: Composition of $C_2H_6N^+$ ion mixture by gas-phase titration

base	PA [kJ/mol]	composition	
		vinylaminium	acetaldiminium
n-tributylamin	985	complete monomodal deprotonation	
2-chloropyridine	897	67	33
meta-toluidine	893	44	56
di-*sec.*butyl ether	874	51	49
methylvinyl ether	868	34	66
3-fluoroaniline	866	35	65
2-phenylpropene	866	> 87	< 13
N-methylformamide	861	no deprotonation	

Two conclusions are drawn from these results. i) None of the $C_2H_6N^+$ product ions is deprotonated by a weaker base than N-methylformamide (PA = 861 kJ/mol) and all product ions are deprotonated by 2-chloropyridine (PA = 897 kJ/mol). This confirms the presence of ions **v** and **a** as the only species present among the product ions. ii) Apparently the $C_2H_6N^+$ product ions are a mixture of ions **a** and **v**, but the composition of this mixture varies with the titration base used for deprotonation. The most acidic component is **v**, and if the titration base is only slightly more basic than vinylamine the apparent amount of **v** in the mixture is much less than by using a stronger titration base.

The explanation of this unexpected result is a *base catalyzed tautomerization within the collision complex of the $C_2H_6N^+$ product ion with the base molecule.* In the collision complex, the base B picks up the proton from the ammonium group of **v** forming a new ion/molecule complex consisting of neutral vinylamine and the protonated base BH^+. This complex can either dissociate into the components or undergo an intramolecular reaction by transferring the proton from BH^+ back to the ß-C-atom of the vinylamine. Since the more stable protonated acetaldimine **a** is

Figure 1

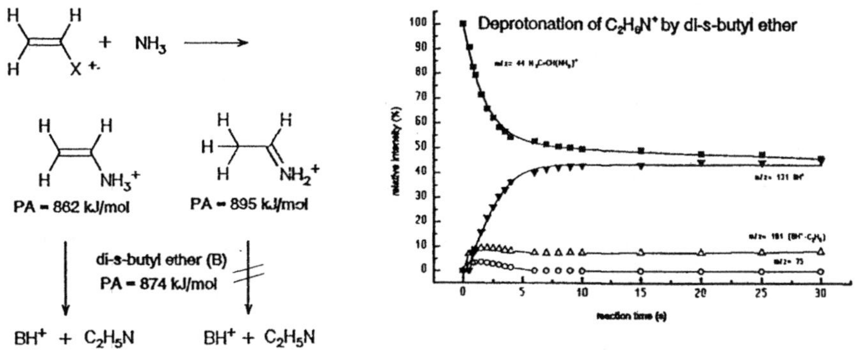

generated in this way, both proton transfers to and from the base B are exothermic for an appropriate base. Thus, the base acts as a proton shuttle and the isomerization corresponds to a *shuttle mechanism*. Further, the efficiency of this base catalyzed tautomerization depends on the branching between dissociation of the complex [vinylamine·BH^+] and the back transfer of the proton: i.e. the lifetime of [vinylamine·BH^+] and the rates of the proton transfer. One factor determining the lifetime of the complex is its internal excess energy, and a rapid dissociation is expected for a complex of high internal energy. Thus, if the first proton transfer from **v** to a strong base is very exothermic, an excited complex [vinylamine·BH^+] is generated by chemical activation and the dissociation of the complex dominates over a proton shuttle. This explains the apparently increasing amount of **v** in the $C_2H_6N^+$ product ion mixture if a relatively strong base is used for the titration experiment.

In fact, likely only the vinylammonium ion **v** is formed initially by the substitution reaction of vinyl halide radical cations with NH_3. This assumption is corroborated by a titration of the $C_2H_6N^+$ product ions with 2-phenylpropene. The PA(2-phenylpropene) = 866 kJ/mol is identical to PA(3-fluoroaniline) but the rates of proton transfer involving carbenium ions are slower than those from onium ions. Hence, 2-phenylpropene is a poor shuttle, and dissociation after the initial proton transfer should prevail in spite of the only slightly exothermic first proton transfer from **v** to 2-phenylpropene. This effect is seen clearly by comparing the amount of isomerization by the proton shuttle mechanism during the deprotonation of the $C_2H_6N^+$ product ions by 3-fluoroaniline and 2-phenylpropene: 3-fluoroaniline induces an isomerization of 65% of the $C_2H_6N^+$ product ions from **v** to **a** (which is not deprotonated by 3-fluoroaniline) and only 35% react as **v** by deprotonation while in the case of 2-phenylpropene more than 87% of the $C_2H_6N^+$ product ions are deprotonated as expected for **v**.

The proton shuttle mechanism presented here for the base catalyzed isomerization of vinylammonium ions **v** to acetaldiminium ions **a** is an interesting example of chemical reactions occurring within collision complexes formed with a definite lifetime during gaseous ion/molecule reactions and of the influence of the detailed reaction dynamics on the course of these reactions. Further, it is expected that proton shuttle and base induced isomerization by proton shifts are general phenomena during the interaction of a base with a protonated molecule exhibiting several basic sites of different basicity. The presence of several protonation sites is quite common for large and complex organic and bio-organic molecules, but the importance of this proton shuttle effect for the experimental determination of the PA of these large molecules by deprotonation and bracketing methods has not yet been considered.

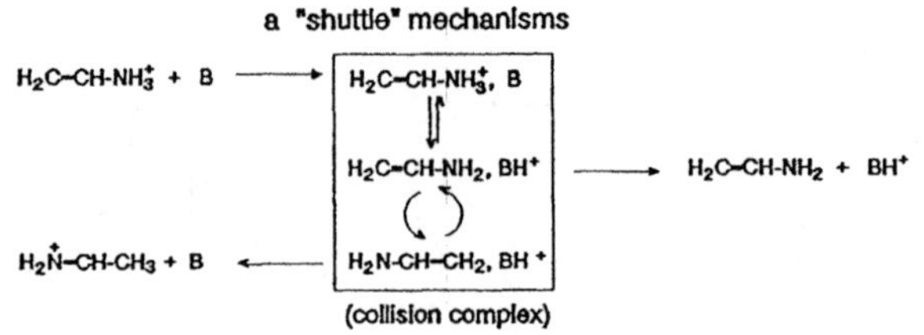

References

1. Wenthold, P. G.; Hu, J.; Squires, R. R.; *J. Am. Chem. Soc.*, **1994**, *116*, 6961.

2. Wenthold, P. G.; Hu, J.; Squires, R. R.; *J. Am. Chem. Soc.*, submitted.

3. Wenthold, P. G.; Squires, R. R.; *J. Am. Chem. Soc.*, **1994**, *116*, 11890.

4. Marinelli, P. J.; Paulino, J. A.; Sunderlin, L. S.; Wenthold, P. G.; Poutsma, J. C.; Squires, R. R.; *Int. J. Mass Spectrom. Ion Process.*, **1994**, *130*, 89.

5. DePuy, C. H.; Bierbaum, V. M.; Flippin, L. A.; Grabowski, J. J.; King, G. K.; Schmitt, R. J.; Sullivan, S. A.; *J. Am. Chem. Soc.*, **1979**, *101*, 6443.

6. Wenthold, P. G.; Squires, R. R.; *J. Phys. Chem.*, **1995**, *99*, 2002.

7. Staneke, P. O.; Groothuis, G.; Ingemann, S.; Nibbering, N. M. M. ; *Int. J. Mass Spectrom. Ion Process.* **1995**, *142*, 83.

8. a) Wojciechowski, K.; *Tetrahedron*, **1993**, *49*, 7277. - b) Danikiewicz, W.; Wojciechowski; and Olejnik, M.; *Tetrahedron Lett.*, **1995**, *36*, 1099.

9. a) Danikiewicz, W.; Wojciechowski, K.; Fokkens, R. H.; Nibbering, N. M. M.; *Org. Mass Spectrom.*, **1993**, *28*, 853; - b) Danikiewicz, W.; Wojciechowski, K.; Fokkens, R. H.; Nibbering, N. M. M.; *Rapid Commun. Mass Spectrom.*, **1993**, *7*, 763.

10. Longevialle, P.; *Mass Spectrom. Rev.*, **1992**, *11*, 157, and references cited therein.

11. Patai, S., Ed. *The Chemistry of Functional Groups: The Chemistry of the Quinonoid Compounds*, Parts 1 and 2 New York: Wiley (1974).

12. Patai, S.; Rappoport, Z., Eds. *The Chemistry of Functional Groups: The Chemistry of the Quinonoid Compounds*, vol. II, parts 1 and 2 New York: Wiley (1988).

13. Harrison, A. G. *Chemical Ionization Mass Spectrometry*, Boca Raton, Florida: CRC Press (1983), Chapter 9.

14. Hunter, E. P. Work in nearing completion.

15. Meot-Ner (Mautner), M. *J. Am. Chem. Soc.* **1984**, *106*, 1257-1264.

16. (a) Ponomarev, D. A.; Takhistov, V. V.; Misharev, A. D.; Orlov, V. M. *Russ. J. Gen. Chem.* **1994**, *64*, 908-911. (b) Brown, P. *Org. Mass Spectrom.* **1970**, *4*, 519-532. (c) Tait, J. M. S.; Shannon, T. W.; Harrison, A. G. *J. Am. Chem. Soc.* **1962**, *84*, 4-8.

17. Dewar, M. J. S.; David, D. E. *J. Am. Chem. Soc.* **1980**, *102*, 7387-7389.

18. Lias, S. G.; Bartmess, J. E.; Liebman, J. F.; Holmes, J. L.; Levin, R. D.; Mallard, W. G. *J. Phys. Chem. Ref. Data* **1988**, *17*, Suppl. 1.

19. a) Thölmann, D.; Grützmacher, H.-F.; *J. Am. Chem. Soc.*, **1991**, *113*, 3281-3287; b) Thölmann, D.; Grützmacher, H.-F.; *Int. J. Mass Spectrom. Ion Proc.*, **1992**, *117*, 415-440; - c) Thölmann, D.; Flottmann, D.; Grützmacher, H.-F., *Chem. Ber.*, **1993**, *124*, 2349-2356. - d) Thölmann, D.; Hamann, S.; Grützmacher; *Int-J.Mass Spectrom. Ion Process.*, **1993**, *137*, 43-54; e) Nixdorf, A.; Grützmacher, H.-F.; *J. Am. Chem. Soc.*, in preparation.

20. Büker, H. H.; Dissertation, Universität Bielefleld 1992; - Büker, H. H.; Grützmacher, H.-F.; European Mass Spectrom.; in preparation.

21. Lias, S.G; Liebermann, J. F.; Holmes, J. L.; Levin, R. D.; Mallard, W. G.; *J. Phys. Chem. Ref. Data*, **1988**, Suppl. 1.
All other PA data cited in this contribution were taken from this compilation.

WORKSHOP ON BIOANALYTICAL APPLICATIONS OF GAS-PHASE ION CHEMISTRY

NICO M.M. NIBBERING
Institute of Mass Spectrometry,
University of Amsterdam,
Nieuwe Achtergracht 129,
1018 WS Amsterdam,
The Netherlands

Introduction

Like the previous NATO Advanced Study Institutes in the field of fundamental gas phase ion physics and chemistry held in Biarritz [1], La Baule [2], Vimeiro [3], Les Arcs [4] and Mont Ste. Odile [5], the meeting in Grainau, Germany at the foot of the "Zugspitze" was both scientifically and socially a great success.
The general trend in mass spectrometry which in the last few years is moving increasingly into the generation of ions from (large) biomolecules by soft ionisation methods, such as matrix-assisted laser desorption ionisation and electrospray, and analysis of these ions by advanced mass spectrometric equipment, such as reflectron time-of-flight and Fourier transform ion cyclotron resonance mass spectrometers, fully justified in the Institute's programme not only tutorial lectures, but also a workshop on bioanalytical applications of gas phase ion chemistry.
Below a summary is given of the ten oral contributions to the workshop, which in an evening session of approximately two hours including discussions were presented in an enthusiastic way. In this summary, an attempt has been made to cover the contributions in coherent subtopics.

1. Hydrogen/Deuterium-exchange of amino acids and peptides in the gas phase

Two contributions were focused on this sub-topic.

E.H. Gur *et al.* (Institute of Mass Spectrometry, University of Amsterdam, Nieuwe Achtergracht 129, 1018 WS Amsterdam, The Netherlands) discussed the parameters which determine the biomolecular reactivity of protonated dipeptides in the gas phase and which may be relevant to analytical applications [6]. To this end a series of alkyl dipeptides with alkyl = H (glycine), CH_3 (alanine) and/or $(CH_3)_2CH$ (valine) and molecules containing only some of the peptide functional groups, such as N-ethylglycyl amide, have been studied using Fourier transform ion cyclotron resonance mass spectrometry. On the basis of hydrogen/deuterium (H/D) exchange of the protonated dipeptides and functional analogues with D_2O three parameters have been uncovered which determine the gas phase reactivity of the protonated dipeptides.

K. R. Jennings (ed.), Fundamentals and Applications of Gas Phase Ion Chemistry, 419–429.

The first parameter is the "local" proton affinity of the most basic group of the peptide. It is observed that it is not the overall proton affinity of the molecule which determines the rate and efficiency of the H/D exchange reaction but rather the nature of the N-terminal side-group. For example the efficiency of the H/D exchange reaction is 23% for the N-terminal glycine peptides, 14% for the N-terminal alanine peptides and 5% for the N-terminal valine peptides. This effect is attributed to the increasing difficulty of moving the proton from the increasingly more basic amino groups (Gly < Ala < Val) to other less basic sites in the corresponding dipeptides and to the longer lifetime of the D_2O/proton dipeptide collision complex of the amino group is less basic.
The second parameter is the total hydrogen bonding capability of the collision complex. The large number of functional groups within the collision complex enables at all times multiple effective hydrogen bonds between all basic sites. This allows all sites to be protonated within the lifetime of the collision complex and effects H/D exchange, as opposed to a situation without effective hydrogen bonding which leaves the proton on the most basic site. This multiple hydrogen bonding eventually leads to what is in the literature termed proton mobility.
The last parameter concerns the restrictions imposed upon the peptide backbone on the orientations of the peptide functional groups. These geometric restrictions limit the effectiveness of hydrogen bonds between peptide functional groups and thus favour protonation of the basic site of the incoming reagent. In the H/D exchange reaction this parameter enables the much less basic D_2O molecule to be protonated in the presence of the very basic peptide amino group.
These three parameters are thought to be valid not only for peptides but also for other polyfunctional molecules.

Subsequently, A.G. Harrison *et al.* (Department of Chemistry, University of Toronto, Toronto, Ontario, Canada M5S 1A1) summarised their results obtained in a study of the H/D exchange reactions of protonated amino acids and small protonated peptides with ND_3 [7]. H/D exchange reactions have been carried out under chemical ionisation (CI) conditions using ND_3 or CH_3OD as reagent gases [8,9], by fast atom bombardment (FAB) using a deuterated matrix such as glycerol-O-d_3 [10] or by electrospray (ES) mass spectrometry using an appropriate deuterated solvent such as CH_3OD [11, 12] to count the number of active hydrogens in samples - in recent years especially in samples of biological interest - by measuring the shift in the m/z ratio of an appropriate ion when the hydrogens are replaced by deuterium through H/D exchange. The methods of counting active hydrogens encounter difficulties for complex molecules because it can be difficult to deconvolute the exchange pattern from the isotopic pattern due to naturally occurring isotopes, particularly if the exchange is incomplete because of conformational effects. An alternative approach, first demonstrated by Cooks and co-workers [13] is to study the exchange reactions of mass-selected ions in low energy gas phase collisions with suitable reagent gases using tandem mass spectrometry.
Similarly, the H/D exchange reactions of the protonated amino acids and small protonated peptides with ND_3 [7] have been studied in the quadrupole collision cell of

a hybrid BEqQ mass spectrometer [14]. The protonated species produce by FAB was mass-selected by the BE stage of the instrument at 6 keV ion energy, decelerated to 1-5 eV ion energy and introduced into the quadrupole collision cell (q) containing ND_3 at $4x10^{-6}$ torr as read by the ionisation gauge attached to the pumping line for the region (the actual pressure in the cell is 100 to 1000 times the indicated pressure). The reaction products were analysed by scanning the final mass-analysing quadrupole. Protonated amino acids $RCH(NH_2)COOH.H^+$, including those with hydroxyl, carboxyl and amine functions in the R group, exchange all active hydrogens readily. The amide hydrogens of protonated glutamine and asparginine exchange less readily as does the N-bonded hydrogen of the imidazole ring of histidine. The phenolic hydrogen of protonated tyrosine and tyrosine-containing peptides does not undergo significant exchange. Protonated arginine undergoes only a limited exchange with ND_3 and arginine-containing peptides are essentially unreactive with ND_3. Although simple protonated di- and tri-peptides exchange all active hydrogens readily, peptides with amide groups in the sidechain exchange the last two hydrogens slowly.
The exchange reaction is believed to proceed by the sequence:

$$MH_n^+ + ND_3 \rightarrow MH_n^+\text{--}ND_3 \rightarrow MH_{n-1}\text{--}HND_3^+ \rightarrow MH_{n-1}D^+\text{--}HND_2 \rightarrow MH_{n-1}D^+ + HND_2$$

The lack of reactivity of protonated arginine and arginine-containing peptides is attributed to the inaccessibility of the $MH_{n-1}\text{--}HND_3^+$ complex because of the high proton affinity of arginine compared to ammonia. In essence, H/D exchange will occur readily only if the two species MH_{n-1} and ND_3 have proton affinities that are within about 25 kcal mol^{-1} of each other.

2. Alkali metal ions and bacterial glycolipids or small nucleotides

Two presentations dealt with the chemical analysis of alkali metal cationised biomolecules:

C.G. de Koster (Unit for Macromolecular Mass Spectrometry, FOM Institute for Atomic and Molecular Physics, Kruislaan 407, 1098 SJ Amsterdam, The Netherlands) presented results of research on extracellular biosurfactants, in particular bacterial glycolipids, which are produced by many bacteria, yeasts and fungi during cultivation to emulsify hydrophobic carbon sources. These substances are of increasing commercial use and rapid and sensitive methods for structure and mixture analysis are required for screening large numbers of bacterial strains. The glycolipid surfactants studied most intensively are sophorose lipids produced by strains from the genus *Candida* or *Torulopsis* and are composed of a sophorose (Glcβ1-2Glc) unit to which a hydroxy fatty acid is ether linked. The compounds may have only one or two acetate groups at the 6' and 6" position of the disaccharide. FAB ionisation of crude sophorolipid preparations show the formation of protonated molecules. Upon collisional activation under high energy (10 keV) single collision conditions, there is a striking difference between the gas phase ion chemistry of activated $[M + Na]^+$ and $[M + H]^+$ glycolipid quasimolecular ions. Collisional activation of protonated glycolipids produced predominantly series of Y ions and B oxonium ions by cleavage of glycosidic linkages between the carbohydrate constituents and the fatty acid-carbohydrate ether bond. These fragmentation reactions afford sequencing of the glycolipid building blocks consuming minimal sample amounts. Collisional activation of sodiated glycoconjugates induces the formation of product ions which are the results of cleavage of glycosidic bonds (Y ions), pyranose/furanose ring C-C bonds of the carbohydrate moieties (A or X ions), and charge remote fragmentations of the glycoconjugate fatty acid alkyl chain. The formation of A/X fragment ions yields sequence information of the building blocks and the position of the glycosidic linkage. The charge-remote fragmentation pattern may be used to localise double bond positions of the fatty acid alkyl chain. The collision-induced decomposition (CID) spectra of the sodiated and the protonated biosurfactant molecules provide sufficient structural information, *i.e.* sequence of the carbohydrate and fatty acid building blocks, 6' and 6" acetate substitutions, linkage positions and double bond positions, for initial screening of large quantities of bacterial culture broth samples.

Subsequently, D. Vollmer (National Centre for Toxicological Research, Division of Chemistry, 3900 NCTR Road, HFT-230 Jefferson, AR 72079, USA) demonstrated the use of cation-exchange resins in FABMS investigations to be a convenient and versatile method for removing or incorporating alkali metal ions into nucleotide analytes. Removal of alkali metal ion contaminants from nucleotide samples in the FABMS experiment is necessary because many problems such as decreased sensitivity or even non-detection of the analyte can occur when the concentration of alkali metal ions is high. Controlled incorporation of metal ions into nucleotides is also important for systematic studies since species of differing metal ion concentrations are known to occur in nucleotide biochemistry, in nucleotide carcinogenesis and as anti-cancer agents.

Removal of Metal Ions. The cation-exchange resins are beneficial for both positive and negative ions, large and small molecular weight nucleotides (within the range 100-2000 Da) and they can still be effective when the relative concentrations of alkali metal contaminants is high. For example, the noise is decreased by a factor of about 4 and the intensities of the peaks due to $[M-H]^-$ and $[M-2H+Na]^-$ are increased by a factor of about 3 in the spectrum of a mixture of 1 μg TTC and 600 μg NaOAc when 75-100 NH_4^+ resin beads are added. For positive ions, the most abundant $[M+H]^+$ ions are obtained when 1 μg of DDT/DTE matrix with 75-100 resin beads (200-400 mesh) in the H^+ form are used. For negative ions, the most abundant $[M-H]^-$ ions are produced from 1 μg of TEA and 75-100 resins in the NH_4^+ form.

Incorporation of Metal Ions. The study of metal ion-nucleotide interactions in the gas phase has been hindered by the ability to desorb, in a controlled fashion, intact metal ion-nucleotide species. Combining metal cation exchange resins with the FAB matrix, however, allows for the study of gas phase interactions of a specific metal ion with a particular nucleotide. This technique produces multiple metal ion species from microgram or sub-microgram quantities of nucleotide sample and it allows for analysis of positive ions or negative ions.

In these investigations, the metal ion-nucleotide species are studied as positive ions because $[M\text{-}nH^{+}+(n+1)Met]^{+}$ ions, in general, gives rise to the most informative fragmentations. Tandem mass spectrometry experiments have been performed for a sufficiently whole set of trinucleotides and different metal ions to permit general conclusions to be drawn concerning the site for metal ion attachment and the relative metal ion affinities for a given nucleotide base. Both protonated and metallated nucleotides fragment to form similar sequence ions; that is, the product ions masses are shifted by mass difference of the proton and the alkali metal ion. Sequence ions indicate that the site of attachment is non-specific (*i.e.* there is more than one site for attachment) and that the cation can reside on the base or the phosphate moiety. When the metal ion resides on the base, the ratio of the product ion abundances of the (metal ion + base)/(metal ion + thymidine) indicates the relative metal ion affinity for each base mimics that of their proton affinities (*i.e.* G > C-A >>T). Multiple metallated species (*e.g.* $[M\text{-}H^{+} + 2Met]^{+}$, $[M\text{-}2H^{+} + 3Met]^{+}$ of the nucleotides also give similar sequence ions that contain one or more metal ions.

3. Electrospray tandem mass spectrometry of DNA-adducts, a new O-linked glycan and intact haemoglobin β-chain.

Three contributions described the use of electrospray ionisation (ESI) tandem mass spectrometry in structural analysis of biomolecules.

E.L.Esmans *et al.* (Department of Chemistry, University of Antwerp (RUCA), Groenenborgerlaan 171, B-2020 Antwerp, Belgium) discussed as part of their DNA research the analysis of the phenylglycidyl ether (PGE) adducts of 2'-deoxycytidine and 2'-deoxyadenosine by liquid chromatograph (LC) ESMS/MS [15-17]. Since modification of DNA after exposure to a chemical carcinogen may lead to DNA misreplication and eventually to tumour formation, a study of the resulting DNA adducts is warranted. Several sensitive methods have been developed for the analysis of carcinogen DNA adducts as immunoassays and ^{32}P post labelling. Unfortunately, these techniques provide no structural information unless some standard compounds are available. This is, however, not often the case. Mass spectrometry is a sensitive technique allowing structural analysis of minute amounts of polar material and is therefore a powerful technique for the analysis of such adducts.

Prior to a study aiming at the isolation and identification of modified DNA constituents obtained after exposure of calf thymus DNA to mutagenic epoxides, a programme was started in which the separation and the mass spectral behaviour of 2'-deoxynucleos(t)ide adducts are studied with the aid of different mass spectral techniques. These experiments provide more insight into the way these epoxides interact with DNA constituents and the resulting adducts can serve as reference compounds when reactions are carried out on oligonucleotides and DNA. This implies the development of a multiple separation method, the determination of a number of PGE units attached to the 2'-deoxynucleos(t)tide, the localisation of the alkylation site and a study of the stability of these DNA adducts.

In view of the aforesaid, the reaction mixtures of PGE with 2'-deoxyadenosine (dAdo) and 2'-deoxycytidene (dCyd) were analysed by both (+) and (-) LC-ESMS. Low energy CID spectra were recorded during LC-MS runs in order to investigate the exact localisation of the PGE unit on the heterocyclic moiety.
The adducts were analysed on a VG "Quattro II" mass spectrometer coupled to a Kontron HPLC system. Separation was done on a RP Select B column (125 x 4 mm) using the following method: isocratic elution for 8 min. (65% 0.01 M NH_4 OAc/ 35% CH_3OH), then to 100% CH_3OH in 10 min. After 12 min. at 100% CH_3OH the initial conditions were resumed. The flow rate was 0.8 ml/min (split 10/1).
The reaction mixture of PGE/dAdo was characterised by the presence of two PGE adducts. The low energy CID spectra of the $[M+H]^+$ ions generated under ES (-) conditions. From these data, it could be concluded that alkylation primarily occurred at position N3. A very small amount of an adduct alkylated at position N4 was detected. This adduct co-eluted with the N3-PGE adduct.

V.N. Reinhold *et al.* (Mass Spectrometry Resource, R806, Boston University Medical Center, Boston, MA 02118, USA) presented the structural details of a new type of O-linked oligosaccharide of 1244 Da which was found on several proteins secreted by the Gram-negative bacterium Flavobacterium meningosepticum [18]. Linkage, sequence and branching of this unusual heptasaccharide was defined by ESMS and CID [19]. The proposed structure was supported by a combination of deuterio-labelling, GC-MS composition and methylation analysis, and the preparation of several chemical analogues and derivatives with each product evaluated by MS and CID. The singly branched structure contained seven residues, including three different uronyl analogues: a methylated rhamnose and mannose, a glucose and a reducing terminal mannose. Only pyranose ring forms were detected: ((2-OMe)Man1→4GlcNAcU1→4GlcU1→4Glc1→4(2-OMe)GlcU1→4[(2-OMe)Rham 1-2]Man).

A. Buzy, *et al.* (Department of Chemistry, University of Warwick, Coventry CV4 7AL, U.K.) presented results of an ESI/MS study of the intact haemoglobin β-chain in order to detect variants. Haemoglobin variants are carried by more than 1 in 800 of the general population and typically are characterised by single amino acid substitutions along either the α-chain or β-chain of the protein [20]. When studied by means of ESI/MS, large peptides often give ions carrying 15-25 charges and the efficiency of collision-induced decomposition (CID) under "low-energy" conditions rises rapidly as the charge on the ion increases resulting in significant fragmentation which gives useful structural information. Frequently, equal amounts of normal and aberrant chains are found, i.e. they occur as $\alpha_2\beta\beta^*$, where β^* indicates an aberrant chain. The mass difference, Δm, between these chains ranges from 1- 129 Da and if ions carry z charges, the peak separation falls to Δm/z in the ESI mass spectrum. The general strategy used was as follows:
0.25μL samples of untreated blood were diluted and centrifuged, acidified with formic acid and injected into a "Quattro II" tandem mass spectrometer (Micromass, UK). The relative molecular mass (RMM) of the protein was found by means of ESI/MS. Since

the accuracy of mass measurement is typically ~ ± 0.3 daltons, variants for which Δm is as little as 1 dalton can usually be detected in this way. The low energy CID mass spectrum of the complete protein was then obtained. For aberrant amino acid residues close to either the N- or C-terminus, a shift in the observed m/z of b or y ions that include the substituted residue may be observable, allowing one to determine the position of substitution. If the aberrant residue is well-removed from either terminus, the ions will be multiply-charged so that the separation on the m/z scale becomes Δm/z. This must be at least 0.3 to be detected in the absence of the normal protein but normally at least 5 if normal and aberrant forms co-exist. Even if Δm is sufficiently high, unequivocal location of the position of substitution may be impossible if the sequence of b or y ions is incomplete. In such cases, cleavage of the peptide is carried out using CNBr followed by separation of the resulting proteins by HPLC. The RMM and CID spectrum of each protein is then determined and if the position and type of substitution is still uncertain, further enzymatic cleavage is carried out on the protein carrying the variant residue.

Two abundant series of multiply-charged ions appeared at m/z = (15,126.4 + z)/z and (15,867 + z)/z respectively where z = the number of protons attached to the sample molecule, corresponding to the α- and β- chains. The β-chain consists of 146 amino acid residues for the central section of which little sequence information can be derived from the low energy CID spectrum: 60 of the 146 amino acid residues (42%) were identified from the CID spectra of the complete β-chain, rising to 89 (61%) from the CID spectra of the two CNBr peptides.

The mass accuracy obtainable compares well with that previously reported [21] which requires isolation of the chains and removal of other components before analysis is carried out. Five samples of human haemoglobin were studied that contain normal and aberrant β-chains having mass differences ranging from 1 - 77 Da. In each case, since the aberrant amino acid residue was not more than 22 residues from one or other terminus, measurement of the RMM and the CID spectrum of the β^{17+} and β^{*17+} ions together for low values of Δm or the CID spectrum of the β^{*17+} ion alone for higher values of Δm enabled one to locate the aberrant amino acid residue precisely or to determine the region of the chain in which it occurred. Difficulties arise if glutamic acid is present since this inhibits the formation of these fragment ions and prevents the position of amino acid replacement being identified with certainty.

4. Low- and high-energy collision-induced dissociation tandem mass spectrometry of glycoalkaloids and their aglycones, sugars and glycosidic antibiotics

Two contributions in this area of research were presented.

M. Claeys *et al.* University of Antwerp (UIA), Department of Pharmaceutical Sciences, Belgium) discussed some recent results of a comparison between low- and high-energy collision-induced dissociation (CID) tandem mass spectrometry in the analysis of glycoalkaloids and their aglycones [22]. The application of tandem mass spectrometry has demonstrable advantages over conventional mass spectrometry in the structural analysis of saponins and related compounds such as the glycoalkaloids. The glycoalkaloids are a group of nitrogenous steroidal glycosides found in common food plants of the *solanum* species, including potatoes, tomatoes and aubergines, which are of biological interest and concern because of their toxicity.

The low energy CID spectra were recorded on a Fisons VG70SEQ mass spectrometer of the hybrid type, whilst the high-energy CID spectra were obtained on a Kratos "Concept" II HH instrument of the four sector type. Low energy CID spectra were also examined in an effort to define which particular fragmentations among the many typically observed during high energy CID, can also be induced by low energy collisions. The results show that both low- and high-energy CID of $[M+H]^+$ ions are useful for the structural characterisation of the aglycone part. Low energy CID favours rearrangement reactions and charge-driven fragmentation of the aglycone rings, whilst high-energy CID spectra are more complex and contain additional ions which appear to result from charge-remote fragmentations, multiple cleavages or complex rearrangements. With respect to the characterisation of the carbohydrate part, the low-energy CID spectra reveal a clear series of Y_n^+ ions (ion nomenclature according to Domon and Costello [23]) which is useful for determining the carbohydrate sequence. The high energy CID spectra are more complex and also show $^{1,5}X_n^+$ and Z_n^+ type ions. It was pointed out that other methods are needed to obtain information on the nature of the individual sugars and specific carbohydrate linkages.

K.R. Jennings *et al.* (Department of Chemistry, University of Warwick, Coventry, CV4 7AL, U.K.) presented results to demonstrate the differences found in the CID spectra of alkali metal cationised and permethylated oligosaccharides. Low energy CID spectra were recorded on a Finnigan-MAT TSQ 70 in which ions were generated by means of FAB ionisation and xenon was used as the collision gas. High energy CID spectra were recorded on a Kratos "Concept" II HH four sector instrument using a collision cell floated 4kV above earth potential, with helium or argon as the collision gas. Both types of CID spectra of the $[M+H]^+$ ions contained peaks due to abundant oxonium ions which give valuable information on sequencing and branching patterns but only limited information on linkages. The corresponding CID spectra of $[M+Na]^+$ ions contain many ions resulting from multiple bond cleavages the relative yields of which are strongly dependent on collision energy and target gas. At low collision energies, peaks arising from Y and C ions are common but peaks arising from ring cleavages are much less common. At high collision energies, using argon as the collision gas, however, the $[M+Na]^+$ ions of the permethylated oligosaccharides gave many fewer Y ions but a considerable number of peaks arising from ring cleavage which were very useful in assigning positions of linkage and, in one case, indicating the position of branching of the oligosaccharide. From this type of observation, it was possible to identify each type of linkage present in several hexoses by the presence or absence of various diagnostic peaks that may arise from the cleavage of two ring bonds [24]. Similar peaks were observed in the high energy CID spectra of the $[M+Na]^+$ of a series of permethylated glycosidic antibiotics when argon replaced helium as the collision gas. There was a considerable increase in overall sensitivity and some linkage information could be derived from the CID spectra [25].

5. Application of low temperature fast atom bombardment mass spectrometry to cryobiology.

This research topic was presented by M.V. Kosevich *et al.* (B. Verkin Institute for Low Temperature Physics and Engineering of the National Academy of Sciences of the Ukraine, 47, Lenin Avenue, Kharkov, 310164, Ukraine) as part of their ongoing study in cryobiology. As stated at the beginning of the presentation, secondary emission mass spectrometric techniques of ion production from solids and viscous liquids applied to cooled and frozen solutions of biomolecules and cryoprotectors are expected to provide information which is valuable for molecular cryobiology. For that purpose, a dedicated low temperature (LT) FAB ion source was designed which allows one to control the temperature of the sample in the range from ambient to that of liquid nitrogen. With the use of the LT FAB technique, the dependences of FAB mass spectra of frozen water solutions of different compounds (alkali metal salts, a cryoprotector glycerol, nucleic acids, nitrogen bases) on concentration and temperature were obtained. An allowance of these mass spectra allowed the investigators to propose a model of ion sputtering the basis of which is related to the strong dependence of the LT FAB mass spectra pattern on the structure of the frozen sample, its surface morphology and phase transitions with temperature variations. It is known that at rates of freezing attainable for macroscopic volumes of water solutions phase separation takes place (which is the main physical mechanism of the damage of biological material during cryo-conservation) resulting in formation of pure water (ice) crystals and eutectics containing solutes. The geometrical dimensions of these structural elements of solid, determined earlier in cryobiological studies, appear to be from two to four orders of magnitude higher than the mean diameter of the zone excited by the impact of a single bombarding particle (10^{-6} - 10^{-4} m as opposed to 10^{-8} m). Due to this fact, ion sputtering occurs independently from separate surface domains of different chemical composition and the resulting net LT FAB mass spectrum represents a superposition of the mass spectrum of neat ice (a characteristic set of water clusters) and the spectrum corresponding to the solute at a given temperature.

Temperature changes cause phase transitions in the samples in accordance with the phase diagrams characteristic of each type of solution. The most interesting examples of phase transitions, which can be monitored by characteristic changes in the FAB mass spectra, are direct sublimation of water from ice crystals (resulting in the formation of the molecular radical ion of water *via* a gas phase FAB ionisation mechanism), melting of eutectic (e.g. glycerol in the case of water-glycerol mixtures) to liquid while water crystals are still solid and dehydration of cryo-crystalline-hydrates.

The revealed sensitivity of LT FAB mass spectrometry to the morphology of the frozen samples and phase transitions with temperature variations forms a basis for further applications of this technique to the study of different effects associated with the freezing of water solutions of cryoprotectors and biomolecules. One of such effects concerns the dehydration of biomolecules during freezing. It is of interest that no mixed clusters of the molecules of different solutes with water were found at subzero temperatures over a wide concentration range. The only case when mixed clusters were sputtered corresponded to the systems in which crystalline hydrates (e.g. cryohydrate $NaCl.2H_2O$) were formed. This observation, in principle, may allow one to estimate the quantity of the so-called strongly-bound water retained by large biopolymers during freezing. The other effect is connected with the significant increase in the concentration of solute due to capture of the water solvent by the growing water crystals under freezing conditions, which may cause the negative changes in the normal functioning of biomolecules. LT FAB allows one to detect corresponding changes in intermolecular interactions. For example, an intense formation and sputtering of clusters of nucleic acids nitrogen bases with alkali metal ions, the latter present in the sample in trace quantities, were observed for frozen water solutions, while for a cryoprotector glycerol solution, which forms a glassy homogeneous solid, only neat protonated bases without metal ligand were observed at the same metal ion concentrations. Further LT FAB studies will be aimed at the elucidation of the peculiarities of intermolecular interactions in analogous systems under low temperature conditions.

References

1. Ausloos, P. (1975) *Interaction Between Ions and Molecules,* Plenum Press, New York.
2. Ausloos, P. (1979) *Kinetics of Ion-Molecule Reactions,* Plenum Press, New York.
3. Almoster Ferreira, M.A. (1984) *Ionic Processes in the Gas Phase,* D. Reidel Publishing Company, Dordrecht.
4. Ausloos, P. and Lias, S.G. (1987) *Structure/Reactivity and Thermochemistry of Ions,* D. Reidel Publishing Company, Dordrecht.
5. Jennings K.R. *(1991)Fundamentals of Gas Phase Ion Chemistry,* Kluwer Academic Publishers, Dordrecht.
6. Gur, E.H., de Koning, L.J. and Nibbering, N.M.M. (1995) The bimolecular hydrogen-deuterium exchange behaviour of protonated alkyl dipeptides in the gas phase, J. *Am. Soc. Mass Spectrom.* **6**, 466-477.
7. Dookeran, N.N. and Harrison, A.O. (1995) Gas-phase H-D exchange reactions of protonated amino acids and peptides with ND_3, *J. Mass Spectrom.* **30**, 666-674.
8. Hunt, D.F., McEwen, C.N. and Upham,. R.A. (1971) Chemical ionization mass spectrometry II. Differentiation of primary, secondary , and tertiary amines, *Tetrahedron Lett.* 4539-4542.
9 Hunt, D.F., McEwen, C.N. and Upham, R.A. (1972) Determination of active hydrogen in organic compounds by chemical ionization mass spectrometry, *Anal. Chem.* **44**, 1292-1294.
10. Sethi, S.K., Smith, D.L. and McCloskey, J.A. (1983) Determination of active hydrogen content by fast-atom-bombardment mass spectrometry following hydrogen-deuterium exchange, *Biochem. Biophys. Res. Commun.* **112**, 126-131.
11. Siegel, M.M. (1988) Hydrogen-deuterium exchange studies utilizing a thermospray mass spectrometer interface, *Anal. Chem.* **60**, 2090-2095.

12. Edmonds, C.G., Pomerantz, S.C., Hsu, F.F. and McCloskey, J.A. (1988) Thermospray liquid chromatography/mass spectrometry in deuterium oxide, *Anal. Chem.* **60**, 23 14-23 17.
13 Ranasinghe, A., Cooks, R.G. and Sethi, S.K. (1992) Selective isotopic exchange of polyfunctional ions in tandem mass spectrometry: methodology, applications and mechanism, *Org. Mass Spectrom.* **27**, 77-88.
14 Harrison, A.G., Mercer, R.S., Reiner, E.J., Young, A.B., Boyd, R.K., March, R.E. and Porter, C.J. (1986) A hybrid BEQQ mass spectrometer for studies in gaseous ion chemistry. *Int. J. Mass Spectrom. Ion Processes* **74**, 13-31.
15. Van den Eeckhout, E, Coene, J., Claereboudt, J., Borramans, F., Claeys, M., Esmans, E. and Sinsheimer, J.E. (1991) Comparison of the isolation of adducts of 2'-deoxycytidine and 2'-deoxyguanosine with phenylglycidyl ether by high-performance liquid chromatography on a reversed-phase column and a polystyrene-divinylbenzene column, *J.Chromatogr.* **541**, 317-331.
16. Lemiere, F.. Esmans. E.L., Van Dongen. W., Van den Eeckhout. E. and Van Onckelen, H. (1993) Evaluation of liquid chromatography-thermospray mass spectrometry in the determination of some phenylglycidyl ether-2'-deoxynucleoside adducts, J. *Chromatogr.* **647**, 211-218.
17. Lemiere, F., Joos, P., Esmans, E., Van den Eeckhout, E. and Claereboudt, J. (1994) Investigation of the adduct formation between phenylglycidyl ether and 2'-deoxy-nucleotides, Proceedings of the 42nd ASMS Conference on Mass Spectrometry and Allied Topics, May29 - June 3, 1994, Chicago, Illinois, USA, p.126.
18. Plummer, T.H., Jr., Tarentino, A.L. and Hauer, C.R. (1995) Novel, specific O-glycosylation of secreted Flavobacterium meningosepticum proteins. Asp-Ser and Asp-Thr-Thr-consensus sites, *J. Biol. Chem.* **270**, 13192-13196.
19. Reinhold, B.B., Hauer, C.R., Plummer, T.H. and Reinhold, V.N. (1995) Detailed structural analysis of a novel, specific O-linked glycan from the proka~ote Flavobacterium meningosepticum, *J. Biol. Chem.* **270**, 13197-13203.
20. Bunn, H.F. and Forget, B.G., "*Haemoglobin: Molecular, Genetic and Clinical Aspects*", Saunders W.B., (1986), pp. 453, 502, 565
21 Shackleton, C.H.L and Witkowska, H. Ewa in "*Mass Spectrometry: Clinical and Biomedical Applications*", Ed. Desiderio D.M., Plenum Press, New York, (1994), **2**, 135
22. Claeys, M., Van den Heuvel, H., Chen, S., Derrick, P.J., Mellon, F.A. and Price, K.R. (1996) Comparison of high- and low-energy collision-induced dissociation tandem mass spectrometry in the analysis of glycoalkaloids and their aglycones, *J. Am. Soc. Mass Spectrom.* **7**, 173-181.
23. Domon, B. and Costello, C.E. (1988) A systematic nomenclature for carbohydrate fragmentations in FAB-MS/MS spectra of glycoconjugates, *Glycoconjugate J.* **5**, 397-409.
24. Lemoine, J., Fournet, B., Despeyroux, D., Jennings, K.R., Rosenberg, R., and de Hofmann, E., (1993) Collision-induced dissociation of alkali metal cationized and permethylated oligosaccharides: Influence of the collision energy and of the collision gas for the assignment of linkage position, *J Am Soc Mass Spectrom.* **4**, 197-203
25. Florencio, M.H., Despeyroux, D., and Jennings, K.R., (1994) Collision gas effects in the collision-induced decomposition of protonated and cationated molecules of carbohydrate antibiotics, *Org. Mass Spectrom.*, **29**, 483-490

WORKSHOP ON „THE APPLICATIONS OF QUANTUM MECHANICAL METHODS IN GAS PHASE ION CHEMISTRY"

J. HRUŠÁK
J. Heyrovský Institute of Physical Chemistry, Academy of Sciences of the Czech Republic, Dolejškova 3, Prague, Czech Republic

1. Introduction

Molecular orbital calculations have been frequently referred in the lectures and discussions during the whole program of the NATO ASI on Fundamentals of Gas Phase Ion Chemistry. It seemed therefore timely to organize a workshop on the applications of quantum mechanical methods. This workshop covered the field of the use of different *ab initio* and density functional (DFT) based methods in the structure and energy profile calculations of various negatively and positively charged systems. Such calculations represent a powerful tool for the interpretation (and understanding) of experimental results. More general aspects of the theoretical treatment of gas phase reactions like the importance of large polarized basis sets and proper inclusion of the correlation energy were discussed previously in the lectures by D. Dixon. In the workshop, five contributions were given and their abstracts will appear hereafter in the order in which there were presented. The main emphasis of the workshop was to demonstrate the different possibilities of electronic structure calculations when applied to specific problems. They were devoted to following topics:

It has been shown by J. Liebman that the different reactivities (proton affinities) in heterocyclic systems can be related to structural features and corresponding changes in the electronic structure of these compounds.

E. Uggerud has demonstrated that the dynamics of gas phase ion reactions can be successfully modeled when the proper shape of the potential energy surface is known.

In his talk R. Squires concentrated on the problem of proper description of the wave function and particularly the role of molecular orbital symmetry in electronic structure calculations. Obviously, this questions are essential for the distonic radical anions and cations. Few examples were given.

K. R. Jennings (ed.), Fundamentals and Applications of Gas Phase Ion Chemistry, 431–444.

G. McGibbon presented an example of a theoretical investigation of a gas phase rearrangements and fragmentations. His mass spectrometric observations can be rationalized based on the shape of the potential energy surface resulting from electronic structure calculations. Further, the good performance of some newly developed methods in the DFT framework were demonstrated.

Some additional examples of the good performance of different density functional schemes for transition metal containing systems were given by J. Hrušák, who has demonstrated that these methods are a useful alternative for smaller systems and remain the only one choice as far as larger open shell systems are concerned.

2. The Proton Affinities of Pyridine, Phosphabenzene and Arsabenzene

J. F. LIEBMAN
Department of Chemistry and Biochemistry, University of Maryland Baltimore County, Baltimore, MD, 21228, USA

P. P. GASPAR
Department of Chemistry, Washington University, St. Louis, MO, 63130, USA.

D. J. BERGER
Department of Chemistry, Virginia Polytechnic Institute and State University, Blacksburg, VA, 24061, USA

It is well-documented [1,2] that the proton affinity of the sp^2 nitrogen-containing heterocyclic base pyridine (**1**, E = N) is greater than that of the sp^3 nitrogen-containing base ammonia. By contrast, the analogous species containing nitrogen's heavier congeners, phosphorus and arsenic, phosphabenzene and arsabenzene[2] (**1**, E = P and As) have lower proton affinities than phosphine and arsine[3,4].

Assisted by PM3 calculations, we recently suggested[5] that the low proton affinities of these phosphorus and arsenic containing heterocycles relative to pyridine arises simultaneously from three factors. As we proceed from E = N, P and As, we find :

a) increasing inductive -withdrawal by the carbon from the hetero atom, the purported basic site

b) increasingly poor C=E overlap in the protonated cations, cf. resonance structure **2**

c) increasing tendency for pyramidalization around the tricoordinated E in the >E-H center (cf. resonance structure **3**)

1

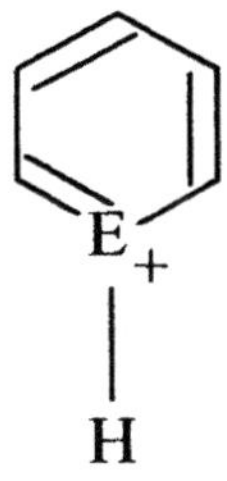

2

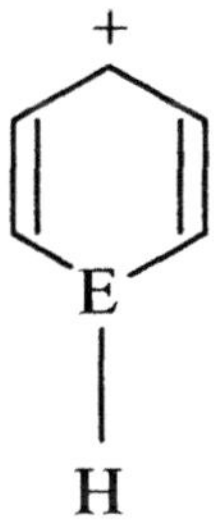

3

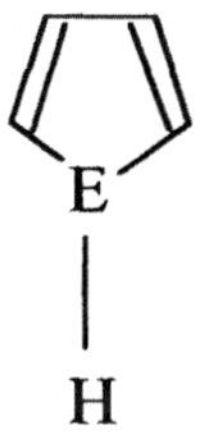

4

The resulting loss of aromatic stabilization in the cation relative to the neutral heterocycle translates as weakened basicity and lowered proton affinity. Correlary to our model are a literature explanation for the measured relative acidities of pyrrole, phosphacyclopentadiene and arsacyclopentadiene[6], species **4** (with E = N, P and As respectively) and a still untested prediction[7] that introduction of another proton into protonated pyridine, phosphabenzene and arsabenzene to form the dication will be increasingly easy (or probably more properly, decreasingly difficult) in that order.

[1] Lias, S. G.; Liebman, J. F.; Levin, R. D.; *J. Phys. Chem. Ref. Data*, **1984**, *13*, 695.

[2] For completeness, these species are also called phosphorin and arsenin respectively.
[3] Hunter, E. P. *The 1995 NIST Proton Affinity scale*, in preparation.
[4] Hodges, R. V.; Beauchamp, J. L.; Ashe III, A. J.; and Chan, W.-T.; *Organometallics*, **1985**, *4*, 457.
[5] Berger, D. L.; Gaspar, P. P.; Liebman, J. F.; *J. Mol. Struc.* (Theochem) in press.
[6] Sunderlin, L. S.; Pamu, D.; Puranik, D. B.; Ashe III, A. J.; Squires, R. R., *Organometallics*, **1994**, *13*, 4732. For completness, the latter two heterocycles are also called 1H-phosphole and 1H-arsole.
[7] Steen Hammerum, unpublished suggestion to the authors.

3. The Loss of H_2 from $CH_3NH_3^+$, $CH_3OH_2^+$ and CH_3FH^+. Reaction Mechanisms and Dynamics from Observation of Metastable Ion Fragmentations and *ab initio* Calculations.

E. LEERE ØIESTAD, Å. LEERE ØIESTAD, H. SKAANE, K. RUUD, T. HELGAKER , E. UGGERUD
Department of Chemistry, University of Oslo, P. O. Box 1033 Blindern, N-0315 Oslo, Norway,

T. VULPIUS
Department of Chemistry, H. C. Ørsted Institute, University of Copenhagen, DK-2100 Copenhagen, Denmark.

Experimental and theoretical results for the following three reactions were presented.

$$CH_3NH_3^+ \rightarrow CH_2NH2^+ + H_2 \quad (1)$$

$$CH_3OH_2^+ \rightarrow CH_2OH^+ + H_2 \quad (2)$$

$$CH_3FH^+ \rightarrow CH_2F^+ + H_2 \quad (3)$$

The distributions of the translational energy (T) released during loss of H_2 from metastable $CH_3NH_3^+$ and $CH_3OH_2^+$ ions were measured using the MIKE technique[1]. For both reactions the most probable T value accounts for approximately 3/4 of the reaction's reverse critical energy. Subject to the same experimental conditions CH_3FH^+ ions do not give rise to any measurable signal for H_2 loss. The relevant parts of the potential energy surfaces of all three reactions were investigated using various *ab initio* quantum chemical computational schemes. *Ab initio* direct dynamics calculations were performed to obtain representative reaction trajectories. Translational energy releases computed at the end of these trajectories (where the fragments have separated) agree with the corresponding experimental figures. The three reactions follow a common polar mechanism which involves an initial transfer of a proton from the most basic center (N, O or F) towards one of the hydrogen atoms of the methyl group. During this stage the proton polarizes the electrons around the methyl hydrogen to give it some hydride character, and in the transition state this has resulted in an embryonic H-H bond. Further electron reorganization during the concerted bond breaking and bond forming process leads to a strong repulsive force along the reaction coordinate as the two fragments depart. This accounts for the highly non-statistical partitioning of the available potential energy into relative translation between the two fragments formed.

[1] Vulpius, T., Leere Øiestad, E., Leere Øiestad, Å. M., Skaane, H., Ruud, K., Helgaker K., and Uggerud, E. (1995) Loss of H_2 from $CH_3NH_3^+$, $CH_3OH_2^+$ and CH_3FH^+. Reaction Mechanisms and Dynamics from Observation of Metastable Ion Fragmentations and *ab initio* Calculations, *European Mass Spectrom.* 1, 121-129.

4. Symmetry-Breaking in Calculations on Distonic Radical-Ions

R. R. SQUIRES
Department of Chemistry, Purdue University
West Lafayette, Indiana USA 47907

Distonic ions may be divided into two distinct classes, depending upon the nature of the charge site. The most commonly studied form of distonic ion possesses a coordinatively- and electronically-saturated charged site, usually of the onium type, such as in the so-called "ylidions" and their homologs [1]. The other type of distonic ion corresponds to an ionized biradical or zwitterion, which possesses two coordinatively-unsaturated sites containing one or three electrons. An important consequence of this latter topology is the possibility for at least two low-lying electronic states of the distonic ion, wherein the odd electron occupies one or the other of the two nominally non-bonding orbitals. This under-appreciated aspect of the ionized biradical class of distonic ion creates some special challenges for their proper theoretical description. Some of these challenges are described in this contribution.

Salem and Rowland [2] define a biradical as a molecule possessing two weakly interacting electrons in degenerate or nearly degenerate molecular orbitals. These MO's are constituted from the two localized non-bonding orbitals, n_1 and n_2, and their electronic interaction may result from direct (through-space) or indirect (through-bond) overlap between n_1 and n_2, or from electron exchange. It is useful to classify biradicals (and ionized biradicals) according to the topological relationship between the two non-bonding orbitals [2]. A *homosymmetric* biradical is characterized by a symmetry element, such as a reflection plane or rotation axis, which relates n_1 to n_2 and allows them to mix, while in a *heterosymmetric* biradical n_1 and n_2 belong to different symmetry representations and, therefore, do not mix. The MO's for homosymmetric biradicals are given by the symmetric and antisymmetric combinations of n_1 and n_2: $\Psi_+ = (n_1 + n_2)$ and $\Psi_- = (n_1 - n_2)$. Usually (but not always) Ψ_+ lies energetically below Ψ_- due to partial overlap between n_1 and n_2. The orbital splittings naturally decrease as the magnitude of the interaction between n_1 and n_2 decreases. It is well known that the totally-symmetric singlet states of homosymmetric biradicals require a minimal two-configuration representation [2].

Occupation of the two valence orbitals Ψ_+ and Ψ_- by three electrons gives rise to a pair of distonic radical anion doublet states with the nominal configurations $\Psi_+^1\Psi_-^2$ and $\Psi_+^2\Psi_-^1$. A two-configuration representation is no longer required since the overall symmetry of each doublet configuration is determined by that of the singly-occupied orbital, and Ψ_+ and Ψ_- will always belong to different symmetry representations. These two distonic ion doublet states will have the same molecular symmetry as that of the homosymmetric biradical from which they are derived, and will possess equal odd-spin and negative charge density at both sites, i.e., they are "delocalized" states. Depending upon the magnitude of the $\Psi_- - \Psi_+$ orbital energy gap, a distonic radical anion derived from a homosymmetric biradical may be subject to second-order Jahn-Teller interactions [3] (vibronic coupling) between the $\Psi_+^1\Psi_-^2$ and $\Psi_+^2\Psi_-^1$ configurations, which can lead to geometric distortion to a structure with reduced molecular symmetry. This low-symmetry state correspond to a "localized" distonic radical anion, with the negative charge and odd-spin density largely constrained to separate sites. When the Jahn-Teller interaction is strong, the higher energy delocalized state (e.g. $\Psi_+^1\Psi_-^2$) becomes a saddle point on a "double-cone" potential energy surface, while the lower energy symmetrical state (e.g. $\Psi_+^2\Psi_-^1$) can be either the global minimum or a transition state between equivalent distorted minima.

However, one must exercise caution in evaluating the significance of radical (and radical ion) states computed to have reduced symmetry because of the well-known "doublet instability" problem [4] that can afflict MO calculations on delocalized radicals that do not provide sufficient correlation between the open-shell electron and the electrons in the next lower-lying orbital. Broken-symmetry solutions to the Hartree-Fock equations for delocalized radicals with otherwise symmetrical structures can arise as an artifact of the inadequacies of a single-configuration representation. Localization of the unpaired electron results as the system acts to minimize Coulombic repulsion between this electron and the electron of opposite spin in the immediately subjacent orbital that spans the same atoms as the singly-occupied orbital. In unconstrained geometry optimizations, doublet instability can spuriously result in unsymmetrical (non-delocalized) structures as low-lying or global minima, and the problem can be aggravated by the use of small basis sets. The same types of unsymmetrical structures that are produced by the real effects of second-order Jahn-Teller interactions can also arise as artifacts of doublet instability. Doublet instability can in some cases be remedied by providing different spatial wave functions for electrons of opposite spin, as in the unrestricted Hartree-Fock (UHF) approach, or by employing MCSCF or CI procedures that allow correlation of the unpaired electron with electrons in the subjacent orbital.

Nevertheless, for some systems artifactual symmetry-breaking can persist even at MCSCF and CI levels of theory [5]. Broken-symmetry multiconfigurational or limited CI wave functions can sometimes result for symmetrical geometries when - correlation is not adequately represented in the MCSCF or CI expansions. This produces a cusp on the potential energy surface at the symmetrical geometry that leads to a discontinuous change in the energy with small symmetry-breaking molecular distortions. Therefore, an unconstrained geometry optimization will erroneously lead to a minimum with a distorted structure. These effects are especially problematic in calculations on charged or heteropolar open-shell systems, and in calculations on systems possessing equivalent, weakly-interacting pairs of orbitals containing one or three electrons [6]. This problem is therefore endemic to calculations involving distonic radical anions and cations derived from biradicals.

[1] Yates, B.F.; Bouma, W.J.; Radom, L. *J. Am. Chem. Soc.* **1984**, *106*, 5805.

[2] Salem, L,; Rowland, C. *Angew. Chem. Int. Ed. Engl.* **1972**, *11*, 92.

[3] Pearson, R.G. *J. Am. Chem. Soc.* **1969**, *91*, 4947.

[4] Huyser, E.S.; Feller, D.; Borden, W.T.; Davidson, E.R. *J. Am. Chem. Soc.* **1982**, *104*, 2956.

[5] Davidson, E. R.; Borden, W. T. *J. Phys. Chem.* **1983**, *87*, 4783.

[6] Borden, W. T.; Davidson, E. R.; Feller, D. *J. Am. Chem. Soc.* **1980**, *102*, 5302; *ibid.* **1981**, *103*, 5725.

5. Structures and Energetics of Isomeric $[CH_4O_3]^{\bullet+}$ Ion-Neutral Complexes. Comparative *ab initio* and DFT Study.

G. A. McGIBBON, J. K. TERLOUW
Department of Chemistry, McMaster University, 1280 Main St. W., Hamilton, Ontario, Canada L8S 4M1

J. HRUŠÁK
J. Heyrovský Institute of Physical Chemistry, Academy of Sciences of the Czech Republic, Dolejškova 3, CZ-182 23 Prague 8 Czech Republic.

H. SCHWARZ
Institut für Organische Chemie der Technischen Universität Berlin, D-10623 Berlin, Germany.

The electron ionization (70 eV) mass spectrum of dihydroxyfumaric acid (DHF), HOOC-C(OH)=C(OH)-COOH contains a sizable peak (40%) at m/z 64, corresponding to a radical cation of the elemental composition $[C,H_4,O_3]$. This ion is clearly the product of an extensive rearrangement reaction but, fortunately, one may entertain only a few realistic structural possibilities for such a species. They comprise the interesting hydrogen-bridged radical cations, $[H_2O{\cdots}H{\cdots}O{=}C\text{-}OH]^{\bullet+}$ (**1**), $[HC(O)O{\cdots}H{\cdots}OH_2]^{\bullet+}$ (**2**) and $[H_2O{\cdots}H\text{-}COOH]^{\bullet+}$ (**2a**), the ylid ion $H_2O\text{-}C(OH)_2^{\bullet+}$ (**3**) and what is ostensibly ionized trihydroxymethane (aka the orthoformic acid), $HC(OH)_3^{\bullet+}$ (**4**).

We have employed quantum chemical techniques to study the above mentioned isomeric $[CH_4O_3]^{\bullet+}$ radical cations. Standard *ab initio* MO methods (i.e. the Moeller-Plesset pertubational theory retained at the second (MP2) and fourth (MP4SDTQ) orders respectively) as well as the Density Functional Theory (DFT) were employed for calculations of relative stabilities and of the energy profile for the individual interconversion reactions. In the DFT calculation we used the „hybrid“ DFT method (HF/DFT), namely the three parameter fit of Becke (B3LYP) as implemented in the GAUSSIAN92 program. Standard polarized basis sets (6-31G**) have been used in both the DFT and *ab initio* calculations. The individual isomers were characterized by the analysis of the corresponding Hessian matrix of second derivatives and the calculated harmonic frequencies were used to calculate the Zero Point Vibrational energy (ZPVE).

The differences in structure and energy between the HF/DFT and MP4SDTQ//MP2 calculations are surprisingly small. Further, the B3LYP calculations results The B3LYP/6-31G** results are presented in Figure 1 below, while the energy profile deduced from the MP4SDTQ/6-31G**//MP2(full)/6-31G** calculation (not given) is almost identical.

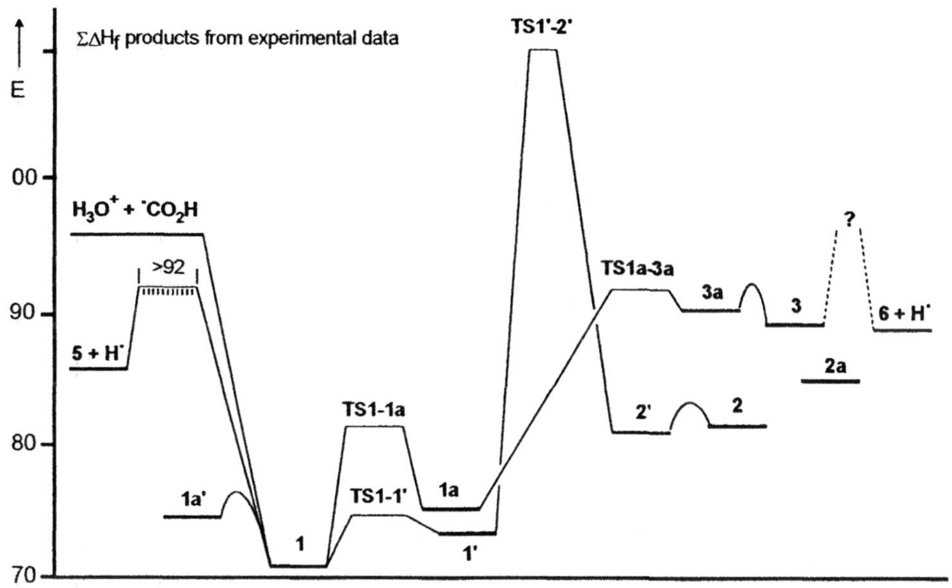

Figure 1 The schematic B3LYP calculated potential energy surface (in kcal/mol). The estimated ΔH_f for the individual isomers are based on the experimental values of the products

Clearly, the most stable ions are the hydrogen-bridged radical cations. These species exhibit substantial conformational flexibility and there are at least four stable conformations of **1** lying within 5 kcal/mol of one another. The ΔH_f of **1** itself was estimated using an empirically based formula to be 71 kcal/mol and its isomers **2** and **3** are calculated to be higher in energy by 11.5 and 19.2 kcal/mol (MP4SDTQ/6-31G**(+ZPVE)//MP2/6-31G**) or 9.4 and 18.1 kcal/mol (B3LYP/6-31G**) respectively. No stable structure corresponding to ion **4** was found. Structurally, ions **1** look more like hydronium ion-carboxyl radical complexes than dihydroxycarbene radical cation-water complexes, a description which applies instead to **3/3a**. Further, an essential barrier (**TS1'/2'**) separates 1 from the remaining part of the potential energy surface.

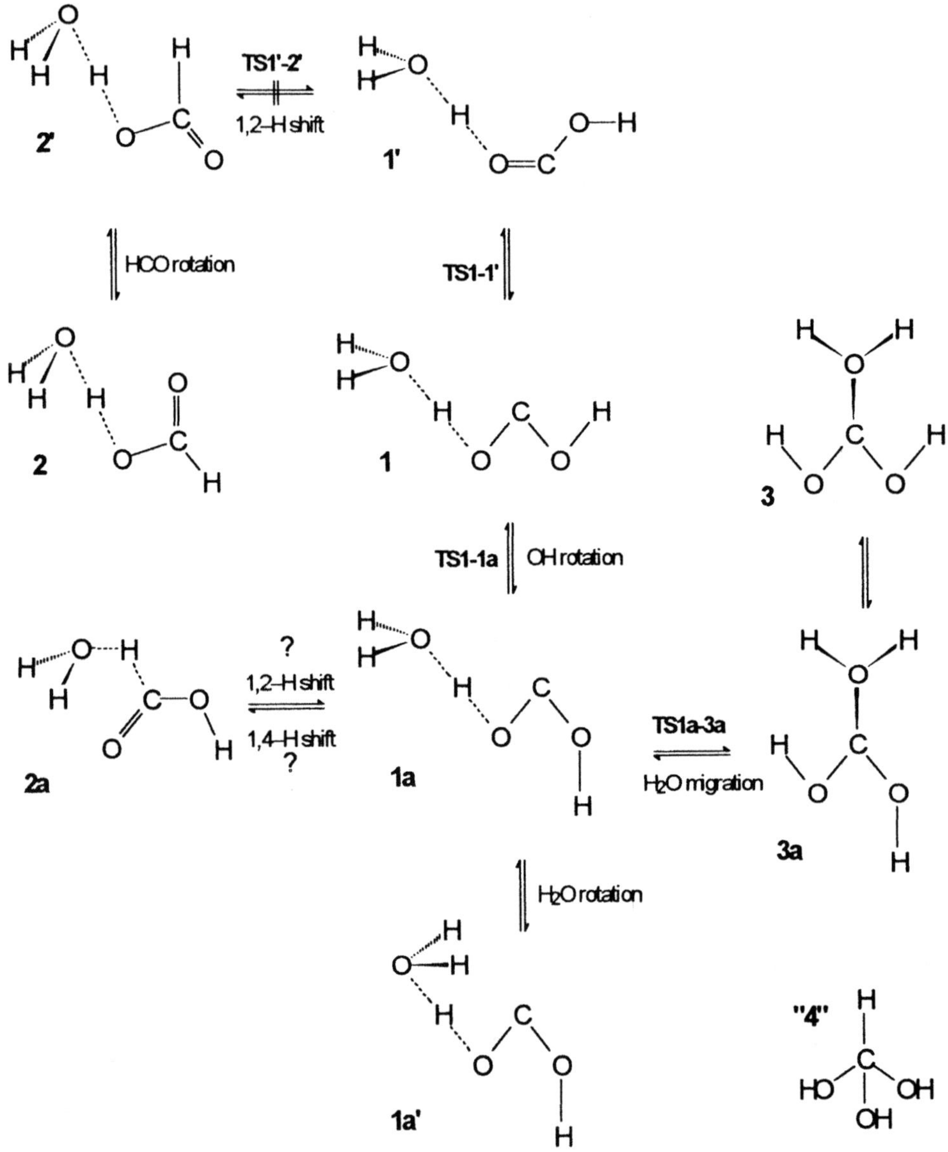

Tandem mass spectrometric experiments (metastable ion, collisional activation and MS/MS/MS) on dihydroxyfumaric acid indicated that two 1,5-hydrogen transfers and sequential expulsions of CO lead, via an intermediate dihydroxyketene-water ion-neutral complex, to the hydrogen-bridged ion **1**. The significant unimolecular dissociations of the latter lead to H_3O^+ + O=C-OH$^\bullet$ and, by $H^\bullet$ loss, to the proton-bound ion $[H_2O\cdots H\cdots O{=}C{=}O]^+$ **5** rather than $(HO)_3C^+$ **6**.

6. The Use of DFT and HF/DFT Methods in the Structure and Energy Calculations of Cationic Transition Metal Containing Systems.

J. HRUŠÁK
J. Heyrovský Institute of Physical Chemistry, Academy of Sciences of the Czech Republic, Dolejškova 3, CZ-182 23 Prague 8 Czech Republic.

Applications of classical *ab initio* methods to organometallic chemistry are connected with very large computational effort. The large number of electrons, the existence of many nearly degenerated electronic states, changes in electronic structure in course of many reactions, and usually the size of the system as such, cause severe problems in the calculations. Conventional high level *ab initio* approaches based on multiconfiguration or coupled cluster techniques accompanied with large flexible basis sets are requested for the correct description of transition metal (TM) containing systems. Such calculations are extremely expensive computational tools and thus just limited to very small (few atoms) systems.

On the other hand, the DFT methods which have been many times successfully applied in the field of organic chemistry offer due to their low computational costs a reasonable alternative. Unfortunately, the amount of experience in using this approach is up to now quite limited. It seems that the basic problem is related to the treatment of the exchange part of the exchange-correlation functional. In open shell transition metal systems this both phenomena (i.e. the exchange as well as correlation energies) have significant influence on the chemical bonding. Very recently appeared systematic studies for small cationic TM containing species have shown, that the hybrid HF/DFT approaches obtain reliable structures and bond dissociation energies [1].

We applied a HF/DFT (namely the Becke3LYP) method in conjunction with the relativistic effective potentials (RECP) and polarized basis sets (i.e. TZP for non-metal atoms and 10s/8p/7d/3f basis for the metals) to a series of cationic closed shell complexes M^+-X (M= Cu, Ag, and Au; X= CH_2, CHF, CF_2, H_2O, NH_3, CO, and C_2H_4) and found a excellent agreement with results of the conventional coupled cluster expansion (CCSD(T)) as well as with experimental data [2]. The Becke3LYP calculated bond dissociation energies are at most 4 kcal/mol of with respect to the CCSD(T) and agree nicely also with the experimental findings. Further, the DFT and HF/DFT converge much faster with the expansion of the basis set as compared to standard *ab initio* techniques [3]. Contrary to the CCSD(T) approach, the changes in geometry and relative energy are negligible small in the DFT and DFT/HF calculations, when an additional *d*-and *f*- polarization functions is added. The results are already converged at the polarized triple zeta level and no higher angular momentum functions are needed.

For the open shell TM-species the situation does not change dramatically. It might be shown that the errors in atomic excitation energies (i.e. the $s^1d^n \rightarrow s^0d^{n+1}$) are typically in the order of 0.2 - 0.3 eV similarly to the CCSD(T) or MCPF approaches. It is also known, that the bond dissociation energies as calculated by the gradient corrected DFT methods are usually overestimated by c. 1 eV. This was traced back to the inadequate description of the HF exchange in the functional and might be compensated by using the „*exact*“ (i.e. Hartree-Fock exchange) [1].

TABLE 1 A comparison of the DFT and CCSD(T) calculated BDE's (in kcal/mol) for some products of the Pd^+ + CH_3X (X = F, Cl, Br, and I) reaction.

System	BP86/DZP	BLYP/DZP	B3LYP/DZP	CCSD(T)/DZPf
$Pd^+ + CH_3$	71,4	65,1	54,5	46,8
$Pd^+ + CH_2F$	75,6	68,6	55,5	49,9
$Pd^+ + CH_2Cl$	81,6	74,4	61,4	55,5
$Pd^+ + CH_2Br$	87,7	79,9	67,1	61,8
$Pd^+ + CH_2I$	95,0	86,5	73,7	68,6
$Pd^+ + CH_3F$	41,1	38,5	35,2	27,5
$Pd^+ + CH_3Cl$	33,7	35,0	30,2	30,4
$Pd^+ + CH_3Br$		45,6		
$Pd^+ + CH_3I$	56,1	50,8	44,1	38,3
$Pd^+ + F$	42,7	38,8	22,6	25,7
$Pd^+ + Cl$	30,5	31,6	12,4	11,5
$Pd^+ + Br$	29,9	24,5	12,1	15,6
$Pd^+ + I$	32,5	27,1	15,1	16,1

As an example may serve the products of the reaction Pd^+ + CH_3X. The results of three different DFT calculations in comparison to the CCSD(T) data are listed in Table 1. It is obvious that the pure DFT schemes (Becke-Perdew and the Becke-Lee-Yang-Parr functionals) reasonably overestimate the covalent Pd-C bond strength in $PdCH_2X^+$ (c. 25 kcal/mol and c. 18 kcal/mol), while the errors for the ionicaly bound species ($Pd-X^+$) and for the ion dipole complexes ($Pd-CH_3X^+$) are significantly smaller (c. 16 kcal/mol and 11 kcal/mol). The B3LYP hybrid DFT method, which consist 20 % of the HF exchange proceed much better than the pure DFT and the errors do not exceed 6 kcal/mol. It should be mentioned that the single point CCSD(T) calculation (at the BP86 calculated geometry) is by a factor of 15 more costly than the corresponding B3LYP calculation. Further, due to the basis set truncation effects the CCSD(T) calculated BDE for $Pd-CH_3^+$ which is the only one known experimentally (58 kcal/mol) is still underestimated, while the B3LYP computed value is pretty good.

As an impressive example of the performance of the DFT approach may serve the calculated relative stabilities of complexes of iron with tetraline, dihydronaphtaline, and naphthalene as experimentally observed in the Fe^+ mediated dehydrogenation reaction:

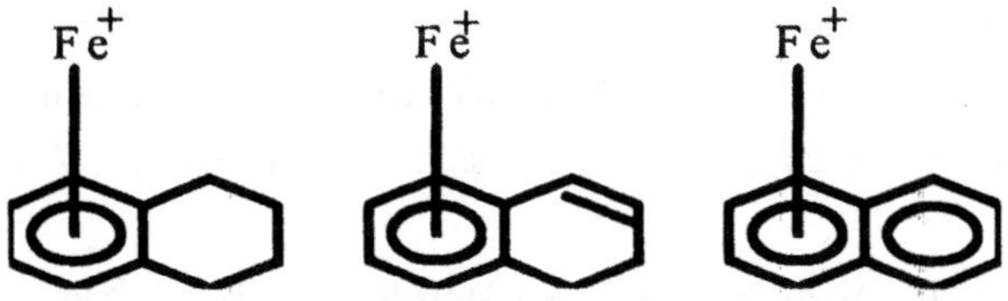

Systems of such size are hardly accessible for traditional *ab initio* CI treatments. Employing the DGAUS program with a relativistic effective core potential and a polarized double zeta basis set we fully optimized the geometries without any symmetry restrictions for both the quartet and sextet states. The sextet electronic configuration was found to be the most stable and we compare the DFT relative stabilities to the experimental values coming from direct ligand exchange measurements in a FT-ICR cell [4]. The data are given in Table 2. It is obvious, that the DFT method is not only able to describe correctly the relative stabilities of the individual isomers, but further, the absolute values of the BDE's correspond nicely (accepting the tendency for overbinding) also to the measured values based on the BDE of the cationic Iron -(benzene) complex.

TABLE 2. The BP/DGAUS/RECP-DZP calculated BDE's (in kcal/mol) in comparison to the experimental values based on relative stabilities from direct ligand exchange.

System	Experimentally		BP/DZP
$Fe^{+} + C_6H_6$	48,6	55 [a)]	57,1
$Fe^{+} + C_{10}H_8$	53,1	59,1	63,2
$Fe^{+} + C_{10}H_{10}$	52,2	58,6	61,8
$Fe^{+} + C_{10}H_{12}$	51,2	57,6	60,3

[a)] Corrected BDE value from ref. [5]

1. Holthausen, M. C.; Heineman, C. Cornehl, H.H.; Koch, W.; Schwarz, H.; *J. Chem. Phys.* **1995**, *102*, 4931

2. (a) Hrušák, J.; Hertwig, R.H.; Schröder, D.; Schwerdtfeger, P.; Koch, W.; Schwarz, H.; *Orgamomettalics* **1995**, *14,* 1284, (b) Schröder, D.; Hrušák, J.; Hertwig, R.H.; Schwerdtfeger, P.; Koch, W.; Schwarz, H.; *Orgamomettalics* **1995**, *14,* 312, (c) Hertwig, R.H.; Hrušák, J.; Schröder, D.; Schwerdtfeger, P.; Koch, W.; Schwarz, H.; *Chem. Phys. Lett.* **1995**, *236,* 194, (d) Hrušák, J.; Schwarz, H.; *J. Chem. Phys.* **1995** (submitted)

3. Hrušák, J.; Hertwig, R.H.; Schröder, D.; Schwerdtfeger, P.; Koch, W.; Schwarz, H.; *J. Phys. Chem.* (accepted)

4. Seemeyer, K.; Hertwig, R.H.; Hrušák, J.; Koch, W.; Schwarz, H.; *Orgamomettalics* **1995** (accepted)

5. Schröder, D.; Sülzle, D.; Hrušák, J.; Böhme D.K.; Schwarz, H.; *Int. J. Mass Spectrom. Ion Proc.* **1991**, *110*, 145

Summary of the Workshop on Fundamentals and Applications in Fourier Transform Ion Cyclotron Resonance Mass Spectrometry

Ron M.A. Heeren
Unit for Macromolecular Mass Spectrometry, FOM Institute for Atomic and Molecular Physics, Kruislaan 407, 1098 SJ Amsterdam, The Netherlands.

This workshop was intended to discuss some of the recent developments in FT-ICRMS and discuss their relevance to various fundamental and practical aspects of mass spectrometry. Prof. Keith Jennings reviewed most of the recent instrumental developments in FT-ICRMS. In his contribution, a strong emphasis was put on the relevance of the dissociative studies of macromolecules which are thought to be realistically possible now. Especially with the advent of ESI-FT-ICRMS, the technique has gained tremendous potential in the study of gas-phase dissociation. The multiply protonated molecular ions of various peptides and proteins are demonstrated to obey very different (lower) energy regimes compared to the singly charged analogues. The contributions of Dr. Heinrich Sievers focused on energy Resolved CID of elementorganic complexes by FT-ICR Spectrometry. Thermochemical data on various was obtained and was used to elucidate the structure of the elementorganic complexes. Dr. Grielof Koster has revisited the photodissociation of C_4H_4 radical cations, by studying the photodissociation of the ions in the ICR cell both immediately after formation and after relaxation as a function of wavelength . It was found that relaxation has no effect on the total ion fraction that can be dissociated as a function of light intensity. Dr. Joseph Schwartz used ligand exchange experiments, in an attempt to bracket the BDE of Fe^+-(N_2). The $Fe(N_2)^+$ ion which lies at the basis of this study is conveniently made accessible by a sequence of ion-molecule reactions carried out inside the ICR-cell. This once again demonstrates the unique capabilities of trapped ion mass spectrometry as a gas-phase reactor. Dr. Ron Heeren elaborated on the work on synthetic polymer characterisation using MALDI-FT-ICRMS. The high resolution capabilities of a 7T external source FT-ICRMS system were used to determine end-group size and structure of medium size polymeric systems. From the same laboratory, Dr. Chris de Koster demonstrated how sustained off-resonance irradiation can be used to probe the lowest energy dissociation pathways in macromolecules. This last lecture clearly demonstrated one possible future in the application of gas-phase ion chemistry in the structural characterisation of macromolecules.

K. R. Jennings (ed.), Fundamentals and Applications of Gas Phase Ion Chemistry, 445–458.

Fundamentals of FT-ICR Instrumentation

Keith R. Jennings, Department of Chemistry,
University of Warwick, Coventry CV4 7AL, U.K.

Basic Ideas Ions are formed either close to the ICR cell or more usually in an external source. In the latter case, they are transported by means of an electrostatic ion guide from the source to the ICR cell which is located in a very strong magnetic field. For analysis, the pressure is very low, typically only a little above 10^{-8} Torr, and ions are detected by absorption of r.f. power which induces a current across the plates of an ICR cell when

$$\omega_c = KB(ze/m)$$

where K is a constant, ω_c is the cyclotron resonance frequency of the ion of mass to charge ratio m/z, and e is the electronic charge. When $B = 3$ Teslas, $\omega_c = 2.6$ MHz for ions of m/z 18 and 15.6 KHz for ions of m/z 3000. Ion reaction and detection are separated by time rather than space as in most other instruments. Electrical pulses are used for ion generation, selective excitation, detection and quenching.

All ions formed are detected simultaneously. A complex time-domain spectrum is subjected to FT analysis and the resulting frequency-domain spectrum is the normal mass spectrum. Current commercial instruments have magnetic field strengths of 3, 4.7, 7 and 9.4 Teslas.

Advantages of FT-ICR-MS The mass resolving power is proportional to $1/m$, B, and the time taken for observation. In order to maintain the coherence of ion motion, the cell pressure must be in the region of 10^{-8} Torr or less and at high values of B, a resolving power of 500 000 has been obtained at m/z 2000. The instrument can cover a wide mass range - best results are obtained below m/z 2500 but electrospray ionisation extends the mass range to over a million. Very high sensitivity can be obtained - one ion of mass approximately 6.25 KDa, carrying 2610 charges, has been observed [1] and since the ions are not destroyed on detection, time-dependent studies can be carried out such as observing the approach to equilibrium or carrying out multiple collision-induced decomposition experiments. The technique is suitable for use with electrospray ionisation and matrix assisted laser desorption ionisation and several different methods of studying collision-induced decomposition of ions have been described, the most widely-used of which is sustained off-resonance irradiation (SORI) [2]. In this technique, precursor ions are irradiated with r.f. radiation slightly removed from the cyclotron resonance frequency so that the radius of their motion is always small and they remain close to the axis of the cell.

Possible Probelms in Using FT-ICR The efficiency of transferring ions from an external source to the ion cell continues to be a cause for concern, especially as the size of the magnet and the strength of the magnetic field increases. The dynamic range normally obtainable is in the region of no more than 1000 since approximately 100 ions are needed for a reasonable signal-to-noise ratio but if there are over 100 000 ions in the cell, Coulombic repulsion between the ions causes them to move away from the cell axis, leading to broadening of the peaks. The dynamic range can be increased to approximately a million by selective ejection of unwanted ions by means of the technique known as SWIFT (Stored Waveform Inverse Fourier Transform) in which all ions except those within a chosen mass range are ejected from the cell by irradiating at all frequencies except those corresponding to the chosen range of ions [3].

Control of the motion of the ions in the cell becomes increasingly difficult as the mass of the ions increases since they drift away from the axis of the cell where control of the ion motion is better than when the ions are away from the axis. For ions of *m/z* above 2500, this results in the degrading of the signal-to-noise ratio and mass resolution. Recently, however, the technique of Quadrupolar Axialisation [4] allows one to confine the ions to the axial region, effecting a considerable improvement in performance at high mass.

Accuracy of mass measurement requires careful control of the number of ions in the cell since although low ion currents avoid diffusion problems, problems arising from ion statistics can cause the isotope pattern to be distorted so leading to an error in the average relative molecular mass. Nevertheless, with the use of an internal standard, accuracies of mass measurements are typically in the region of 2 ppm for molecules having relative molecular masses in the region of 10 000 - 20 000.

Concluding remarks. Despite these difficulties, FT-ICR-MS is now established as the high resolution, high mass instrument of choice having excellent sensitivity, resolving power and mass range, with good performance as a tandem mass spectrometer.

References

[1] R.D. Smith, X. Cheng, J.E. Bruce, S.A. Hofstadler and G.A. Anderson, *Nature*, (1994), **369**, 137.

[2] A.J.R. Heck, L.J. de Koning, F.A. Pinske and N.M.M. Nibbering, *Rapid Com.. Mass Spectrom.*, (1991), **9**, 406.

[3] S.L. Mullen and A.G. Marshall, *J. Am. Chem. Soc.*, (1988), **110**, 1766.

[4] L. Schweikhard, S. Guan and A.G. Marshall, *Int. J. Mass Spectrom. Ion Proc.*, (1992), **120**, 71.

Energy Resolved CID of Elementorganic Complexes by FT-ICR Spectrometry

H. L. Sievers and H.-F. Grützmacher
Department of Chemistry, University of Bielefeld, Germany

Previous studies [1] have shown, that under CI conditions and by electron impact ionization group V EI_3 compounds react with several n- and π-donor ligands building [EI_2 ligand]+ complexes. The bimolecular gas phase reaction

$$\mathbf{EI_3^+ + ligand \longrightarrow [EI_2\ ligand]^+ + I}$$

E = P, As, Sb and Bi

ligand = benzene, propene and acetone

is exergonic in case of E = As, Sb, and Bi. The efficiency of the corresponding PI_3^+ reaction with acetone is close to the collision limit, but too low to observe the reactions with benzene or propene at the ICR pressure and time scale.
Energy resolved CID by FT-ICR spectrometry [2] of EI_3^+ and [EI_2 ligand]$^+$ complexes was applied to gain thermochemical data for interpreting the reactivity of EI_3^+ and the structure of [EI_2 ligand]$^+$ complexes.

Appearance energies of the [EI_2 ligand]$^+$

AE (eV)	PI	AsI	SbI	BiI
iodine	1.88	1.15	1.22	0.78
propene		1.13	1.06	0.87
benzene		1.06	1.04	0.86
acetone	1.32	1.25	1.24	

AEs were derived from the experimental threshold curves assuming a linear increase of fragment ion intensity above AE and fitting the Doppler broadening corrected ideal cross section function [3] to the data. A Bruker Spectrospin CMS 47 FT-ICR spectrometer equipped with an external EI or CI source with the Bruker Infinity™ Cell was used for these investigations. The energy resolved CID procedure was checked by determining the AE of I loss from iodobenzene radical cation indicating an accuracy of better than ± 0.2 eV.

An additional MIKE spectrometry study of [EI_2 benzene]$^+$ complexes, not presented here, and measurements of the ligand exchange reaction in [AsI_2 ligand]$^+$ do not show a significant reverse activation barrier (< 0.1 eV). Hence the ligand bond dissociation energies of the [EI_2 ligand]$^+$ complexes are obtained by adding the volume term nRT to the corresponding AE. The ligand BDEs vary from 87 kJmol^{-1} for [BiI_2 propene]$^+$ and [BiI_2 benzene]$^+$ to 125 kJmol^{-1} for [EI_2 acetone]$^+$ complexes indicating a very weak interaction of the carbene analogues EI and the organic ligands.

In case of PI_3^+, the AE of PI_2^+ cannot be correlated with the PI_2^+-I BDE. The formation of [PI_2 acetone]$^+$ is exergonic despite the significant lower AE of PI_2^+ from this complex. Hence, there is an activation barrier for the PI_2^+-I bond cleavage. In general, the iodo BDE of the EI_3^+ must be below the lowest corresponding EI_2^+-ligand BDE.

The AEs determined in this study by energy resolved CID open the possibility for a more detailed theoretical analysis of the reaction mechanisms and bonding type of heavy element organic compounds.

References

[1] (a) H. Grützmacher, D. Thölmann, H.-F. Grützmacher in *Fundamentals of Gas Phase Ion Chemistry*, NATO ASI Series C; (K. R. Jennings Ed.); Kluwer Academic Publ. Dordrecht, The Netherlands, *1991*, Vol. 347, pp 144-145. (b) H. L. Sievers,H.-Fr. Grützmacher, H. Grützmacher, S. Pitter, *J. Am. Chem. Soc.* **1995**, *117*, 2313-2320.

[2] (a) C. E. C. A. Hop, T. B. McMahon, G. D. Willett, *Int. J. Mass Spectrom. Ion Proc.* (1990), **101**, 191-208. (b) C. E. C. A. Hop, T. B. McMahon, *J. Am. Soc. Mass Spectrom.* (1994), **5**, 2774-2781 and ref. cited therein. (c) A. R. Katritzky, C. H. Watson, Z. Dega-Szafran, J. R. Eyler, *J. Am. Chem. Soc.* (1990), **112**, 2471-2478; (d) A. R. Katritzky, C. H. Watson, Z. Dega-Szafran, J. R. Eyler, *J. Am. Chem. Soc.* (1990), **112**, 2479-2484.

[3] Chantry, P. J. *J. Chem. Phys.* (1971), **55**, 2746-2759.

Photodissociation of C_4H_4 Radical Cations

G. Koster and W.J. van der Hart
Leiden Institute of Chemistry, Gorlaeus Laboratories, Leiden University, P.O. Box 9502, 2300 RA Leiden, The Netherlands

In a previous paper from this laboratory [1], it was shown that relaxation of the internal energy is important in the photodissociation of $C_4H_4^{+\bullet}$ fragment ions from, e.g., 1,5-hexadiyne. In this earlier work it was concluded that immediately after formation, the ions have an internal energy sufficient for dissociation after absorption of a single photon, whereas after relaxation photodissociation no longer takes place. The photodissociation thus was assumed to be determined by a competition between photodissociation and relaxation. This model explains why at low light intensities a much smaller fraction dissociates than at high light intensities, even at long irradiation times. One of the unsolved questions, however, was why two-photon processes were not observed. For this reason the photodissociation of the ions both immediately after formation and after relaxation was restudied as a function of wavelength.

In contrast to the conclusions from [1], it was found that relaxation has no effect on the total ion fraction that can be dissociated as a function of light intensity. Clearly, the original model was too simple. The new experiments at 514 (2.41 eV) nm and 488 nm (2.54 eV) appeared to be in very good agreement with the reaction scheme shown in figure 1 in which it is assumed that the whole process is a competition between one- and two-photon dissociation, relaxation by infrared emission and photon induced isomerization from a light-absorbing structure A, probably vinyl acetylene, to a structure B, probably dimethylene cyclopropene, which does not absorb visible light. The corresponding energy diagram is shown in figure 2.

At 458 nm (2.71 eV) the ions photodissociate by a one-photon process. It follows that the dissociation limit in figure 2 is 2.6-2.7 eV above the ground state of structure A.
In an additional poster contribution to this advanced study institute it has furthermore been shown that the conclusions from both our experiments and those obtained by McLafferty and coworkers from collision induced dissociation measurements [2] are in excellent agreement with ab initio calculations on the isomerization pathways of C_4H_4 radical cations.

References

[1] W.J. van der Hart, (1988), *Org. Mass Spectrom.* **23**, 187
[2] M.-Y. Zhang, B.K. Carpenter and F.W. McLafferty, *J. Am. Chem. Soc.*, (1991), **113**, 9499.

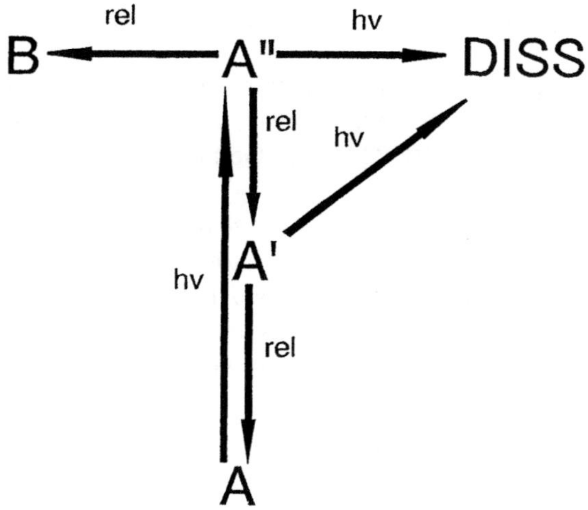

1 Reaction scheme for the photodissociation and photoisomerization of C_4H_4 radical cations. See figure 2 for the distinction between ions A, A' and A".

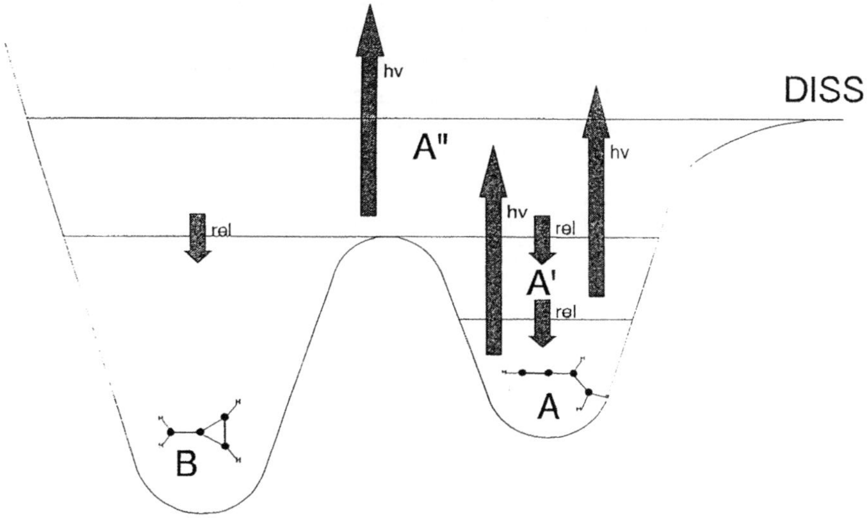

2. Energy diagram for the photodissociation and photoisomerization of C_4H_4 radical cations.

Generation and Structural Characterization of $Fe(N_2)^+$

Joseph Schwarz, Christoph Heinemann, and Helmut Schwarz*
Institut für Organische Chemie der Technischen Universität Berlin, Strasse des 17 Juni D-10623 Berlin, Germany.

The $Fe(N_2)^+$ ion is conveniently accessible by the sequence of ion-molecule reactions depicted in Scheme 1. Fe^+ reacts with ß-butyrolactone under the formation of $Fe(CO_2)^+$ as the major product ion (55 %) [1], the minor products are $FeC_3H_4^+$ (3 %), $FeC_3H_6^+$ (3 %), $FeC_2H_4O^+$ (35 %) and $FeC_3H_6O^+$ (4 %). In the subsequent reaction step N_2 is introduced into the ICR cell via pulsed valve to yield $Fe(N_2)^+$ via a ligand exchange reaction [2].

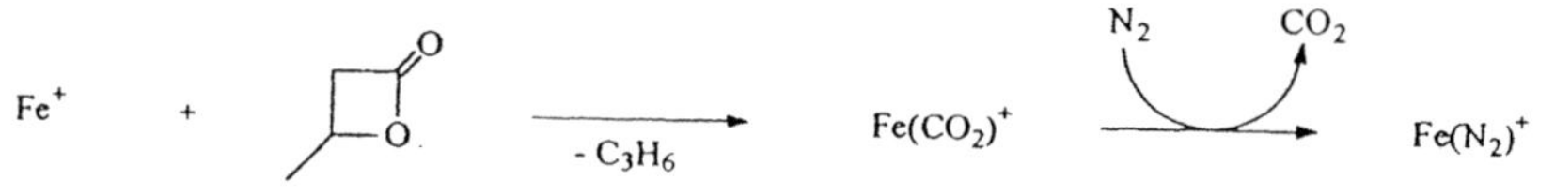

Due to the very high bond dissociation energy (BDE) of the dinitrogen triple bond (D_0 = 228.4 kcal/mol) and the generation process of the $Fe(N_2)^+$ ion, we did not expect the insertion of the transition metal ion into the NN triple bond to generate $NFeN^+$. In fact, when $Fe(N_2)^+$ is mass selected and subjected to collisional activation with argon, over the entire energy regime applied, the complex exclusively decomposes to Fe^+ and N_2. By employing ligand exchange experiments, we have tried to bracket the BDE of Fe^+-(N_2). Ligands L (with L = C_2H_6, CH_4, and Xe) which were introduced via a pulsed valve replace N_2 from the thermalized $Fe(N_2)^+$ complex, while in the reaction of $Fe(N_2)^+$ with Ar and Kr (also pulsed-in) no displacement is observed.

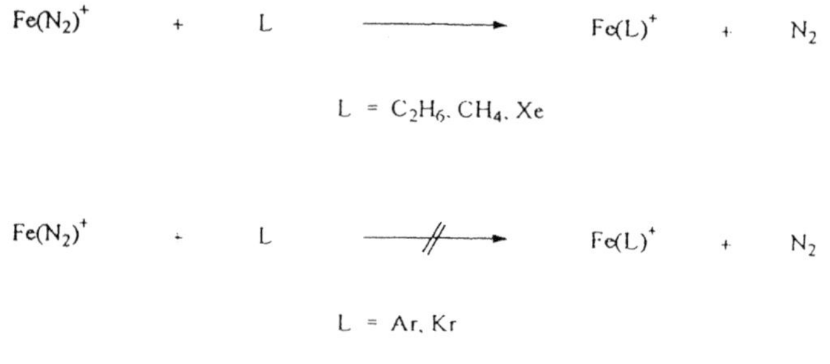

In view of the very small range of bond dissociation energies for the Fe^+-(X) complexes (BDE(Fe-$(CO_2)^+$) = 8 ± 2 kcal/mol, BDE(Fe-$(Xe)^+$) = 9 ± 2 kcal/mol) and the possibility to observe slightly endothermic reactions when reactants were introduced via pulsed valves we have performed additional experiments to determine the BDE of $Fe\text{-}(N_2)^+$. In order to check the upper limit for BDE($Fe^+\text{-}(N_2)$), isolated $Fe(Xe)^+$ (generated from $Fe(CO_2)^+$ and pulsed-in Xe) was reacted with N_2 in a pulse experiment and formation of $Fe(N_2)^+$ is observed. The occurrence of the exchange reaction $Fe(N_2)^+ + Xe$ $Fe(Xe)^+ + N_2$ in both directions implies (*i*) the BDEs of $Fe(N_2)^+$ and $Fe(Xe)^+$ are very similar and (*ii*) bracketing the upper limit for the BDE($Fe^+\text{-}(N_2)$) by isolating $Fe(N_2)^+$ and pulsing-in Xe is not the method of choice. Rather, equilibrium measurements should be performed. However, the presence of ß-butyrolactone which also reacts with either $Fe(N_2)^+$ and $Fe(Xe)^+$ complicates this procedure and, in addition, causes serious intensity problems. Therefore, we have chosen a more indirect method for the determination of the relative BDEs of $Fe^+\text{-}(N_2)$ and $Fe^+\text{-}(Xe)$. The rate constants of the reaction of $Fe(N_2)^+$ with Xe (k_{Xe}) and of $Fe(Xe)^+$ with N_2 (k_{N2}) were determined independently and then converted into the equilibrium constant K_{eq}. For the exchange process $Fe(N_2)^+$ with Xe the rate constant amounts to $k_{Xe} = 1.6 \cdot 10^{-10}$ cm^3 molecule^{-1} s^{-1}. In the analogous reaction of $Fe(Xe)^+$ with N_2 we determined a rate constant of $k_{N2} = 1.0 \cdot 10^{-10}$ cm^3 molecule^{-1} s^{-1}. From this, one obtains the equilibrium constant $K_{eq} = k_{Xe}/k_{N2} = 1.6 \pm 0.4$ for the reaction

$$Fe(N_2)^+ \quad + \quad Xe \quad \underset{k_{N_2}}{\overset{k_{Xe}}{\rightleftharpoons}} \quad Fe(Xe)^+ \quad + \quad N_2$$

with the assumption of 10% errors in the *relative* rate constants. The standard free enthalpy ΔG for this process, $\Delta G = \Delta H - T\Delta S$ (eq. 1), can be obtained from the measured equilibrium constant *via* the Gibbs-Helmholtz-equation, $\Delta G = -RT \ln(K_{eq})$. Accordingly, one arrives at the conclusion that substitution of Fe^+-coordinated Xe by molecular N_2 is endoergic by 0.3 ± 0.1 kcal/mol.

However, the problem of the exact coordination of N_2 to Fe^+ *side-on* (C_{2v}) versus *end-on* ($C_{\infty v}$) cannot be solved experimentally with the methods presently available to us. Therefore, we have initiated theoretical investigations of the $Fe^+\text{-}(N_2)$ complex which were carried out within the restricted open-shell coupled cluster approximation. For iron, we chose the 8s7p6d4f2g contraction of a 21s15p10d4f2g primitive set according to Widmark *et al* [3] which has recently been successfully employed for the description of weakly as well as strongly bound cationic iron complexes. The ANO contraction for the nitrogen atoms was of the type (14s9p4d3f)/[6s5p3d2f]. The calculations indicate that the ground state of $Fe(N_2)^+$ is of quartet multiplicity with the dinitrogen molecule coordinated *end-on* to cationic transition metal centre ($^4\Sigma^-$ state, principal 3d config. $3d\sigma^2$, $3d\pi^2$, $3d\delta^2$). Together with the results of our previous studies of cationic iron/noble-gas systems [4] these data can be used to calculate the entropy change ΔS (translational, rotational, and vibrational contributions) of the forward reaction and so the term $T\Delta S$ of this process. These calculations indicate that N_2 is at least 1 kcal/mol more strongly bound to Fe^+ and the fact that at Xe is capable to replace N_2 in the $Fe(N_2)^+$ complex 300 K is a result of an entropy-driven reaction. The BDE(Fe^+-(N_2)) amounts to 11.9 $\pm$ 2.3 kcal/mol.

[1] Schwarz, J.; Schwarz, H. *Organometallics*, (1994), **13**, 1518.
[2] For a more detailed description of the experiments and theoretical calculations see: Schwarz, J.; Heinemann, C.; Schwarz, H. *J. Phys. Chem.*, (1995), **99**, 11405.
[3] (a) Widmark, P.-O.; Perrson, B. J.; Roos, B. O. *Theor. Chim. Acta*, (1991), **79**, 419. (b) Pou-Amerigo, R.; Merchan, M.; Widmark, P.-O.; Roos, B. O. manuscript in preparation.
[4] Heinemann, C.; Schwarz, J.; Koch, W.; Schwarz, H. *J. Chem. Phys.*, in press.

Characterisation of synthetic polymers with FT-ICR-MS

R.M.A. Heeren, G.J. van Rooij, C.G. de Koster and J.J. Boon

Unit for Macromolecular Mass Spectrometry, FOM Institute for Atomic and Molecular Physics, Kruislaan 407, 1098 SJ Amsterdam, The Netherlands.

Mass-spectrometric characterisation of commercial polymer formulations is used to determine molecular weight distribution, to identify monomeric units and end-groups, sequence and structure of copolymers as well as additives and impurities. In particular high resolution high speed, FTICR-MS analysis will shown to be invaluable in polymer characterisation. Here, a few examples will be presented of direct polymer analyses with MALDI FTICR-MS for both molecular weight distribution determination and end-group identification of medium size synthetic polymers. Additionally it will be shown how temperature programmed "in-source" pyrolysis FTICR-MS can be used as an indirect method for the analysis of polymer additives.

Direct mass spectrometric analysis of synthetic polymers has received much attention since the advent of Matrix Assisted Laser Desorption and Ionisation. This technique in combination with Time-Of-Flight Mass Spectrometry has been shown to facilitate the measurement of entire molecular weight distributions of synthetic polymers up to molecular weights of 1MDalton. Interpretation of these measurements, even at lower molecular weight is generally difficult because of the lack of resolution. Although "delayed extraction" has dramatically improved the resolution of modern TOF instruments, still high resolution measurements are required to positively identify monomeric composition and end-group functionality. We have used MALDI on an external ion source FTICR-MS to identify the monomeric composition of an Ethyleneoxide-Propyleneoxide co-polymer as well as its end-group structure [1,2]. In figure 1 a broadband MALDI-FTICR mass spectrum of such an EOPO copolymer is presented. The resolution (m/dm)50% of 40000 is more than sufficient to resolve the individual isotopic peaks and hence determine both the number of EO and PO monomeric units, as well as the end group structure, as indicated in the figure. The spectrum was obtained after signal-averaging 25 transients of 128 kwords. The delay between firing of the laser and trapping of the ions was optimised to 1050 μs, such that the entire molecular weight distribution was trapped and analysed. After a 2 ms delay the ions were excited to approximately 95% of the cell radius before the image current induced on the detection electrode was measured.

The analysis of polymer additives and impurities is generally difficult because of the wide variety of available compounds and the low additive/impurity concentration (<1-5% w/w) compared to polymer and filler levels. However, direct temperature resolved pyrolysis mass spectrometry has already proven to be an analytical tool for fast analysis of unknown mixtures of polymers and e.g. the presence of fire-retarding additives. In a previous publication we have shown how fast, high-resolution temperature resolved "in-source"

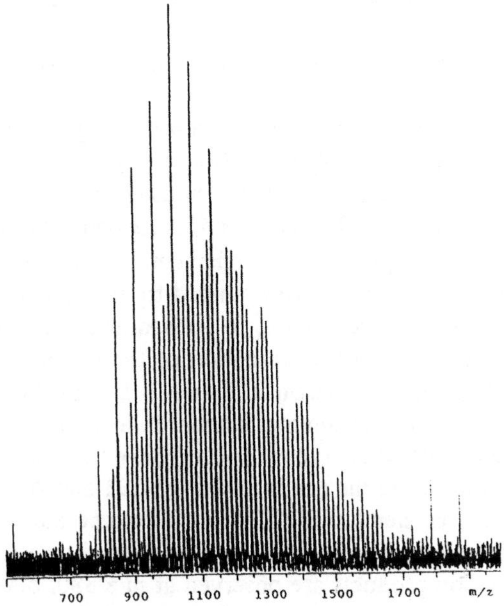

Figure 1 MALDI-FTICR spectrum of an EOPO copolymer

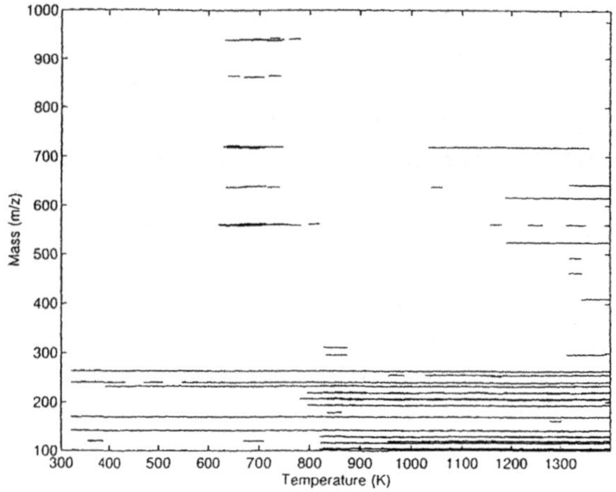

Figure 2 Temperature resolved spectra of a fire-retarded polymer

pyrolysis FTICR-MS can be used to identify fire-retarding additives [3]. Here we present measurements on a fire-retarded polymer system using high speed, temperature resolved broadband screening. A sample of the polymer housing of a TV-set was freezer-milled for 3 grinding periods of 2 min each. The powder was suspended in toluene and subsequently 1-2 μl of this solution was applied to the filament of a hair-pin type filament pyrolysis probe. The temperature of this probe was ramped from 300 K to 1200 K inside an 70 eV electron impact ionisation source leading to subsequent desorption of the volatile fire-retarding additives and the thermal degradation of the polystyrene matrix. The resulting temperature resolved spectra are shown in figure 2. During this run PFTBA was leaked into the source as to provide an internal standard. In figure 2 there are three different processes to be observed, of which typical spectra have been shown as insets in the figure. In the whole temperature range the peaks at m/z 69, 131 and 263 are visible indicative for the presence of the internal standard PFTBA. in the temperature region around 700 K the desorption of decabrominated diphenylethers is indicated by the isotopic cluster at m/z 960. Thermal degradation products are observed at m/z 880, 800, 720 and 640. Finally at elevated temperatures the polystyrene degradation products at m/z 78, 91, 104, 117, 193 and 207, are observed. One of the great advantages of this methodology is the combination of analysis speed and resolution. The high analysis speed allows the study of thermal desorption kinetics of large molecules. Direct temperature resolved "in-source" pyrolysis combined with MALDI on an FTICR-MS proves to be a very useful tool in the exact analysis of complex polymeric systems.

[1] C.G. de Koster, M.C. Duursma, G.J. van Rooij, R.M.A. Heeren and J.J. Boon, Rapid Comm. in Mass. Spec. (1995), **9**, 957-962.

[2] G.J. van Rooij, M.C. Duursma, R.M.A. Heeren, J.J. Boon and C.G. de Koster, accepted for publication in J. Am. Soc. Mass Spec.

[3] R.M.A. Heeren, C.G. de Koster and J.J. Boon, Anal. Chem, (1995) **67**, 3965-3970.

Identification of the lowest energy dissociation pathway of collisionally activated O-methyl oligosaccharide $[M + Na]^+$ cations by SORI-FTMS

C.G. de Koster, R.M.A. Heeren and J.J. Boon
Unit for Macromolecular Mass Spectrometry, FOM Institute for Atomic and Molecular Physics, Kruislaan 407, 1098 SJ Amsterdam, The Netherlands,

Collisional activation mass spectrometry of oligosaccharide quasimolecular ions is widely applied for sequencing of the primary structure of a carbohydrate. Several mechanism for glycosidic bond cleavage and fragmentation of the furanose/pyranose ring of the oligosaccharide cations have been proposed to rationalize the gas phase ion chemistry of collisionally activated oligosaccharide ions. In spite of a growing number of mechanistic studies about unimolecular fragmentation of oligosaccharide ions, no thermochemical data are known. A further improvement of gas phase carbohydrate ion chemistry requires understanding of the potential energy hypersurface of these ions. In this study we have probed the lowest dissociation limit of sodium cationized maltoheptose per methyl ether molecules using multiple low energy collisions in an ICR.

Under conditions of MALDI ionization in an external ion source O-methyl ethers of maltooligosaccharides produce abundant $[M+Na]^+$ cations. These ions are trapped into the ICR cell. Mass selected ions monoisotopic 12C and 113C $[M+Na]^+$ quasimolecular ions are subjected to multiple low energy collisions with Ar. The mass selected m/z 1497 / 1498 $[M+Na]^+$ maltoheptose cations ions are off resonance excited ($\Delta f = -1500$ Hz) for a period of two seconds. Sustained off resonance excitation (SORI) leads to an oscillation of the ion orbit radius as a function of the excitation time. Ion translational energy calculations show that the amount of internal energy available on collision of the ion with neutral argon is 0.35 eV at maximum ion velocity and 0.1 eV at the top of the ion orbit amplitude. Upon excitation there are approximately 545 collisions per second of the cation with Ar neutrals at an ICR cell Ar pressure of 1.10-5 mbar. The time interval between two collisions is approximately 2 milliseconds. This is a long period in comparison with the vibration frequencies of the cation. In this time interval redistribution of the acquired centre-off mass energy takes place over the vibration modes of the ion. In addition to activating collisions, many collisions may also deactivate the ions. Finally, the time scale for effecting dissociation is long (~ 2 s), facilitating slow, low-energy rearrangements with decomposition occurring by the lowest energy pathway. Thus, by employing a long duration off-resonance pulse ions are slowly collisionally activated by sequential inelastic collisions and, as a consequence, the lowest energy pathway for unimolecular dissociation of carbohydrate ions should be populated by this method.

The CID spectrum of SORI activated m/z 1497 $[M+Na]^+$ O-methyl maltoheptose cations is given in Fig.1. Collisional activation of the ions under multiple low energy conditions leads predominantly to the formation of fragment ions at m/z 1279, 1075, 871, 667 and

463 by losses of $C_{10}H_{18}O_5$ + n. $C_9H_{16}O_5$ (n = 0,1,2,3 and 4) neutrals, respectively. The ions are formed by glycosidic bond cleavage with hydrogen transfer from the non-reducing end to the glycosidic bond oxygen atom which becomes structural part of the product ion (Scheme 1). The positive charge (the sodium cation) is retained on the reducing end of the carbohydrate. This fragmentation reaction is denoted as Y-ion formation using the nomenclature of Domon and Costello. In contrast with high energy collisional induced dissociation experiments under single collision conditions (data not shown) formation of Y ions is a highly selective reaction. The most important conclusion is that these experiments clearly demonstrate that formation of Y ions is the lowest energy decomposition pathway for sodium cationized maltoheptose O-methyl ethers.

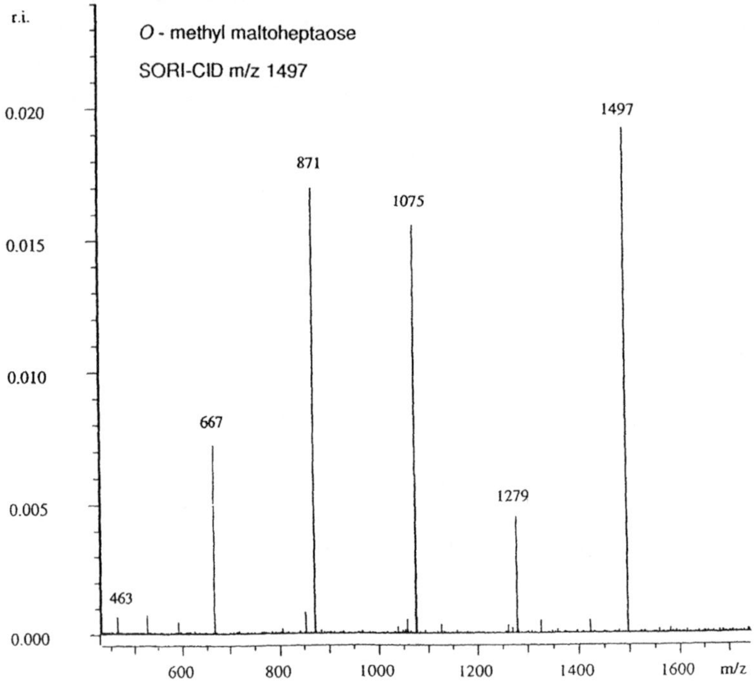

Figure 1 CID spectrum of SORI activated O-methyl maltoheptose

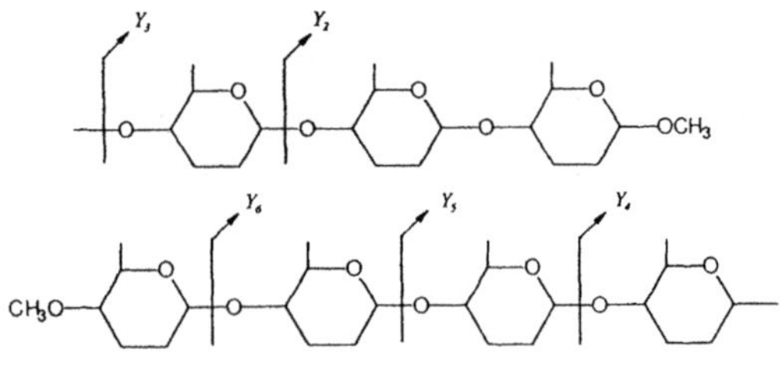

Scheme 1 SORI m/z 1497 $[M+Na]^+$ O-methyl maltoheptose

FUNDAMENTALS OF GAS PHASE ION CHEMISTRY: SUMMARY OF THE PANEL DISCUSSION

T.B. McMahon
Department of Chemistry, University of Waterloo, Waterloo, Ontario, Canada, N2L 3G1

A panel discussion on fundamental aspects of gas phase ion molecule reactions was held in which 7 presentations were made. The phenomena discussed covered a broad range of topics including kinetic, mechanistic and structural aspects of gaseous ions. The contributions of 4 of the participants are presented below.

IONIC AND NEUTRAL COMPLEXES OF BORANE (1)

Ivan Černušák[a], Carol A. Deakyne[b], Hans Lischka[c], Joel F. Liebman[d], Adriana Pappová[a,e] and Anne Skancke[f]
[a]*Dept. of Physical Chemistry, Comenius University, SK-84215 Bratislava, Slovak Republic*
[b]*Dept. of Chemistry, Eastern Illinois University, Charleston, IL 61920, USA*
[c]*Institute of Theoretical Chemistry and Radiation Chemistry, University of Vienna, A-1010, Austria*
[d]*Dept. of Chemistry and Biochemistry, University of Maryland Baltimore County, Baltimore, MD, 21228, USA*
[e]*1995 Summer Research Student, Dept. of Chemistry and Biochemistry, University of Maryland Baltimore County, Baltimore, MD, 21228, USA*
[f]*Dept. of Chemistry (IMR), the University of Tromsø, N-9037, Norway*

The current theoretical project arose from the confluence of several disjoint, and published, theoretical studies. One dealt with neutral boron-containing carbonyl (Lewis acid/Lewis base) complexes with the empirical formula H_nBCO for n = 0 - 3, and their isomers.[1] While the n = 3 species is long-known, isolated and well characterized and the n = 0 species was recently observed in a matrix, the n = 1 and 2 remain species "merely" invoked as reaction intermediates. The second dealt with cyanoborane complexes, their isomers and derivatives[2], most notably $HBCN^-$ and H_2BCN. The diatomics CO and CN^- are isoelectronic to each other and therefore so are their respective borane(1) complexes HBCO and $HBCN^-$. Saying "complexes" suggests structural diversity and inherent complicatons: we recall the third key paper[3] that discussed at some length the relative Lewis acidity of the valence isoelectronic BF_3 and SO_3. Such diversity and complication is already seen for HBCO and $HBCN^-$. Both species "enjoy" classical chain structures, the former[1] as a triplet and the latter[2] as isomeric singlets with cyano and isocyano connectivities H-B-C-N and H-B-N- C[4].

K. R. Jennings (ed.), Fundamentals and Applications of Gas Phase Ion Chemistry, 459–468.

Both HBCO and $HBCN^-$ also form rings without imaginary frequencies: these classical and strained species are found with aromatic π^2 electronic configurations.

CF^+ is yet another isoelectronically related diatomic molecule. What can be said about $HBCF^+$ and its isomers? Two factors are operative in determining the relative stabilities within the series, $HBCF^+$, HBCO, and $HBCN^-$: the Lewis acid strength of the HB fragment and the backbonding effect. Provided the former is the dominant effect, the stability should increase in the order $HBCF^+$, HBCO and $HBCN^-$ since the nonbonded pair of electrons in the CF^+ fragment is (based on electrostatic and electronegativity) reasoning more tightly bound than the carbon of CO which in turn is more tightly bound than the CN^- carbon. Backbonding of B-H bonds into empty π^* orbitals of CO is recognized as a key stabilizing mechanism for H_3BCO. If also important for HBCO, it is presumably less important[5] for $HBCN^-$ and H_3BCN^- and more important for $HBCF^+$ and H_3BCF^+. As such, stability in the order $HBCN^- <$ HBCO $< HBCF^+$ is expected. Definitive calculational conclusions are not available because of missing runs and differing basis set and/or correlation energy levels in studies 1 and 2. However, we have already found minima corresponding to acyclic $HBCF^+$ as well as to a ring structure[6].

Finally, we recall that the gas phase ion chemistry of CN^- and the also isoelectronic NO^+ are both quite rich. What will be found for CF^+? Studies here are even more embryonic.

References

1. Skancke, A. and Liebman, J.F. (1994) *J. Phys. Chem.* **98** 13215.
2. Ćernušák, I., Beck, S and Bartlett, R.J. (1992), *J. Phys. Chem.* **96**, 10284; Ćernušák, I. and Lischka, H. (1995) *Chem. Phys. Lett.* **241** 261.
3. Deakyne, C.A. and Liebman, J.F. (1991) *J. Mol. Struct. (Theochem)* **235** 343.
4. In "study 2" the possibility of a triplet species was not considered, while in "study 1" the possibility of the isomeric isocarbonyl "H-B-O-C" was likewise not discussed.
5. This follows from the charge on the "new, carbon-containing" acceptor as well as expected σ, π and π* MO compositions in terms of coefficients on C and on N, O and F AOs in the diatomic.
6. Preliminary findings on $HBFC^+$ show a short B-F bond and no bonding to carbon, suggestive of formation of HBF^+ by fluorine abstraction by boron.

ISOMERIZATION OF C_6H_6 RADICAL CATIONS: 1,3-HEXADIEN-5-YNE

W.J. van der Hart
Leiden Institute of Chemistry, Gorlaeous Laboratories, Leiden University, P.O. Box 9502, 2300 RA Leiden, The Netherlands

In a recent paper [1] from this laboratory it was shown by ab initio calculations that the barrier for hydrogen shifts in the benzene radical cation is about 50 kcal mol^{-1}. Once the internal energy is sufficient for a 1,2-hydrogen shift, the moving hydrogen can go to any position in the ring. The barrier for carbon scrambling via an isomerization to the fulvene structure was calculated to be 17 kcal mol^{-1} higher than that for carbon scrambling.

According to previous semi-empirical calculations [2], the lowest barriers that can be reached from the structures involved are those leading to the 1,3-hexadien-5-yne radical cation and to non-classical ion structures with a five-membered ring and an exocyclic CH-bond. In the present work the semi-empirical results were used as a starting point for ab initio calculations on these isomerizations using the Gaussian 92 and 94 and GAMESS-UK program packages. The geometries of the stable ion structures and transition states were optimized at the ROHF/6-31G** level. For these optimized structures single-point multireference configuration interaction (MRCI) calculations were done with the Table-CI option of GAMESS-UK. The resulting relative energy values (kcal mol^{-1}) are summarized in Fig. 1. Here it should be noted that the value marked with an asterisk is still a bit uncertain.

The values in Fig. 1 show that:

1. The heat of formation of most non-classical ions involved is lower than that of the classical 1,3-hexadien-5-yne radical cation.
2. The barrier for isomerization of the 1,3-hexadien-5-yne radical cation is about 16 kcal mol^{-1} which explains the previous experimental conclusion [3] that its value should be lower than 1 eV. Interestingly, the results indicate that there is a possibility that a (very) small fraction of the 1,3-hexadien-5-yne radical cation does not isomerize to the benzene structure but is trapped in a non-classical ion structure with a five-membered ring.

References

1. van der Hart, W.J. (1995) *J. Am. Soc. Mass Spectrom.* **6** 513.
2. van der Hart, W.J. (1994) *Int. J. Mass Spectrom. Ion Processes* **130** 173.
3. van de Guchte, W.J. and van der Hart, W.J. (1993) *Int. J. Mass Spectrom. Ion Processes* **123** 7.

ION/MOLECULE ASSOCIATION REACTIONS

P.S. VINOGRADOV, D.M. BORISENKO, O.V. DMITRIEV
Institute for Energy Problems of Chemical Physics, Russian Academy of Sciences, Leninsky 38, B1.2., Moscow 117829, Russia

Ion/molecule association reactions are the processes that produce the ion clusters that play an important role in ion chemistry of the Earth's atmosphere at the altitudes below 85 km. Association reactions are not elementary processes and usually are presented by a simplified two stage model

$$A^+ + B \underset{k_d = 1/t}{\overset{k_f}{\rightleftharpoons}} [A^+ B^*] \xrightarrow{(+M),\ k_{st}} A^+B + M$$

The value of the excited collisional complex lifetime controls the association rate at low pressure, being a factor in the expression of the third order (termolecular) rate constant. At high concentration of M the rate of the process becomes independent of [M], and the reaction appears to proceed by second order kinetics. The idea of the estimation of t consists in comparing the lifetime with the time between collisions and can be obtained from the dependence of the association rate vs M pressure. The results of such studies performed in our group obtained by the flowing afterglow technique were reported. In the course of our experiments 3 types of kinetic dependencies were studied: on the reaction time, on reactants' flow and on the pressure of the buffer gas. The latter dependence was the basis for the procedure of estimation of the lifetimes of complexes.

Table 1 shows data obtained in our group for atmospheric reactions. It follows from the table that the attachment of the fourth and the fifth water molecules $H^+(H_2O)_n$ ion clusters appears to be a second order process at pressures in the range Torr while for $n < 4$ the process proceeds by third order kinetics. Longer complex lifetimes for a greater number of particles in the complex seems to be consistent with RRKM theory. The lifetimes of collisional complexes with 4 and 5 H_2O molecules are estimated to be about 10 ns.

Nevertheless, the $O_2^+ + O_2 \rightarrow O_4^+$ shows a significant second order kinetics contribution for the association process at pressures of about 1 Torr although the number of particles in the complex is only four. One of the possible explanations of the phenomenon may be a non-adiabatic transition to a binding (electronically excited) state. Note that the probability of the transition is low. If one compares the second order reaction rate constant and the collision rate constant the probability can be estimated as one per half a million. But once the system has got into the nonadabatic channel with low probability, the same low probability is expected to get out, i.e. the system should "wait" there for half a million vibrational periods. That results in the lifetime estimation of 10 ns.

Table 1

Reaction			Rate Constant
1)* $O_2^+ + O_2$ (+M)	→	O_4^+ (+M)	4.5 10^{-15} cm^3/s
2) $O_2^+ + N_2 + N_2$	→	$O_2N_2^+ + N_2$	3.0 10^{-33} cm^6/s
3) $O_2N_2^+ + O_2$	→	$O_4^+ + N_2$	2.5 10^{-12} cm^3/s
4) $O_4^+ + N_2$	→	$O_2N_2^+ + O_2$	1.0 10^{-14} cm^3/s
5) $O_2N_2^+ + N_2$	→	$O_2^+ + N_2 + N_2$	1.5 10^{-13} cm^3/s
6) $O_2^+ + CO_2 + N_2$	→	$CO_4^+ + N_2$	3.0 10^{-31} cm^6/s
7) $H_3O^+ + H_2O + Ar$	→	$H^+(H_2O)_2 + Ar$	2.5 10^{-27} cm^6/s
8) $H^+(H_2O)_2 + H_2O + Ar$	→	$H^+(H_2O)_3 + Ar$	4.4 10^{-27} cm^6/s
9)* $H^+(H_2O)_3 + H_2O + Ar$	→	$H^+(H_2O)_4 + Ar$	5.0 10^{-12} cm^3/s
10)* $H^+(H_2O)_4 + H_2O + Ar$	→	$H^+(H_2O)_5 + Ar$	1.0 10^{-12} cm^3/s
11) $H^+(H_2O)_5 + Ar$	→	$H^+(H_2O)_4 + H_2O + Ar$	4.0 10^{-14} cm^3/s
12) $H^+(H_2O)_4 + Ar$	→	$H^+(H_2O)_3 + H_2O + Ar$	4.0 10^{-14} cm^3/s
13) $H^+(H_2O)_3 + Ar$	→	$H^+(H_2O)_2 + H_2O + Ar$	4.0 10^{-14} cm^3/s
14) $H^+(H_2O)_2 + Ar$	→	$H_3^+O + H_2O + Ar$	4.0 10^{-14} cm^3/s

Note to Table 1. Termolecular reactions that obey the second order kinetics are marked by *.

As another example of the second order association process at Torr pressure CH_3^+ + H_2O reaction can be mentioned. The value of the second order reaction rate constant is estimated as 10^{-9} cm^3/s. Note that the similar (by number of particles) $H_3O^+ + H_2O$ reaction is a third order process (see Table 1). In the former case, a great area of the potential energy surface ("valleys" corresponding to different products) is available for the system at given initial energy.

Table 2 shows some data on the reactions of simple ions and water ion clusters with acetonitrile molecules. The latter type of reaction is interesting for stratospheric ion chemistry. Recently the measurements of ionic composition in the stratosphere have shown that ion clusters containing acetonitrile appeared to be an abundant component of detected positively charged species. The structures of combined water-acetonitrile ion clusters were studied by R. Squires by the CID technique. As yet, quantitative data on the rates of the reactions were absent, so one of the goals of these studies was to measure their rate constants.

Table 2 includes data on reactions of attachment of acetonitrile to ions and ion clusters. The reactions appear to be second order at Torr pressures. Note that the formation of protonated acetonitrile trimer (reaction k_3) was observed, yet its effective rate constant is one order of magnitude lower than that for the protonated dimer which has a more stable structure.

TABLE 2. Rate Constant Values Obtained at High Concentration of Water

Reaction	$k/10^{-9}\ cm^3s^{-1}$	$k/10^{-26}\ cm^6s^{-1}$	Lit. Values
$HW^+ + W \rightarrow HW_2^+$ *	0.0214	0.0665**	$6.65 \cdot 10^{-28}\ cm^6/s$ 1
$HW_2^+ + W \rightarrow HW_3^+$	0.055 ± 30%	0.171	$1.51 \cdot 10^{-27}\ cm^3/s$ 2
$HW_3^+ + W \rightarrow HW_4^+$	0.051 ± 30%	0.164	$1.5 \cdot 10^{-27}\ cm^6/s$ 3
$HW_5^+ + W \xrightarrow{K_4,He} HW_4^+$ **	$K_P = 1.1 \cdot 10^4\ atm^{-1}$	$K = 1.1 \cdot 10^4\ atm^{-1}$	4
$HW_3^+ + A \rightarrow HW_2A^+ + W$	3.7 ± 15%		5
$HW_2A^+ + A \rightarrow HWA_2^+ + W$	2.9 ± 15%		6
$HWA_2^+ + A \rightarrow HWA_3^+$	$0.25^{+100\%}/_{-50\%}$		7
$HW_2^+ + A \rightarrow HWA^+ + W$	4.5 ± 15%		8
$HWA^+ + A \rightarrow HA_2^+ + W$	1.9 ± 15%		9
$HW^+ + A \rightarrow HA^+ + W$ ***	5 ± 10%		$4.7 \cdot 10^{-9}\ cm^3/s$ 10
$HA^+ + A \rightarrow HA_2^+$	1.7 ± 10%	5.3	$4.3 \cdot 10^{-9}\ cm^3/s$ 11
$HA_2^+ + A \rightarrow HA_3^+$	0.007 ± 50%		12
$HA_2^+ + W \rightarrow HWA_2^+$	$0.002^{+100\%}/_{-50\%}$		13
$HW_4^+ + A \rightarrow HW_3A^+ + W$	3.5 ± 20%		14
$HW_3A^+ + A \rightarrow HW_2A_2^+ + W$	3.1 ± 20%		15
$HW_2A_2^+ + A \rightarrow HWA_3^{++} + W$	$2.4^{+100\%}/_{-50\%}$		16
$HWA_3^+ + W \rightarrow HW_2A^{2+} + A$	$0.0066^{+100\%}/_{-50\%}$ ****		17
$HWA_3^+ + He \rightarrow HWA_2^+ + A$	$3.3\ 10^{6\ +100\%}/_{-50\%}$		18
$HWA_2^+ + W \rightarrow HW_2A_2^+$	$0.012^{+100\%}/_{-50\%}$		19
$HW_5^+ + A \rightarrow HW_3A^+ + W_2$	1.2 ± 20%		20
$HA_3^+ + W \rightarrow HWA_3^+$	$3^{+100\%}/_{-50\%}$		21

* $W = H_2O$, $A = CH_3CN$ **, *** used for calibration, **** $k_{16}/k_{17} = 366$

As for attachment reactions of water/acetonitrile protonated clusters with acetonitrile, they appeared to be less effective in comparison with the reaction of a change of a ligand. So for stratospheric ion chemistry such reactions seem to be the dominating way of combined water/acetonitrile ion clusters formation.

Table 3 shows the results of the studies of chemical reactions in $Xe^+ + CH_3CN$ system. The most interesting result is the absence of effective charge transfer reactions, even though the ionization potential of Xe is only by 0.01 eV greater than that for acetonitrile and practically resonant transfer was expected. Perhaps, the low charge transfer efficiency takes place due to strong "lock in" effect arising in the course of attraction of the Xe ion and the large dipole moment (~4 Debye units) of the acetonitrile molecule. For a locked geometry of the complex Xe^+ -- NC -- CH_3 the transfer of an electron from bonding NC orbital to a *p* orbital of Xe appears not to be effective.

TABLE 3. Reactions in $Xe^+ + CH_3CN$ System
$A = CH_3CN$; $B = CH_2CN$

Reaction	$k / 10^{-9}$ cm³/s	Extrapolated low P value of k/cm^6s^{-1}
$Xe^+ + A \rightarrow XeA^{+*}$	$4.0^{+100\%}/_{-50\%}$	$5.7 \cdot 10^{-26}$
$XeA^+ + A \rightarrow XeA_2^+$	$0.9^{+100\%}/_{-50\%}$	$1.0 \cdot 10^{-25}$
$XeA^+ + A \rightarrow HA^+ + B + Xe$	$0.021^{+50\%}/_{-20\%}$	
$XeA_2^+ + A \rightarrow XeA_3^+$	$0.03^{+150\%}/_{-50\%}$	
$HA^+ + A \rightarrow HA_2^+$ ***	$1.1 \pm 30\%$	$5.1 \cdot 10^{-26}$
$XeA_2^+ + A \rightarrow HA_2^+ + B + Xe$	$0.09^{+50\%}/_{-20\%}$	
$HA_2^+ + A \rightarrow HA_3^+$ **	0.007	
$Xe^+ + A \rightarrow Xe + A^+ + 0.01$ eV		$k < 3\ 10^{-11}$ cm³/s
$Xe^+ + A \rightarrow XeH^+ + B + 0.036$ eV		$k < 1\ 10^{-11}$ cm³/s
$Xe^+\ \overset{+\Delta}{N}\equiv\overset{-\Delta}{C}$-CH	D=4 Debye units	

Yet, the attachment reaction appears to be a very efficient process at Torr pressures though the number of atoms in the collision complex is 7. The attachment of the second and even the third acetonitrile molecule to Xe^+ was observed.

The estimations of the collision complex lifetimes made previously from the pressure dependencies of the association rate was based on the simple model mentioned above. If the dependence is measured in a large enough pressure range one can hope to obtain more detailed information from the data processing (e.g. in representing complex formation and stabilization rate constants as a product of the probability of the process and the collisional rate constant there is a temptation to obtain also the corresponding probabilities). Yet, if at least two types of complexes with different lifetimes and probabilities can be formed no processing can yield any reliable information. So other direct methods for estimations of lifetimes are necessary.

CALCULATION OF PROTON AFFINITIES OF STRONG BASES USING HIGH LEVEL AB INITIO METHODS: A TEST FOR THE PROTON AFFINITY SCALE?

J.F. GAL
FT-ICR Group, Laboratoire de Chimie Physique Organique, University of Nice-Sophia Antipolis
06108 Nice Cedex 2, France

During the last 5 years, several papers have proposed changes in the proton affinity (PA) scale, particularly in the upper basicity range[1-5]. Smith and Radom used ab initio calculations at the G2 level to differentiate between the proposed changes[3]. The G2 PAs were found to be consistently within 10 $kJ.mol^{-1}$ of the scale compiled by Lias and co-workers (LBLHLM scale in the following),[6] except for isobutene. The isobutene PA was shown simultaneously to be in error[5] and a new value in excellent agreement with the G2 PA was put forth. The G2 PAs agree even better with the latest experimental results of Szulejko and McMahon[4]. In order to further compare the LBLHLM scale and the calculated PAs, ab initio calculations on stronger bases than those studied by Smith and Radom were foreseen. From the computational point of view, small molecules with a minimum of flexible groups should be selected. We choose as potentially strong, but sufficiently small, bases: formamidine, acetamidine and guanidine (see "The Chemistry of Amidines and Imidates", in the Patai series, Wiley, N.Y., 1991). The G2 method as well as less computer-time consuming codes were used. Simultaneously, we tried to obtain experimentally their PA by proton transfer equilibrium measurement, using FT-ICR mass spectrometry. The neutral bases were liberated from their hydrochlorides. Unfortunately, we were not able to isolate and use formamidine (probably because of its rapid decomposition: hydrolysis, polymerization?): the FT-ICR mass spectra showed only the presence of formamide and ammonia in the spectrometer. The two other neutral bases exhibit some affinity toward H_2O and CO_2, but are stable enough to be introduced in the spectrometer and their pressure measured. Proton transfer equilibria with several reference bases were measured and the reported experimental results are referenced to the LBLHLM scale. The PA for these imino-bases, calculated at the G2 level (table 1), show the same trend between theory and experiment to that reported by Smith and Radom for the weaker amino-bases[3]: the calculated values are always larger, but by less than 10 $kJ.mol^{-1}$.

TABLE 1. G2 ab initio calculations of PAs for strong bases ($kJ.mol^{-1}$)

CH_2=NH (formaldimine)	862.3	(exp. 854.0)[b]
H_2N-CH=NH (formamidine)	943.1[a]	(exp. not avail.)
H_2N-C(CH_3)=NH (acetamidine)	969.9	(exp. 962.7)[b]
$(H_2N)_2$C=NH (guanidine)	980	(exp. 974.9)[b]

(a) ref. 8; (b) ref. 6

Acetamidine and guanidine (table 1) are stronger bases than dimethylamine (the strongest base calculated by Smith and Radom) by 40-50 $kJ.mol^{-1}$, formaldimine being close to ammonia. The difference ΔPA = PA(calc) - PA(exp) does not seem to inflate with the currently accepted absolute PA's. The guanidine PA is similar to that of triethylamine and quinuclidine. To obtain even stronger monofunctional bases, alkylation at the nitrogen atoms of amidines or guanidine is required.[9] Such molecules are not readily calculated at the G2 level. The efficiency of less time-consuming codes: G2(MP2), Density Functional theory (DFT), and a composite G2(DFT) corresponding to DFT optimized geometrics (6 31G*), have been tested. The results obtained on acetamidine are presented in table 2.

TABLE 2: Comparison between G2 and less expensive PA calculations on acetamidine ($kJ.mol^{-1}$); for details on computational methods, see refs. 7 and 8

Experimental	962.7	$\Delta = 0$
G2 (for comparison)	969.9	7.2
G2 (MP2)	969.9	7.2
G2 (BLYP)	966.9	4.2
G2 (SVWN)	966.9	4.2
BLYP/6-311+G(3df,2p)	968.6	5.9
SVWN/6-311+G(3DF,2P)	956.9	-5.8

Interestingly, the G2 and G2(MP2) give the same result, higher than the experimental value by 7.2 $kJ.mol^{-1}$. Using DFT optimized geometrics, G3 gives slightly lower PAs, but again higher than the experimental one. The two DFT methods give also PA's in agreement with experiment, but we note that the BLYP PA is close to the G2 PA, and that the SVWN scheme is the only one to give a PA lower than the experimental value.

In conclusion, according to our G2 calculations, the necessity of a large correction for the PA of bases situated at the upper part of the scale is not obvious. Whether a 10 $kJ.mol^{-1}$ upward adjustment in this range is needed or not cannot be decided from calculations (even the best available) only.[10] It seems that PA can be calculated to within 10 $kJ.mol^{-1}$ using DFT, with the possibility to study reasonably large molecules.

1. Acknowledgements

I thank Dr. Hab. E.D. Raczyńska, SGGW, Warsaw, for valuable suggestions.

2. References

1. Meot-Ner, M. and Sieck, L.W. (1991) *J. Am. Chem. Soc.* **113**, 4448-4460.
2. Szulejko, J.E. and McMahon, T.B. (1991) *Int. J. Mass Spectrom. Ion Process* **113**, 279-294.
3. Smith, B.J. and Radom, L. (1993) *J. Am. Chem. Soc.* **115**, 4885-4888.
4. Szulejko, J.E. and McMahon, T.B. (1993) *J. Am. Chem. Soc.* **115**, 7839-7848.
5. Keister, J.W., Riley, J.S. and Baer, T. (1993) *J. Am. Chem. Soc.* **115**, 12613-12614.
6. Lias, S.G., Liebman, J.E. and Levin, R.D. (1984) *J. Phys. Chem. Ref. Data* **113**, 695-808; Lias, S.G., Bartmess, J.E., Liebman, J.E., Holmes, J.L., Levin, R.D. and Mallard, W.G. (1988) *J. Phys. Chem. Ref. Data* **17**, Suppl. No. 1.
7. González, A.I., Mó, O., Yáñez, M., Léon, E., Tortajada, J., Morizur, J.-P., Leito, I., Maria, P.-C. and Gal, J.-F. submitted.
8. Tortajada, J., Léon, E., Luna, A., Mó, O. and Yáñez, M. (1994) *J. Phys. Chem.* **98**, 12919-12926.
9. Decouzon, M., Gal, J.-F., Maria, P.-C. and Raczyńska (1993) *Rapid Commun. Mass Spectrom.* 7, 599-602; Raczyńska, E.D., Maria, P.-C., Gal, J.-F. and Decouzon, M. (1994) *J. Phys. Org. Chem.* 7, 725-733, and reference 13 cited therein.
10. A new analysis of the PA data base is presented by Ed Hunter (NIST, Maryland), elsewhere in this book.

METAL ION CHEMISTRY WORKSHOP: DISCUSSION SUMMARY

PETRA A. M. VAN KOPPEN
Department of Chemistry
University of California
Santa Barbara, CA 93106

Coordinatively unsaturated transition metal complexes are important intermediates in both catalytic and organometallic reactions. These metals are extremely versatile in their chemistry because the nd and (n+1)s orbitals are close in energy, which gives rise to a relatively high density of low lying excited states. Systematic studies of their reactivity and thermochemistry have given insight into the energetics and mechanisms of these reactions. A primary focus has been to determine the factors that underlie σ–bond activation by the metal center. These factors are crucial for understanding how transition metals function as catalysts.

State selected studies have shown that the metal ion reactivity is a strong function of electronic state configuration. Electronically excited state metal ions are often significantly more reactive than the corresponding ground state ions. Ligating the metal provides an alternate means of changing the electronic configuration and reactivity of the transition metal center. Recent studies of cluster assisted σ–bond activation indicate that the addition of two H_2 or two CH_4 ligands to first row transition metal ions provides a sufficiently strong field to induce σ–bond activation of H_2 and CH_4. Because the strength of the ligand field is a function of both the number of ligands and the type of ligands, the effect of the ligand field on the metal ion reactivity can be studied. This is a new and exciting area of research, presently being explored using experimental techniques in conjunction with theory.

This workshop covered a variety of aspects of this field, including organometallic chemistry of both transition metal and main group metal ions. Investigations involving methods such as equilibrium, time of flight, collision induced dissociation, photo-fragmentation and metastable ion kinetic energy measurements were reported. A summary of the contributions follows.

1. Cluster Assisted Sigma Bond Activation at Transition Metal Centers

Petra A. M. van Koppen, Paul R. Kemper, John E. Bushnell, Michael T. Bowers
Department of Chemistry, University of California, Santa Barbara, CA 93106, U.S.A.

Our approach to studying the effect of the ligand field on the electronic structure and reactivity of transition metal centers has been to measure equilibria between transition

K. R. Jennings (ed.), Fundamentals and Applications of Gas Phase Ion Chemistry, 469–482.

metal ions and a number of simple ligands.[1-4] In this abstract we will focus on the mechanism and driving force for cluster assisted σ–bond activation observed for both Sc^+/nH_2 and Ti^+/nCH_4 reactions.

In the equilibrium experiment,[3] product/parent ion ratios were measured as a function of reaction time. As the drift time was increased the product/parent ion ratios became constant, indicating equilibrium had been reached. The ratios were converted to equilibrium constants and standard free energies.[3] A plot of ΔG^o versus T yielded straight lines for all systems with the intercept equal to ΔH^o_T and a slope equal to ΔS^o_T Standard statistical thermodynamic methods were used to obtain ΔH^o_0 from ΔH^o_T and ΔS^o_T.

Cluster assisted σ–bond activation is a new class of reactions first observed for Sc^+ reacting with nH_2.[1] This system displayed quite complicated kinetics in establishing the "simple" equilibrium shown in Reaction (1), where $Sc(H)_2^+$ represents either the $Sc(H_2)^+$ adduct or the inserted HSc^+H product.

$$Sc^+ + H_2 \rightleftharpoons Sc\,(H)_2^+ \qquad (1)$$

For all first row transition metal ions, the equilibrium for the addition of a single H_2 to form the $M(H_2)^+$ adduct was established very quickly, except for Sc^+. In this case, a very slow approach to equilibrium was observed and was modeled to yield a rate constant of 3 x 10^{-17} cm^3/sec at 350K, an efficiency of ~10^{-7}. This appears to rule out formation of a simple $Sc(H_2)^+$ electrostatic complex that should be formed with essentially unit efficiency. In addition, a negative temperature dependence was observed for this rate constant, which implies that the inefficiency of this reaction cannot be due to an energy barrier along the reaction coordinate.

Electronic structure calculations[5,6] indicate that the inserted HSc^+H product has $^1A_1$ symmetry which correlates directly to a component of the ^1D, $3d^2$ excited state of Sc^+. Thus, ground state Sc^+ (^3D, $4s^13d^1$) reacts with H_2 via a spin-orbit coupled crossing to form the inserted HSc^+H product. Theory predicts a substantial barrier (~20 kcal/mol)[5] associated with this crossing. From our experimental data, however, a process involving an energy barrier is ruled out.

The details of the mechanism proposed to explain the experimental observations have been published.[1] The Sc^+ ion does insert into the H–H bond but not in a simple single collision mechanism. The driving force for the H–H σ–bond activation is summarized as follows. Because the d-orbitals for atomic transition metal ions are degenerate, the ground electronic configurations are high spin. Ligating the metal ion splits this degeneracy. The ligand field and the d-orbital splitting increase with the number of ligands and with the field strength of the ligand. At a certain point electron pairing becomes energetically favorable and a spin change will occur. For Sc^+ reacting with nH_2, two H_2 ligands provide a strong enough field to allow a spin change from a triplet to a singlet surface and σ–bond activation is observed (reaction (2)).

$$Sc(H_2)_2^+ + H_2 \rightleftharpoons (H_2)_2Sc^+\!\!\begin{smallmatrix}H\\H\end{smallmatrix} \rightleftharpoons Sc^+\!\!\begin{smallmatrix}H\\H\end{smallmatrix} + 2H_2 \qquad (2)$$

The possibility of a cluster assisted σ–bond activation mechanism inspired us to look for other systems to study. Methane seemed to be a likely candidate because for all first row transition metal ions dehydrogenation of methane is endoergic.[7,8] However, for third row transition metal ions, multiple dehydrogenation of methane leading to the oligomerization of methane is observed at thermal energy.[9] Here we show that cluster assisted σ–bond activation by Ti^+ is the first step to oligomerization of methane by a first-row transition metal ion.

For Ti^+ injected into a high pressure of CH_4, a strong signal corresponding to $(CH_4)_2Ti^+{=}CH_2$ is observed. Reaction of $(CH_4)_2Ti{=}CH_2^+$ with methane initiates the oligomerization process, reaction (3), converting methane to ethylene.

$$(CH_4)_2Ti^+{=}CH_2 + CH_4 \rightarrow (CH_4)_2Ti^+\!\!\begin{smallmatrix}CH_2\\|\\CH_2\end{smallmatrix} + H_2 \qquad (3)$$

In our experiment, with the addition of the third methane to Ti^+, both $(CH_4)_2Ti(H)(CH_3)^+$ and $Ti(CH_4)_3^+$ are observed. These clusters are clearly distinguishable even though they have the same mass. At low temperatures (255-340 K) addition of CH_4 to $Ti(CH_4)_2^+$ comes quickly into equilibrium ($t << 500$ μs) indicating simple adduct formation is occurring (reaction (4a)). At high temperatures (500-580 K) addition of CH_4 occurs very slowly and equilibrium is obtained only at very long times ($t > 3000$ μs) indicating σ–bond activation is occurring (reaction (4b)).

$$Ti(CH_4)_2^+ + CH_4 \rightarrow Ti(CH_4)_3^+ \qquad \Delta H = -6.6 \text{ kcal/mol} \qquad (4a)$$

$$Ti(CH_4)_2^+ + CH_4 \rightarrow (CH_4)_2Ti^+\!\!\begin{smallmatrix}H\\CH_3\end{smallmatrix} \qquad \Delta H \sim -16 \text{ kcal/mol} \qquad (4b)$$

The rate of the slow approach to equilibrium has a positive energy dependence, and modeling yields a barrier to insertion of 8 kcal/mol. The fact that two distinct linear regions are observed in the ΔG^o versus T plot allows bond dissociation energies to be obtained for both reaction (4a) and (4b). This analysis indicates σ–bond activation is favored over adduct formation by ~10 kcal/mol as shown in Figure 1.

We are very excited about the possibilities of studying the effect of ligation on the electronic structure and reactivity of the transition metal center. There appears to be an opportunity to "tune" the σ–bond activation process by selecting the right ligand.

Experiments are underway to test simple ligands for these properties. From these studies we hope to gain further insight into the factors which underlie σ–bond activation and the role transition metals play as catalysts.

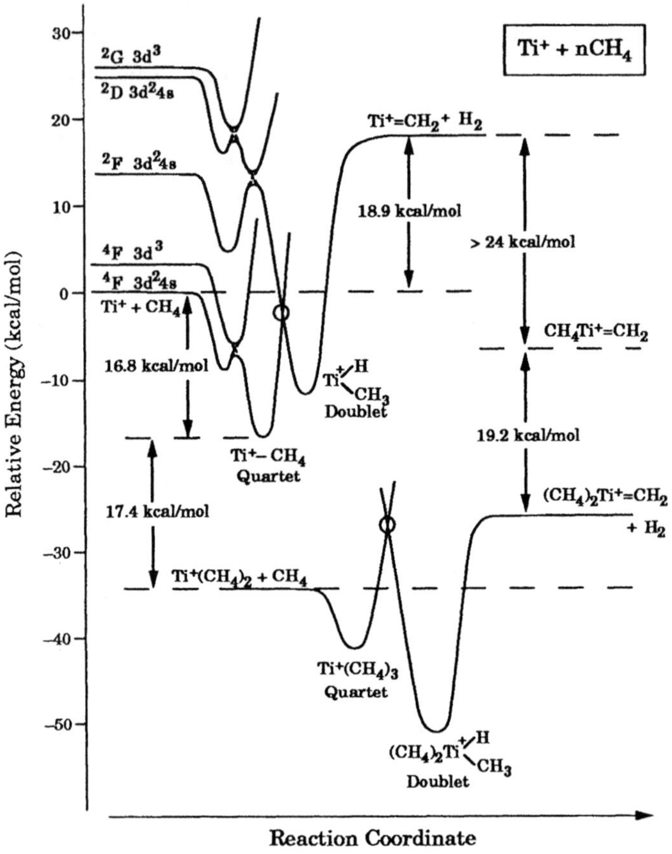

Figure 1. Reaction coordinate diagram for $Ti^+ + CH_4$ (top part) and $Ti(CH_4)_2^+ + CH_4$ (bottom part). State designations are for the Ti^+ atom.

2. Biomolecular Collisions in Real Time Crossed-beam Reactions of Ni^+ and Co^+ with Hydrocarbons

Robert J. Noll, Sung Soo Yi and James C. Weisshaar

Department of Chemistry University of Wisconsin-Madison, Madison, WI, 53706, U.S.A.

We report on a single-collision, crossed-beam experiment that measures the evolution of *bimolecular* collision complexes into product ions *directly* in real time, with

100 ns resolution.[10] Reactions of Co^+ and Ni^+ with propane and n-butane were studied (reactions (5) and (6)).

$$M^+ + C_3H_8 \rightarrow MC_2H_4^+ + CH_4 \quad (5a)$$
$$MC_3H_6^+ + H_2 \quad (5b)$$
$$MC_3H_8^+ \text{ (adducts)} \quad (5c)$$

and

$$M^+ + n\text{-}C_4H_{10} \rightarrow MC_2H_4^+ + C_2H_6 \quad (6a)$$
$$MC_3H_6^+ + CH_4 \quad (6b)$$
$$MC_4H_8^+ + H_2 \quad (6b)$$
$$MC_4H_{10}^+ \text{ (adducts)} \quad (6d)$$

Reactions of Co^+ with C_3D_8 and n-C_4D_{10} were studied as well. A neutral beam of transition metal atoms, a pseudo-skimmed molecular beam of alkane, and a 7-ns pulsed dye laser beam cross in field-free space in the extraction region of a time-of-flight mass spectrometer. Reactant translational and internal energy is carefully controlled. At $t = 0$, the laser pulse creates a packet of M^+ cations in the ground spin-orbit level by resonant two-photon ionization.[11] The reactant ion is either $Co^+(3d^8, {}^3F_4)$ or Ni^+ $(3d^9, {}^2D_{5/2})$, The collision energy is either ~ 0.01 eV (merged beam configuration) or 0.2 eV, controlled by the angle between the metal ion and hydrocarbon beams. The internal energy distribution of the alkane beam is cooled to 50-100 K by mild expansion. Immediately after photoionization, ion-molecule collisions commence. After an adjustable reaction time Δt_{rxn} of 0-8 ms, a pulsed electric field accelerates all ions for mass analysis by time-of-flight over a 1-*m* field free path. Such mass spectra provide time-resolved snapshots of the evolution of bimolecular collision complexes to elimination products.

For $Co^+ + C_3H_8$ with Δt_{rxn} = 8 μs and 0.2 eV collision energy, the mass spectrum of products includes 40% adducts (5c); the balance of products are the CH_4 and H_2 elimination channels (5a) and (5b), in 1:3 ratio. The $Ni^+ + C_3H_8$ products under the same conditions include 73 % adducts (5c); in addition CH_4 and H_2 elimination occur in 6:1 ratio. Deuteration increases the fraction of adducts to 93 % for $Co^+ + C_3D_8$. When the collision energy is reduced to 0.01 eV, the fraction of long-lived adducts increases to 83 % for Co^+ and 96 % for Ni^+. These results suggest a small potential energy barrier for the Ni^+ $({}^2D_{5/2}) + C_3H_8$ reaction; Co^+ $({}^3F_4) + C_3H_8$ reactants apparently lie just above some threshold along the path to elimination products. We have also used a retarding field in the 1-m flight tube to separate the $CoC_3H_8^+$ peak into *very* long-lived adducts (lifetime greater than 20 μs) and the fragments of metastable decay to $CoC_3H_6^+ + H_2$ elimination products, which form on a 2-10 μs time scale.

For Co^+ + C_4H_{10} under the same conditions (Δt_{rxn} = 8 μs and 0.2 eV collision energy), the mass spectrum of products exhibits only 16 % adducts (6d); the dominant elimination component branches among the C_2H_6 (5a), CH_4 (5b) and H_2 (5c) channels in the 10:1:3 ratio. Surprisingly, decreasing the collision energy to 0.01 eV significantly decreases the fraction of adducts to 6 % while leaving the branching ratios among elimination channels essentially unchanged. Deuteration to n-C_4D_{10} increases the fraction of adducts to 45 %; $CoC_2H_4^+$ + C_2D_6 dominate the elimination products. The product branching for Ni^+ + n-C_4H_{10} is remarkably similar to that of Co^+ + n-C_4H_{10}.

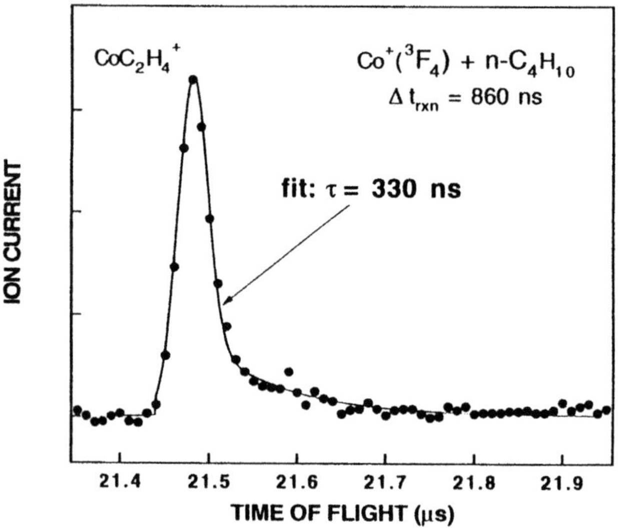

Figure 2. $CoC_2H_4^+$ from Co^+ + n-C_4H_{10} for 0.2 eV, Δt_{rxn} = 860 ns.

In both M^+ + n-C_4H_{10} reactions, the experiment reveals that the ethane elimination channel (6a) occurs on a range of time scales from 350 ns to 10 μs, and possibly even longer. The 350 ns time scale becomes evident when Δt_{rxn} is decreased to 860 ns. Under those conditions the $CoC_2H_4^+$ peak shows a marked tailing towards longer times (Figure 2), the characteristic signature of metastable complex decay in the extraction region of the mass spectrometer.[12] The ions in the tail accelerate initially as $CoC_4H_{10}^+$ adduct ions, then fragment *during* extraction, and complete their acceleration as less massive $CoC_2H_4^+$ elimination products. We have successfully modeled the tail as a single exponential decay of the complex with lifetime ± 100 ns. Similar tails are observed for both the Co^+ and Ni^+ reactions with n-butane at both 0.01 and 0.2 eV. Surprisingly, the lifetime is essentially independent of the identity of the metal and of collision energy over the range studied. However, at 0.2 eV deuteration to n-C_4D_{10}

increases the lifetime of the complex to 800 ± 150 ns for Co^+ and to 700 ± 150 ns for Ni^+. The tail for $Co^+ + n\text{-}C_4D_{10}$ deviates significantly from that predicted by single exponential decay. In addition, some of the longer-lived complexes that survive extraction subsequently fragment to Co^+ + *n*-butane or $CoC_2H_4^+ + C_2H_6$ on a time scale of 2-20 μs, as revealed by the retarding field technique. A substantial fraction of the adduct ions have not yet fragmented when they reach the detector at $t = 25$ μs.

In summary, the experiment shows that in each reaction the same H_2 or alkane elimination product evolves on a wide range of time scales, even though the collision energy and internal energy of reactants have been carefully controlled. We are exploring whether or not statistical rate theory in conjunction with energetic guidance from *ab initio* calculations can explain all the features of the data. The range of time scales may arise from the broad distribution of orbital angular momenta among complexes formed at different impact parameters within the Langevin circle.

3. Electron Affinity and Catalytic Activity of Iron Porphyrins

Douglas P. Ridge
Department of Chemistry, University of Delaware, Newark, DE 19716, U.S.A.

The importance of metalloporphyrins as biological redox catalysts suggests that these species might be useful as catalysts for industrial oxidation processes. The catalysis group at Sun Co. has, in fact, investigated a number of metalloporphyrins as catalysts for the oxidation of alkanes to alcohols.[13] They found that catalytic activity increased with the number of electronegative substituents on the catalyst and with the redox potential of the catalyst. This suggested that catalytic activity might increase with the gas phase electron affinity of the catalyst, and hence we undertook to measure the electron affinities of a number of metalloporphyrins.

The first series of metalloporphyrins examined were halogenated iron tetraphenyl porphyrins.[14] A number of these species had been found to catalyze the oxidation of isobutane to tertiary butyl alcohol by molecular oxygen. The metalloporphyrins were introduced into an FT-ICR spectrometer on a heated probe. Reference electrophiles were introduced through a batch inlet. Ionization was accomplished by low energy electron impact and by capture of electrons scattered from the pulsed electron beam. Charge transfer reactions of the metalloporphyrin and reference electrophile approached equilibrium in the ion trap. Equilibrium constants or limits on equilibrium constants were determined and electron affinities deduced from the resulting free energy changes and from the known electron affinity of the reference electrophile.

Comparison of the electron affinities of the halogenated iron tetraphenyl porphyrins with their reduction potentials suggested that the latter were subject to differential solvent effects. Hence the electron affinities were a better measure of the intrinsic electron affinity of the catalysts.

We can now add the electron affinities of iron octaethyl porphyrins to our original results and make a more significant test of the quantitative relationship between electron affinities and catalytic activity.[15] The octaethyl porphyrins were substituted with different numbers of nitro groups at the meso positions to increase the ligand electronegativity. We have electron affinity and catalytic activity (turn-over numbers) for two halogenated iron tetraphenyl porphyrins and two nitrated octaethyl porphyrins. A plot of the log of the turn-over number against the electron affinity for these four points is linear with a correlation factor of 0.99. Such a correlation is useful in and of itself for substantiating patent claims.

The mechanistic basis for the correlation between electron affinity and catalytic activity has to do with the ability of the metalloporphyrin to bind oxygen at the metal. In the catalytic mechanism molecular oxygen is pulled apart by two metalloporphyrins each binding an oxygen atom to their metal centers. These mono-oxo species then react with the alkane to make alcohols. Factors that enhance the reactivity of the oxo intermediate enhance the effectiveness of the catalyst. We hypothesize that as the electronegativity of the ligand increases the metal-oxygen bond in the oxo species decreases. Hence increasing ligand electronegativity gives increasing oxo reactivity.

The reactivity of the metalloporphyrins with oxygen atom donors reflects the relationship between ligand electronegativity and ability to bind an oxygen at the metal. Both anions and cations of iron porphyrins without halogen or nitro substituents or with only a few electronegative substituents react with NO_2 by O atom transfer. Species with more electronegative ligands are unreactive. Species with axial ligands are also unreactive verifying that the metal is the oxidation site.

This work is an example of the potential that exists for the application of the study of ion molecule reactions to quite practical problems in mechanistic chemistry. Realizing this potential will require collaborative work between gas phase ion chemists and imaginative industrial chemists.

4. Photochemical Processes on Transition Metal Atoms and Small Clusters

D.A. Kirkwood, J.F. Winkel and A.J. Stace
School of Molecular Sciences, University of Sussex, Falmer, Brighton BN1 9QJ

We report the first results of a new approach to studying the chemistry of metal clusters. Bare metal atoms and clusters are generated from volatile organometallic compounds and complexed with one or two reactant molecules (either ethanol or propanol) in an ion trap where they are excited with radiation from a tunable infrared laser.[16] Although a single photon of the latter is not sufficient to break a chemical bond, because the clusters spend ~ 1 ms within the radiation field, they absorb 15 - 20 photons (1.5 - 2eV) which, in some cases, is sufficient to drive the chemical reaction. For the most part, photoexcitation results in the loss of an alcohol molecule; however, in some instances the complexes lose either water or a small unsaturated hydrocarbon. For example, with small iron/ethanol cluster ions reactions (7) and (8) are observed.

$$Fe(CH_3CH_2OH)_2^+ + nh\nu \rightarrow Fe(CH_3CH_2OH{\bullet}CH_2CH_2)^+ + H_2O \quad (7)$$

or

$$Fe_{2,3}(CH_3CH_2OH)_2^+ + nh\nu \rightarrow Fe_{2,3}(CH_3CH_2OH)^+ + CH_3CH_2OH \quad (8)$$

Photofragment yields exhibit a very pronounced dependence on laser wavelength and at least one infrared absorption feature has a peak center very close to that seen for alcohol molecules adsorbed on to the surfaces of metals and metal oxides. An example of an absorption profile recorded for a complex between iron and *n*-propanol is shown in Figure 3. These data have been recorded by summing over the photofragment yields as a function of laser wavelength.

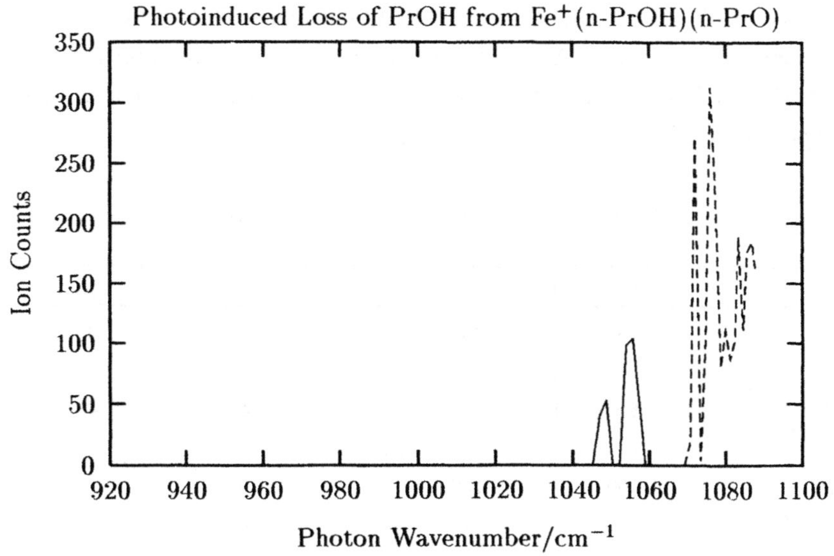

Figure 3. Infrared absorption profile recorded for the complex $Fe^+{\bullet}$(n-PrOH)(n-PrO) as a function of CO_2 laser wavelength.

5. A New Approach to Study Metal Ion Reactions

Jinsong Ni and Alex G. Harrison
Department of Chemistry, University of Toronto, Toronto, Canada M5S 1A1

Fast atom bombardment (FAB) has been shown to be a convenient source of singly-charged metal ions.[17-22] The present work explores the usefulness of studying reactive collisions of singly-charged transition metal ions so-produced with organic substrates in the quadrupole collision cell of a hybrid BEqQ mass spectrometer.[23] There has been one brief report[19] of the reactions of mass-selected metal ions with a 2-methylpropane in the quadrupole cell of a triple quadrupole mass spectrometer but no previous studies employing a hybrid sector-quadrupole instrument.

In our initial studies, the transition metal ions Sc^+ through Cu^+ were mass-selected and introduced into the quadrupole cell at 3 eV incident ion energy to react with cyclopentene under multiple collision conditions. Table 1 records the product distribution observed. The Mn^+ ion is largely unreactive with cyclopentene. The major primary product for Sc^+, Ti^+, V^+, Fe^+, Co^+ and Ni^+ is $M(C_5H_6)^+$ arising by reductive H_2 elimination from the $M(C_5H_8)^+$ complex. Cr^+ and Cu^+ show only clustering to form $M(C_5H_8)^+$. This difference has been rationalized[24] in terms of the energetics of insertion of the metal ion into a C-H bond. The $M(C_5H_6)^+$ product reacts further with c-C_5H_8 to produce $M(C_5H_5)_2^+$, presumably the ionized metalocene, except for $Co(C_5H_6)^+$ which also produces $Co(C_5H_6)_2^+$ and $Ni(C_5H_6)^+$ only clusters with a further cyclopentene. Where comparison is possible the results are in good agreement with results obtained in earlier studies by ICR[25,26] indicating that the tandem mass spectrometric method is a viable approach to the study of metal ion reactions.

Table I. Products from M^+/Cyclopentene Reaction

Product	Sc	Ti	V	Cr	Mn	Fe	Co	Ni	Cu
M^+	26	73	40	64	100	29	9	4	23
$M(C_5H_5)^+$		10	10		1	2			
$M(C_5H_6)^+$	8	12	5			4	7	4	6
$M(C_5H_8)^+$				12	6				16
$M(C_5H_5)_2^+$	100	100	100			100	73		
$M(C_5H_5)(C_5H_6)^+$		16	37			7			
$M(C_5H_6)_2^+$							100		
$M(C_5H_5)(C_5H_8)^+$					5				
$M(C_5H_6)(C_5H_8)^+$								100	8
$M(C_5H_8)_2^+$				100	20				100

Because of the possibility of excited metal ion states and the unknown kinetic energy of the reacting ions, the most likely application of this approach is the use of metal ion reactions to distinguish among isomeric molecules. Figure 4 shows the spectra observed for the reaction of Ti^+ with six C_5H_8 isomers under multiple collision conditions. A rich chemistry is evident and a ready distinction among the isomers is possible.

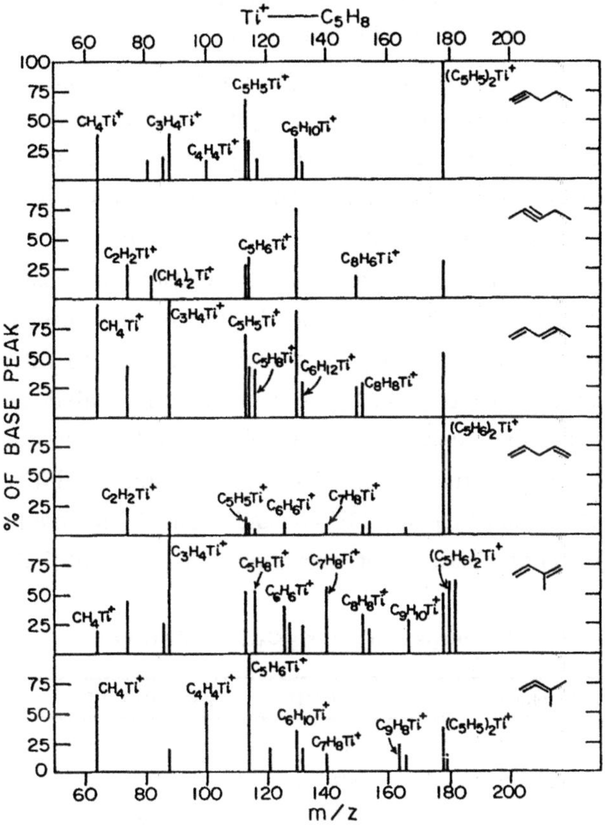

Figure 4. Reaction of Ti^+ with C_5H_8 Isomers

6. Gas Phase Ion/Molecule Complexes of AsI_2^+ with C_7H_8 Isomers

H. L. Sievers, C. Heide, and H.-F. Grützmacher
Department of Chemistry, University of Bielefeld, Germany

The gas phase ion chemistry of C_7H_8 radical cations is something like an "evergreen" of mass spectrometry. The radical cation of cyclohepta-*1,3,5*-triene ($CHT^{+\bullet}$), norborna-2,*5*-diene ($NBD^{+\bullet}$), quadricyclane ($QUAD^{+\bullet}$), and toluene isomerize prior to unimolecular decomposition,[27] and loss of $H^\bullet$ and C_2H_2 are the only significant decomposition reactions. The rearrangement of $C_7H_7^+$ formed from the toluene$^{+\bullet}$

precursor is still a matter of investigation.[28] The transition metal induced isomerization of QUAD to NBD is of general interest, because it releases 27 kcal/mol of energy stored by the photoisomerization of NBD to QUAD. A study[29] of gas phase $[M\ C_7H_8]^+$ (M = Fe, Cu) complexes showed that isomerization of QUAD to NBD was followed by a retro Diels-Alder reaction. In the product ion acetylene was replaced reversibly by C_2D_2. Hence, in these complexes the chemical energy storage system is "destroyed". The most abundant ion in the CAD spectra of [M toluene]$^+$ and [M CHT]$^+$ (M = Fe, Cu) is $C_7H_7^+$ and these complexes are best described by a single $(C_7H_7)M(H)^+$ structure.

A previous investigation[30] of $[AsI_2$ ligand$]^+$ complexes, formed under CI conditions and by EI ionization of a mixture of AsI_3 and several n- and π-donor ligands, demonstrated the capability of these main group metal complexes of reversible ligand exchange. The interactions between hydrocarbons and cationic fragments EX_2^+ of group-15 elements are of special interest since these fragments are isolobal to carbenes. In addition, the number of valence electrons of the EX_2^+ entity may change reversibly from 6 to 8 during complexation, resembling the change from 16 to 18 valence electrons in transition metal complexes, which is the key to their manifold catalytic properties.

MIKE spectra were recorded with a VG AutoSpec EBE mass spectrometer and exchange reactions were performed with a Bruker Spectrospin CMS 47x FTMS. MIKE spectra of $[AsI_2$ NBD$]^+$ and $[AsI_2$ QUAD$]^+$ are similar and differ from those of the other isomers (Table II). Hence, there are at least three isomers identified by their MIKE spectra.

Table II. Comparison of $[AsI_2\ C_7H_8]^+$ MIKE Spectra

	$[AsI_2$ toluene$]^+$	$[AsI_2$ NBD$]^+$	$[AsI_2$ QUAD$]^+$	$[AsI_2$ CHT$]^+$
$C_7H_7^+$	19%	51%	47%	59%
$AsC_7H_7^+$	3%	3%		
$AsIC_7H_7^+$	6%	29%	40%	17%
$AsI^{+\bullet}$	12%			
AsI_2^+	60%	11%	13%	24%

Additional information about structure and reactivity of $[AsI_2\ C_7H_8]^+$ complexes were obtained from ligand exchange reactions (9) - (11).

$$[AsI_2\ NH_3]^+ + NBD \rightleftarrows NH_3 + [AsI_2\ NBD]^+ \quad (9)$$

$$[AsI_2\ NH_3]^+ + QUAD \rightleftarrows NH_3 + [AsI_2\ NBD]^+ \quad (10)$$

$$[AsI_2\ NH_3]^+ + toluene \leftrightarrows NH_3 + [AsI_2\ NBD]^+ \quad (11)$$

In $[AsI_2\ C_7H_8]^+$ complexes, norbornadiene and quadricyclane are bonded stronger than toluene, confirming the presence of different $[AsI_2\ C_7H_8]^+$ isomers. In contrast to the transition metal complexes mentioned above, the entire C_7H_8 ligand is replaced reversibly. Thus, if there is any insertion of AsI_2^+ into a C–C or C–H bond, these reactions must be reversible. According to the MIKE spectra, AsI_2^+ triggers the valence isomerization of QUAD to NBD. Most probably, a consecutive retro Diels-Alder reaction, driven by the exergonic isomerization, is avoided by loosing I from the collision complex.

References

1. Bushnell, J.E.; Kemper, P.R.; Bowers, M.T. *J. Am. Chem. Soc.*. **1994,** *116* , 9710.
2. van Koppen, P. A. M.; Kemper, P.R.; Bushnell, J.E.; Bowers, M.T. *J. Am. Chem. Soc.*. **1995,** *117* , 2098.
3. Kemper, P.R.; Bushnell, J.E.; van Koppen, P.A.M.; Bowers, M.T. *J. Phys. Chem.* **1993,** *97,* 1810.
4. Bushnell, J.E.; Kemper, P.R.; Bowers, M.T. *J. Phys. Chem.* **1993,** *97,* 11628.
5. Rappe, A. K.; Upton, T. H. *J. Chem. Phys.* **1986,** *85,* 4400.
6. Alvarado-Swaisgood, A. E.; Harrison, J. F. *J. Phys. Chem.* **1985,** *89,* 5198.
7. Armentrout, P.B. in *Gas Phase Inorganic Chemistry*, Russel, D.H. (Ed.) Plenum Press: New York, **1989**.
8. Bauschlicher, C.W.; Partridge, H.; Sheehy, J.A.; Langhoff, S.R.; Rosi, M. *J. Chem. Phys.* **1992,** *96,* 6969.
9. a) Irikura, K. K.; Beauchamp, J. L. *J. Phys. Chem.* **1991**,*95*, 8344. b) Wesendrup, R.; Schroder, D.; Schwarz, H. *Angew. Chem. Int. Engl.* **1994,** *33,* 1174.
10. Noll R. J.; Weisshaar, J. C. *J. Am. Chem. Soc.*. **1994,** *116*, 10288.
11. Sanders, L; Sappey, A. D.; Weisshaar, J. C. *J. Chem. Phys.* **1986**, 85, 6952. Sanders, L.; Hanton, S. D.; Weisshaar, J. C. *J. Chem. Phys.* **1990**, 92, 3485. Hanton, S.D.; Noll, R. J.; Weisshaar, J. C. *J. Chem. Phys.* **1992,** 96, 5165.
12. Werner, A. S.; Baer, T. *J. Chem. Phys.* **1975**, 62, 2900. Proch, D.; Rider, D. M.; Zare, R. N. *Chem. Phys. Lett.* **1981**, 430.
13. Ellis, P. E., Jr.; Lyons, J. E. *Catal. Lett.* **1991** 3, 389.
14. Chen, H. L.; Ellis, P. E., Jr.; Wijesekera, T.; Hagan, T. E.; Groh, S. E.; Lyons, J. E.; Ridge, D. P. *J. Am. Chem. Soc.* **1994,** 116, 1086.
15. Jin, L.; Ellis, P. E., Jr.; Wijesekera, T.; Shaikh, S. N.; Lyons, J. E.; Ridge, D. P. *Cat. Lett.*, submitted.

16. Winkel, L. F.; Woodward, C. A.; Jones, A. B.; Stace, A. J. *J. Chem. Phys.* **1995,** *103,* 5177.

17. Rabrenovic, M.; Ast, T.; Beynon, J. H. Int. J. *Mass Spectrom. Ion Processess.* **1984,** *61*, 31.

18. Freas, R. B.; Campana, J. C. *J. Am. Chem. Soc.* **1985,** *107*, 6202.

19. Mestdagh, H.; Morin, N; Rolando, C. *Tetrahedron Lett.* **1986,** *27*, 33.

20. Drewello, T.; Eckhart, K.; Lebrillo, C. B.; Schwarz, H. *Int. J. Mass Spectrom. Ion Processess.* **1987,** *76*, R1.

21 McCullough, S. M.; Jones, A. D.; Lebrillo, C. B. *Int. J. Mass Spectrom. Ion Processess.* **1991,** *107*, 545.

22. Chen, L.Z.; Miller, J. M. *J. Am Soc. Mass Spectrom.* **1991**, *2*, 120.

23 Harrison, A. G.; Mercer, R. S.; Reiner, E. J.; Young, A. B.; Boyd, R. K.; March, R. E.; Porter, C. J. *Int. J. Mass Spectrom. Ion Processess.* **1986,** *74,* 13.

24. Ni, J.; Harrison, A. G. *Int. J. Mass Spectrom. Ion Processess.* in press.

25. Jacobson, D. B.; Freiser, B. S. *J. Am. Chem. Soc.* **1983**. *105*, 7492.

26. Lech, L. M.; Freiser, B. S. *Organomet.* **1988**. *7*, 1948.

27. See for example, Rabrenovic, M; Brenton, A. G.; Ast, T. *Org. Mass Spectrom.* **1983**, 18, 587 and references therein.

28. Lifshitz, C. *Acc. Chem. Res.* **1994**, 27, 138 and references therein.

29. MacMillan, D. K.; Hayes, R. N.; Peak, D. A.; Gross, M. L. *J. Am. Chem. Soc.* **1992**, 114, 7801.

30. Sievers, H. L.; Grützmacher, H.-Fr.; Grützmacher H.; Pitter. S. *J. Am. Chem. Soc.* **1995**, 117, 2313.

WORKSHOP OF COLLISION-INDUCED DECOMPOSITION OF IONS

Keith R. Jennings,
Department of Chemistry,
University of Warwick,
Coventry CV4 7AL, U.K.

Introduction

Rather than have a formal lecture on collision-induced decomposition of ions, it was decided that a workshop would be more appropriate to cover various aspects of the subject. There continues to be interest at the fundamental level as well as in its use for structure determination and analytical applications. Furthermore, the process is now used in a wide variety of different types of instrumentation, each of which requires the use of somewhat different experimental techniques, some of which lead to collision-induced decompositions at high translational energy whereas others are used primarily for the study of low energy collision processes. In keeping with the tutorial aims of the Institute, the workshop began with a general introductory lecture which was followed by a number of oral presentations, the contents of which are summarised below.

1. Fundamentals of Collision-Induced Decomposition of Ions

All collision-induced decomposition experiments consist of the same series of steps:

(1) isolation of a particular ion of interest, the precursor ion, which has been formed in ion source,
(2) giving the ion translational energy of typically 30-100 eV in a low energy experiment or of several keV in a high energy experiment,
(3) allowing the ion to collide with an inert gas such as helium or argon as a result of which a little of the translational energy is converted into internal energy
(4) which results in fragmentation and the formation of product ions which make up the CID spectrum.

The type and extent of fragmentation will depend both on the amount of internal energy deposited in the precursor ion as a result of the collision and on the structure of the ion. The assumption is usually made that if a collision-induced decomposition experiment is carried out in a particular, closely-controlled manner, any differences in the CID spectra of two compounds arise from differences in structure, that is to say, effects of differences in internal energy can be neglected. Although there are several well-documented cases of internal energy effects being observable for relatively small precursor ions such as CH_3CN, the assumption that they can normally be neglected for larger ions is quite

K. R. Jennings (ed.), Fundamentals and Applications of Gas Phase Ion Chemistry, 483–498.

justifiable. It follows that for this assumption to be valid, one must compare like with like since the CID spectrum given by a particular precursor ion is usually very dependent on the way in which it is produced.

High energy collision processes are normally studied in a magnetic sector instrument in which precursor ions typically possess 5 - 10 keV of translational energy. Collisions usually occur in a collision cell placed between the electric and magnetic sectors of a two sector instrument or between the second and third sector of a four-sector instrument. Product ions of a chosen precursor ion are obtained by use of a linked scan of the magnetic and electric sector field strengths, B and E, such that B/E is constant throughout the scan. Post-acceleration of the ions produced in the collision cell of a four sector instrument is normally desirable to minimise loss of ions by scattering and this requires the scan to be modified somewhat. Under these conditions, electronic excitation is thought to predominate for ions of modest mass although there is increasing evidence that for heavier ions, such as larger peptides and carbohydrates, vibrational excitation is also important. Helium has traditionally been used as the collision gas but for ions of m/z greater than 1500, argon is now more usual with xenon being used for ions of still higher m/z. The amount of energy deposited is thought to peak at 2-3 eV but there appears to be a long high energy "tail" which gives rise to higher energy fragmentation processes.

Low energy processes are normally studied in triple quadrupole instruments, ion traps and ion cyclotron resonance (ICR) instruments in which the energy of singly-charged precursor ions is rarely greater than 100 eV and vibrational excitation occurs. In the triple quadrupole instrument, the precursor ion is selected by the first mass analyser, collision occurs in the rf-only central section and product ions are collected by scanning the second mass analyser. The amount of energy deposited is thought to peak at about 40% of the maximum available based on centre of mass considerations and higher energy processes are less commonly observed. When used with electrospray ionisation, however, which may produce a precursor ion carrying 20 or more charges, it is important to recognise that a potential difference of 50 eV produces translational energy of 1 keV so that higher energy processes may then be observed. Since scattering losses are much reduced, the collection efficiency and hence the sensitivity of this type of experiment can be very high. Furthermore, in ion traps and ICR instruments, successive CID experiments can be carried out, often referred to as $(MS)^n$ where n is the number of CID steps carried out.

Although CID was widely used to investigate the structures of ions in order to provide information on the mechanisms of fragmentation processes, its main use in recent years has been to investigate the CID of MH^+ ions produced by one of the several "soft" ionisation methods currently in use for work with biological samples. These include fast atom bombardment (FAB) ionisation, matrix assisted laser desorption ionisation (MALDI) and electrospray ionisation (ESI) which with larger molecules produces (M

+ nH)$^{n+}$ ions where n may typically range from 10 to over 50. Whereas for a singly-charged precursor ion, the product ions are always observed at low values of m/z, it is quite common for the fragmentation of a multiply-charged precursor ion to give a product ion of higher m/z if a small, charged fragment is lost.

It is difficult to deposit sufficient energy in ions of m/z above about 2500 to produce sufficient fragmentation to obtain full structural information and for large peptides and proteins, cleavage of the molecule on the C-terminal side of a methionine residue by means of CNBr is commonly employed followed by a tryptic digest which frequently produces smaller peptides of m/z of 1000 - 2000 which can be successfully investigated by means of CID.

In addition to its use for structural purposes, CID is also of value in the detection of a specified component in a complex mixture. The normal mass spectrum obtained by electron or chemical ionisation contains many peaks the origins of which are difficult to assign. Providing that the component of interest produces an ion of relatively high mass, this ion may be chosen as the precursor ion and its CID spectrum obtained for comparison with the CID spectrum of an authentic sample. The specificity of the method and the absence of "chemical noise" much improves the sensitivity of detection of trace components in a mixture, e.g. the explosive RDX in a biological matrix.

Implicit in the applications of CID spectra mentioned above is the relationship of the spectrum to the structure of the molecule from which the precursor ion is derived. Marked differences in the spectra are a good indication of differences in the structures of the sample molecules but minor differences must be treated with caution since they could arise from differences in the way the experiment was carried out, producing for example, minor internal energy effects. Care must also be taken to recognise that molecules of very similar structures, such as positional isomers, may often give very similar CID spectra so that additional evidence may be necessary before a particular sample can be positively identified.

Other less-frequently used methods of making use of CID are the precursor ion scan in which a chosen product ion formed from a variety of precursors are collected and the constant neutral loss scan in which all product ions formed by the loss of a chosen neutral species are collected. The precursor ion scan is particularly useful for identifying compounds belonging to a particular class, all of which fragment to give a specific fragment ion, e.g. the m/z 149 ion given by all phthalates. The constant neutral loss scan requires both mass analysers of a triple quadrupole to be scanned simultaneously, the second set to transmit ions lower in mass by that of the chosen neutral species lost and is relatively little used.

2. Directed Fragmentation in an Ion Trap

R.E. March, (Department of Chemistry, Trent University, Peterborough, Ontario, Canada K9J 7B8)

The process by which ion dissociation is effected by collisions with buffer gas atoms in a quadrupole ion trap differs substantially from that of high-energy or low-energy collisions in pressurised Faraday cage-type or r.f.-only quadrupole collision cells, respectively. In an ion trap, ions are excited resonantly throughout an activation period which is some 1 000 times the flight time in an r.f.-only collision cell; during this period, each ion suffers some 600 collisions with helium buffer gas. Yet ions experience, on average, some 50 r.f. cycles of the drive or storing potential between successive collisions so that ion kinetic energy is maintained throughout the activation period. The rate of ion activation is controlled by the amplitude and frequency of the applied supplementary or "tickle" voltage.

When the ion trap is operated as a tandem mass spectrometer it is advantageous, on occasion, to increase tandem mass spectrometric specificity even at the expense of sensitivity. Such specificity enhancement usually requires that ion fragmentation be directed preferentially to a specific fragmentation channel. For example, in the determination of polychlorinated dibenzo-p-dioxins by tandem mass spectrometry, enhanced specificity is achieved by monitoring fragment ions formed by the loss of $COCl^{\cdot}$ rather than those formed by the loss of $Cl^{\cdot}$ or Cl_2.

The quest for directed fragmentation of mass-selected ions in an ion trap is facilitated by the change in ion secular frequency with axial displacement from the centre of the ion trap; such frequency changes are due to non-linear resonance effects in the ion trap. Thus the rate of ion activation which is the dominant factor in directing ion fragmentation can be varied over a substantial range by temporal variation of the "tickle" voltage amplitude of single frequency excitation, multiple frequency excitation and r.f.-modulated single-frequency excitation. Preliminary results were presented for the tandem mass spectrometric determination of polychlorinated biphenyls wherein fragmentation of non- and mono-*ortho* chlorine substituted biphenyls of relatively high toxicity was directed to the loss of Cl_2, while fragmentation of di-, tri- and tetra-*ortho* chlorine substituted biphenyls (of relatively low toxicity) was directed to the loss of $Cl^{\cdot}$.

3. Low-Energy Collision-Induced Dissociation of Triatomic Ions: Competing Kinematic Models

Anil Shukla, Ron Tosh, Rahul Chawla and Jean Futrell, (Department of Chemistry and Biochemistry, University of Delaware, Newark, DE 19716, U.S.A.)

Collision-induced dissociation (CID) plays an ever increasing role in characterising high mass ions (singly and multiply charged) produced by soft ionisation techniques which result in little fragmentation and limited structural information. Even though the technique is widely used, the mechanisms of energy transfer and dissociation are not well understood. Several recent studies [1,2] of the CID of high mass ions have reported very large, physically implausible amounts of kinetic energy losses that could not be explained by the generally accepted two step model of collisional activation and subsequent dissociation. Uggerud and Derrick [3] have developed an impulsive collision model to explain such large energy losses in CID experiments. According to this model, the collision between the ion and a neutral particle involves only a single atom or a small part of the ion (and not the whole ion), such that energy and momentum transfer take place between the neutral and that part of the ion. To reconcile their model with experimental observations, they further assume that scattering is small and measured energy losses represent averaging over several collisions with different atoms in the ion.

The impulsive collision theory (ICT) developed by Uggerud and Derrick is similar in many respects to the spectator stripping model [4] used to describe ion-molecule association reactions. The knock-out model of Mahan and co-workers [5], proposed for low energy CID, is an elaboration of the spectator stripping model explicitly applied to CID processes; this model, however, was rejected by them on energetic grounds in comparison with experiments. In this report we demonstrate that CID of CS_2^+ and CO_2^+ ions at low collision energies is well-described by an impulsive, knock-out model we have developed [6].

The crossed-beam experiment [7] used for this study consists of a high pressure ion source to generate relaxed ions, a deceleration lens to reduce their kinetic energy to the desired collision energy, a supersonic beam of neat argon colliding at 90°, and a rotatable detector assembly comprised of a cylindrical energy analyser, quadrupole mass filter and electron multiplier. The neutral beam was chopped mechanically for signal averaging. The laboratory angular and energy distributions of fragment ions were measured at a series of laboratory angles with respect to the collision centre. Some experiments were performed at low source pressure using a different instrument [8] with similar basic features. The experimental data were converted to Cartesian velocity contour plots using methods which have been described elsewhere [9].

Figure 1 shows the centre of mass (CM) velocity contour plot of the S^+ fragment ions at 17.8 eV collision energy. A Newton diagram showing velocity vectors of both reactants in the laboratory and CM reference frames and the elastic scattering circle (ESC) are superimposed on the contour plot. Even though the intensity maximum of the contour plot is located within the ESC (as expected for an endothermic process), a significant fraction of the S^+ signal lies at laboratory scattering angles which are just outside the ESC. As a matter of fact, the peaks of fragment ion velocity distributions measured at angles > 28° lie outside the ESC. Following the conventional two-step CID mechanism, these data imply that energy transfer decreases with increasing angle and becomes superelastic at larger angles. This counter intuitive result has not been observed previously in CID processes. Moreover, the lifetimes of non-dissociating excited states ($A^2\Pi_u$ and $B^2\Sigma_u^+$) of CS_2^+ are too short for them to be present in the reactant ion beam; further, ions formed in the high pressure source are likely relaxed to the ground state by several hundred collisions within the source. We therefore suggest that the S^+ atoms found at angles outside the ESC are, in fact, evidence for inelastic scattering following different kinematics.

A careful examination of Figure 1 and the kinetic energy distributions at individual laboratory scattering angles shows that the intensity outside the ESC is part of a ridge that includes the intensity maximum which is located well within the ESC. Figure 2 shows this effect very clearly where the maximum of each energy distribution (solid bar representing peak maximum and probable error) has been potted as a function of laboratory scattering angle. These intensity maxima are not located on a circle whose centre is the CM of the Ar/CS_2^+ but rather they are centred on a different point on the relative velocity vector, corresponding approximately to the CH of Ar/S^+ interaction. A circle drawn about the Ar/S^+ CM passes through both the intensity maximum and the large angle intensity ridge extending beyond the ESC as shown in Figure 1 (dashed circle) and close to most of the intensity maxima in Figure 2 (marked S^+ recoil).

The results described above strongly suggest that collisions between CS_2^+ ions and argon neutrals involve only one of the sulfur atoms in the ion and that sufficient energy is transferred in the collision process to break the $CS\text{-}S^+$ bond. Momigny and co-workers [10] have shown that the dissociation path involves an avoided crossing between the B state and the lowest energy dissociation asymptote. This mechanism is entirely consistent with our observations if the non-adiabatic transition from the CS_2^+ ground state to the B state occurs as the reactants approach and subsequent coupling of the $^4\Sigma^-$ asymptote reduces the force between S^+ and CS so that Ar and S^+ may undergo a two-body collision with no momentum transfer to CS (a similar hypothesis was suggested by Mahan and co-workers [5] in their earlier work). If this is the case, the energy required (4.8 eV) for the dissociation is transferred into CS_2^+ in the approach trajectory. This effectively reduces the reactant relative velocity vector and shifts the S^+/Ar CM towards the CS_2^+/Ar CM. A circle drawn from this shifted origin actually demonstrates a better match to our experimental data, as shown by the dashed circle of Figure 2.

Although the calculation of energy transfer into internal modes and kinetic energy release for the process described above differs from that done for the two-step model for CID, energy and momentum conservation still provide the necessary relationships. This is easily illustrated by means of a Newton diagram analogous to those used to rationalise two-step CID processes. Figure 3 shows the recoil velocities after collision in which argon strikes only S^+ within the CS_2^+ ion. Since the momentum of CS is unchanged by the collision, a measurement of the recoil vector of S^+ is sufficient to determine the kinematics of this CID process completely. In the case of CID, energy is transferred from relative kinetic energy into internal energy and hence the recoil vectors are reduced as shown in Figure 3. The recoil vector of S^+ is measured in the experiment and that of CS remains the same as initial CS_2^+. These two vectors then give the position of the "virtual" location of the excited CS_2^{+*} ions with respect to the original CS_2^+ CM after collision. The difference in the relative energy of CS_2^+/Ar and CS_2^{+*}/Ar enables us to deduce the kinetic energy transferred into internal modes.

The method described above has been used to calculate the energy transferred into the CS_2^+ ions leading to S^+ fragment ions. A circle centred at the centre of the S^+/Ar CM and drawn through the solid bars in Figure 2 defines the radius of the S^+ recoil circle and gives the kinetic energy transfer into CS_2^+ ions to be 4.5 ± 0.8 eV. This is in very good agreement with the thermochemical threshold of 4.8 eV for this dissociation process.

The CM velocity contour plot for CS^+ fragment ions at 17.8 eV collision energy also shows an analogous ridge of intensity centred about the CS^+/Ar CM, suggesting that the two fragmentation processes in CS_2^+ follow complementary pathways. In the case of CS^+, however, the peak intensity maximum does not follow the same circle as the extended ridge. A plausible interpretation is that there are two or more competing reaction pathways for this process. Further research now in progress to define this phenomenon experimentally may provide an acceptable description of this dissociation process.

Similar scattering behaviour has been observed in the CID of CO_2^+ ions to O^+ and CO^+ fragment ions at 23.8 eV collision energy, as shown in Figure 4 for O^+ ions. Similar to the $^+$ ion, the maximum intensity peak within the ESC and the associated ridge do not fall on the same circle of symmetry, suggesting two competing mechanisms with different kinematics. If we assume that the intensity maximum and the higher angle intensity ridge arise from two-step and impulsive collisions, respectively, then the enrgy transferred into internal modes for the two processes corresponds to 5.9 ± 0.5 eV and 4.5 ± 0.7 eV respectively, comparable to the threshold energy for this process, 5.3 eV.

The possibility that a combination of two mechanisms is operative is supported by the measurements of the unreactive scattering of CO_2^+ ions in collision with argon. Figure 5 shows the velocity contour plot of inelastically scattered CO_2^+ ions at 23.8 eV collision energy. The intensity maximum near zero degrees corresponds to an elastic collision process where no energy is transferred from kinetic into internal modes of the CO_2^+ ions. As the angle increases, however, scattering becomes increasingly inelastic, reaching a maximum energy deposition of about 4.8 eV at laboratory scattering angles of about 15°. The angular dependence of the velocity contours between 0° and 15°exhibits the kinematics of CO_2^{+*} recoil that would occur if argon were to collide with one of the O atoms. Thus our results suggest that scattering of CO_2^+ ions is dominated by O^+/Ar collisions until the dissociation threshold is reached. At that point, the two-body inelastic collisions of CO_2^+ ion with argon centred on the CO_2^+/Ar CM take over the process, with the most probable energy transfer into internal excitation equal to 4.8 eV. If this interpretation is correct, two kinematically distinct mechanisms deposit internal energy into CO_2^+ up to the limit of the lowest energy dissociation process. It is interesting to note that a similar experiment on scattering CS_2^+ ions at 17.8 eV collision energy exhibited completely different behaviour: almost all CS_2^+ ions were found in the ESC, suggesting that there is no energy transfer into internal modes for this similar molecular ion.

In all the experiments described above, the primary ion beams were produced in a high pressure ion source; thus reactant ions were in the ground electronic state with very little (if any) vibrational excitation. When CO_2^+ are formed in a low pressure electron impact ion source, a significant shift in the energetics contour plot for O^+ scattering is observed. Figure 6 shows the CM velocity contour plot for O^+ fragment ions at 49 eV collision energy. The intensity maximum is found just outside the ESC and a ridge of intensity follows the O^+ recoil circle for O^+/Ar collisions. This can be rationalised only by assuming that CO_2^+ ions formed in the ion source retain sufficient internal energy that CID can proceed with little or no energy transfer. This quite a surprising result since the radiative lifetimes of the A and B states of CO_2^+ (of the order of 120 nsec.) are much shorter than the ion transit time from the source to the collision centre and the C state is fully predissociated. Our results nevertheless suggest the possibilities that a long-lived excited state of CO_2^+ exists which has not been observed previously or that a bottleneck exists for energy redistribution from a specific vibronic level. We are extending the range of collision energies and varying pressure in the high pressure ion source to obtain further information on this phenomenon.

We gratefully acknowledge financial support from the National Science Foundation, Grant No. CHE-9021014.

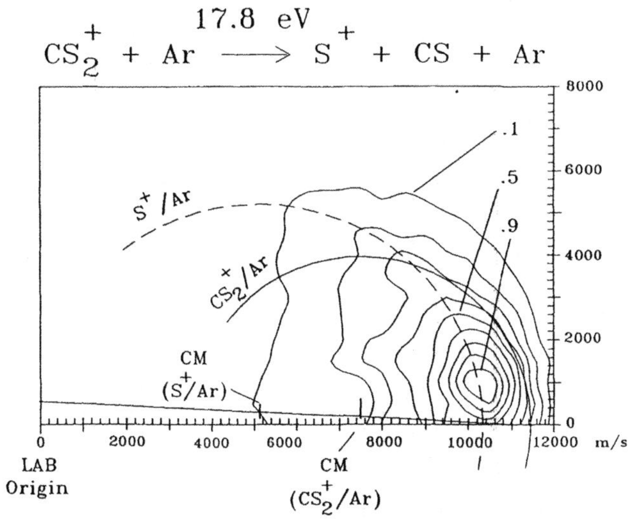

Figure 1. CM velocity contour plot for the CID of CS_2^+ ions to S^+ ions in collision wit argon at 17.8 eV collision energy. A Newton diagram showing velocity vectors of reactants has been superimposed on the contour plot to show the laboratory origin and the CM. The circle marked CS_2^+/Ar represents the ESC for the two-body collision process with the origin marked Cm (CS_2^+/Ar). The circle marked S^+/Ar has been drawn to pass through the intensity maximum and the lower intensity ridge with CM (S^+/Ar) as the centre. The numbers marked on the contours represent their relative intensities.

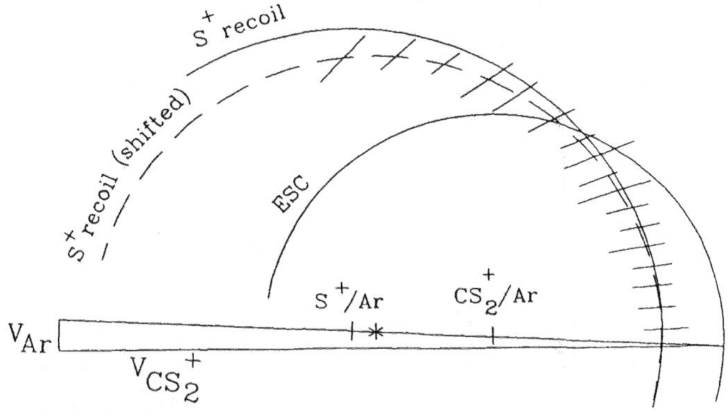

Figure 2. Plot of experimental data (solid bars), representing the most probable range of S^+ velocity at each scattering angle, drawn on the Newton diagram. The solid circle marked S^+ recoil is centred on the S^+/Ar CM while the dashed circle is centred on the S^+/Ar CM (marked by the asterisk on the relative velocity vector) which has been shifted towards the CS_2^+/Ar CM to account for energy transfer equivalent to the dissociation energy.

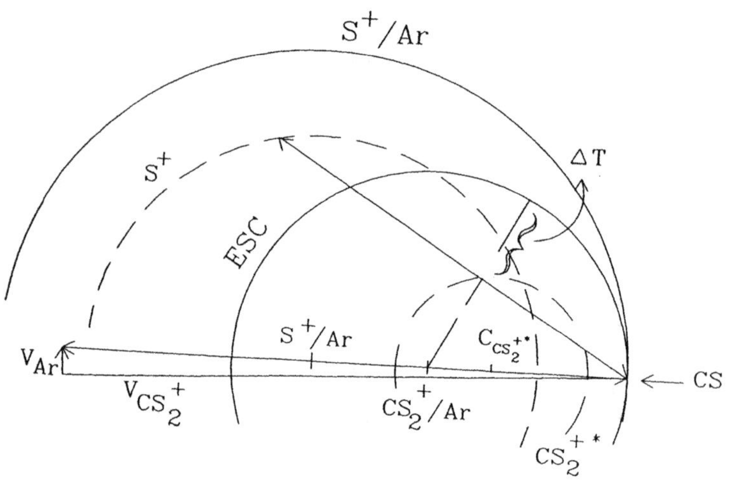

Figure 3. Newton diagram depicting the impulsive knock-out model for the collision of CS_2^+ ions where the argon atom interacts only with the S^+. The circles mark ESC (CS_2^+/Ar) and ESC (S^+/Ar) represent the zero energy transfer limits for the CID proceeding via CS_2^+/Ar and S^+/Ar collisions, respectively. The dashed circles marked S^+ and CS_2^{+*} represent the recoil vectors for the S^+ formed by inelastic S^+/Ar collisions and corresponding location for internally excited CS_2^{+*} ions that dissociate to form S^+. The energy transferred into internal modes in the impulsive knock-out model is represented by ΔT as shown here.

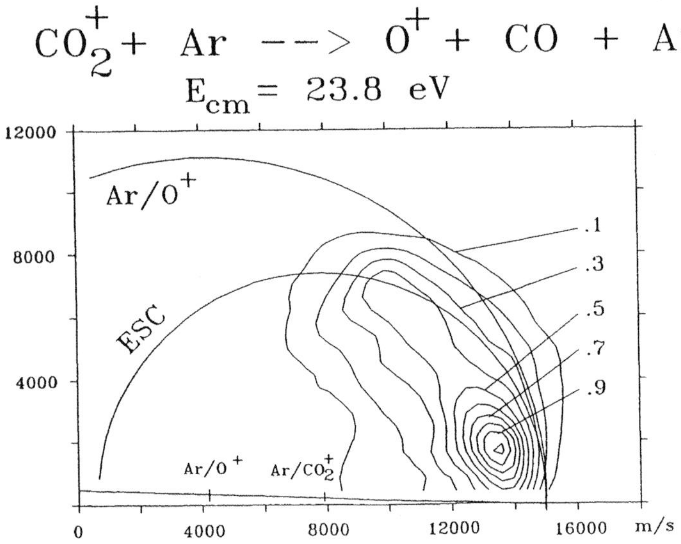

Figure 4. Semilogarithmic CM velocity contour plot for the CID of CO_2^+ ions to O^+ ions in collision with argon at 23.8 ev collision energy. The circles marked ESC and Ar/O^+ correspond to elastic scattering circles for the CID proceeding via the two-step or impulsive knock-out process. Relative intensities for three of the contours are indicated.

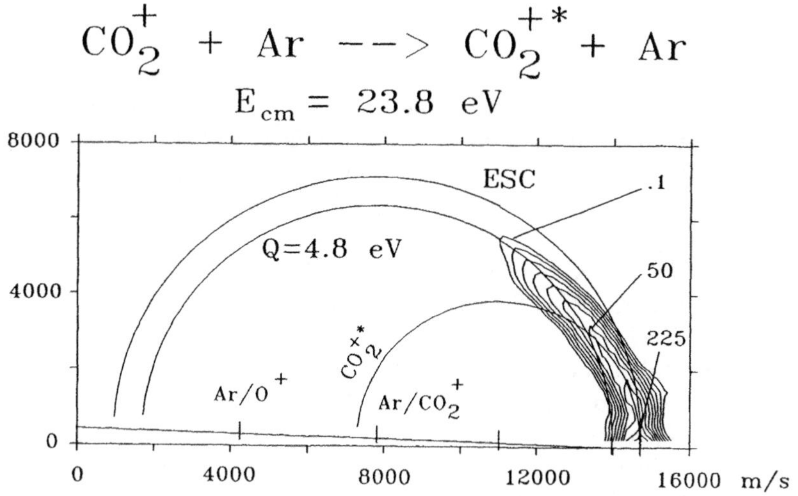

Figure 5. CM velocity contour plot for inelastic scattering of CO_2^+ ions in collision with argon at 23.8 ev collision energy. The circles marked ESC and Q=4.8 eV correspond to elastic scattering and energy transfer of 4.8 eV from kinetic into internal modes of CO_2^+ ions, respectively, for the collisions involving the whole CO_2^+ ion. The circle marked CO_2^{+*} (O^+/Ar) represents the CO_2^{+*} recoil vector of argon collides with only the O^+ ion. The numbers marked on the contours represent their relative intensities.

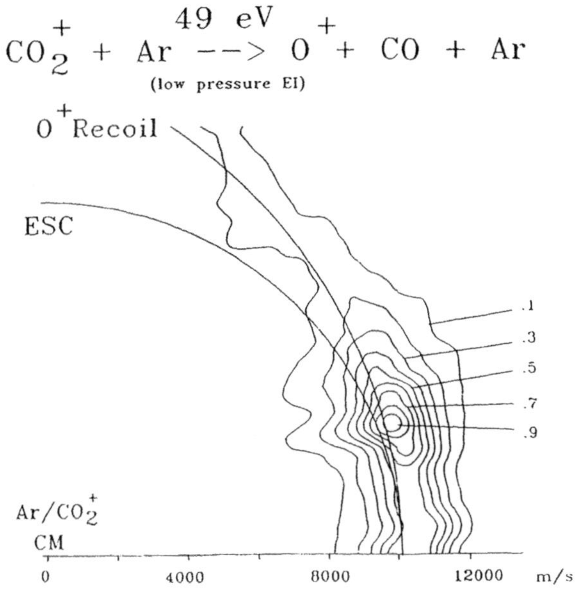

Figure 6. CM velocity contour plot for the CID of CO_2^+ ions to O^+ ions on collision with argon at 49 eV collision energy. CO_2^+ ions were produced by electron ionisation in a low pressure source. The circles marked ESC and O^+ recoil present elastic scattering circles for the CID proceeding via the two-step and O^+/Ar collisions, receptively.

References

1. R.G. Gilbert, M.M. Shiel and P.J. Derrick, *Org. Mass Spectrom.*, (1985), **20**, 439
2. A.J. Alexander, P. Thibault and R.K. Boyd, *J.Am. Chem. Soc.*, (1990), **112**, 2484
3. E. Uggerud and P.J. Derrick, *J. Phys. Chem.*, (1991), **95**, 1430
4. A. Henglein, K. Lacmann and B. Knoll, *J. Chem. Phys.*, (1965), **43**, 1048
5. M.H. Cheng, M.H. Chiang, E.A. Gislason, B.H. Mahan, C.W. Tsao and A.S. Werner, *J. Chem. Phys.*, (1970), **52**, 6150
6. R.E. Tosh, A.K. Shukla and J.H. Futrell, *J. Phys. Chem.* (in press)
7. C.R. Blakely, P.W. Ryan, M.L. Vestal and J.H. Futrell, *Rev. Sci.Instrum.*, (1976), **47**, 15.
8. A.R.Shukla, S.G. Anderson, S.L. Howard, K.W.Sohlberg and J.H. Futrell, *Int. J. Mass Spectrom.Ion Proc.*, (1988), **86**, 61
9. A.K.Shula and J.H. Futrell, *Mass Spectrom. Rev.*,(1993), **12**, 211
10 J. Momigny, H. Wankenne, G. Mathie, J.F. Flamme and M.A. Almoster Ferreira, *Adv. Mass Spectrom.*, (1974), **6**, 923

4. Collision Induced Decay of Water Ion Clusters

P.S.Vinogradov, O.V. Dmitriev and I.N. Veretenikov (Institute for Energy Problems of Chemical Physics, Leninsky Prospect 38, Building 2, Moscow 117829, Russia).

Some results on the CID of water ion clusters formed in a flow tube are reported. In our flowing afterglow apparatus, designed for studies of the kinetics of ion-molecule reactions, a sector magnet is used for mass analysis. At a high enough pressure in the high vacuum part of the apparatus, the so-called "metastable peaks" were observed.

An equipotential cell was placed after the skimmer in the path of the ion beam to allow the study of the spontaneous decay of ion clusters produced in the nozzle and the CID of ion clusters at sufficiently high pressures in the cell, as shown in Figure 1. The CID experiments can be considered as "modelling experiments" for measurement of decay times of ion/molecule and ion cluster/molecule complexes produced in the nozzle, based on metastable peak analysis.

If the potential applied to the cell is varied, both the time of flight *via* the cell and the position of a metastable peak (corresponding to the decay inside the cell) are varied. Note that no preliminary mass separation of ionic species is necessary and that the process $A^+ \rightarrow B^+ + C$ is identified by the position of a metastable peak.

For a model of a decay process of a CL^+ cluster,

$$CL^+ \ (+\ He) \xrightarrow{k_1} CL^{+*} \xrightarrow{k_2 = 1/\tau} CL_{-1}^+ + H_2O \qquad (1)$$

the ratio of integral intensities is expressed as

$$[CL_{-1}^+]/[CL^+] = \frac{k_1 k_2[\tau \exp(-t/\tau) - (1/k_1)\{\exp(-k_1 t)\}]}{(k_1 - k_2)\{\exp(-k_1 t)\}} \qquad (2)$$

where t is the time of flight *via* the cell, τ is the decay time of the CL^{+*} ion. Figure 2 shows the dependencies of the daughter ion/parent ion ratios *versus* the cell potential. Figure 2(a) shows that no significant disturbance of the alignment of the ion beam occurs as ΔU is varied, yet it is interesting to note that the decay efficiency is increased by a factor of about 1.8 as one water molecule is added to each cluster.

Figure 2(b) shows that the increase in the time of flight inside the cell results in the increase in the fraction of the ion clusters that undergo decay outside the cell (not in the field free region nor at other places where they have a different velocity). The time of flight inside the cell is given by $t = 1.3\ cm/[2(3kV - \Delta U)e/m_0]^{1/2}$. If we assume that the collision cross section is not dependent on the energy, we can estimate the lifetime of the clusters excited by collision. The estimated lifetime is in the region of $\tau = 10^2 - 10^{-4}$ nsec.

It appears, however, that the shapes of the fitted curves obtained from equation (2) are not close to those obtained experimentally. It is possible that the excitation energy of the cluster (electronic, vibrational and rotational) depends on the translational energy. Some information on the excitation energy can be obtained from the kinetic energy

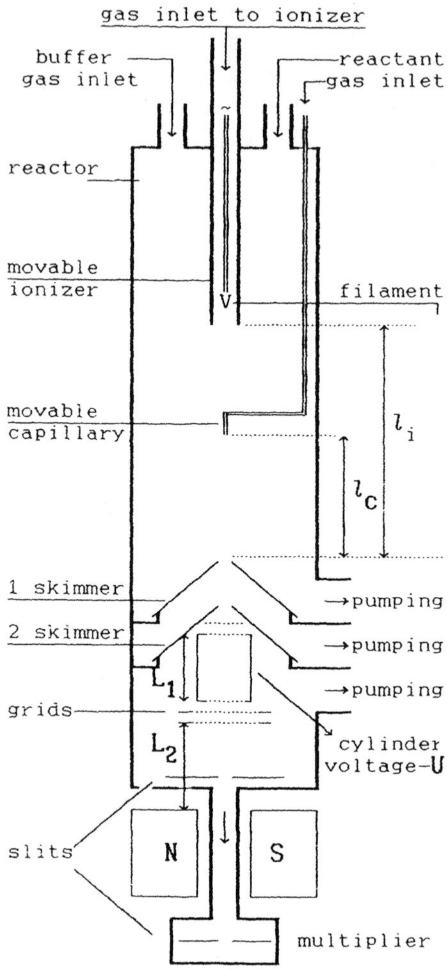

Figure 1. The experimental time-of-flight apparatus showing the collision cell (Cylinder at voltage U).

distribution. Figure 3 shows the shapes of the metastable peaks observed, from which information on the distribution of the kinetic energy release may be obtained.

It appear, however, that a large number of excited states of the clusters can arise, giving rise to a wide range of different decay times, so that the estimate given above is necessarily somewhat crude.

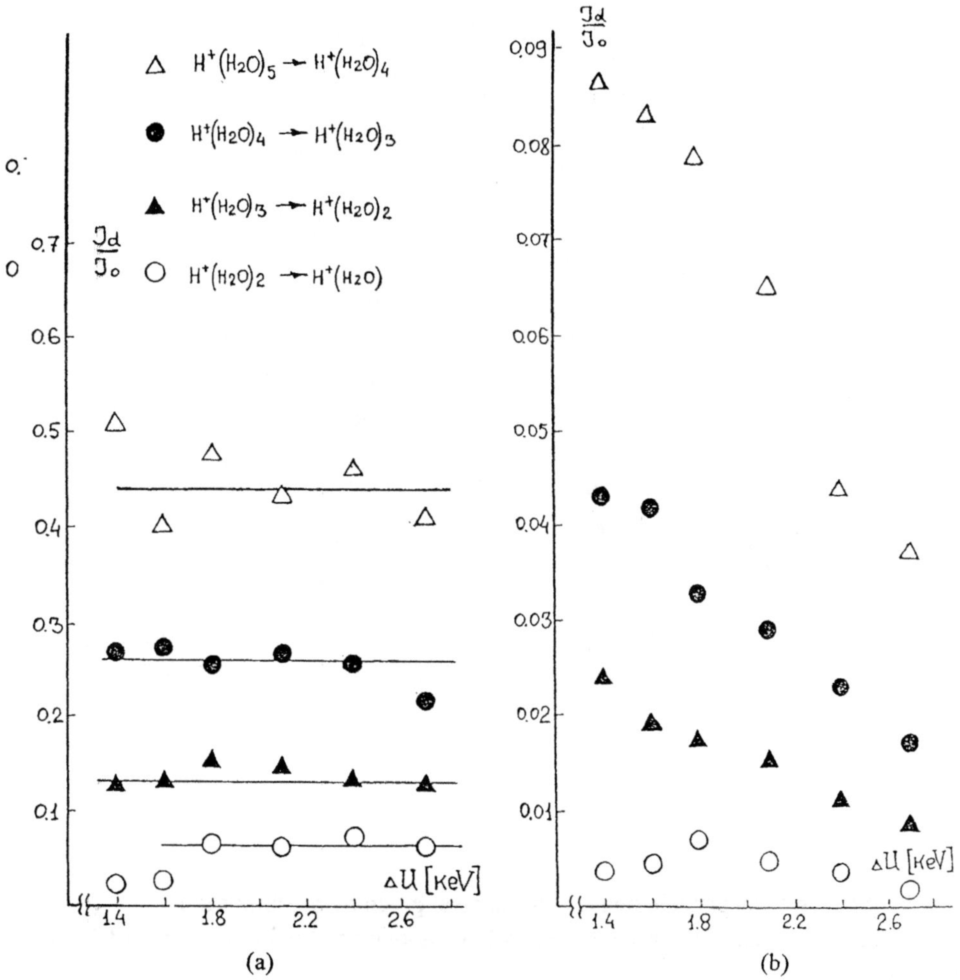

Figure 2. CID of water ion clusters. (a) The decay of cluster ions after acceleration to a translational energy of 3 keV; the variation of the potential of the collision cell does not disturb the alignment of the beam. Decay efficiency varies by a factor of about 1.8 when one water molecule is added to a cluster. (b) Ratios of the integrated intensities of the parent ion and metastable ion plotted against the potential of the collision cell. From the increase in the number of decompositions of ion clusters within the cell with increase in time of flight, one may estimate the lifetimes of clusters.

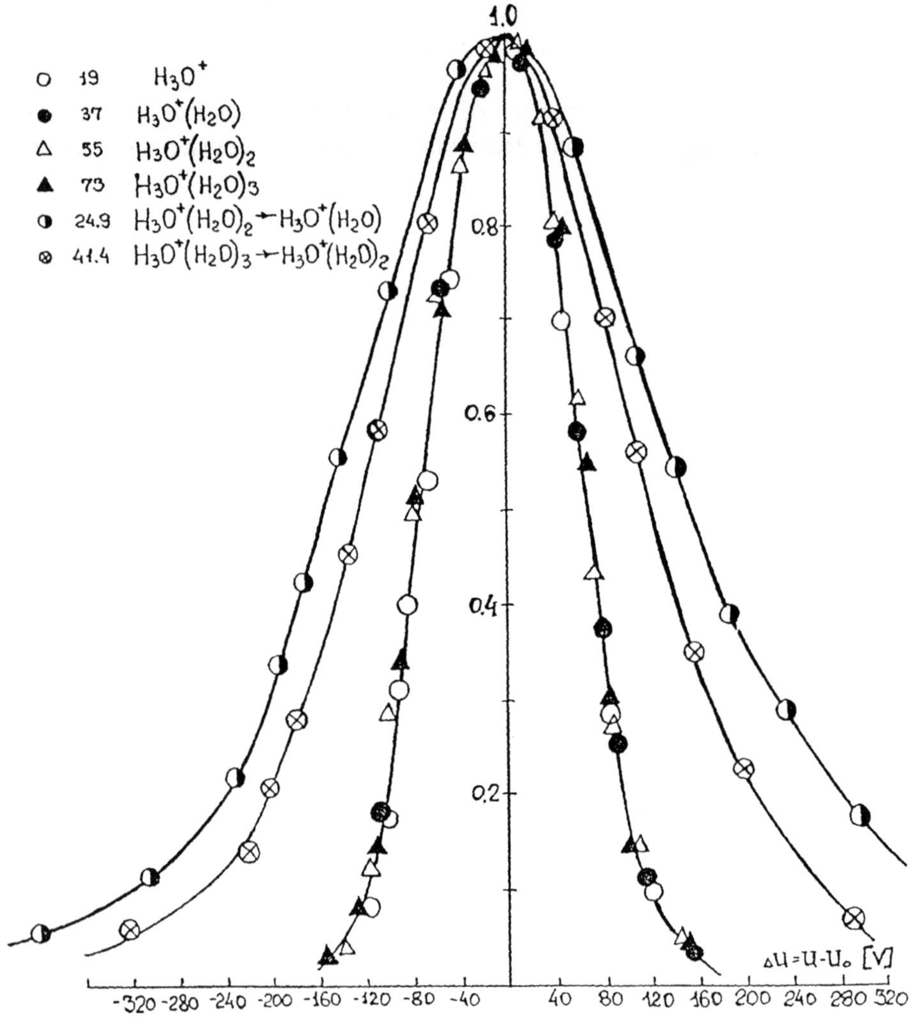

Figure 3. Shapes of metastable peaks and the peak given by precursor ion. $\Delta U = U - U_0$ is the scan of the accelerating voltage above and below the standard value of 3kV.

INDEX

GPSR Compliance

The European Union's (EU) General Product Safety Regulation (GPSR) is a set of rules that requires consumer products to be safe and our obligations to ensure this.

If you have any concerns about our products, you can contact us on ProductSafety@springernature.com

In case Publisher is established outside the EU, the EU authorized representative is:

Springer Nature Customer Service Center GmbH
Europaplatz 3
69115 Heidelberg, Germany

Batch number: 10755248

Printed by Printforce, the Netherlands